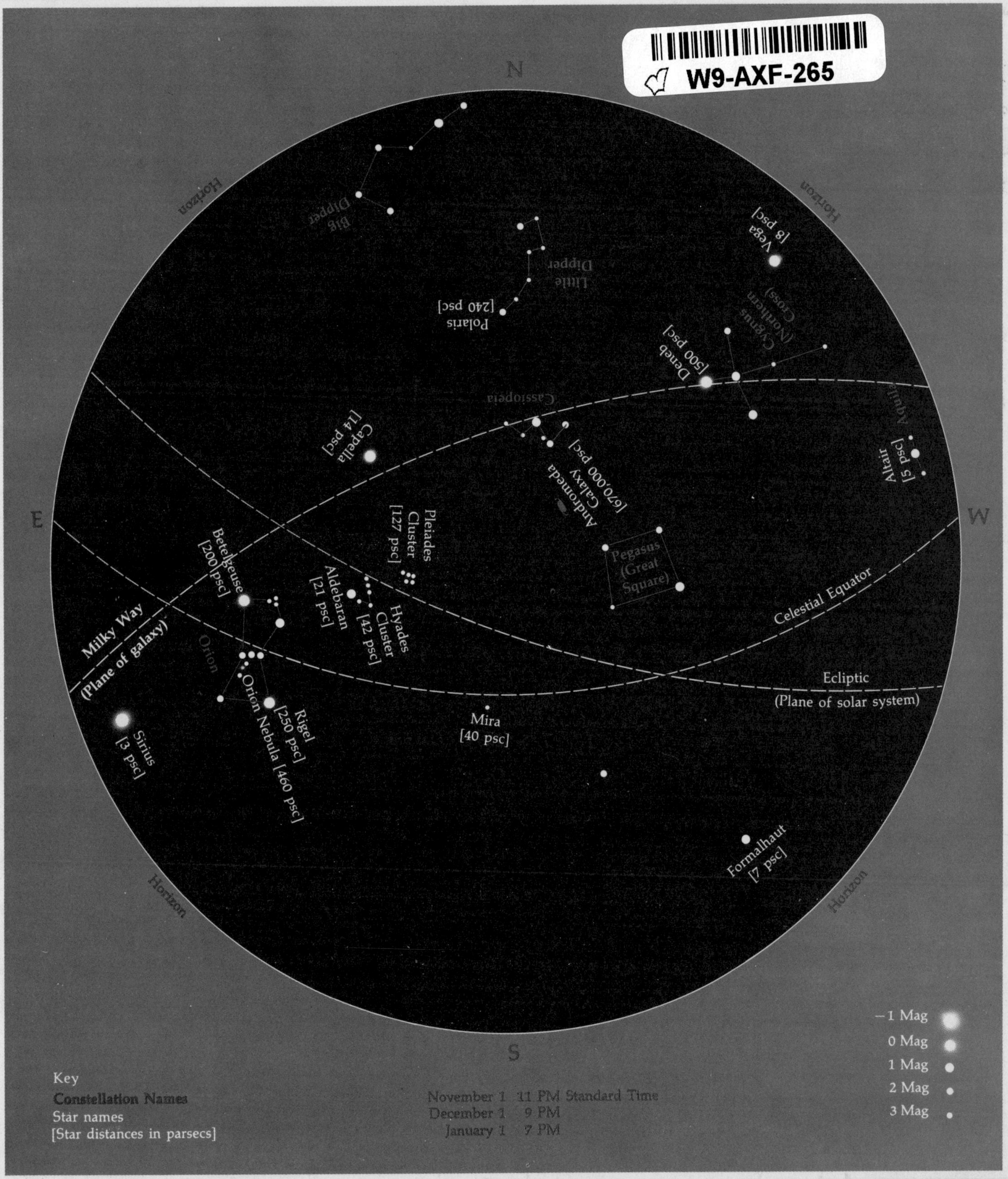
N
S
E
W
Horizon
Horizon
Horizon
Horizon
Big Dipper
Little Dipper
Polaris [240 psc]
Vega [8 psc]
Cygnus (Northern Cross)
Deneb [500 psc]
Cassiopeia
Capella [14 psc]
Andromeda Galaxy [670,000 psc]
Altair [5 psc]
Aquila
Pleiades Cluster [127 psc]
Pegasus (Great Square)
Betelgeuse [200 psc]
Aldebaran [21 psc]
Hyades Cluster [42 psc]
Milky Way (Plane of galaxy)
Orion
Celestial Equator
Ecliptic (Plane of solar system)
Orion Nebula [460 psc]
Rigel [250 psc]
Sirius [3 psc]
Mira [40 psc]
Formalhaut [7 psc]
−1 Mag
0 Mag
1 Mag
2 Mag
3 Mag
Key
Constellation Names
Star names
[Star distances in parsecs]
November 1 11 PM Standard Time
December 1 9 PM
January 1 7 PM

The endpapers of this book show four star maps of the evening sky during the four seasons of the year. These maps have been simplified to show only the brightest stars and the better-known constellations. The stars are identified by name and distance. The reader may use the distances to visualize the night sky as an array of stars in three-dimensional space. For example, Aldebaran appears as a fairly luminous star in the foreground, and adjacent Betelgeuse as a much more luminous star at a greater distance. In addition to the star data, three important imaginary circles are shown: the celestial equator; the ecliptic, marking the plane of the planets' orbits; and the Milky Way, the plane of our galactic disk. Monthly star charts showing fainter stars appear in some of the publications listed in Appendix 3.

Astronomy:
The Cosmic Journey

William K.
Hartmann

Wadsworth
Publishing Company, Inc.
Belmont, California

Astronomy Editor: *H. Michael Snell*
Developmental Editor: *Autumn Stanley*
Production Editor: *Larry Olsen*
Designer: *Dare Porter*
Copy Editor: *Kevin Gleason*
Technical Illustrators:
Darwen and Vally Hennings
Catherine Brandel and Victor Royer
Mark Schroeder
Layouts: *Nancy Benedict and Dare Porter*
Photo Research: *Brenn Lea Pearson*
Editorial Assistance: *Katherine Head*

Printed in the
United States of America
3 4 5 6 7 8 9 10—
82 81 80 79

Spaceman photo on p. 1 courtesy of
NASA. Eta Carinae photo on pp. xii–1
courtesy of Cerro Tololo Observatory,
Chile, and Kitt Peak National Observatory.
Poem on p. 213 from "This is outer space"
in SPACEWALKS by Burgert Roberts.
Copyright © 1971 by Burgert Roberts.
Reprinted by permission of Harper & Row,
Publishers, Inc. Cover photo of
Stonehenge by Autumn Stanley.

Library of Congress Cataloging in
Publication Data

ISBN 0–534–00546–2

Library of Congress Cataloging in
Publication Data
Hartmann, William K.
    Astronomy.

    Bibliography: p. 507
    Includes index.
    1. Astronomy.   I. Title.
QB45.H32   520   77–21807
ISBN 0–534–00546–2

Preface

With human footprints on the moon, radio telescopes listening for messages from alien creatures (who may or may not exist), technicians looking for celestial and planetary sources of energy, and a spacecraft on its way out of the solar system, an astronomy book published today enters a world different from the one that greeted books a few years ago. Astronomy has broadened to involve our basic circumstances and our enigmatic future in the universe. With eclipses and space missions broadcast live into our homes, astronomy has become an adventure for all people, an outward exploratory thrust that may one day be seen as an alternative to wars, mindless consumerism, and ideological bickering among nations.

Today's astronomy students no longer accept a mere listing of current facts; they ask, as people have asked for ages, about our basic relationships to the rest of the universe. They may study astronomy partly to seek points of contact between science and other human endeavors: philosophy, history, politics, environmental action, even the arts and religion.

In designing this book, the Wadsworth editors and I have tried to respond to these developments. Rather than jumping at the start into the murky waters of cosmology, I have begun with the viewpoint of ancient people on earth and worked outward across the universe. This method of organization automatically (if loosely) reflects the order of humanity's discoveries about astronomy and provides a unifying theme of increasing distance and scale, illustrated in our *Contents in Brief*. In view of the wealth of recent discoveries, this earlier tradition of starting with our feet on the solid earth may also give a clearer view of the universe.

The arrangement of this book, then, aims to give an unfolding, ever-expanding panorama of our cosmic environment. We hope it unfolds like a story in which each chapter provides not only a new facet, but also a growing understanding of the relationships among the elements of the whole.

The subtitle refers to three separate cosmic journeys that we undertake simultaneously. First, we travel through historical time, where we see how humans slowly and sometimes painfully evolved our present picture of the universe. Second, we journey through space, where we see how our expanding frontiers have revealed the geography of the universe. Beginning with an earth-centered view, we study the earth–moon system, the surrounding system of planets, the more distant surrounding stars, our own vast galaxy, and the encompassing universe of other galaxies. Finally, we travel back through cosmic time. Familiar features of the earth are typically only a few hundred million years old. The solar system is about 4.6 billion years old. Our galaxy may be 12 or 13 billion years old. The universe itself began (or began to reach its present form) perhaps 12 to 20 billion years ago.

Because astronomy touches many areas of life and philosophy, I have freely considered a wide range of topics of common interest, including space exploration, financing of science, cosmic sources of energy, the bookstall's barrage of astrology and other pseudo-science, and the possibility of life on other worlds, as well as the conventional "hard science" of astronomy. This variety of topics shows how basic scientific research touches all areas of life—I hope in a way that lets readers ponder the relation between science and priorities in our society.

The arrangement of text material into eight modules and twenty-five chapters should give instructors some flexibility in tailoring a course according to their interest. For example, those who are not much interested in historical development could omit or pass lightly over *Module A*.

Each module gives some historical background, describes recent discoveries and theories, and then discusses advances that might occur if society continues to support research. This more or less chronological approach has several purposes. Since there is often a certain logic to the order in which discoveries were made, historical emphasis may help readers remember the facts. Second, historical discussion allows us to introduce basic concepts in a more interesting way than by reciting definitions. Third, there is a widespread fallacy that the only progress worth mentioning is that of the last few decades. Astronomy, of all subjects, shows clearly that, to paraphrase Newton, we see as far as we do because we stand on the shoulders of past generations. Exploration of the universe is a continuing human enterprise. As we try to maintain and improve our civilization, that is an important lesson for a science course to teach.

Another philosophy we have followed is to treat astronomical objects in an *evolutionary* way, to show the sequence of development of matter in the universe. Stars, pulsars, black holes, and other celestial bodies are linked in evolutionary discussion, rather than listed as different types of objects detected by different observational techniques. Further, we try to treat astronomical objects as real *places* in the universe rather than as points of light at the end of earthbound telescopes.

As a result of Wadsworth's extensive market surveys during preparation of the text, we have adopted several noteworthy design concepts. Mathematics is reduced to a minimum and discussed primarily as applications of five commonly used basic equations. These are described *qualitatively* in the main text, so that the book can be used in an entirely nonmathematical way. The five equations are also described more *quantitatively* in optional boxes with blue-screened background. For a more mathematical approach, instructors can emphasize these (and discuss the *Advanced Problems* at ends of chapters, which are mostly applications of the five basic equations). We have also included supplemental *Enrichment Essays* that can be used or omitted at will. These follow the color photo section, *The Universe in Color*, which thus conveniently divides the basic text from the supplements. To reduce costs, we have limited our color photographs primarily to examples that illustrate the role of color in understanding astronomical phenomena.

More specifically, teaching aids incorporated in the book include:

Text

1. Photos reproduced by a duotone two-ink process to hold the richest possible blacks and contrasts. In addition to the classic large telescope photos, I have included three other categories:

a. *Photos from recently published research papers.*

b. *Photos by amateurs with small and intermediate instruments, often used to show sky locations of well-known objects in the large-telescope photos. These can help readers to visualize and locate these objects in the sky, a difficult task if based on classic large-telescope photos alone. Photographic data provided with many of these pictures may be used in setting up student projects in sky photography.*

c. *Scientifically realistic paintings show how various objects might look first hand to observers in space. Discussion of features shown in the paintings illustrates a synthesis of scientific data from various sources.*

2. Key concepts are shown in **boldface** type. These are repeated in *Concepts* lists at the ends of chapters as aids to review. Definitions of key concepts are included in an expanded *Glossary* at the end of the book.

3. The five basic equations are introduced in the text as needed, but are set off in boxes by color screen for optional use. The five boxes discuss:

I. *The Small Angle Equation,* useful for calculating apparent sizes of objects at known distances.

II. *Calculating Circular and Escape Velocities,* useful for deriving speeds or masses in co-orbiting systems (planetary, binary star, galactic).

III. *Measuring Temperatures of Astro-nomical Bodies: Wien's Law,* which shows how radiation measurements can reveal the temperatures of distant objects.

IV. *The Doppler Effect: Approach and Recession Velocities,* which shows how spectral measures can reveal radial velocities of distant objects.

V. *The Stefan-Boltzmann Law: Rate of Energy Radiation,* which shows how temperature and luminosity measurements can reveal sizes of radiating sources.

4. Limited numbers of references to technical and nontechnical sources appear in the text. They are there partly to help students and teachers find more material for projects, and partly to help instructors emphasize that statements should be verifiable. These sources are included in an expanded *References* section at the end of the book.

End of Chapter Materials

1. *Chapter Summaries* review basic ideas of the chapter and sometimes synthesize material from several preceding chapters.

2. *Concepts* lists include the important concepts appearing in **boldface** in the text. Reviewing the *Concepts* lists is a good way for the student to review the content of each chapter.

3. *Problems* are aimed at students with nonmathematical backgrounds.

4. *Advanced Problems* usually involve simple arithmetic or algebra and are usually applications of the five basic equations. These can be omitted in nonmathematical courses.

5. *Projects* are intended for class use where modest observatory or planetarium facilities are available. The intent is to get students to do astronomical observing or experimenting.

Supplementary Material

1. *Enrichment Essays* can be used or not as instructors wish. These include essays on:

A. *Telescopes and Observing.*
B. *The Pseudo-Science of Astrology, UFOs, Ancient Astronauts, and Astro-Catastrophes.*
C. *Astronomical Coordinates and Time-Keeping Systems.*

2. *Appendices* are included on:
1. *Powers of Ten.*
2. *Units of Measurement.*
3. *Supplemental Aids in Studying Astronomy.*

3. The *Glossary* defines most terms included in the *Concepts* lists, as well as other key terms.

4. The *References* section includes all sources mentioned in the text and others used in compiling the text. Non-technical references useful for student term papers are starred (*); widely available journals and magazines are emphasized in this group, including most astronomy articles appearing in *Scientific American* in recent years.

5. Two indexes are included, an *Index of Names* and an *Index of Terms.*

6. *Star Maps* for the four seasons are featured on the endpapers. Since more detailed, larger maps are usually available in classrooms or laboratories, these have been simplified, emphasizing the plane of the solar system and the plane of the galaxy, and major constellations mentioned or illustrated in the text.

Acknowledgments

My thanks go to many people who helped produce this book. I have tried to incorporate as many of their suggested corrections and improvements as possible, although final responsibility for weaknesses remains mine.

Included among two groups of reviewers who criticized parts of the text for research comprehensiveness and teaching potential were: Helmut A. Abt, Richard A. Bartels, Bart J. Bok, Lee Bonneau, David R. Brown, Warren Campbell, Robert J. Chambers, Kwan-Yu Chen, Clark G. Christensen, Dale P. Cruikshank, Robert J. Doyle, Benjamin C. Friedrich, F. Trevor Gamble, Owen Gingerich, Hubert E. Harber, George H. Herbig, Ulrich O. Herrmann, J. R. Houck, William W. Hunt, Gerald H. Newsom, Terry P. Roark, John M. Samaras, Myron A. Smith, Richard C. Smith, Michael L. Stewart, Kjersti Swanson, James E. Thomas, Scott D. Tremaine, Peter D. Usher, Walter G. Wesley, Ray J. Weymann, Richard M. Williamon, R. E. Williams, and Warren Young.

I especially thank Dr. Gingerich for his time in sending me articles illuminating various problems of ancient astronomy, and Dr. Christensen for suggesting end-of-chapter problems. I also thank Clark R. Chapman, Gayle G. Hartmann, and Floyd Herbert for discussions and criticisms that affected the presentation of material; and, finally, a wide circle of colleagues, especially in Tucson and at the University of Hawaii, for helpful discussions.

In locating and providing photographs, Dale Cruikshank, Clarence Custer, John Eddy, Walter Feibelman, Steven Larson, Alan Stockton, Don Strittmatter, and Dave Webb were very helpful. Charles Federer (*Sky and Telescope*), David Moore (Kitt Peak National Observatory), and Holly Chaffee (Arizona State Museum) gave kind assistance with their institutional files. I especially thank James Hervat, Ron Miller, and Adolf Schaller for stimulating discussion about their paintings, and Chesley Bonestell for his hospitality and for discussions about his paintings. Floyd Herbert and Mike Morrow helped me in making several wide-angle sky photos with clock-driven mounts. Uncredited photographs are my own.

I thank Joanne Metcalfe, Linda Mata, Gayle Hartmann, and Anne Hartley for assistance with typing.

The staff at Wadsworth Publishing Company made an enormous effort to ensure a useful, creative, and beautiful product. I thank Mike Snell for overseeing with humor and determination the complexities of reviewing and production; Don Dellen for his faith in initiating the project; Autumn Stanley for developmental work that stimulated me to write more clearly and interestingly; Larry Olsen for excellent management of the final editing and production; Dare Porter for his creativity in design and interest in the subject; Jenny Sill and Susan Snell for their hospitality and assistance; and Kevin Gleason for many helpful editorial suggestions.

William K. Hartmann
Tucson, May 1977

module a

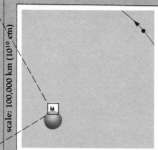

scale: 100 km (10^7 cm)

The Early Discoveries
Prehistoric Advances:
Phenomena in the Sky •
Historic Advances:
Worlds in the Sky •

module b

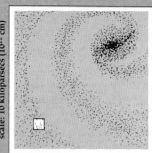

scale: 100,000 km (10^{10} cm)

**Exploring the Earth–Moon
System** The Earth as a
Planet • The Conquest of
Gravity and Space • The
Moon •

module c

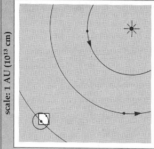

scale: 1 AU (10^{13} cm)

The Solar System Dis-
covering the Solar
System • Mercury and
Venus • Mars • The Outer
Planets • Comets, Meteors,
Asteroids, and Mete-
orites • Origin of the
Solar System •

Contents in Brief

*Astronomical distance
and scale are illustrated
in this brief overview
of the Contents*

module d

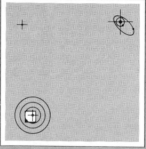

scale: 1 parsec (10^{18} cm)

Stars and Their Evolution
The Sun and Its Radiation
• Measuring the Basic
Properties of Stars •
Nearby Stars and Stellar
Evolution • The Births
of Stars • The Deaths
of Stars •

module e

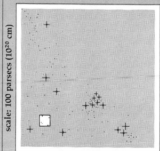

scale: 100 parsecs (10^{20} cm)

**Environment and
Groupings of Stars**
Interstellar Atoms, Dust,
and Nebulae • Binary and
Multiple Star-Systems •
Star Clusters and
Associations •

module f

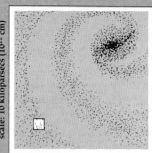

scale: 10 kiloparsecs (10^{22} cm)

Galaxies The Milky Way
Galaxy • The Local
Galaxies • The Expanding
Universe of Distant
Galaxies •

module g

scale: 1 megaparsec (10^{24} cm)

Frontiers Cosmology:
The Universe's Structure
• Cosmogony: A Twentieth-
Century Version of the
Creation • Alien Life
in the Universe •

Contents

*Module
d

Stars
and Their
Evolution
213*

*Module
e

Environment
and Groupings
of Stars
307*

Module i

Galaxies

355

Module 9

Frontiers

421

**Module
h
Enrichment
Essays**
477

For
Erdys, Ernie, Grace, and Harry
with thanks

Astronomy:
The Cosmic Journey

Invitation to Astronomy Today

If you awoke one day to discover that you had been put on a strange island, your first project after getting food and water would be to try to find out where you were. This is just what has happened to all of us. We have all been born on an island in space: the earth. We are all passengers on a cosmic journey. Astronomy is the process of finding out where we are.

This definition of astronomy would not always have satisfied people, because at one time in history we thought we knew where we were: at the center of the universe, with the sun, stars, planets, and satellites all revolving around us. The definition is not fully satisfactory now, either, because astronomy is mingling with other sciences; but it reminds us that we want to know where we are in space and time.

Changing Definitions of Astronomy

Paradoxically, the more we learn, the harder astronomy is to define. When astronomy was restricted to observations by earthbound observers, it was easily defined as "the science of objects in the sky." But now that we have actually begun to explore space, astronomy loses its convenient boundaries. Other disciplines become involved. Is the astronaut who chips a rock sample off the moon practicing astronomy? What about the researcher who studies the geology of the sample once it reaches earth? What of the nuclear physicist who measures properties of nuclear reactions going on in the centers of stars? What about the biologist interested in life on Mars? Their work is included in our definition of astronomy.

There is a popular misconception that as various scientists pursue new research, knowledge proliferates into an endless complex of different specialized disciplines. Indeed, students are usually taught this way, with advanced courses probing into narrower and narrower specialties. According to this idea, research is like climbing a tree, starting near the trunk, and proceeding to ever more specialized branches of knowledge.

But historically, science has been more like the process of working our way *down* the tree of knowledge. At first we see a bewildering variety of seemingly unrelated phenomena—twigs of knowledge. After sorting through these twigs, or miscellaneous observations, we see that they meet in a branch. As workers map details on the branch, someone with vision may discover that the first branch is joined to another branch. By grasping the major branches closer to the trunk, we can control and understand the more specialized phenomena represented by the higher branches. As we work our way toward the trunk, more and more branches join. This is why different specialties are connecting with astronomical research; astronomy is becoming more general.

A historical example illustrates this process. Around 1600, Galileo and others conducted many experiments to understand seemingly unrelated types of motion, such as the motion of falling objects and the motion of the moon. The wide variety of motions and accelerations might have seemed impossible to encompass in a simple way. However, by about 1680, Isaac Newton realized that they were connected by simple relationships between force, acceleration, and gravitational attraction. Suddenly, as a result of Newton's insight, ordinary students could understand motions better than the Galileos and Aristotles of the past.

In the same way, today's students can grasp the relationship between the earth and the cosmos better than the pioneer explorers who mapped astronomical frontiers in the past.

Our Definition of Astronomy

This book treats astronomy not as a narrow set of academic observational results or abstruse physical laws, but as a voyage of exploration with practical effects on humanity. A broad view of discoveries of the last years, the last decades, and even the last centuries is at least as important as the latest factual detail, because the broad view reveals where research will be going in the future and how it may continue to affect us. In an era of

energy crises, food shortages, and environmental threats, when badly applied technology has created such horrors as biological weapons and hydrogen bombs, educated people are called on to make judgments about their own actions—the kinds of jobs they have, the materials they consume, and the actions of organizations they work for. In short, we may be called on to judge what kind of civilization should continue. If so, we all need to know how our world is affected by its cosmic surroundings, and to understand the scientific and social procedures for obtaining and extending that knowledge.

Therefore, this book is organized according to the following definition of **astronomy:**

Astronomy is the study of all matter and energy in the universe, emphasizing the concentration of this matter and energy in evolving bodies such as planets, stars, and galaxies, and fully recognizing that we observers— humanity—are part of the universe, and that our home, the earth, is only the point from which our voyage of exploration has started.

The reason for emphasizing humanity is that astronomy influences how we think about ourselves and our role in the universe. Our ecological experience of the sixties and seventies has shown that we *must* look at the earth as a single astronomical body in order to survive. As the poet Archibald MacLeish said after one of the Apollo flights, we have now seen the earth as "small and blue and beautiful," and ourselves as "riders on the earth together."

A Survey of the Universe

The chapters of this book start on earth and move outward through space. However, before beginning, we give a brief preview of the universe, starting at the largest scale.

The **universe** is everything that exists; to the best of our knowledge, the universe consists of untold thousands of clusters of galaxies. **Galaxies** are swarms of billions of stars. Galaxies differ in form. Some are football shaped; some, irregular. Our galaxy, like many others, is disk-shaped with about a hundred billion stars arrayed in curving spiral arms. Most galaxies, including ours, are surrounded by a halo of **globular star clusters,** each a spheroidal mass of hundreds of thousands of stars. The center of our galaxy (and others) is a puzzling, violent place, as we will see in later chapters. If we made a model of our galaxy big enough to cover North America, the earth would not be as big as a basketball, or even a BB. It would be about the size of a large molecule.

Scattered inside our galaxy, mostly in the spiral arms, are groupings of stars called **open star clusters** and clouds of dust and gas called **nebulae.** Scattered at random are the individual stars, called **field stars.** Each **star** is an enormous ball of gas, mostly hydrogen. In the center of a typical star, atoms are jammed together so closely that their nuclei interact to create the heat and light of the star. Some stars, mostly faint, also contain non-gaseous or non-hydrogen material.

Three hundred million billion kilometers (200,000,000,000,000,000 miles) from the center of the galaxy and 40 thousand billion kilometers from the nearest star is the **sun.** A hydrogen sphere more than a million kilometers in diameter, the sun is the center of the **solar system**—the small system made up of the sun and its family of rocky or icy planets and satellites. About 150 million kilometers from the sun is the **earth.** The closest two planets approach the earth within about 50 million kilometers. The **moon** is only 384,000 kilometers away. The earth itself is nearly 13,000 kilometers across, a tiny speck among myriads of larger and smaller specks in the universe.

A Word about Mathematics

The preceding survey of the universe shows how hard it is to express astronomical quantities in ordinary units. The distances and masses are too great to be grasped conveniently. For this reason the system of expressing numbers as **powers of ten** is useful. For example, the distance from the sun to the galactic center becomes simply 3×10^{17} kilometers; the distance from earth to sun is 1.5×10^8 kilometers. A reader uncertain about this system should consult Appendix 1 at the end of the book. We will use this system occasionally, as convenience dictates.

When using any equation to calculate physical quantities, you must remember to use consistent sets of units. For example, the English system of inches, ounces, etc. must not mix in the same equation with the metric system of centimeters, grams, etc. Because the metric system, long used by scientists, is becoming universal, we use it here. Specifically, we will show some simple calculations in the **cgs system,** meaning that lengths will be expressed in centimeters, masses in grams, and time intervals in seconds.

Facing the Universe

The universe is immense; we are perhaps audacious even to undertake such a study. Yet something noble in the human spirit urges us to seek knowledge of the universe and our place in it. Throughout history, scientific discoveries in the heavens have conflicted with other (religious, metaphysical, or philosophical) conceptions. Persecutions, legal action, even wars have resulted from the challenge of new knowledge to old beliefs. In this book we present the latest and best scientific theories, and the evidence for them, with the understanding that our knowledge is limited and ever-changing. Indeed astronomy unifies humanity by showing us that we are all in this together, facing the unknown—a word that aptly describes much of the universe, its past and its future.

The universe is no abstract concept; it consists of real, physical places. Suppose that we were able to explore these places, and that in all these vast reaches, among all the stars and galaxies, we found no one else—no living creatures anywhere, except ourselves.

Or suppose that we discovered nonhuman intelligent life—alien civilizations, perhaps advanced and incomprehensible, perhaps very simple. Suppose the universe is full of civilizations, and that our contacts with them turn out to be a million times as radical as the first contacts between the American Indians and Europeans, in 1492.

These are two wildly different possibilities. One or the other is likely to be true. Either boggles the mind. We are just barely entering the century in which we may be able, if research continues, to decide between them.

Concepts

astronomy

universe

galaxy

globular star cluster

open star cluster

nebula

field star

star

sun

solar system

earth

moon

powers of ten

cgs system

Problems

1. Scientific discoveries are sometimes described as conflicting with religion or philosophy. Before continuing this book, examine your beliefs and comment on which views best match your own:

a. *There are concepts in astronomy that conflict with my own religious or philosophical views.*

b. *The two areas of thought turn out to be consistent; there is no problem.*

c. *If there is a conflict, scientific hypotheses will evolve and eventually become consistent with my religious or philosophical views.*

d. *Science measures the reality only of the physical world; there is no need for this to be consistent with an inner psychological or spiritual reality.*

e. *Science tries to reflect how all nature behaves; thus science and religious philosophy of daily life should be consistent with each other, to produce a well-integrated personality.*

1

The Early Discoveries

To paraphrase Brahms in his reference to Beethoven, we hear the tramp of the giant behind us.

Sir Fred Hoyle,
commenting on the builders
of Stonehenge

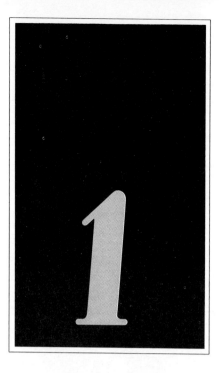

Prehistoric Advances: Phenomena in the Sky

In 1906 the American astronomer Percival Lowell wrote:

> *Smoke from multiplying factories . . . has joined with electric lighting to help put out the stars. These concomitants of an advancing civilization have succeeded above the dreams of the most earth-centered in shutting off sight of the beyond, so that today few city-bred children have any conception of the glories of the heavens which made of the Chaldean shepherds astronomers in spite of themselves.*

This observation seems even more apt today. It is difficult to realize how important the sky once was in daily affairs, and how much we still depend on basic, seemingly obvious ideas that actually were developed only through centuries of struggle to understand nature.

There are several reasons to survey the ancient discoveries. First, they were often the most basic discoveries —so basic that they provide a foundation for current exploration. We forget how hard they were to recognize. For example, we think it obvious that the earth moves round the sun, but early scholars were severely criticized for saying so.

Second, discoveries were often made just when society reached a point where they *could* be made. Therefore, a historical approach provides a good framework for learning important concepts in a logical order.

Third, many common ideas today come from astronomical traditions of the past. For example, although we favor a decimal numerical system, we have 24 hours in a day instead of ten, and 360 degrees in a circle instead of 100. The reason comes from ancient astronomy and mathematics. To understand contemporary ideas and conventions, it is useful to review the past.

The Earliest Astronomy: Motives and Artifacts (ca. 30,000 B.C.)

Archeology and anthropological studies of present-day primitive tribes[1] shed light on the earliest astronomical practices. Imagine yourself an intelligent hunter of 30,000 years ago. Agri-

---

[1]Time grows short for these studies. The last few pockets of primitive culture are being discovered and integrated into their surrounding cultures by scientific expeditions and even guided tours. Frontiers on this planet are nearly gone.

culture is not yet practiced. You have to depend on hunting and gathering for food. Celestial phenomena are not your most immediate concern, but you know that certain celestial cycles are important. You want to know when berries on the mountain will ripen, or when migratory birds will arrive at a nearby lake. Developing such abilities would require cumulative day-counts and knowledge of seasons, best found in the sky. When you travel far, you want to be able to find your way home. Celestial objects are your only reliable guides in unfamiliar landscapes.

If you are a woman, you also want to know in what season the child you may carry will be born, or when your next menstrual period is due. Early women undoubtedly used the 29.5 day cycle of the moon's phases to keep track of their 27 to 30 day menstrual periods. Some experiments reportedly indicate that menstrual cycles of women sleeping in the light of an artificial "moon" become synchronized to its cycle of phases, and some sci-

Figure 1·1 **Paleolithic evidence suggests that ancient people tabulated the phases of the moon. The configuration shown here, called "the old moon in the new moon's arms," is seen in the evening sky a few days after a new moon, when the portion not lit by the sun is dimly lit by light reflected off the earth.**

Figure 1·2 **The sun rises and sets at a different horizon point each day, with a flattened shape caused by atmospheric distortion. These phenomena exemplify the combination of atmospheric and astronomical effects not distinguished by early people.**

entists have even speculated that the human menstrual cycle evolved to match the period of the moon's phases (Luce, 1975).

Early people probably dealt with these problems by keeping records of events in the sky.

There may be actual physical evidence of the earliest practice of astronomy. Curious notations have been found on thousands of artifacts dating as far back as 30,000 B.C., scattered over wide areas in Europe, Africa, and the Soviet Union. Typical examples are bone tools with clusters of carefully scribed lines. Microscopic examination shows that the lines were not carved all at once, but in groups at different times. Some archeologists believe the markings may be the beginnings of calendar systems—perhaps counts of the number of days between phases of the moon, such as new moon (the thinnest crescent) and full moon (fully illuminated disk). The makers of these tools were not the brutish cavemen of some stereotypes; their craftsmanship is demonstrated

by the magnificent paintings of their prey—mammoths, boars, reindeer—that they left on European cave walls.

This interpretation is supported by contemporary **calendar sticks** made by modern aborigines. For example, a lunar calendar stick carved by Indian Ocean islanders shows a series of grooves almost identical to the Paleolithic examples of 30,000 years earlier. And North American Indian calendar sticks similarly record historical events, phases of the moon, and other natural events by sequences of carvings (see Figure 1·3).

Figure 1·3 **Portion of a calendar stick carved by Papago Indians of Arizona. The stick, begun in 1841 and passed on to the carver's son, carries a record of social and natural events, including an earthquake, for the years 1841 to 1939. Since the Papagos had no written language, the symbols can be read only by the carver. In such a way the earliest records of astronomical events may have accumulated, allowing subsequent discoveries of recurrent cycles.** (Calendar stick from collection of Arizona State Museum, University of Arizona)
▼

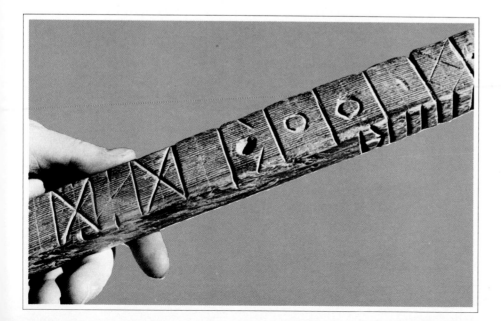

Thus, proto-astronomical record-keeping probably began as early as 30,000 B.C. By 10,000 B.C., people in various cultures were probably familiar with time-keeping by cycles of the moon, and perhaps with longer cycles as well.

Calendar Refinements (10,000–3000 B.C.)

Around 10,000 B.C., the domestication of animals and the cultivation of crops began. This agricultural revolution brought two new incentives to understand the sky. First, to know when to plant, people had to be able to *predict* seasonal changes. Second, in the stable villages made possible by agriculture, people kept records; and the mere existence of records gave scholars an incentive and opportunity to seek seasonal sequences and other subtle celestial cycles.

Exactly this stage of development is shown in historical records of Indians of the American Southwest. The Papago Indians, for example, named each of twelve lunar cycles (months)

that made up the year—among them were a "Frost Moon" (October), "Green Moon" (March, plants leafing), "Hungry Moon" (May, no wild foods), and the like. This was obviously the beginning of a calendar system. But since each cycle of lunar phases is about $29\frac{1}{2}$ days, 12 such cycles would be only 354 days, not a full year. Thus, the lunar-based calendar would have to be readjusted each year.

People perhaps began to get a better sense of the yearly cycle by counting days or moon cycles between certain dramatic annual events. In Egypt, for instance, the Nile flooded at a certain time of year. At high northern latitudes people looked eagerly for the sun to return to the northern sky after the cold, short days of winter. In some early societies, day-counts of this cycle gave estimates of about 360 days, later refined to 365 days and then $365\frac{1}{4}$ days.

Accumulated observational records would have revealed the correlation between this annual cycle of seasons and the apparent movements of the sun. The sun does not follow the same path across the sky every day. At northern latitudes in summer it is high in the midday sky; on winter middays it is low in the southern sky. Thus, careful observations of the sun's motion relative to other features of the sky would have helped determine the exact number of days in the year and the exact date of the observation itself.

Organized observations of this kind were made throughout the world thousands of years ago. Mesopotamian calendars were being designed by 4000 to 3000 B.C. In recent decades, anthropologists have found modern pre-technological people using similar means to create sophisticated systems of time measurement.

Other Early Discoveries

Before going on to describe ways in which these principles were utilized and extended, we pause to describe other aspects of the sky discovered by early people and still encountered today. These concepts are an important aid in describing many astronomical phenomena.

North Celestial Pole Anyone who sleeps outside at night notices that the stars slowly wheel around a single point in the sky. In the northern hemisphere that single point is called the north celestial pole; in the southern hemisphere, the south celestial pole. The wheeling of the stars is due to the earth's rotation on its axis, and therefore one complete circuit takes about 24 hours. The **celestial poles** can be thought of as the *projections of the earth's axis on the sky,* or the spots targeted by vertical searchlights at the north and south poles. Recognizing the celestial pole helped in navigation; if you can locate the north celestial pole in the sky, you know which way is north.

North Star A bright star, **Polaris,** happens to lie near the north celestial pole. Thus, it seems to stand still all night above the northern horizon, as other stars wheel around it. It thus serves as an excellent reference beacon. For nearly 1,000 years, the star Polaris has been located within a few degrees of the north celestial pole, but in other times, other stars have been near the pole. Any star located near the north celestial pole is called the **north star.**

Zenith and Meridian Wherever an observer stands, the point directly overhead is called the **zenith.** The **meridian** is an imaginary north–south line from horizon to horizon through the celestial pole and the zenith. This line marks the highest point that each star reaches above the horizon on any night; it is the dividing line between rising stars and setting stars. For this reason it is especially useful in time-keeping. In the daytime, the moment when the sun crosses the meridian is called noon; in very early times, therefore, people used this astronomical concept to divide the day.

Celestial Equator The imaginary circle lying 90° from either the north or the south celestial pole is the **celestial equator.** It is the projection of the earth's equator onto the sky and is parallel to the daily east-to-west paths of the stars as they wheel around the celestial poles. The celestial equator helped divide the sky into easily recognizable portions.

Figure 1·4 **General properties of the sky as defined by the earth's daily rotation.**
▼

The 360° Circle Our division of the circle into 360 units, or degrees, follows Babylonian practice of the late pre-Christian era. Their division probably comes, in turn, from the ancient estimate of 360 days for one cycle of the "sun around the earth." Their division of 1 degree into 60 minutes of arc, and 1 minute of arc into 60 seconds of arc, may also have been favored because the numbers 360 and 60 are easily divided into a variety of equal parts (for instance, both numbers are divisible by 2, 3, 4, 5, and 6). The decimal (ten-based) system of counting and writing numbers later replaced the sexagesimal (sixty-based) system in most uses *except* the ancient studies of angles and time.

The Ecliptic By observing the relative positions of sun and stars at dawn or dusk, one can establish that the sun appears to shift nearly a degree to the east each day, relative to the stars. (This can be confirmed by noting that the sun "moves" 360° in 365 days.) Ancient observers found that the sun

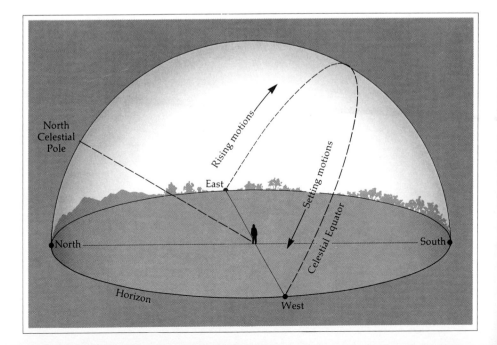

traces the same path among the stars year after year. The path differs from the celestial equator (being tipped to it by $23\frac{1}{2}°$ and crossing it at two points). *The path of the sun among the stars is called the* **ecliptic** (see Figures 1·5 and 1·6).

The Planets Ancient observers— priest-astronomers and wakeful shepherds alike—found that five of the brighter starlike objects were not fixed like the rest. They came to be known in the Western world as planets—from the Greek word for wanderers.[2] Not until after the telescope was invented, in about 1610, was there any proof that these **planets** were globes like our own earth, or that three more faint ones had gone undiscovered. For this reason the ancients spoke of five planets, whereas we speak of nine (counting our own).

The Zodiac Ancient observers tracking the paths of the planets from night to night among the stars discovered that the planets never stray far from the ecliptic. In fact, *the planets move all the way around the sky in a zone about 18° wide, centered on the ecliptic.* This zone is called the **zodiac.** After the stars were grouped in easily remembered patterns called constellations. those constellations located within the 18° came to be known as the signs of the zodiac.

[2]Many of our technical terms have Greek roots because the modern Western tradition has a direct link to Greek civilization through the Roman and medieval worlds. However, many of the original concepts discussed here were discovered in *pre-Greek* times and passed down to them, only to receive Greek names. Later writers, often unaware of the earlier work, revered the Greek thinkers and kept their terms.

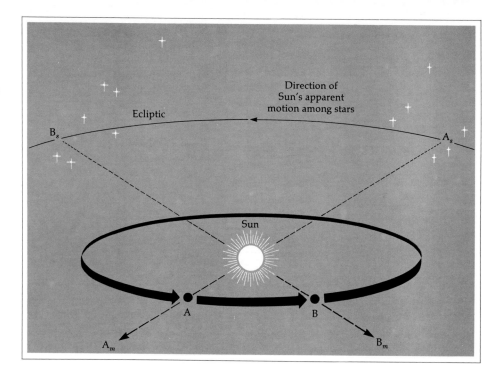

Figure 1·5 **A model of the earth's revolution around the sun.** As the earth moves from point A to point B in its orbit, the sun first appears among background stars B_s, then against stars B_s. Similarly, the midnight sky, opposite the sun, first contains stars in direction A_m, and later contains those in direction B_m.

Figure 1·6 **Consequences of Figure 1·5 for an earthbound observer.** At dawn on date A, the sun can be seen near certain bright stars. A few days later (date B) the sun has shifted eastward along the ecliptic, and the stars appear farther from the sun at dawn.

Heliacal Risings and Settings Any given star rises and sets slightly earlier each night. The **heliacal** (he-LIE-ah-cal) **rising** of a star occurs *on the first day each year when the star can be seen just before dawn.* (*Heliacal* means "near the sun," from the Greek *helios,* sun.) One day earlier, the star would rise a few minutes *after* the sky gets too bright. On the day of heliacal rising, the star glimmers near the horizon for just a few minutes before that hap-

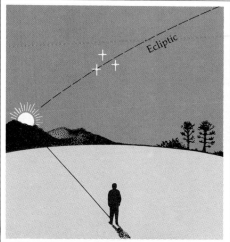

Dawn, date *A*

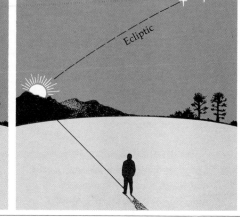

Dawn, date *B*

pens. On the day after, the star rises slightly earlier and can be seen for several minutes before the sky lightens. Similarly, **heliacal setting** *occurs on the last day when the star can be seen at dusk.* On the next day, by the time darkness falls, the star has already set.

Why would anyone living in, say, 4000 B.C., care about the heliacal rising or setting of a certain star? The answer is that each heliacal rising or setting corresponds to a certain date in the year. Depending on the star, that date might be the first day of spring, or the day when planting should start. In the absence of clocks and calendars and morning newspapers on the doorstep, heliacal risings and settings would tell the date with an accuracy of a day or two. It is scarcely surprising that the ancient Egyptian calendar began with the heliacal rising of Sirius, which marked the beginning of the Nile's annual flood.

Agriculture in Java was once scheduled according to the heliacal rising of Orion's belt. Australian aborigines begin their spring when the Pleiades, a prominent star cluster, rises in the evening sky. Papago Indians of Arizona also divide their agricultural year according to the Pleiades (Castetter and Bell, 1942):

> Pleiades rising in summer, start
>   planting;
> At the zenith [directly overhead]
>   at dawn, too late to plant more;
> Past the zenith, time for corn
>   harvest;
> One quarter down from the zenith,
>   time for deer hunting;
> Setting, time for harvest feast.

Origin of the Constellations

As memory-aids to help them learn and locate stars, early observers named groups of stars for their resemblances to familiar animals, objects, or mythical characters—Orion (the hunter), Leo (the lion), Taurus (the bull), and so on. These groups are **constellations.** People in different cultures saw different patterns, often derived from their own mythologies; for example, the ancient Chinese constellations differ from Western constellations.

Our own familiar constellations came out of the Near East. As early as 3300 B.C., Leo and Taurus often appear in struggle on Mesopotamian artifacts. Star symbols sometimes shown on the lion's shoulders reveal that the artisans had the constellation in mind rather than a real lion. The struggle motif may refer to Leo's chasing Taurus out of the sky: as Leo rises, Taurus starts setting. Although these images originated when the risings and settings of Leo and Taurus coincided with equinox and solstice dates, the image gained a mythic importance of its own and persisted in art until A.D. 1200 (Hartner, 1965).

The English astronomer Michael Ovenden (1966) did some astronomical sleuthing to find out who formalized most of our present constellations, when, and why. The most ancient constellations fill only the northern sky, leaving a puzzling empty zone around the south celestial pole.[3] Evidently, the ancient constellation-makers couldn't see this southern zone from their northern latitude. As shown in Figure 1·7, for an observer

at any given northern latitude, L, the sky is divided into three zones: the *north* **circumpolar zone,** L° from the north celestial pole, containing stars that never set; the **equatorial zone,** containing stars that rise and set; and the *south* **circumpolar zone,** L° from the south celestial pole, containing stars that *never rise.* The size L of the southern circumpolar zone is the clue to the northern latitude L of the ancient observers. Using this and other clues, Ovenden obtained a latitude of 36°N $\pm\frac{1}{2}$° for the ancient constellation-makers. He estimated the date of their activity by using a more complicated fact about the sky. Because of lunar and solar forces acting on the earth, the north and south celestial poles and the celestial equator show a **precession:** that is, their positions slowly drift in a small circle among the stars. During the 26,000 years required for one such circuit, different stars become the north star. The ancient constellations show a rough symmetry around the star Alpha Draconis, about 25° from Polaris. Since this star was the north star around 2600 B.C., Ovenden concluded that most constellations were designed about 2600 B.C. ±800 years.

These findings clarify the purpose of the constellations. Many people have supposed the constellations were dreamed up by uneducated shepherds or sailors who amused themselves by finding in random star groupings their favorite mythical heroes and heroines, familiar animals, and objects from their daily lives. It now appears that the original constellations were more carefully designed to help identify the celestial pole, equator, and coordinates of about 2600 B.C. ±800 years. The purpose may have been to help teach navigation. This idea accounts for the seemingly forced nature of some constellations. For example, Hydra, a

[3]Constellations now filling this southern zone, such as the Southern Cross and the Telescope, were not charted until European navigators sailed these waters in the 1600s and 1700s. In 1930 the International Astronomical Union revised the boundaries of all constellations for easier record-keeping. Ancient names were kept, but the old, irregular boundaries were replaced by neater, geometric ones.

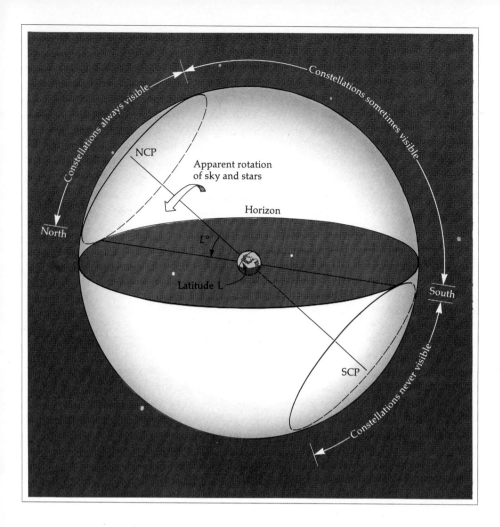

Figure 1·7 For an observer at latitude L in the
northern hemisphere, no stars and
constellations within L° of the north
celestial pole (NCP) ever set. Stars in a
middle zone rise and set, and a southern
zone is never seen.

Figure 1·8 How to determine your latitude ▶
astronomically. Angle *L* is the latitude, by
definition. The elevation θ of the north
celestial pole above the horizon equals the
latitude L.

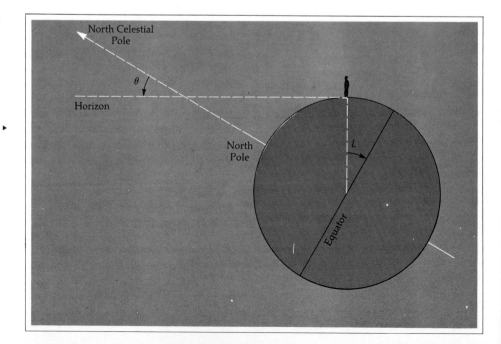

Figure 1·10 **From polar latitudes the celestial equator is so near the horizon that the sun may spend entire days above or below the** horizon. **This series of multiple exposures shows "the midnight sun"—the sun skimming the horizon but not setting.** (Photographed in northern Europe by Aleko E. Lilius, courtesy *Sky and Telescope*)

▼

peculiar, elongated constellation of 25 faint stars, lies accurately along the celestial equator of about 3000 B.C. Certain additional evidence suggests the designers were seafarers.

What seafaring civilizations lay near 36°N in the period 2600 B.C. ±800

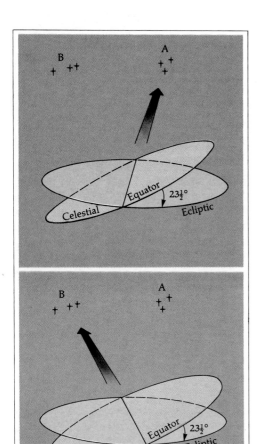

Figure 1·11 **The effect of precession, or rotation of the plane of the celestial equator. At different dates, A and B, during the 26,000-year precessional cycle, the intersection of the celestial equator and ecliptic is seen in different positions, A and B, among the stars.**

years? The Minoan sailors of Crete (35°N) are perhaps the best possibility. Vestiges of Minoan culture were transmitted to the Greeks, and thence to us, which is consistent with this theory. *Many constellations, therefore, may be Minoan creations handed down to us from around 2600 B.C.*, with still earlier elements incorporated in them. We should not assume that "it all started with the Greeks."

Solstices and Equinoxes

We can understand the tremendous importance of seasonal changes—and especially the first days of spring and summer—to primitive or outdoor peoples. Such people could use the constellations as a background to help map the annual (360° per year) movement of the sun along its path, the ecliptic. They could see that the sun's apparent motion along the ecliptic produced four important dates during the year, the **solstices** and **equinoxes.** For simplicity, we will describe these from the viewpoint of a northern observer:

Spring Equinox, about March 21; First Day of Spring Sun crosses the celestial equator moving north.[4] Rises due east, sets due west. An important day to ancients in the northern hemisphere because it marked the return of the sun to "their" sky, bringing warmth.

Summer Solstice, about June 22; First Day of Summer Sun reaches point farthest north of celestial equator. Rises and sets farthest north. Most hours of daylight, ensuring

---

[4]As a memory-aid, recall that the sun as it crosses the equator is *equidistant* from the north and south celestial poles.

warm weather. Most important day of year in many ancient northern calendars.

Autumn Equinox, about September 23; First Day of Autumn Sun again crosses celestial equator, moving south. Colder weather on the way.

Winter Solstice, about December 22; First Day of Winter Sun farthest south of celestial equator. Rises and sets farthest south. Fewest hours of daylight.

Once they understood equinoxes and solstices, the ancients could make the systematic observations necessary to maintain accurate calendars. They could also determine equinox and solstice dates, often marked by major ceremonies. For example, on equinox dates the sun would rise due east, 90° around from the north point marked by the north star. A post or monument on the eastern horizon could mark the spot behind which, seen from a fixed vantage point, the sun would rise on this special day. Once permanent villages were established, special "observatory" buildings could be constructed with built-in markers.

Astronomical Orientation of Ancient Buildings

Cultures with at least some astronomical awareness often aligned buildings along north–south and east–west lines. But such alignments alone do not establish great sophistication on the builders' part, because the four cardinal directions can be determined in a single night's observation of the rising points, meridian crossing, and setting points of stars.

Figure 1·12 **Movement of the sunset position along the horizon during an eight-month period. Moving south in A, the sun reaches winter solstice in B. Here the setting position reverses direction and proceeds north again in C, reaching the vernal (spring) equinox in E. Summer solstice is reached in H. The author's former home, from which these pictures were made, happened to occupy the focus of a "natural Stonehenge," where the two solstice positions were marked by prominent mountain peaks. Such situations may have triggered the prehistoric idea of building artificial solstice markers, to reset calendars twice a year.**
▼

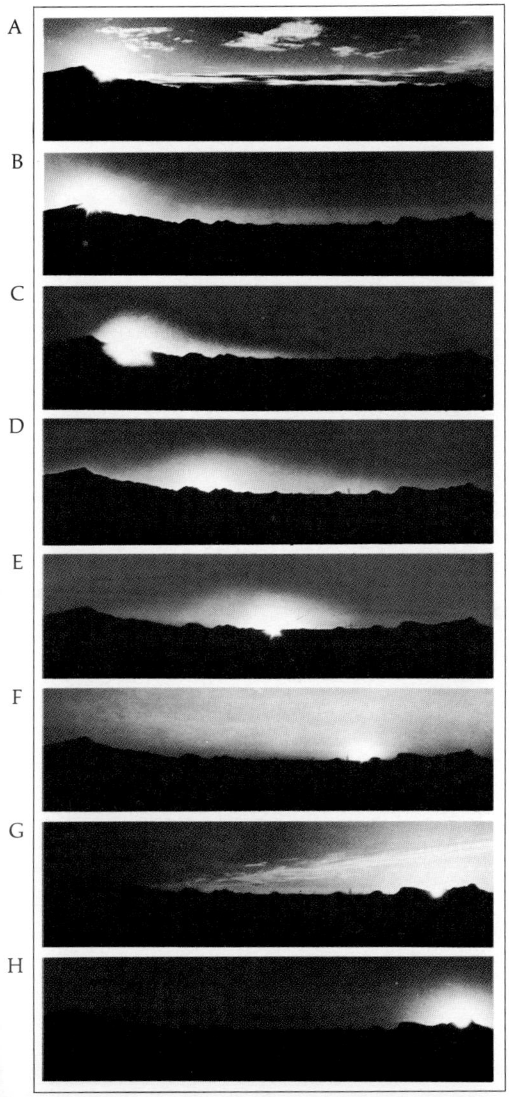

An east–west building alignment is called the **equinoctial orientation,** because the sun rises and sets exactly on the east–west line only on the equinox dates. Archeologists have found equinoctial buildings, including major temples, all over the world. The earliest examples, many thousands of years old, indicate that ancient builders early recognized north, east, south, and west from astronomical observation. Magnetic compasses could *not* have been involved, because magnetic north is usually slightly off true north, and because the earliest magnetic compass was apparently made no earlier than 1400–1000 B.C., by the Olmecs, in Mexico. Magnetic materials were not discovered by the Greeks

before 600 B.C., and compasses were first recorded in China around 300 B.C. to A.D. 100 (Carlson, 1975).

Some equinoctial buildings reveal much knowledge and skill, both astronomical and mechanical. A famous example is the Great Pyramid of Khufu in Egypt, built around 2600 B.C. The sides are aligned north–south and east–west with an average error of only a fraction of a degree. The pyramid's dressed-rock base deviates from a true plane by only 0.004 percent, measured by the four corners. A 380-foot tunnel in the north face points toward the point where the north star of that era, Alpha Draconis, crossed the meridian below the north celestial pole. (Other astronomical or pseudo-astronomical purposes postulated for the pyramid are now dismissed by most Egyptologists.)

Ancient astronomers also designed buildings with more subtle **solstitial orientation,** or alignment toward a point on the horizon marking sunrise or sunset on the summer or winter solstice. To explain the importance of

Figure 1·13 **The Stonehenge principle. The sun's position at solstice is 23½° from the celestial equator. The rising (or setting) position on this date can be marked by a prominent pillar. Other markers in a ring could allow observation of other selected risings and settings during the year.**
▼

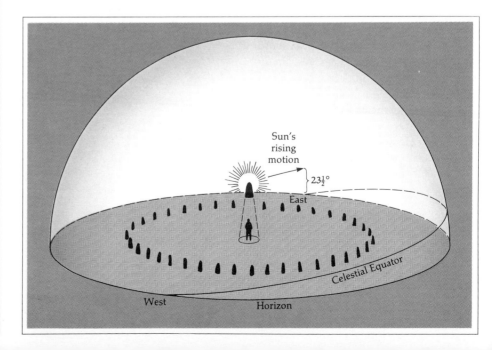

solstices to ancient astronomical observers, let us reconsider solstices by putting ourselves in the ancients' place.

If we watch the sun set on several succeeding days, always looking from the same observing post, we see the sun touch the horizon at different points on different days (Figure 1·12). If we had started our observations on the first day of spring (spring equinox), we would find the sun setting due west. The next day it would set farther north, and each day set still farther north, until, by the first day of summer (summer solstice), it would have reached its northernmost point, *rising and setting farther north than on any other day of the year.* By the next day it would have slipped back a little to the south. In other words, at the summer solstice, the sunset point appears to *reverse* its progression. We could set up a marker to tell us which point marked the solstice day.

Similarly, at winter solstice, the setting and rising points reverse their southerly progression and begin to slip back to the north. Because of the sun's unique reversal, these dates, easily observed each year, were ideal dates on which to reset the calendar.

As early as 2700–2600 B.C., certain massive stone buildings were apparently oriented in order to "clock" the sun's motion as a solstice drew near.[5] Instead of facing east, where the sun rose at the equinoxes, these buildings were designed with sighting devices aimed toward one or more solstice points on the horizon.

---

[5]Wooden structures of this kind were almost certainly built, probably well before the stone ones we know about, but they have not survived.

The Case of Stonehenge

Among the most famous solstitially oriented structures in the world is **Stonehenge,** in southern England (see Color Photos 1 and 2). Until recently it posed a complete mystery: Why should people living around 2600 B.C. have built a ditch, bank, and ring of pits more than twice the diameter of

Figure 1·14 **Aerial plan of the original Stonehenge construction around 2600 B.C., consisting of an outer ditch and mound, an avenue toward summer solstice sunrise, and a ring of posts or stones. Selected moonrises could also be observed.**
▼

the dome of Mount Palomar Observatory? Why should they have set up a 16-foot-high rock—the heel stone—outside the ring on a broad straight avenue running $\frac{1}{2}$ km out of the circle? Why should other people more than 500 years later have brought 30- to 50-ton stone blocks some 30 km to use in massive structures in the center of the circle, not to mention hauling five-ton stones about 380 km (240 miles) over sea and land to be arranged also in a circle around the center of Stonehenge?

As long ago as 1740 the English scholar Dr. W. Stukely reported that the avenue and the heel stone are almost perfectly aligned on the sum-

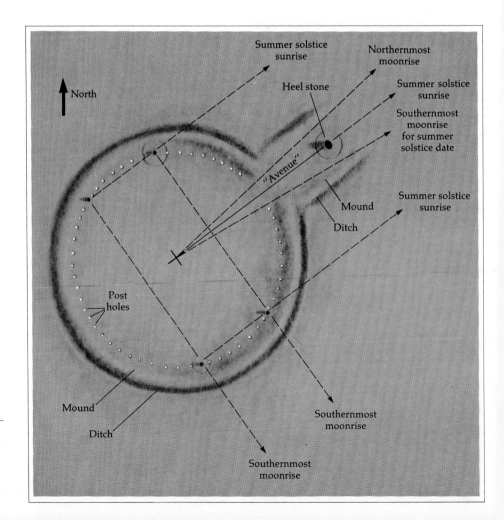

mer solstice sunrise. Yet when the American astronomer Samuel Langley in 1889 and the English astronomer J. N. Lockyer in 1894 spoke of Stonehenge as as astronomical observatory, scholars said ancient people could not have been sophisticated enough to orient large structures this way.

Then in the early 1960s, Smithsonian astronomer Gerald Hawkins and others drew attention to other astronomical alignments at Stonehenge. The large number of stones suggests an immense number of lines, which Hawkins sorted by computer. He found that not only the most northerly (midsummer) sunrise, but

▲
Figure 1·15 **Sketch of summer solstice sunrise as observed from Stonehenge after massive trilithons were constructed inside the original monument, around 2000 to 1500 B.C. The dark stone just below the sun is the heel stone, or solstice marker.**

Figure 1·16 **A view of Stonehenge. The largest stone structures (right) are among the latest (2000 B.C. or later). The fence at left marks an earlier earth bank surrounding the structure, and the white spots mark post or stone holes; these features and an associated**
▼

avenue leading toward summer solstice sunrise (far side of monument) are believed to date from about 2600 B.C.

also the other midsummer and midwinter sunrise and sunset positions were marked by stones. Moreover, Hawkins and an English researcher, C. A. Newham, independently concluded that special sighting stones had been designed to allow observations of the *moon's* motions as well, including its northernmost and southernmost points of rising and setting.

All of these stones and directions belonged to the oldest part of Stonehenge, dating from about 2600 B.C. according to the radiocarbon dating system, as revised in the later 1960s. The larger, later stones (as recent as 2000 to 1500 B.C.) revealed no significant alignments—indicating an intellectual decline in society, as though the later builders, though marvelous en-

Figure 1·17 England and the French coast contain many other ancient stone ring monuments besides Stonehenge, such as this ring in England. Astronomical analyses of these rings are still being made; some have shown astronomical alignments.
▼

Figure 1·18 Lockyer's illustration of the ruins ▶ of the long hall in the Temple of Amon-Re, aligned toward the summer solstice sunset position.

gineers, had lost track of the monument's original purpose.[6] (Similarly, in modern times Stonehenge has been used for magical rites by Druid cultists.)

Today there is little doubt that Stonehenge was carefully designed and used for observations or ceremonies connected with the sun's rising on summer solstice, and probably also for observations of the moon. Many other prehistoric stone circles in England and northern Europe also have sight-lines with astronomical significance (see Figure 1·17).

[6]The redating of Stonehenge upset the old diffusionist idea that civilization originated solely in the Near East and then spread slowly into Europe. Stonehenge, once dated later than the high Mediterranean civilizations, was considered a crude northern copy of Eastern ideas. The new date of 2600 B.C. suggests more direct exchanges of cultural ideas.

Observatories, Temples, and Cathedrals

Because astronomical alignments in prehistoric sites have only recently been established, archeologists are now on the lookout for sites where early peoples observed solstices or other events. Some examples follow:

Temple of Amon-Re, Karnak, Egypt (ca. 1400 B.C.) One of the largest temples ever built, it covered twice the area of St. Peter's in Rome. Its central hallway, about 370 m (1,200 feet) long, was aligned within half a degree of the summer solstice sunset position (Figure 1·18). Other Egyptian temples had similar orientations, whereas still others were oriented toward heliacal risings and settings of certain stars. Because of precession, the rising and setting points of stars shift slowly over the centuries, and some of these temples show evidence of rebuilding as if to correct for such shifts, even though the rebuilding often ruined the symmetry of the temple itself by changing the angle of the central hall. The Egyptians are known to have made astronomical observations: Chicago's Oriental Institute has part of a device for charting star movements, made by the young Pharaoh Tutankhamen himself around 1350 B.C.

Angkor Wat, Cambodia (ca. A.D. 1150) The axis of the Angkor Wat temple is east–west within a fraction of a degree; parallel walls have equal lengths within 0.1 percent; there are reportedly more solstitial, equinoctial, and lunar alignments than would be expected by chance. A Chinese merchant who visited the site in 1296 reported that, just like the Chinese, the Khmer builders understood astronomy and could predict eclipses.

Figure 1·19 An aerial view of geometric figures constructed around A.D. 200 (?) by clearing pebbles in the Peruvian coastal desert. Somewhat more than a random ▼ percentage of the figures appear astronomically aligned, and may have served a calendric purpose.

Tihuanaco, Bolivia (A.D. 375–725) At the ruined city of Tihuanaco, a rectangular Temple of the Sun was built with walls diverging less than a degree from the N–E–S–W points. From the exact center of the west wall an observer could see the solstice sunrise just at the outer edges of the corner pillars of the opposite wall. Legends of solar worship or observation are associated with the site. Possible astronomical alignments have been found in nearby Peruvian desert areas (Figure 1·19).

Caracol Temple, Yucatan (ca. A.D. 1000) Many examples of possible astronomical alignment are among the ruins of Mexico and Guatemala. Caracol is a cylindrical tower located at the famous ruined city of Chichen Itza. Sighting lines defined by window design, and other alignments, mark the horizon positions of solstice sunsets and sunrises, as well as the most northerly and southerly points reached by Venus, a planet whose positions were carefully recorded in Mayan manuscripts still existing today.

Casa Grande, Arizona (ca. A.D. 1350) The same trade routes that carried cultural influence northward from central Mexico in the first millennium of the Christian era probably carried astronomy as well. An unusual four-story adobe building at the Casa Grande site in Arizona has been identified as a probable observatory similar in conception to the Caracol. Oddly shaped upper-floor windows can be used to sight solstice positions and certain significant lunar rising and setting positions. (See Color Photo 4.)

American Indians in Historic Time (ca. 1540) Spanish explorers recorded Indian astronomical practices. Inca ceremonies (in Cuzco, Peru) determined the solstices, and runners were sent out to spread the news of the start of the new year through the Inca empire. Because the sunrise position on the horizon slowed and stopped at the solstice point, this point was known in the Inca empire as the "sun's hitching post." Montezuma (in what is now Mexico City) reportedly proposed tearing down and realigning the wall of a building to bring it into better accord with the sun's position at equinox. In the 1800s, solstice observations similar to those proposed at Casa Grande were actually recorded at Indian pueblos in the Southwest.

Figure 1·20 **Astronomical alignments at the Medicine Wheel, suggested by J. A. Eddy, 1974.**
▼

Medicine Wheel, Wyoming (A.D. 1500–1760) The closest American analog to Stonehenge is a circular stone ring with cairns marking summer solstice sunrise and sunset positions, as well as certain bright star positions on the night horizon (Figure 1·20). It resembles in more massive form the plan of the sundance lodge commonly used in the 1800s by Plains Indians for summer solstice ceremonies (Eddy, 1974).

St. Peter's Cathedral, Rome (ca. A.D. 330–1506; 1506–) The ancient Mediterranean tradition of orienting temples for practical astronomical purposes may have carried over to modern times as a tradition whose rationale is forgotten. The old (330–1506) and new (1506–) cathedrals of St. Peter were oriented so accurately

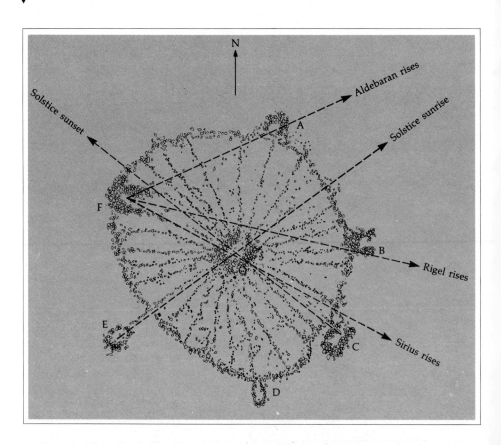

to the east that on the morning of the vernal equinox (Lockyer, 1894):

> . . . the great doors were thrown open and as the sun rose, its rays passed through the outer door, then through the inner door, and penetrating straight through the nave, illuminated the High Altar.

The pre-Christian temple in Jerusalem reportedly had the same orientation. On the morning of vernal equinox, priests carried out rituals in a shaft of sunlight in the innermost temple. In related tradition, some Christian cathedrals and missions were oriented toward sunrise on the day of the saint to whom they were dedicated. All this may go back to prehistoric calendric observations.

Eclipses: Occasions for Awe

So far we have treated the moon and the sun as independent bodies that could be tracked for practical purposes. Cycles of the moon divide the year into about twelve equal months; solstices and equinoxes, defined by sunrise positions, divide the year into four quarters. Ancient people who discovered these facts may have gained a sense of well-being: At least *some* things in their environment could be counted on from year to year. Calendars could be made and agriculture regulated. The predictable, friendly sun could be worshipped as a deity, always providing light and warmth.

What, then, could be made of **eclipses,** the sporadic occasions when the sun, or more commonly the moon, disappears while above the horizon? In the few minutes of a total solar eclipse the sky turns dark, stars can be seen in daytime, and an uncanny chill and gloom settle over the land. During the hours of a total lunar

eclipse, the moon turns blood red. Small wonder that the Greek root of eclipse, *ekleipsis,* means abandonment, or that the ancient Chinese pictured a solar eclipse as a dragon trying to devour the sun. Eclipses must have seemed an unpredictable menace to the scheme of things.

We know eclipses awed early people. In Greece, the poet Archilochus, observing the solar eclipse of April 6, 648 B.C., was moved to write:

> Nothing can be sworn impossible . . . since Zeus, father of the Olympians, made night from mid-day, hiding the light of the shining sun, and sore fear came upon men.

▲
Figure 1·21 **The March 1970 total solar eclipse observed in the village of Atatlan, Mexico.**

Figure 1·22 **Local Indians on the steps of the church moments after the total solar eclipse of March 1970 in Atatlan, Mexico.**
▼

Herodotus reports that a war between the Lydians and Medes stopped when an eclipse surprised their competing armies during a battle in the 580s B.C. A hasty peace was cemented by a double marriage of couples from the opposing camps.

Knowledge of what causes eclipses would allay people's fears of them. And if some few people knew how to predict eclipses, these few would become very powerful. To cite a historical example, in February 1504, Christopher Columbus was having trouble convincing Jamaicans that they should feed him and his crew. Seeing that his almanac foretold an eclipse of the moon on February 24, Columbus

warned the Indians that the Christian god would punish them by turning the moon to blood. The scoffing natives changed their attitude when the eclipse actually began on schedule.

This event may have inspired Mark Twain, whose Connecticut Yankee established his authority in King Arthur's court by predicting a total solar eclipse, and then "commanding" the eclipse to end. This imaginative scene suggests the political prestige perhaps gained by some of the first priest-astronomers who discovered how to predict eclipses.[7]

---

[7]Even animals are affected by eclipses. I have seen roosters crowing and birds flocking to trees at midday because of the gloom during a solar eclipse, causing the creatures to believe evening has fallen. There is a story about an eclipse in Colorado during which an astronomer arrived too late to get a prime observing site. He hurriedly set up his equipment in an empty chicken coop to protect his instruments from the wind, and then spent most of the eclipse trying to shoo away the chickens, who dutifully reported to the roost when darkness fell.

Cause and Prediction of Eclipses

Whatever their emotions and motives, some early people learned how to predict eclipses—and this was an early step toward recognizing that we live on a world among worlds orbiting in space.

To understand how eclipses can be predicted, we should first understand the **causes of eclipses.** We must temporarily abandon the earth-centered viewpoint of ancient people and consider the geometry that would be witnessed by an observer in space during an eclipse.

Two important things to remember about eclipses are: (1) they happen when the shadow of one celestial body falls on another; and (2) what you see during an eclipse depends on your position with respect to the shadow. If you see an eclipse at all, you are either in the shadow (a celestial body has come between you and the source of light) or you are looking at the shadow from a distance as it falls on some surface.

Earthbound observers see two types of eclipses: solar and lunar. **Solar eclipses** occur when the moon, on its $29\frac{1}{2}$-day journey around the earth, happens to pass between the earth and the sun (Figure 1·25). A **total solar eclipse** occurs if the moon completely covers the sun, as seen by an earthbound observer. A **partial solar eclipse** occurs if the moon is "off center" and covers only part of the sun. By coincidence, the moon happens to have the same angular size as the sun— about $\frac{1}{2}$ degree, or a little less than the angular size of a little fingernail at arm's length.[8] Therefore, the moon usually just covers the sun during a

---

[8]If this concept is unclear, see the small angle equation, p. 38, for further discussion of angles.

Figure 1·23 **Temperatures recorded during the total eclipse of the sun in Mexico, March 1970. The peculiar cooling would have been another of many awe-inspiring effects of eclipses observed by ancient witnesses.**

▼

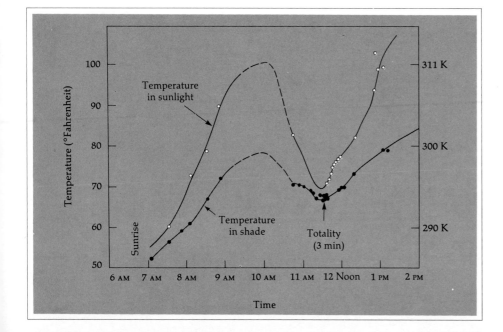

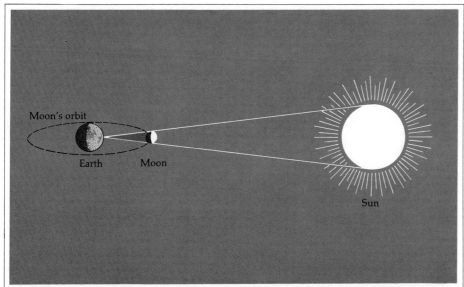

▲
Figure 1·24 **The totally eclipsed sun photographed with a 4.4-inch aperture telescope-filter assembly specially designed to enhance coronal details. The glowing structure is the hot gas of the sun's outer atmosphere, or corona, visible to the naked eye only during eclipses. The planet Venus appears nearby, being on the far side of the sun during this particular eclipse. This eclipse of November 1966 was photographed at 13,000 feet in the Bolivian Andes.** (National Center for Atmospheric Research)

◄ *Figure 1·25* **Geometry of an eclipse of the sun (not to scale). The shadow of the moon falls on the earth.**

Moon's orbit

Earth   Moon

Sun

solar eclipse. But if the moon happens to be at the farthest point in its orbit, it has a smaller angular size than usual and doesn't quite cover the sun. This causes an **annular solar eclipse,** in which the observer sees a ring (Latin *annulus*) of light, which is the rim of the sun, surrounding the moon.

Lunar eclipses occur when the moon, on its journey around the earth, passes through a point exactly on the opposite side of the earth from the sun. This point lies in the shadow cast by the earth (Figure 1·27). It takes the moon a few hours to pass through the earth's shadow, during which time the moon turns an astonishing copper-red from sunlight passing through the earth's atmosphere.

Eclipses can happen in other parts of the solar system. For example, Figure 1·28 is a photo of a solar eclipse as seen from the moon's surface.

Figure 1·26 **The "diamond-ring effect," with diffraction spikes produced by optical effects in the lens of the 35 mm camera used to record the March 1970 eclipse. The effect occurs when the first or last brilliant solar rays pass through mountain valleys at the edge of the moon, as the moon moves across the sun's face.**

Figure 1·27 **Geometry of an eclipse of the ▶ moon (not to scale). The moon passes through the earth's shadow.**

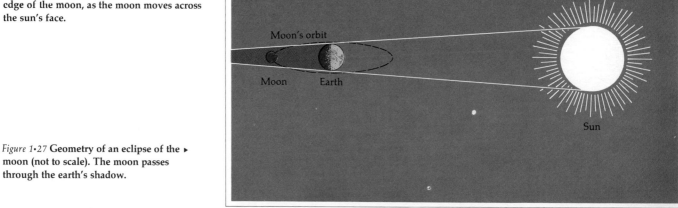

Moon's orbit

Moon   Earth

Sun

Umbral and Penumbral Shadows

In solar and lunar eclipses, the shadows of the earth and the moon are not sharply defined. Because the sun, as seen from the earth, has an angular size of half a degree, sunlight is slightly diffuse; its rays come from slightly different directions.

Thus, shadows—including eclipse shadows—cast by sunlit objects near the earth are not sharply defined. The inner, "core" area of a shadow—the part that receives no light at all—is named from the Latin word for shadow, **umbra.** The outer, fuzzy boundary is called the **penumbra.** An observer in the umbra sees the sun entirely obscured; an observer in the

A

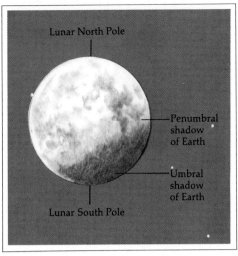

B

Figure 1·29 **A. A photograph of a partial lunar eclipse. B. A diagram of same, showing umbral and penumbral shadows.**

Figure 1·28 **An eclipse of the sun by the earth, photographed from the surface of the moon. This television image, made by Surveyor III in 1967, shows a ring of light that is the earth's atmosphere, back-lit by the sun, which lies behind the black disk of the earth. The brightest light is coming through the atmosphere over eastern Asia; local clouds affect the amount of light transmitted. (NASA)**
▼

penumbra can see part of the sun's disk. Halfway into the penumbra, the observer would see half the sun's disk obscured. For example, if you hold your hand at eye level above smooth ground in sunlight, its shadow will have a central dark umbra and a penumbra about a centimeter wide. If you hold your hand high and spread your fingers, their shadows will be indistinct. An ant in the umbra would see a total solar eclipse; an ant in the penumbra would see a partial solar eclipse.

The relative sizes of umbra and penumbra depend on the distance between the shadow-casting body and the surface on which the shadow appears. During a lunar eclipse, the earth's shadow on the moon has an umbra several times the moon's di-ameter, and a considerably larger penumbra. *Total* lunar eclipses, with the moon in the umbra, can last up to $1\frac{3}{4}$ hours.

On the other hand, the moon's umbral shadow on the earth is 167 miles wide at most (Figure 1·30). Due to the motion of this shadow across the earth, total solar eclipses cannot last more than $7\frac{1}{2}$ minutes. In an annu-lar eclipse, the moon is too far away to produce any umbral shadow, and the eclipse has no true total phase.

Frequency of Eclipses

To be able to predict eclipses, we must understand the intervals between them. Discovery of these intervals several thousand years ago marks the beginnings of our ability to predict seemingly mysterious celestial events. To explain the technique, we again abandon the earthbound point of view for a viewpoint in space. If the moon's orbit around the earth lay exactly in the plane of the earth's orbit around the sun, the moon would pass exactly between the earth and the sun every time around, as can be seen from Figure 1·25. It would also enter the earth's shadow on every pass, and there would be both a solar and a lunar eclipse every month.

But the moon's orbit is tipped by an angle of 5° out of the earth's orbital plane. Therefore the moon is likely to pass "above" or "below" the earth-sun line and the earth's shadow. Since the two orbital planes must, by simple geometry, intersect in a line, there is a line (NN' in Figure 1·32) that contains the **nodes,** the only two points where the moon passes through the earth's orbital plane. This line between the two points is called the **line of nodes.** To produce an eclipse, the moon must be at one of the nodes, and even then, *an eclipse will occur only if the line of nodes is pointing at the sun as the moon passes through one node.*

When is the line of nodes pointed at the sun? If no external forces acted on the moon's orbit, the orbital plane would stay fixed with respect to the stars, and the line of nodes would line up with the sun twice a year, as can be seen in Figure 1·33. Eclipses *could* occur at these two possible times each year, but *only* if the moon moved through points N and N′ at these moments. Predicting eclipses, then, would require alertness during only two intervals per year.

But there is one last complication. The moon's orbit is not fixed with respect to the stars. It is disturbed by various forces. Therefore, the moon's orbit shows a **precession of nodes:** the orbit swings slowly around, always keeping its 5° tilt to the earth's orbital plane. This precession of nodes is analogous to the precession of the earth's axis. The line of nodes, NN′, rotates slowly with respect to the stars, taking 18.61 years to complete one rotation. Therefore, the line of nodes aligns itself with the sun twice every 346.6 days rather than twice every year. *Eclipse year* is the name given to this 346.6-day period.

Thus several simultaneous cycles must be exactly in phase to produce an eclipse: the 29.5-day cycle of lunar revolution with respect to the sun; the 1-year cycle of the earth's revolution; and the 18.6-year lunar precessional

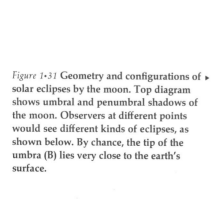

▲
Figure 1·30 **Model of the states of Colorado and Wyoming, showing the size of the umbral shadow of the moon falling on the earth during the total solar eclipse of July 29, 1878.** (John A. Eddy, National Center for Atmospheric Research)

Figure 1·31 **Geometry and configurations of ▶ solar eclipses by the moon. Top diagram shows umbral and penumbral shadows of the moon. Observers at different points would see different kinds of eclipses, as shown below. By chance, the tip of the umbra (B) lies very close to the earth's surface.**

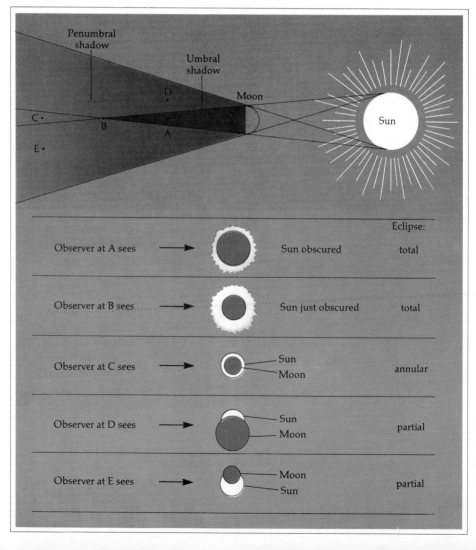

cycle. These overlapping cycles cause subtle periodicities in the occurrence of eclipses.

Discovery of the Saros Cycle (ca. 1000 B.C.?)

From the information given above we can understand one periodicity among eclipses. The moon passes nearest the sun's direction every 29.5 days (month), and the node N lines up with the sun every 346.6 days (eclipse year). An eclipse occurs if the two cycles coincide. How often do they coincide? If figures more exact than those given above are used, 223 lunar months equal 6,585.321 days, while 19 eclipse years equal 6,585.781 days. Thus, the two cycles come almost exactly into phase with each other (to within only 0.46 days) every 6,585 days, or 18 years, 11 days. This interval is called the **saros cycle** (a Greek name, from an earlier Assyrian-Babylonian word). Early astronomers discovered that if in a given year, a particular sequence of eclipses occurred, a similar sequence would probably occur after one saros.

Shorter periodicities, such as 41 and 47 months, also exist but are less accurate for predicting eclipses. Various periodicities were discovered and used by early astronomers to predict eclipses.

Because the moon's umbral shadow on the earth is so small, a fixed observer has only a small chance of seeing any given solar eclipse. But because the earth's umbra is large, and because the moon can be seen from the whole night hemisphere of the earth, half the earth will see every lunar eclipse (barring cloud cover).

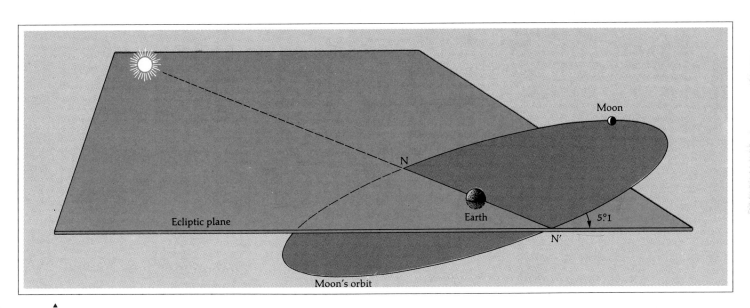

Figure 1·32 **The moon's orbit lies out of the ecliptic plane. Thus, only at points N and N' can the earth, moon, and sun be aligned to produce an eclipse.**

Figure 1·33 **As the earth moves around the sun, there are only two periods each year when the nodal line NN', of the moon's orbit, aligns with the sun so that eclipses can occur.**

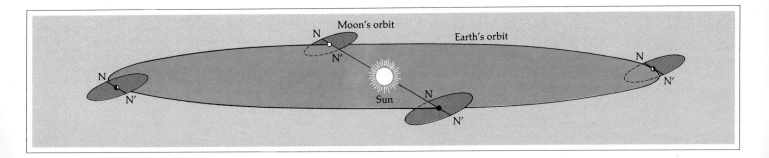

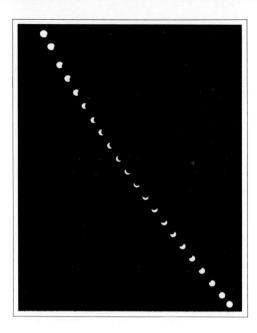

▲
Figure 1·34 **Multiple exposures showing the sun's movement across the sky during a partial solar eclipse on July 10, 1972. At maximum, 72 percent of the sun was covered by the moon. Photos were made six minutes apart with a 35 mm camera and a dark filter.** (NASA photo by A. K. Stober)

Total lunar eclipses are therefore relatively common for any observer, and they were thus easier than solar eclipses for ancient astronomers to predict using the saros and other cycles. Modern astronomers use computers and orbital theory, not cycles, and accurately predict all eclipses.

In summary, on the average a given city may witness a lunar eclipse nearly every year and a partial solar eclipse nearly every other year, but a total solar eclipse only about once every four centuries. Many of the dates are related by cycles.

Let us now return to the viewpoint of ancient peoples. Once records were made and kept, astronomers could benefit from their community's past observations of eclipses. If they discovered the periodicities, they could

begin to make rough predictions of eclipses. We know this happened by 700 B.C. in the Mediterranean area and probably by about A.D. 500 in Central America because records found in Assyrian libraries and Mayan cities discuss prediction of eclipses.

Three Examples of Ancient Eclipse Prediction

Thales (ca. 580 B.C.) Historical records indicate that one of the first known astronomers, Thales of Miletus, predicted the eclipse that stopped the battle between warring Greek factions in the 580s B.C. Thales may have made this prediction by knowing of the saros from Mesopotamian records, or he may have discovered a useful 3-saros periodicity of 669 months, which would have been prominent in the eclipse records of his region for the preceding 125 years. You can repeat his discovery of this cycle by studying these records in Table 1·1. Similar sequences can be seen in Table 1·2, which lists total solar eclipses for the rest of this century.

Stonehenge (ca. 2000 B.C.) Archeological data indicate much earlier prehistoric discoveries of eclipse phenomena. The builders of Stonehenge may have been trying to predict eclipses. Excavations show evidence of arrays of posts or sighting lines aligned to mark the most northerly *moon*rise. If ancient observers used these to track the moon's motion and measure when it was at the most northern (or southern) part of its orbit, then they would know that nodal passage—or potential eclipse days—would come one-fourth of a month later. Many ancient stone monuments in Great Britain contain similar lunar sighting lines and may be lunar obser-

vatories, possibly related to eclipse calendars (Thom, 1971). Astrophysicist Fred Hoyle comments that we don't know for certain whether ancient people used Stonehenge to predict

Table 1·1
Partial Solar Eclipses Observed in Miletus, 710–580 B.C., Showing Repeated Intervals of 669 Months[a]

Date (B.C.)	Number of Months (Lunations)
711 Mar. 14[b]	0
704 Oct. 19	94
700 Aug. 6[b]	141
689 Jan. 11[b]	270
687 Nov. 10	305
662 Jan. 12[b]	604
661 Jun. 27[b]	622
657 Apr. 15[b]	669
651 Jun. 7[b]	745
650 Nov. 21[b]	763
648 Apr. 6[b]	780
646 Sep. 8	810
641 Nov. 11	874
637 Aug. 29	921
636 Aug. 19	933
635 Feb. 12[b]	939
610 Sep. 30	1256
608 Feb. 13	1273
607 Jul. 30	1291
603 May 18	1338
597 Jul. 9	1414
596 Jun. 28	1426
594 May 9	1449
588 Jul. 29[b]	1526
587 Dec. 14	1543
585 May 28[b]	1561
584 May 18	1573
582 Sep. 21[b]	1602
581 Mar. 16	1608

[a]Thales may have had essentially this table for his predictions of eclipses. He may have discovered the 669-month cycle (noted by brackets) or other cycles. You can find other periodicities, for example, 47 months. A *lunation* is one complete cycle of lunar phases.

[b]Eclipses nearly total in Miletus.
Source: Data from Hartner (1969).

eclipses, but we know that *we* could use it to predict eclipses:

It is admittedly astonishing that they should have done so, but our astonishment in this respect could really be a measure of our ignorance rather than a reflection on the intellectual capacity of stoneage man.

Mayan Eclipse Astronomy (ca. A.D. 400)

To say that ancient people may have tracked the moon to make predictions and calendars based on eclipses is not idle speculation. We know that Mayan astronomers of Mexico and Guatamala did just that. One of three priceless Mayan manuscripts left after the Spanish conquest is a record of observed and predicted solar eclipses as well as other astronomical information such as the motions of Venus. Tragically, because the Spaniards burned 27 other Mayan manuscripts in 1562, we may never know the extent of other Mayan knowledge that was almost preserved. While the Mayan calendric system had a 365-day yearly count and a so-called "long count" or cumulative tally of days since a starting date about 3,000 years ago, it also had a cyclic count useful for predicting eclipses. A combination of word-names and number-names for days (such as our usage *July* 4) produced a 260-day cycle called the **Mayan Sacred Round.** In other words, the day names repeated every 260 days. As we have seen, the nodes of the moon's orbit are aligned with the sun (in position for potential eclipses) once every 173.31 days. Eclipses could thus occur three times in 519.93 days, scarcely two hours short of two Mayan Sacred Rounds. The Sacred Rounds thus let Mayan priests keep in step with the sequence of possible eclipses.

Why did the Mayans (and perhaps others) develop an eclipse-based calendar? One motivation may have been that *five* solar eclipses were visible in the Mayan area during the 13 years from A.D. 331 to 344: a total eclipse in 331, a partial eclipse in 335, near-total eclipses in 338 and 344, and an annular eclipse in 342 (Harber, 1969). Such a cluster of eclipses may have impelled the Mayan priest-astronomers toward keeping records so they could predict the sun's next disappearance. In fact, the first astronomical observations in the Mayan manuscripts date from about this time. Interest in eclipses may have been further spurred in A.D. 495, when two partial solar eclipses were visible in Central America only 30 days apart, with a lunar eclipse between them (Owen, 1975). Mayan astronomy was supported by the state and highly organized as shown by a convention of prehistoric Mesoamerican astronomers at Copan, Honduras (probably on May 12, 485) to discuss the calendar system.

Although the Mayan priests probably could not predict every eclipse accurately, once they got their calendric system working they could have had the best of both worlds. When they correctly predicted an eclipse, their power and knowledge were

	Table 1·2 Total Solar Eclipses Until A.D. 2000[a]		
Date	Duration of Totality (min.)	Selected Regions where Totality Visible
1976 Oct. 23	4.9	Australia
1977 Oct. 12	2.8	Venezuela
1979 Feb. 26	2.7	NW U.S., Canada, Greenland
1980 Feb. 16	4.3	Africa, India
1981 Jul. 31	2.2	Siberia
1983 Jun. 11	5.4	New Guinea
1984 Nov. 22	2.1	New Guinea, Chile
1987 Mar. 29	0.3	Atlantic, Africa
1988 Mar. 18	4.0	Philippines
1990 Jul. 22	2.6	Finland, N. Asia
1991 Jul. 11	7.1	Hawaii, Central America, Brazil
1992 Jun. 30	5.4	S. Atlantic
1994 Nov. 3	4.6	Chile, Brazil
1995 Oct. 24	2.4	Iran, India, Vietnam
1997 Mar. 9	2.8	NE Asia
1998 Feb. 26	4.4	Central America
1999 Aug. 11	2.6	Europe, India

[a]Brackets show pairs of eclipses separated by one saros interval.

Table 1· 3 Selected Lunar Eclipses
(Eclipses at least partly visible in the United States)

1979 September 6

1981 July 17

1982 July 6

1982 December 30

1983 June 25

1986 April 24

1987 October 7

manifest; and when predicted eclipses did not occur they could attribute this happy omen to the power of their rituals. The successes and failures of Mayan astronomy may thus have worked together in a way that nourished the cultural traditions and helped sustain the society for centuries.

1

Summary

Unaided by telescopes, unknown geniuses of prehistoric time made many of the most basic astronomical discoveries. They (1) recognized and tracked five planets; (2) discovered the celestial poles and N–E–S–W coordinates defined by daily star motions; (3) learned to use heliacal and solstitial risings and settings to formulate calendars; (4) recognized the ecliptic and the zodiac as the paths of the sun and planets, respectively; (5) may have recognized some effects of precession, which causes stars to shift their positions relative to the celestial poles and the celestial equator.

Stable societies encouraged astronomy, and vice versa. Astronomy probably contributed to civilization by creating calendars and encouraging the keeping of records.

Around 2600 b.c. there may have been a "golden age" in the Mediterranean world, when modern constellations were designed on or near Crete, the great pyramid was carefully oriented in Egypt, and the Stonehenge solstitial observatory was built in England. By around 1400 b.c. building temples with astronomical alignments was a well-developed art in Egypt.

A similar sequence under way in America since the first millennium of the Christian era was ended by European invasion. Efforts to understand eclipses resulted in still more sophisticated astronomical knowledge, recording of cycles over long periods of time, and invention of calendar subtleties.

Concepts

calendar sticks

celestial poles

Polaris

north star

zenith

meridian

celestial equator

ecliptic

planets

zodiac

heliacal risings

heliacal settings

constellations

circumpolar zones

equatorial zone

precession

solstices

equinoxes

equinoctial orientations

solstitial orientations

Stonehenge

eclipse

causes of eclipses

solar eclipse

total solar eclipse

partial solar eclipse

annular solar eclipse

lunar eclipse

umbra

penumbra

nodes

line of nodes of moon's orbit

precession of nodes of moon's orbit

saros cycle

Mayan Sacred Round

Problems

1. Suppose you are standing facing north at night. Describe, as a consequence of the earth's rotation, the apparent direction of motion of each of the following:

a. *A star just above the north celestial pole (or NCP).*
b. *A star just below the NCP.*
c. *A star to its left.*
d. *A star to its right.*

2. Answer the same questions above, but for an observer in the southern hemisphere facing south and looking near the south celestial pole.

3. What is your latitude? What is the approximate elevation of Polaris above your horizon?

4. What is the radius of the north circumpolar zone of constellations, seen from your latitude?

5. How many degrees from the south celestial pole is the southernmost constellation that you can see from your latitude?

6. Solar eclipses are slightly more common than lunar eclipses, but many more people have observed lunar eclipses than solar eclipses. Why?

7. Is there an umbral shadow of the moon on the earth during an annular solar eclipse? Why or why not?

8. How would a lunar eclipse look if the earth had no atmosphere? Compare the earth's appearance from the moon during such an eclipse with its appearance from the moon in reality with its actual atmosphere.

9. Why would lunar and solar eclipses each occur once per month if the moon's orbit lay exactly in the ecliptic plane?

10. Compare ancient European and American cultures of around A.D. 200 to 1000.

    a. *How much did each know about eclipse phenomena?*
    b. *How many years apart did they achieve a comparable ability to predict eclipses?*
    c. *To understand the causes of eclipses?*
    d. *During this period, were the two cultures' rankings in technology similar to their astronomical rankings?*
    e. *Do you think that either of these two rankings (or the two combined) offer a valid measure of cultural achievement?*

Advanced Problems

11. Using the diagram below, prove that the angle of elevation, λ, of the north celestial pole above the horizon equals the latitude of the observer. Why does the angle from celestial equator to zenith equal the latitude?

Projects

1. Starting on the date of new moon (predicted on many calendars) observe the sky at dusk and record whether the moon is visible. Repeat each evening for several weeks and record the moon's appearance. Repeat at next new moon. How many days is the moon visible between new moon and first quarter? Between first quarter and full moon? How might these counts relate to clusters of grooves, such as 3, 6, 4, 8, . . . found on prehistoric tools? Can you *prove*, or only speculate, that the prehistoric records are lunar calendars?

2. From the preceding project, determine how much later the moon sets or rises each night.

3. If a planetarium is available, arrange for a demonstration showing:

    a. *Position of north celestial pole.*
    b. *Position of celestial equator.*
    c. *Daily motion of stars.*
    d. *Prominent constellations.*
    e. *Daily motion of sun with respect to stars.*
    f. *Position of ecliptic.*
    g. *Planetary motions.*

4. Measure the angle λ of elevation of the North Star above the horizon. A protractor can be used as shown to make the measurement. Compare the result with your latitude.

5. Identify a bright planet and draw its position among nearby stars each night for several weeks. (A sketch covering about 10° × 10° should suffice.) Does the planet move with respect to the stars? How many degrees per day? (The latter result will differ from one planet to another and from one week to another.)

6. Find a point with a clear western horizon and determine the date of heliacal setting for some bright star or star group. If several students work independently on the same star, compare results. With how many days' uncertainty can the heliacal setting date be identified?

7. From a point with a clear western horizon chart sunset positions with respect to distant hills, trees, or buildings for several days and demonstrate the motion of the sunset point from day to day. Do the same for a few days around winter or summer solstice and demonstrate the reversed drift of the sunset position. How accurately can you measure solstice date in this way?

8. Using a light bulb across the room for the sun, a small ball for the moon, and your eye for a terrestrial observer, simulate total, partial, and annular solar eclipses.

9. Using the same props, demonstrate a total eclipse of the moon. Show why a lunar eclipse occurs only during a full moon.

10. If an eclipse of the moon occurs while you are taking this course (see Table 1·3), observe it. Compare visibility of surface features on the moon before the eclipse, in the penumbra, and in the umbra. Confirm that the curved shadow of the earth defines a disk bigger than the moon. What color is the moon, and why? Would the moon be lighter or darker during

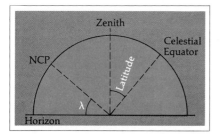

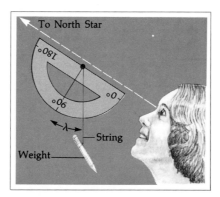

eclipse if there were an unusually large number of storm clouds around the "rim" of the earth, as seen from the moon?

Historic Advances: Worlds in the Sky

In their epic *Gilgamesh*, the Mesopotamians forecast that their own works would vanish as the wind. As they predicted, their names were lost and many of their discoveries degenerated into myth. But some of their knowledge of nature reached the Greek world. The Greeks, with their blend of practicality and imagination, expanded this knowledge rapidly. Historical records give the names and sometimes the biographies of many Greek astronomers. Primarily because of their work, we now accept without question the earth's roundness, the nearness of the spherical moon, and other commonplace but subtle ideas. If you doubt the sophistication of these early scientists, try to measure the diameter of the earth and the relative distances between earth, moon, and sun without any electronic devices. *They* did.

Because discoveries often depend on an underlying conceptual or philosophical framework, we pause in this chapter to look at the attitudes toward nature that the Greeks inherited and developed.

Figure 2·1 **An Egyptian conception of the universe. Stars are distributed in different shells around the universal center. The shells are separated by boundaries depicted as goddesses. This conception may have carried over to Greek and medieval times, when the planets and stars were also pictured as distributed in concentric shells.**
▼

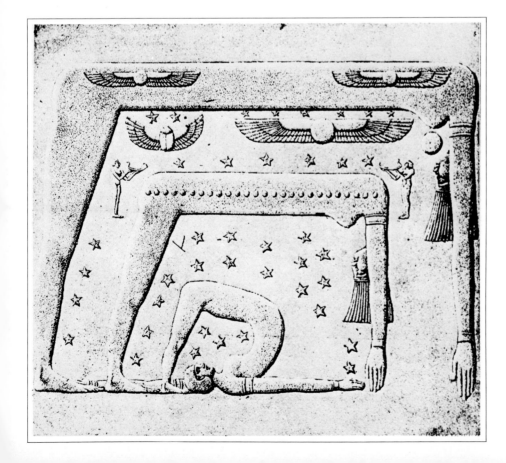

From Early Cosmologies to Abstract Thinking (2500–100 B.C.)

Neither Greek scholars nor their systematic observations burst upon the Mediterranean scene from nowhere. The *idea* of thinking about abstract physical concepts can be traced back to early **cosmologies:** theories about the origin and nature of the universe. One important group of early cosmologies were those of the Egyptians, who identified separate natural forces and assigned personalities to each one. Various cosmological theories developed, assigning different roles to different godlike personages who interacted with the real world. For example, the *Memphite Theology* (ca. 2500 B.C.) spoke of an intelligence that organized the "divine order" of

the universe. According to Wilson (1951), this theory's

insistence that there was a creative and controlling intelligence, which fashioned the phenomena of nature and which provided, from the beginning, rule and rationale, was a high peak of pre-Greek thinking.

This kind of thinking stepped toward astronomical science, because it assigned different gods, intelligences, or forces to different kinds of natural events, seeking relations between them. Early naturalists expressed these relations in myths; later naturalists expressed them in "laws," or generalizations derived from repeated observations.

It is interesting to trace how different systems of thought evolved. Many ideas seem to have carried over directly from the world of 3,000 years ago to us today. For instance, the dominant god among all Egyptian gods came to be Amon-Re, to whom the solstitial temple at Karnak was dedicated (see Chapter 1). Around 1350 B.C. the revolutionary Akhenaten, husband of the famous Queen Nefertiti, created a new religion based on the idea of a single god, Aton, the disk of the sun. Akhenaten's heresy was shortlived because he was overthrown by priests of the old religion. They installed a new young pharaoh, Tutankhamen, whose treasure-filled tomb astonished the world when it was discovered in 1922. Out of this world, around 1300 B.C., escaped a tribe of nomads whose book of monotheistic religious thought is the core of Western religious principles today. Some scholars believe their Psalm 104 may be a direct transcription of a hymn to the sun written by Pharaoh Akhenaten himself and still preserved (Pritchard, 1955). The Pharaoh wrote:

*How manifold it is, what you have
made! . . . You created the world
according to your desire . . . All
men, cattle, and wild beasts. . . .*

The Hebrews, in turn, proposed a
cosmology to explain the phenomena
they saw in nature:

*In the beginning God created the
heavens and the earth. The earth was
without form. . . .*

This cosmological theory asserted that
the earth was created in six stages, or
days: (1) light, day, night; (2) sky;
(3) dry land, ocean, plants; (4) sun,
moon, stars; (5) sea creatures, birds;
(6) humans.

Cosmological theories of this type,
which were especially common in the
Middle East, stimulated new questions
about relations of phenomena in the
universe. Out of this thinking came an
important new idea. Regardless of
the question of gods, facts could be
learned about nature by systematic
observations and experiments from
which repeatable results could be
obtained. A naturalist in Alexandria
could get the same results as an earlier naturalist in Athens. **Science** (from
Latin, to know) is simply the process
of learning about nature by applying
this technique: questions are formulated that can be answered by observation or experiment, and then the
observations or experiments are carried
out.

This system of scientific observation
and recording of nature was developed
most highly in Greece.[1] In addition
to richly provocative Egyptian and
Hebrew sources, the Greek world
received a legacy of astronomical concepts, such as the ecliptic, zodiac,

solstices, saros cycle, the sexagesimal
(six-based) system of measuring time
and angles, and generations of astronomical observations of eclipses and
planet motions from other Mesopotamian (and perhaps European)
sources.

Figure 2·2 **English painter William Blake's
version of the Western creation myth
reflects the idea that the universe was
created according to rational principles.
This idea contributed to the birth of
modern science, for it allowed philosophers
to study the evolution of the universe
without considering sudden changes in
natural laws that might be caused by a
pantheon of capricious gods.** (Photograph
courtesy of the Bettmann Archive, Inc.)
▼

Early Greek Astronomy (ca. 600 B.C.)

Around 600 B.C. the Greeks began
vigorously applying this new way of
learning about the universe. They
talked more of tangible physical "elements" and less of metaphysical relations. They realized that geometric
principles could be used to measure
cosmic distances as well as farmyards.

One of the first known Greek
thinkers was Thales of Miletus (a
Greek-dominated town in present-day
Turkey). Living about 636–546 B.C.,
Thales was a noted statesman, geom-
eter, and astronomer. He is best

---

[1]Along with this scientific development
around 1000 B.C. arose the pseudo-science
of astrology, described in more detail in
Enrichment Essay B.

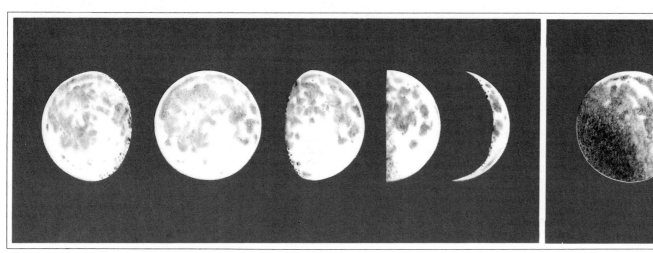

A   B

▲

Figure 2·3 **The phases of the moon (A) indicated to some Greeks that the moon is not a disk but a sphere illuminated by the sun. The earth's curved shadow on the moon (B) during every lunar eclipse also suggested that the earth is spherical.**

known for predicting the peacemaking solar eclipse mentioned in Chapter 1. He probably knew some Mesopotamian astronomical concepts—perhaps the saros cycle, the lengths of seasons, and the daily-changing position of the sun among the constellations of the ecliptic. Thales also reportedly speculated that the sun and stars were not gods, as was then usually thought, but balls of fire. Of course, Thales could not prove his idea, but he got other Greeks thinking in terms of tractable, physical ideas.

Thales' school produced several notable thinkers. Anaximander (ca. 611–547 B.C.) made astronomical and geographical maps, and speculated on the relative distances of the sun, moon, and planets from earth. He said that the matter from which things were made was an eternal substance:

> . . . the source from which existing
> things derive their existence is also
> that to which they return at their
> destruction. . . .

Heraclitus (ca. 535–ca. 475 B.C.) made this remarkable comment:

> This ordered cosmos, which is the
> same for all, was not created by any
> one of the gods or by mankind, but it
> was ever and shall be ever-living Fire,
> kindled in measure and quenched in
> measure. . . . The fairest universe is
> but a dust-heap piled up at random.

*The Pythagoreans:
A Spherical,
Moving Earth (ca. 500 B.C.)*

Pythagoras (flourished 540–510 B.C.), famous for his theorem on right triangles, was also one of the first experimental scientists. Pythagoras proposed the then unusual idea that the earth is spherical. He may have gotten this idea by studying the phases of the moon. The lunar **terminator** (the line separating the lit side from the unlit side) changes its curvature as the moon's phases progress, thus revealing that the moon is spherical rather than a flat disk (Figure 2·3). By analogy, then, the earth, the sun, and other bodies might also be spherical.

In southern Italy, Pythagoras founded a school that had wide influence around 450 B.C. It is unclear, however, which thinkers should be credited with which ideas in this school. Pythagoras himself put the earth at the center of the universe, but later Pythagoreans proposed that it moved, like the moon and the planets, around a distant center. The universe was spherical with a central "fire" containing a force that controlled all motion. Around it, in order outward from the center, moved the earth, the moon, the sun, the five planets, and the stars. This system predates by more than 2,000 years Copernicus' revolutionary model of the planets moving around the sun. The idea of a spherical earth persisted among some Greeks, though it was not universally accepted.

Anaxagoras (ca. 500–ca. 428 B.C.) is credited with deducing the true cause of eclipses. Thereafter, the observed roundness of the earth's shadow on the moon undoubtedly helped to establish the theory that the earth itself is a spherical body. After 30 years' residence in Athens, Anaxagoras was charged with impiety and banished for saying that the sun was an incandescent "stone" larger even than Greece.

Plato (ca. 400 B.C.)

Though known primarily as a philosopher, not a scientist, Plato (ca. 427–347 B.C.) reasoned that astronomy contributed to the civilization of humanity. In *Timaeus*, he said that *philosophy came from astronomy:*

> *. . . had we never seen the stars, the sun, and the heavens, none of the words we have spoken about the universe would have been uttered. But now the sight of day and night, and the revolutions of the years, have created number and given us a concept of time as well as the power of enquiring about the nature of the universe: and from this source we have derived philosophy. No greater good ever was or will be given by the gods to mortal man.[2]*

Aristotle: The Earth Again at the Center (ca. 350 B.C.)

The most influential Greek scientist–philosopher was Aristotle (384–322

---

[2]In the 1800s Ralph Waldo Emerson extended this thought with the lines:

> *If the stars should appear one night in a thousand years, how would men believe and adore, and preserve for many generations the remembrance of the city of God?*

Science fiction writer Isaac Asimov ironically reversed this thought in a 1941 story called "Nightfall." He imagined a planet in a multistar system with six suns. Only once in 2050 years did the orbits of all the suns bring them all to one side of the planet, plunging the other side into darkness so that the stars came out. Instead of inspiring worship, this event so frightened the inhabitants that they burned everything around them to give them light. The result was that civilization on this planet was burned to the ground every 2050 years and had to start over again.

B.C.). His views, built on earlier knowledge, were biased in favor of absolute symmetry, simplicity, and an abstract idea of perfection. Aristotle's universe was spherical and finite, with the earth at the center. Planets and other bodies moved in a multitude of spherical shells centered on the earth. The shells were supposed to turn with varying rates, which explained the observed changeable motions of the planets.

Aristotle is credited with founding modern scientific investigation. His school at Lyceum (a grove near Athens) contained a library, a zoo, and lavish physical and biological research equipment paid for by his onetime pupil Alexander the Great, then ruler of Greece. Aristotle became the indisputable Greek master, his authority unchallenged for 2,000 years. Therefore, his rejection of the Pythagorean idea and his placement of the earth instead of the sun at the center, turned out to delay progress in astronomy. However, there was little reason for him to choose a sun-centered over an earth-centered system, since *either view was consistent with observations known in his time.*

Aristotle was right in several important astronomical ideas.

1. He thought the moon was spherical.

2. He argued that the sun was farther away than the moon because:
   a. *the moon's crescent phase shows that it passes between earth and sun;*
   b. *the sun appears to move more slowly in the sky than the moon (this second argument is not rigorous, but the first is).*

3. He thought the earth was spherical because:
   a. *the curvature of the moon's terminator rules out a disk, and if the moon is a sphere, the earth is likely to be one, too;*
   b. *as a traveler goes north, more of the northern sky is exposed while the southern stars sink below the horizon—a circumstance that would not arise on a flat earth.*

The *apparent* motions of the sun, moon, and stars around the earth could be explained, said Aristotle, either by their actually moving around us or by movement of the earth. But Aristotle concluded that the earth is stationary, and gave a very powerful argument. If the earth were moving, we ought to be able to see changes in the relative configurations of the various stars, just as, if you walk down a path, you see changes in the relative positions of nearby and distant trees. If you line up a tree in the middle distance with a very distant tree, and then step to the right or left, the nearby tree will seem to shift left or right with respect to the distant one. Such a shift in position due to your own motion is called **parallax,** or a **parallactic shift.** If the earth were moving in a straight line we would see a continuous parallactic shift of the nearer stars with respect to more distant stars; and if the earth were moving around some distant center we would see a periodic parallactic shift back and forth among the stars. But a visual survey of the stars and the constellations, from year to year and season to season, showed no evidence of such a shift. So, reasoned Aristotle, the earth must not move.

The reasoning was sound, but the observations were not precise enough to reveal the parallactic shifts, which actually exist. It took years of observation by telescope to discover real parallactic motions among the stars.

Aristotle died shortly after being forced to leave Athens for allegedly teaching that prayer and sacrifices to the Greek gods were useless.

Aristarchus: Relative Distances and Sizes of Moon and Sun (250 B.C.)

Aristarchus (ca. 310–230 B.C.) of Samos (an island off present-day Turkey) cleverly extended the Greek methods of seeking quantitative data. His only surviving work is "On the Sizes and Distances of the Sun and Moon," although his other astronomical works are quoted by other Greek authors.

To estimate the relative distances of the sun and moon from earth, Aristarchus imagined a triangle, *E* (earth), *M* (moon), *S* (sun) (Figure 2•4). He chose a first- or third-quarter moon because the angle *EMS* is then a right angle. The angle *MES* could be measured from earth, or better yet determined by observing the arc traveled by the moon from first quarter to third quarter, which is 2 × *MES*. Then if we call any side of the triangle, for example *ES,* one unit in length, the triangle can be laid out and the *relative* (but not the absolute) distance measured. Aristarchus concluded that the sun was 18 to 20 times as far away as the moon. His method was essentially correct, but he could not measure angle *MES* accurately enough. The sun is really about 380 times as far away as the moon. Aristarchus also formulated a way to measure the relative sizes of the earth and moon. We can reconstruct his calculations, which used what he knew about eclipses. Assuming that a total eclipse of the moon was in progress (Figure 2•5), he drew a circle to represent the earth. Because he evidently knew the sun's angular diameter was $\frac{1}{2}°$, he could represent the earth's umbral shadow by lines leading away from the earth, converging at an angle of $\frac{1}{2}°$. Aristarchus knew, too, that the moon is three-eighths as big as the earth's shadow, so that wherever he placed the moon in the earth's shadow, its diameter would have to be three-eighths the distance across the shadow at that point. But where to place the moon's circle? Aristarchus knew that the moon, like the sun, has an angular diameter of $\frac{1}{2}°$, as seen from earth. These two criteria specified the position and size of the moon, as trial and error with sketches will show (Figure 2•5).

What about the sun? By guessing that the sun was 18 to 20 times as far away as the moon, Aristarchus applied the observation that the sun's angular diameter is $\frac{1}{2}°$, like the moon's, thus fixing its relative size. Aristarchus concluded that the moon was one-third as big as the earth, and that the sun was about 7 times as big as the earth. The correct figures are closer to $\frac{1}{4}$ and 100, but Aristarchus was on the right track.

Aristarchus made still another contribution. Because he thought the sun was much bigger than the earth, he guessed (without much supporting observation, in this case) that the sun, not the earth, must be the central body in the system. (For this an outraged critic declared he should be indicted for impiety.)

Although Aristarchus made some quantitative errors, he was nonetheless far in advance of later scholars who thought the earth was flat. Aristarchus correctly visualized the moon in orbit about a round earth, the earth in orbit around the sun, and the method of measuring interplanetary distances.

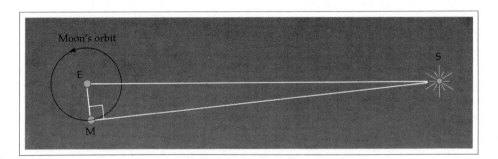

▲ *Figure 2•4* **The geometry by which Aristarchus estimated the relative distances of the moon and sun, around 250 B.C. (not drawn to true scale).**

Figure 2•5 **The geometry by which Aristarchus estimated the relative sizes of the moon and the earth (not drawn to true scale).**
▼

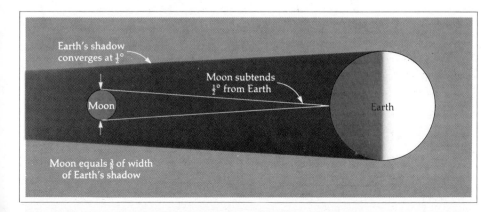

These ideas were not confirmed for another 2,000 years!

Eratosthenes: Earth's Size (200 B.C.)

As Greece declined and Rome prospered, Greek scholars became resident intellectuals in many parts of the Mediterranean world. Eratosthenes (ca. 276–ca. 192 B.C.) was a researcher and librarian at the great **Alexandrian library** in Egypt. He reportedly completed a catalog of the 675 brightest stars, and measured the inclination of the earth's polar axis to the ecliptic plane as shown in Figure 2·8. This inclination, $23\frac{1}{2}°$, causes our seasons,

pointing the northern hemisphere toward the sun in summer and away from the sun in winter.

Eratosthenes is most famous for using geometric relations to measure the earth's size. Told that at summer solstice the sun shone directly down a well near Aswan, he noted that the sun's direction was off vertical by $\frac{1}{50}$ of a circle on the same **date** at Alexandria (Figure 2·9). He realized this difference had to be due to the curvature of the earth, and concluded that the earth's circumference was 50 times the distance from Alexandria to the site of the well. His estimate was probably within 20 percent of the right answer. This Greek master clearly

understood the shape and approximate size of the earth 1,700 years before Columbus.

Hipparchus: Star Maps and Precession (ca. 130 B.C.)

From his observatory on the island of Rhodes, Hipparchus (ca. 160–ca. 125 B.C.) observed the positions of astronomical bodies as accurately as possible and compiled a catalog of some 850 stars. His exhaustive observations—all done, of course, without a telescope—enabled him to predict with reasonable accuracy the position of the sun and moon for any date. Hip-

---

Optional Math Equation I · The Small Angle Equation

To realize that the same simple geometry used in laying out a pasture could be used to map interplanetary distances was a major breakthrough. It required familiarity with the difference between **linear measure** (measurement of the linear distance between two points, commonly measured in meters or kilometers) and **angular measure** (measurement of the difference in direction between two points as seen from a certain specified point, commonly measured in degrees).

It is thus important in astronomy to be able to use angular measure, based on the following system:

$$\text{full circle} = 360 \text{ degrees } (360°)$$
$$1° = 60 \text{ minutes of arc } (60')$$
$$1' = 60 \text{ seconds of arc } (60'')$$

Some idea of angular sizes can be gained by remembering that the angular size of the sun and moon is about $\frac{1}{2}°$ or 30'. This is roughly the size of your little fingernail viewed at arm's length. The smallest objects whose shape you can detect, or resolve, is roughly 2' (or 120''), depending on your eye's keenness. The disks of the brighter planets are around 20'' to 60'' across, and hence cannot be seen by the naked eye.

Angles and linear measures can be combined in an extremely useful and simple equation called the **small angle equation,** which involves the angular size of an object, its linear size, and its distance. If any two of these quantities are known, the third can be calculated. Let us call the angular size α'', to be expressed in seconds of arc. Let the diameter of the object be d and its distance, D. Then the small angle equation is:

$$\frac{\alpha''}{206265} = \frac{d}{D}$$

The number 206265 is called a *constant of proportionality;* it stays the same in all applications of the equation.

Consider an example. Suppose a friend who is 2 m tall is standing across a field, where he subtends an angle of $\frac{1}{2}°$, or 1800'', as shown in Figure 2·6. How far away is he? We want to solve the equation for D. Rearranging the equation, we have

Figure 2·6 An application of the small angle equation. If your friend is 2 m tall and subtends an angle of $\frac{1}{2}°$ (or 1800 seconds of arc), his distance D is 230 m.

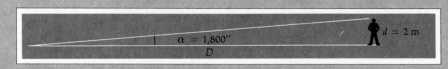

$$\alpha = 1,800''$$
$$D$$
$$d = 2 \text{ m}$$

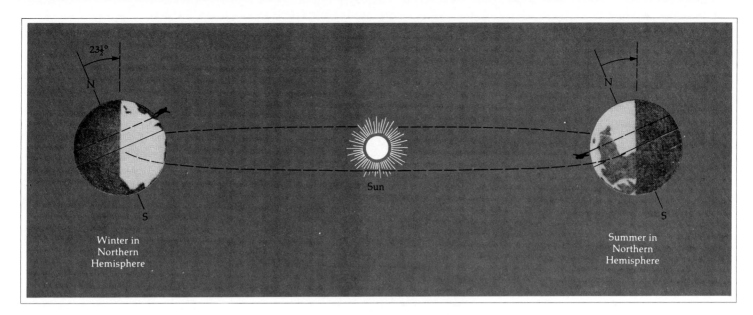

▲
Figure 2·8 The seasons are caused by the earth's inclination of $23\frac{1}{2}°$ to its plane of orbit around the sun. Eratosthenes measured the difference between the sun's noontime elevation in summer and winter at a fixed latitude (dashed straight line). The angular difference between these two elevations is twice the earth's inclination.

$D = 206265\, d/\alpha''$. Using cgs units, we would write $d = 200$ cm. Thus, expressing the equation in cgs units and powers of ten, we would have

$$D = \frac{206265\, d}{a''}$$

$$= \frac{2.06\ (10^5)\ 2\ (10^2)}{1.8\ (10^3)}$$

$$= 2.3 \times 10^4 \text{ cm} = 230 \text{ m}.$$

Your friend is about one-sixth of a mile away.

As the Greeks realized, exactly the same geometry can be used to investigate astronomical distances. The Greeks could not get good measurements of the moon's diameter, but only its angular size, α, which is roughly $\frac{1}{2}°$ or 1800''. If we use the modern knowledge that the moon is about 3,500 km in diameter, we can estimate its distance just as you did

for your friend's distance above. In cgs units, d would be 3.5×10^8 cm. The equation would read:

$$D = \frac{206265\, d}{\alpha''} \simeq \frac{2.06\ (10^5)\ 3.5\ (10^8)}{1.8\ (10^3)}$$

$$\simeq 4 \times 10^{10} \text{ cm} \simeq 4 \times 10^5 \text{ km},$$

or about 400,000 km.

Several mathematical notes should be observed. First, the symbol $\simeq$ means "approximately equal to"; it is useful whenever approximate values (such as $\frac{1}{2}°$) are involved.

Second, this calculation shows the economy of writing the numbers as powers of ten, for example 3.5×10^8 cm instead of 350,000,000 cm.

Third, the answer is given only to an accuracy of one significant figure. **Significant figures** are the number of digits *known for certain* in a quantity. For example, π (pi) to one significant

figure is 3; to three significant figures it is 3.14. Generally, an answer to a calculation should have no more significant figures than does the poorest-known or least accurate number in the equation.* In this case that number was the angular size of the moon, $\frac{1}{2}°$ or 0.5°, with only one significant figure.

A fourth note is that, following our rule given in the prologue, we converted data to cgs units. (In our small angle equation this would have been unnecessary, since the dimensions of d and D are the same and cancel out; but it is a useful habit.)

---

*This statement is especially important to users of electronic hand calculators, which blindly print answers with seven or eight significant figures, even when the accuracy of the input is perhaps only two significant figures.

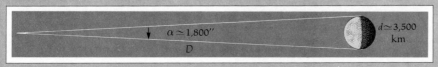

◀ *Figure 2·7* The same geometry as in Figure 2·6 can be applied to measure interplanetary distances, such as the distance to the moon.

parchus has been called antiquity's greatest astronomer, though much of his raw material probably came from Babylon.

The most important discovery attributed to Hipparchus is *precession* (though astronomers of earlier centuries may have been aware of its effects—see Chapter 1). Comparing his own measurements of star positions with materials handed down to him from centuries before, Hipparchus found that, with respect to the background stars, there had been curious shifts in the positions of the north celestial pole, the vernal and autumnal equinoxes, and other coordinates. The whole celestial equator was oriented somewhat differently with respect to the stars! Could the old maps be wrong? Hipparchus concluded instead that the whole coordinate system of celestial equator and poles was drifting slowly with respect to the distant stars. This drift came to be known as *precession,* or **precession of the equinoxes.**

In modern terms, *precession is the result of a wobble of the spinning earth, due to forces produced by the sun and moon.* Just as a spinning top describes a conical wobble because of the earth's gravity, the earth's polar axes describe a conical wobble with respect to the

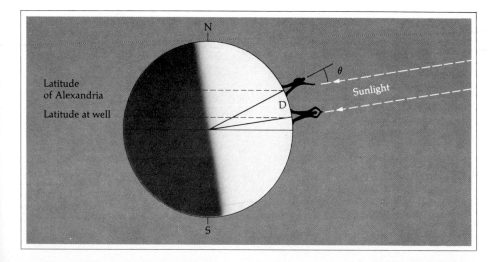

A

fixed stars; hence each celestial pole describes a circle among the stars (Figure 2•10). The circle, $23\frac{1}{2}°$ in radius, is centered on the ecliptic poles. Thus, as mentioned in the last chapter, in different millennia, different stars become the north star; a complete cycle takes about 26,000 years. All star coordinates, which are measured with respect to the celestial equator, therefore change slightly each year.   •

Hipparchus also contributed to the description of solar and planetary motions. Although a few of Hipparchus' contemporaries accepted Aristarchus' idea that the earth moved around the sun, Hipparchus and most other astronomers thought this unnecessarily complex. His own observations showed that the sun's apparent motion with respect to the stars was not uniform from day to day; he therefore

◄ *Figure 2•9* **The geometry of Eratosthenes' measurement of the size of the earth. When the sun was directly over a certain well, Eratosthenes measured the angle** θ **at Alexandria, a known distance (D) away, and found D was $\frac{1}{50}$ of the way around the earth.**

◄ *Figure 2•10* **A. This ten-minute exposure shows stars in the north polar region, trailed by their daily motion around the north celestial pole near Polaris, the bright star in the lower right. B. Map of stars in the same region shows the Little Dipper (prominent in photo A) and the changing path of the north celestial poles in previous centuries. Because of precession, the north celestial pole 2,000 years ago did not lie near a prominent star. About 4,600 years ago, Alpha Draconis (lower left) was the north star. The north celestial pole moves in a circle around the ecliptic pole (top).**
▼

concluded that the sun moved around the earth in a circular orbit whose center was *not* the earth, but was slightly offset.

Like so many ancient works, most of Hipparchus' writings have disappeared except in others' reports—according to which Hipparchus also studied the relative distance from earth of the moon and the sun. He calculated that the moon was $29\frac{1}{2}$ earth diameters away, close to the correct value of

about 30. Hipparchus apparently realized that the sun must be much farther away than Aristarchus' estimate of 20 times the lunar distance.

Ptolemy:
Planetary Motions (A.D. *150*)

Claudius Ptolemy (flourished ca. A.D. 140) was another scholar associated with the Alexandrian library. His

B

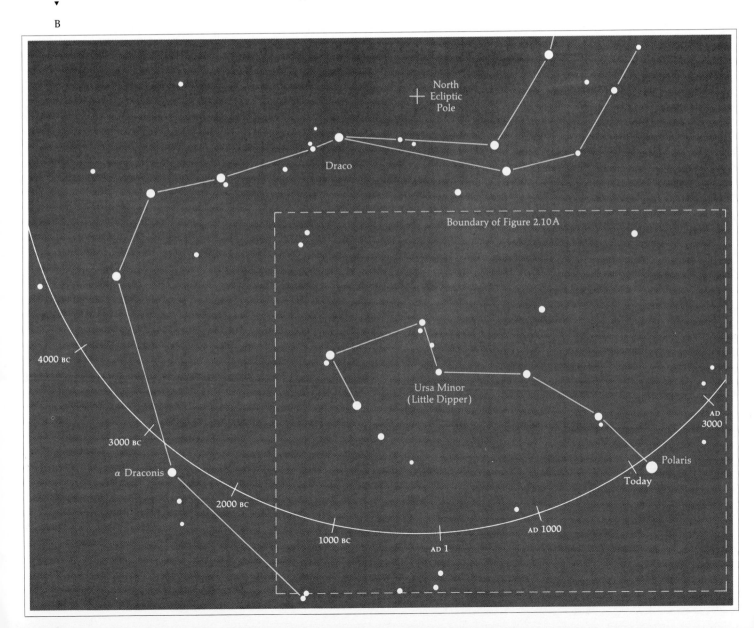

fame as an astronomer is based on a 13-volume work *The Mathematical Collection.* Passed to the Arabs after the destruction of the library, the work became known as *al Megiste* (the greatest), and eventually in Europe as the *Almagest,* a name known for a thousand years.

So slow was the progress of science and so great the influence of the earlier Greeks, that Ptolemy refers to Hipparchus (who died perhaps 240 years earlier) almost as a contemporary, and it is sometimes hard to separate their contributions. A tra-

dition arose that Ptolemy merely synthesized earlier work, especially that of Hipparchus. However, a good part of Ptolemy's work is now seen as original and puts him among the important astronomers of all ages.

Ptolemy extended the star catalog to 1,022 entries, correcting older reported star positions to compensate for precession.[3] But his best-known contribution was a *method for predicting the positions of the sun, moon, and planets,* called the **epicycle theory.** Following the tradition of his day, Ptolemy assumed that the earth was at the

center. In *circular* orbits, in order outward from the earth, he placed the moon, Mercury, Venus, the sun, Mars, Jupiter, and Saturn. To explain why planets apparently do not move at uniform rates, Ptolemy devised combinations of circular motions. Each planet, he said, moved in a circle called the deferent, whose center was offset from the earth. The planet did not move strictly on this circle, but in a smaller circle called an epicycle, whose center moved along the deferent just as the moon moves around the earth while the earth moves around the sun. Astronomers who applied this theory and its later changes during the Middle Ages generally predicted within a few degrees the actual positions of planets.

Ptolemy has been criticized for abandoning Aristarchus' earth-centered system, and for biasing his theory toward a supposed "perfection" of the circle, thus delaying the introduction of the true elliptical orbits. But the system fitted the observations available in Ptolemy's time. Aristotle's old argument still seemed true: The earth could not move around a distant center, because no one had observed parallax. Certainly no one had *observed* elliptical motions. And the system worked. The trouble with Ptolemy's choice of Aristotle over Aristarchus lies not in its being a "mistake," but in its historical effects: The *Almagest* became the epitome of ancient astronomy, and the earth-centered system held sway for 1,400 years.

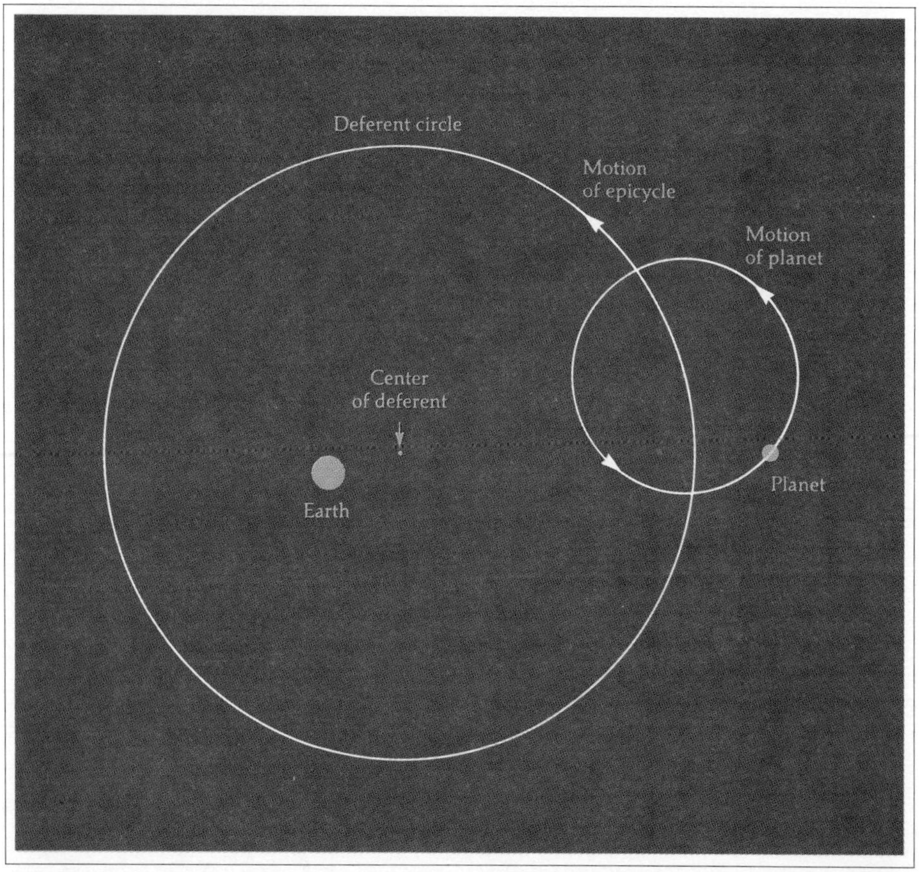

The Loss of Greek Thought (ca. A.D. 500)

Alexandria, where Cleopatra first fascinated Julius Caesar and Marc Antony around 47 B.C., was the world's intellectual center by A.D. 250. With

---

Figure 2·11 **Ptolemy's system for explaining planetary motions. Ptolemy thought each planet moved in a circular epicycle, whose center moved in a circular orbit (deferent) around a point near the earth.**

▼

[3]Correction for precession continues today. For precise setting of a large modern telescope, a star's coordinates published for "epoch 1950" or any other year must be corrected for the current date, or the star could be missed.

Deferent circle

Motion of epicycle

Motion of planet

Center of deferent

Earth

Planet

Rome's fall and the world in disorder in 410, maintaining the great library of ancient discoveries was becoming more and more difficult.

Among the last guardians of the old knowledge in Alexandria was one of the first known woman astronomers, Hypatia (ca. A.D. 375–415). She wrote a commentary on Ptolemy's work and invented astronomical navigation devices, but was murdered by a mob during one of the riots that plagued Alexandria during the decline of the ancient world. In A.D. 640, after a 14-month seige by the Arabs, Alexandria fell and the library was reduced to shambles. The buildings were burned and the best collection of Greek knowledge was irrevocably lost. In the absence of printing, there were few other reference books. The rate of new discoveries declined, and Europe slipped into the Dark Ages.

Al-Battani and the Arab Astronomers (ca. A.D. 900)

The Arabs inherited Greek science from Alexandria and from other parts of the world. For example, they were influenced by Indian astronomical-mathematical writings, called siddhantas, incorporating Greek ideas that had reached India as early as A.D. 450. In 773, the Arab court was visited by an Indian who could predict eclipses, and the caliph ordered the Indian books translated. The tables contained the system of numerals (including zero) we now call arabic numerals.

The next recorded measurement of the earth's circumference was made under the auspices of the Caliph Abdullah al-Mamun about 820, near Baghdad. The result was only 3.6 percent too large. About the same time, Muhammad al-Khwarizmi published the first Arab astronomical tables, whose Indian origins are revealed by their reference longitude, at an observatory in central India. Al-Khwarizmi's tables and his work on arithmetic were widely used in Europe centuries later, and his name survives in the word *algorithm:* an arithmetic formulation.

The Arab astronomer best known in medieval Europe was Muhammad al-Battani (ca. 858–929), known in the West as Albategnius. Al-Battani measured the length of the year to within a few minutes. Following the Ptolemaic conception, he remeasured many quantities related to solar and planetary movements. Among these was a value for the eccentricity (noncircularity) of the earth's orbit (he would have said the sun's orbit) only 4 percent too big. His originality is shown by several corrections he made to Ptolemy's results.

The Moslem empire spread to Spain, an important event because it brought both ancient learning and contemporary Arab science to Europe. About A.D. 1000, Al-Khwarizmi's astronomical tables were published in Spain with the reference longitude corrected to Cordoba, Spain. Astronomical practice then spread in many directions. In the 1200s a grandson of Genghis Khan started an observatory in Persia, while a new set of planetary tables was compiled in Spain. In the 1400s Ulugh Beg, a grandson of the

Table 2·1 Astronomical Discoveries of the Greeks		
Observation	Inference	Observer Commonly Quoted
Curved lunar terminator	Moon round	Pythagoreans
Round shadows during lunar eclipses	Earth round	Pythagoreans
Crescent phases of moon	Moon between earth and sun	Aristotle
Different stars at zenith at different latitudes	Earth round	Aristotle
No evident stellar parallax observed by naked eye	Distance earth moves is small compared to distances to stars	Aristotle
Relative sizes and angles of moon and earth's shadow	Moon smaller than earth, and sun bigger than earth	Aristarchus
Angle from first quarter to last quarter moon = 180°	Sun tens of times farther away from earth than moon	Aristarchus
Relation between angular shift of zenith position among stars and linear distance traveled on earth	Calculable circumference of earth	Eratosthenes
North celestial pole's shift with respect to constellations	Precession	Hipparchus

Mongol conqueror Tamerlane, established a famous observatory at Samarkand, in what is now the southern Soviet Union. Ulugh Beg became the first post-Greek astronomer to compile a new star catalog that was not merely a copy of Ptolemy's.

The Arab astronomers improved the old observations and constructed beautiful astronomical instruments now seen in many museums. According to Pannekoek (1961), some, such as al-Battani, related astronomy to their practice of religion in finding their place in the universe. However, Pannekoek notes:

> Their aim, however, was not to further the progress of science—this idea was lacking throughout. . . . The tables were needed for astrological purposes. Hence the astronomers had to compute and publish them ever anew.

In any case, Arab mathematicians and astronomers were instrumental in transmitting and improving scientific practice.

The Universality of Astronomy

Although it is the Mesopotamian–Greek–Arab stream of astronomy that most clearly feeds modern Western conceptions, other astronomical centers, partly independent in their development, did influence the West. Early developments in India, for example, almost certainly influenced the Mesopotamians; perhaps future work will show they were even more important than we know. Other cultures' astronomy remained isolated. The next sections highlight some examples of non-Western astronomy.

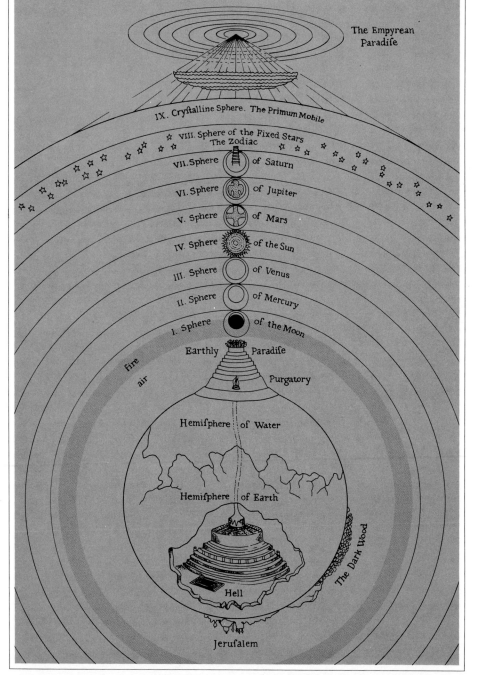

◄ Figure 2·12 A diagram of the universe, after a medieval sketch based on the description by Dante in the *Divine Comedy*, ca. 1310. This diagram shows the persistence of Ptolemy's conception 1,160 years after he did his work. The heavens are arranged in seven concentric planetary spheres with the earth at the center. Beyond lies the starry zodiac and an outer "crystalline sphere" marking the boundary of the universe.

*Astronomy in
India: A Hidden Influence*

Astronomical practices in India reportedly date back to about 1500 B.C. The first known astronomy text, describing planetary motions and eclipses, and dividing the ecliptic into 27 or 28 sections, appeared around 600 B.C. By this time India had contact with the Mesopotamian and Greek worlds, and influences probably traveled both ways. Texts dating from around A.D. 450 use Greek computational methods and refer to the longitudes of both Alexandria and Benares, a major Indian astronomical center. Arabs who later visited India wrote of Brahmagupta (ca. 1050) as one of the greatest Indian astronomers, who reportedly helped introduce Graeco-Indian astronomy to the Arabs.

Unfortunately, most records of this fertile early period of Indian astronomy were destroyed during invasions in the 1100s. The great center at Benares was destroyed in 1194, and various university libraries of Buddhist and other ancient literature were burned in religious wars. A massive observatory—one of the world's five major observatories by the 1700s—was reestablished at Benares in later centuries (and damaged again by invading religious fanatics).

*Astronomy in
China: Modern Ancients*

Tradition that Chinese astronomers were predicting eclipses before 2000 B.C. is legend; scholars put the appearance of such techniques in China closer to 1000 B.C. Thus, Chinese astronomy flourished about the same time as the main Greek flowering, probably influenced by the same

Middle Eastern cultures. The excellent Chinese observations include the world's best lists of the mysterious "guest stars," or novae and super-novae, recorded from 100 B.C. to the present, and still consulted by modern astronomers.

Chinese conceptions of the universe were strikingly modern. As early as

Figure 2·13 **An interpretation of ancient Oriental observations of a comet. A teacher is discoursing on the significance of the**
▼

120 B.C. a Chinese text stated (quoted by Needham, 1959):

> *All time that has passed from antiquity until now is called* chou; *all space in every direction . . . is called* yü.

The Chinese term for the universe, *yü-chou,* was thus similar to the modern

celestial visitor. (Painting by Chesley Bonestell)

concept of "space–time," a four-dimensional continuum encompassing all that has existed.

Another Chinese statement of this period is apt (quoted by Needham, 1959):

> *The Earth is constantly in motion, never stopping, but men do not know it; they are like people sitting in a huge boat with the windows closed; the boat moves but those inside feel nothing.*

Contrast this with Aristotle's view, which, for most Westerners, put a stationary earth at the universe's center. Unfortunately, these advanced ideas had little influence on Western astronomy until after the Renaissance.

The Chinese astronomer Yü Hsi reportedly discovered precession independently, and in A.D. 336 stated, "I think that the heavens are infinitely high." At that time in the West, Ptolemaic groundwork was being laid for the medieval conception that planets moved in as many as 10 separate physical shells surrounding the earth. By the time Western scholars visited China, many Chinese ideas were closer to the truth than Western concepts. In 1595, for example, Matteo Ricci listed among "absurdities" he found in Chinese astronomy (quoted by Needham, 1959),

> *They say that there is only one sky and not ten skies; that it is empty and not solid. The stars are supposed to move in the void, instead of being attached to the firmament. . . . Where we say there is air between the spheres, they affirm there is a void.*

These quotes, of course, indicate only selected speculations and not data confirmed by experiments or observations. But experimental data were not yet available anywhere.

Astronomy in Polynesia: Stars as Guideposts

Though they had cosmological myths, the Polynesians were not so much interested in understanding the relationship of worlds in space as in plotting the relationships of islands in the trackless Pacific. Driven by this need, Polynesian navigators developed amazing skills. Lacking the magnetic compass, they steered toward chosen directions by learning where certain bright stars rise and set on the ocean horizon. Those directions remain constant for each star (except for the very slow changes of precession during several generations). Modern Polynesians can sail such a constant setting that they can accurately report subtle nighttime shifts in wind direction on the open ocean even though they themselves are moved by the wind (Lewis, 1973).

Polynesian navigators found their latitude by noting which stars passed directly over their masts. For example, in A.D. 1000, if Aldebaran passed overhead, the boat was about 5° south of Hawaii's latitude; if Arcturus, about 100 miles north of that latitude. This method was accurate to about 0.2 to 0.5 degrees, or about 12 to 30 miles. To reach, say, Hawaii from Samoa, one could simply sail north to the correct latitude, then east until the island was sighted.

American Indian Astronomy: Science Cut Short

Many people still underestimate the sophistication of American Indian civilization. American natives, as was noted in Chapter 1, built astronomically aligned observatories, tracked and recorded planetary positions, and devised calendars based on eclipse cycles. European explorers found them conducting astronomical observations of solstice dates, planetary positions, and so on. Much of this art developed rapidly between A.D. 200–1200, but after 1300 Mesoamerican influence declined and "frontier" cultures in the Mississippi valley and Southwest virtually collapsed. It is interesting to speculate how far American civilization might have evolved had it escaped European influence for one more millennium.

2

Summary

We have seen how all varieties of cultures, preliterate and technological alike, moved toward certain concepts of the earth as a world among other worlds in space. Some of these concepts were purely practical; some, abstract. These movements came in fits and starts, and were scattered over various parts of the world. Progress toward knowledge has not been continuous. Cultures have advanced and then slid back, depending on their stability and vigor.

Many of the discoveries reviewed here dealt with the relation of the earth, the moon, and the planets. Table 2·1 furnishes a good review of key observations by the Greeks. Their advances, among all those of antiquity, were most important in influencing Western scientific thought. Although conceptual models of the universe differed from culture to culture, all cultures moved toward discovering astronomical relationships.

Concepts

cosmologies

science

terminator

parallax, parallactic shift

Alexandrian library

linear measure

angular measure

small angle equation

significant figures

precession of the equinoxes

epicycle theory

Problems

1. How critical can we be of early theorists who believed the earth was at the center of the universe? Explain why or why not by considering the following questions.
   a. *Did they have any basis for not putting the earth at the center?*
   b. *Did either theory fit the available observations?*
   c. *Did any of the Greeks prove that any celestial bodies revolve around a central earth or do not do so?*

2. As the moon goes through its phases:
   a. *Why is its terminator usually curved?*
   b. *At what lunar phase is it straight?*
   c. *At what lunar phase is it not seen?*

3. Contrast the types of astronomical observations available to the Greeks with those available today.

4. Do you agree with the quotations from Plato and Muhammad al-Battani implying that astronomical events may have influenced the beginnings of philosophizing about the universe?

Could astronomical phenomena have influenced religious concepts?

5. Do you think Aristotle's faith in symmetry and "perfection" helped or hindered his investigations of the universe?

6. Why is the moon's umbral shadow on the earth much smaller than the moon itself? (Hint: See Figures 2·4 and 2·5.)

7. How does Hipparchus' discovery of precession prove the existence of earlier careful astronomical records or stars' positions?

8. Do you think destructive events such as the pillaging of the Alexandrian library or the sacking of the observatory and library at Benares are significant or insignificant in world history? (The answer, of course, requires that you explain your concept of "significant.")

9. Does the sun pass through the zenith every day on the equator? If not, on what dates does it do so?

Advanced Problems

10. At latitude $+40°$ will the sun ever pass through the zenith? At latitude $+23\frac{1}{2}°$? If so on what date(s)?

11. The angular diameter of the sun is roughly $\frac{1}{2}°$ and its distance from earth is about 150 million km.
    a. *Use the small angle equation to estimate the sun's diameter.*
    b. *Could Aristarchus determine the ratio of the sun's diameter to its distance?*
    c. *Why could the Greeks, such as Aristarchus, not determine the sun's linear diameter or distance?*

12. A backyard telescope (a few inches in aperture) can reveal features with angular diameter 1″. The moon is about 400,000 km away. What size

lunar crater could you see with such a telescope?

13. If you landed on a small satellite and found that walking in a 1-km straight line caused the stars in front of you to rise 1° farther above the horizon (while stars overhead also shifted by 1°), what would be the satellite's circumference? What would be its diameter?

Projects

1. Using a distant, strong light source such as the sun or a light bulb, a small ball to represent the moon, and your eye to represent a terrestrial observer, show that crescent phases of the moon prove that it passes between the earth and the sun.

2. If you travel far enough during vacation to change your latitude, compare measurements of the elevation of the north star (or the sun during daytime, taking care not to stare directly at it) made from various locations. A sighting device like that described for the problems in Chapter 1 can be used. (Point the device at the sun by watching its shadow, without looking directly at it.) How accurately can latitude be determined in this way? Measure the number of kilometers or miles corresponding to your change of latitude, and estimate the circumference of the earth (this method is similar to Eratosthenes'). The project can be done as a class effort, with different people's reports of elevation angles plotted against their latitude to give a curve showing how elevation changes with latitude. Coordinate with your instructor.

3. With a camera and fairly fast black-and-white film, such as Tri-X, make time exposures of the night sky. Try different exposures, such as 1

minute, 5 minutes, and 1 hour. Make
one series including the north star,
one toward the east or west horizon,
and one toward the south horizon.
Explain the patterns made by the
trails.

4. During a camping trip or late-
evening outing, pick an equatorial
constellation in the sky and follow its
motion. Make a series of sketches at
different hours, showing its position
with respect to the horizon. Do the
same for a circumpolar constellation
and contrast the results.

5. Using star maps, trace out the
position of the celestial equator in the
sky. Compare this to the position of
the sun's path, the ecliptic.

b

Exploring the Earth–Moon System

The earth is the cradle of humanity, but one cannot live in the cradle forever.

Konstantin Tsiolkovsky, 1899

The Earth as a Planet

More than 2,000 years have elapsed since some of the Greek naturalists realized that the earth is a moving spherical world in space. Yet only during the last 400 years have most people agreed that the world is round. Only during the last few decades have people begun to realize that the earth is only one of many *places* in space, with landscapes, days, nights, seasons, weather, and geological processes. Only the last years have brought us actual photographs of these other places with their lonely lava flows, rugged craters, dusty winds, and scudding clouds.

The visits of our spaceships to other planets have promoted a new conception of the earth. The earth is not the one, universal, stable landscape that geologists and geographers once imagined. It is just one planet out of many, one laboratory in which one set of chemical and meteorological conditions has acted to produce one set of rock types, one kind of climate, a certain set of life-forms. We easily forget that earth is simply the closest planet. If we can understand its rocks, climates, terrain, magnetism, and other properties, we can better understand what we find on other planets. Similarly, as we explore other planets, we encounter other laboratories for planetary evolution, where we can learn more about how earth evolves and how its evolution may be affected by our own activities.

To place the earth in an astronomical context, the plan of this chapter is fourfold: first, to describe some consequences of earth's rotation; second, to determine the origin and age of the earth; third, to describe the interior of the earth; and finally, to show that earth is not a homogeneous, static ball, but a dynamic world affected both by internal processes and by cosmic processes.

Earth's Rotation

That the earth is a rotating ball is obvious to us; we can view satellite films of the earth rotating. However, early evidence that the earth is a rotating body like other astronomical bodies involved several discoveries still of everyday interest.

Earth's Oblate Shape

Magellan's voyage around the world in 1522 settled that the earth is round, but measurements showed that the distance corresponding to one degree of latitude, measured by Eratosthenes' method, varies at different latitudes. Results showed that the earth is not exactly spherical, but bulges at the equator. The *equatorial diameter* of 12,756 km (about 8,000 mi) is about 43 km (27 mi, or 0.3%) longer than the *polar diameter*.[1]

The bulged shape of the earth is called an **oblate spheroid**, not a sphere. The unexpected bulge was first related to the spin around 1680 when Isaac Newton showed that only a non-spinning earth would be spherical. He found that a spinning earth would have an equatorial bulge because of centrifugal effects, and the amount of the bulge would depend on the strength and mass distribution of material inside the earth. Thus, measurement of the oblateness of the earth or any other planet can be used to analyze the properties of the material inside the planet.

Irregularities in Earth's Rotation

Originally, our time-keeping system was based on the earth's rotation, which was assumed to be constant. Modern careful time-keeping, based on very regular standards such as atomic vibrations, has revealed that earth's rotation is not exactly constant. A gradual slowdown of about 0.0008 seconds per day is caused by lunar gravitational forces, and the earth once turned much faster. Smaller, irregular changes result from shifts of

---

[1]Even the system of kilometers and meters is related to the earth's shape. In 1791, the French Academy of Sciences decided to replace the haphazard English units—foot, inch, pound, ounce—with a more straight-forward system. They defined the fundamental unit of length, the meter, as one ten-millionth of the distance from the equator to the pole. (More recently, greater accuracy has been achieved by defining the meter in terms of wavelengths of light, but it is still about one ten-millionth of the equator–pole distance.)

material inside the earth. Official clocks are adjusted for those changes.

Coriolis Drift

Another effect of earth's rotation is seen in the cyclonic storm patterns familiar from satellite photos (Figure 3·3). Consider an air parcel moving

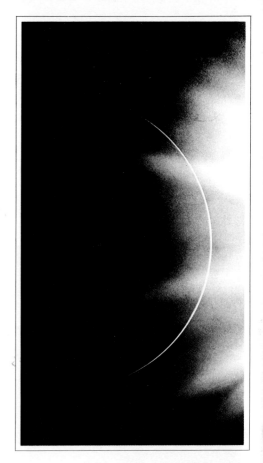

Figure 3·1 **The earth as a planet. The crescent ▶ earth was photographed by Apollo astronauts. It is backlighted by the sun nearby, just out of the frame to the right.**

Figure 3·2 **Geophysical and other observations from space have provided new information about the earth. This view by Apollo 16 astronauts reveals the cloud- and snow-covered polar regions, and an unusually cloud-free United States. Baja California, prominent beneath a few clouds, is a land mass splitting off from North America by sea-floor spreading in the Gulf of California. (NASA)**
▼

northward off the equator toward a low pressure region in the atmosphere. Because the earth is turning, the equatorial air mass would make one trip around the circumference, 40,000 km, in 24 hours. Thus, regardless of any other motions, it has an eastward motion of nearly 1700 km/hr, whereas material to the north, near the pole, has no such eastward drift. Thus, as the mass moves northward, it finds itself moving eastward faster than the ground or local air. Relative to the ground, it is drifting eastward as shown in Figure 3·4. Similarly, an air mass drifting south toward the low pressure zone would drift westward. Any such eastward and westward deflections due to the planet's rotation are called **coriolis drift.** Coriolis drift means that rotation is set up in the vicinity of low pressure storm systems, as Figure 3·4 shows; rotation is coun-

Figure 3·3 **The spiral rotational pattern of a storm off New Zealand demonstrates the coriolis effect (see Figure 3·4) caused by the earth's rotation.** (NASA photo by Skylab 4 astronauts)
▼

terclockwise in the northern hemisphere, clockwise in the southern. The reader can demonstrate on Figure 3·4 that opposite rotation occurs as air moves away from high pressure systems. Coriolis drifts explain the scroll-like cloud patterns so prominent in space photos of earth. No coriolis drifts, hence no scroll patterns, would occur on non-rotating planets.

Foucault Pendulum

In 1851, French physicist Jean Foucault set up a 60-pound pendulum on a 200-foot cord in the dome of the

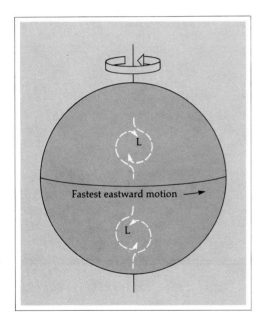

▲
Figure 3·4 **Due to the coriolis effect, air masses moving toward or away from the equator are deflected east or west relative to the ground. Air moving into low pressure areas (L) produces cyclonic storm patterns; the sense of rotation is opposite in the two hemispheres, and is a proof of the earth's rotation.**

Pantheon in Paris and made a remarkable demonstration of the earth's rotation. When he set the pendulum in motion, its direction of swing slowly shifted from north–south toward east–west. The same display of the **Foucault pendulum** can be seen in many planetariums today. To understand what is happening, consider an imaginary experiment: at the north pole we set up a massive pendulum on a universal pivot that allows it to swing in any direction. We draw back the weight carefully, lined up on a certain star. As we let go, the pendulum starts swinging ponderously back and forth on a straight path aligned with the star. If neither earth nor stars moved, the pendulum's path would stay fixed with respect to the ground and the star. If the stars were turning around an unmoving earth, the target star ought to move out of line with the pendulum's path, which would remain fixed relative to the ground. But if the earth is turning under the pendulum, then the pendulum would keep swinging toward the star but the turning of the earth would shift the pendulum's path relative to the ground. The last effect is the one that actually occurs. A similar effect occurs at other latitudes, as Foucault showed.

Earth's Rotation and Relativity

Earth's oblate shape, rotation irregularities, coriolis drifts, and the Foucault pendulum seemed to prove (if there were any doubt left over from Greek times) that the apparent daily motion of the sun and stars is not due to their motion around the earth, but is due to rotation of the earth itself. But have we been too glib in claiming this proof? Assume a universe composed of only two planets in *relative motion,*

meaning simply that an observer on object *A* would perceive motion relative to object *B,* and vice versa. No one could say which object is "really" moving, because there would be no background frame of reference. This is the **principle of relativity:** only *relative* motion is definable. Don't we then have a conflict between our proof that the earth rotates and the principle of relativity?

It is interesting that consideration of a seemingly mundane topic such as the daily rotation of earth can lead directly to the frontiers of theory. Such questions led Einstein to his *general theory of relativity,* which is a treatment of non-linear or accelerated motions, such as rotation. Einstein was fond of arguing by means of "thought experiments," and posed the following one. Suppose an observer lives not on the earth but on a spinning disk, like a giant phonograph record, isolated among the stars. He would experience a certain gravitational pull toward the disk's center, but he would also experience a tendency to keep moving in a straight line, tending to throw him off the disk. We call the latter tendency "centrifugal acceleration," and attribute it to the rotation of the disk. (If the rotation were fast enough, this centrifugal effect would exceed the disk's gravity, and the observer would slide off.)

But suppose the observer on the disk is determined to base his physical interpretation of the universe on the idea that his disk is not turning, that the universe is rotating around him. If he assumes he is at rest, then he must attribute the centrifugal force he experiences to the properties of matter in the universe. Einstein (1961) says,

> . . . on the basis of the general principle of relativity he is justified

in doing this. The force acting on himself, and in fact on all other bodies which are at rest relative to the disk, he regards as the effect of a gravitational field. Nevertheless, the space-distribution of this gravitational field is of a kind that would not be possible in Newton's theory of gravitation.

The observer *could,* however, describe this unusual "gravity" mathematically. If there were satellites nearby, the observer could describe their motions consistently with this mathematical formulation, although the problem would be inconveniently complex.

In short, if someone absolutely insisted on formulating his understanding of the universe from the premise that the earth is stationary and nonrotating, it would be valid for him to do so, but his physics and mathematics would become more complicated than our familiar and simple descriptions, based primarily on Isaac Newton's conception. He would also have to accept the earth as virtually the only nonspinning body in the known universe (since the sun, other planets, and stars demonstrably spin at different rates, and have satellites moving around *them*). We prefer the Newtonian conception of a spinning earth because it is simpler than a conception of a still earth and a spinning universe. All of this illustrates that science claims only to find simple, *workable descriptions* of nature, and not its "ultimate reality," which is considered an insoluble problem.

*Earth's Rotation
and Celestial Coordinates*

Earth's rotation has a practical application in allowing us to define coordinate systems to locate things both on the ground and in the sky.

For example, on earth's surface the north pole is the pole of rotation; latitude measures angular distance from the equator; longitude measures angular distance parallel to the equator. In the same way, as shown in Figure 3·5, astronomers project the latitude and longitude lines out onto

the sky to provide coordinates to locate directions of objects as seen from earth. In the last chapter we described how the north and south celestial poles and the celestial equator are located in this way. The celestial equivalent of latitude is called **declination;** declination measures angular

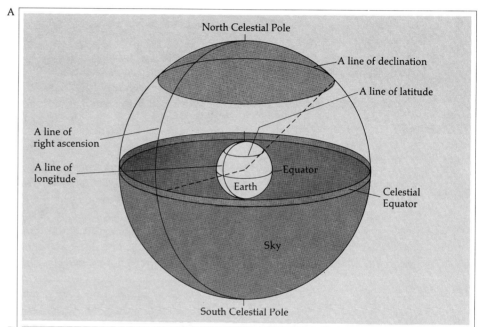

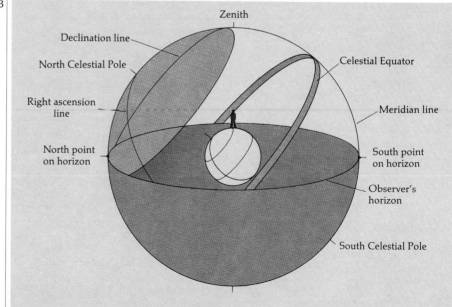

distance from the celestial equator. Thus the celestial equator is at declination 0°; the north celestial pole is at declination 90°; the south celestial pole is at declination −90°. The celestial equivalent of longitude is called **right ascension.** Just as longitude can be expressed either in degrees or in hourly time zones, astronomers use time zones to measure right ascension; that is, the celestial equator is divided into 24 hours, instead of 360°. A certain object's location might be described as R.A. 3^h 20^m, declination +53°.

Further details of this system are discussed in Enrichment Essay C; it is used by astronomers all over the world in pointing telescopes at celestial objects.

◄ *Figure 3·5* **A. The diagram shows lines of right ascension and declination obtained by projecting longitude and latitude lines from the center of the earth onto the sky. B. The diagram has been tipped to represent the perspective of an observer at an intermediate northern latitude. Various points** *above the observer's horizon* **are identified. In both A and B, the celestial equator is shown as a solid band for emphasis.**

Earth's Age and Origin

If the preceding facts were all we knew about the earth, we might picture it as a homogeneous, rigid, spinning sphere—like an unchanging billiard ball suspended in timeless space. But a closer look reveals that this is no billiard ball. It has mountain ranges, erupting volcanoes, surging oceans, sliding glaciers, a tempestuous atmosphere. What forces make it

evolve? How did it come into being? How old is it?

A variety of astronomical and geological evidence indicates that the earth was formed about 4.6 billion (4.6×10^9) years ago from particles orbiting around the sun, and that the earth displays many ongoing evolutionary processes, caused by both internal and external forces. Chapter 11 will return to details of the process that formed the earth and all the other planets; here we will emphasize processes that relate to the earth's early history and that will help us understand the conditions we find on other planets.

Debates about Earth's Age

Our view of earth as an evolving planet of great age has not come easily. Prior to a few centuries ago, the earth's age was hardly even considered, except for vague ideas that it was very old. Buddhist traditions, for example, called for a cyclical universe, with an indeterminate age for earth. In 1646, English scholar Sir Thomas Brown wrote that determining earth's age "without inspiration . . . is impossible and beyond the Arithmetick of God himself." Around 1650, several other scholars hypothesized that the earth's age and history could be deciphered from references in ancient scriptures. Archbishop James Ussher used this method and calculated that the whole cosmos was formed on Sunday, Oct. 23, 4004 B.C., and that humanity was created on Friday the 28th.

However, in the same century, naturalists began to realize that if sediments accumulated at the rate measured at the mouths of European rivers, many millennia would have been required to accumulate the sedi-

ment deposits actually observed. Even as evidence for earth's great age accumulated, the idea of geological evolution, or change, was as controversial as the later question of biological evolution. As late as the 1700s some people argued that fossils and strata deposits interpreted by geologists as signs of evolving climates and landforms were merely "devices of the Devil" put in rocks to mislead us. Even today some Fundamentalists argue that fossils and other geologic features were created and put in the ground all at once. By the same logic one might imagine that all "ancient" artifacts and history books were fabricated and placed in museums in, say, 1836. Such a hypothesis cannot be proven wrong, or even tested, but scientists reject such views on the grounds that they have little predictive or productive consequence; they don't lead us toward new observations or possibilities for new discovery.

Scientists who studied our planet from the 1500s to the 1800s were strongly influenced by medieval conceptions that the earth started in an Eden-like state (Davies, 1969):

> *a magnificient, lush estate which God had stocked with everything necessary for human well-being before admitting man as a freehold tenant. . . .*

Those who conceived of earth's evolution felt that it had been a process of shriveling, as when an apple dries out. Mountains were evidence of earth's decline in old age, ". . . even as warts, tumours, wenns, and excrescencies are engendered in the superficies of men's bodies." This view even affected emotional perceptions of the landscape. When John Dennis crossed the Alps in 1693, he reported not the majestic panoramas that impress us, but "only a vast, but horrid, hideous, and ghastly ruins." The theory of

mountain formation by simple shriveling of the earth persisted into our
century, and was proven wrong only
in recent decades (see Figure 3·6).

Measuring Rock
Ages by Radioactivity

All questions of earth's age and
evolution were clarified by discovery
of a reliable technique to determine
ages of rocks. This subject might seem
like a far cry from astronomy, but we
must remember that earth's rocks, like
lunar rocks and meteorites, are cosmic
materials. We live on a large rock.

In 1896 the French physicist Antoine
Becquerel accidentally left some photographic plates in a drawer with some
uranium-bearing minerals. Later he
opened the drawer and found the
plates fogged. Being a good physicist,
he did not dismiss the event but in-
vestigated. He found that the uranium
emitted rays which, like Roentgen's
X-rays of 1895, could pass through
cardboard. The "rays" turned out to
be energetic particles emitted by
unstable atoms. Radioactivity had
been discovered.

A **radioactive atom** is an unstable
atom that spontaneously changes into
a stable form by emitting a particle
from its nucleus. The original atom
thus becomes either a new element
(changes in the number of protons in
the nucleus) or a new *isotope* of the
same element (no change in the number of protons but a change in the
number of neutrons). The original
atom is called the **parent isotope** and
the new element is called the **daughter
isotope.** The time required for half
of the original parent atoms to disintegrate into daughter atoms is called
the **half-life** of the radioactive element.
If a billion atoms of a parent isotope were present in a certain min-

A

eral grain in a rock, a half-billion
would be left after one half-life, a
quarter billion after the second half-
life, and so on. As examples, half
the rubidium 87 atoms in any given
sample change into strontium 87 in
50 billion years; uranium 238 changes
into lead 206 with a half-life of 4.51
billion years; potassium 40 changes
into argon 40 with a half-life of 1.30
billion years; and carbon 14 decays
into nitrogen 14 with a half-life of
5,570 years.

Early in the twentieth century,
physicists realized that here was a way
to determine the ages of rocks. Suppose we determine the *original* numbers of different parent and daughter
isotopes in a rock. In part, this can
be done by measuring numbers of
stable isotopes, which usually occur in

certain ratios to the unstable isotopes
in a given fresh mineral. Then, if we
simply count the *present* numbers of
parent and daughter isotopes in the
rock, we can tell how many parent
atoms have decayed into daughter
atoms, and hence tell how old the
rock is. If half the parent atoms have
decayed, the age of the rock equals
one half-life of the radioactive parent
element being studied. This technique
of dating rocks is called **radioisotopic
dating,** since radioactive atoms that
decay are called radioisotopes.

It is important to note that the
quantity being determined in radioisotopic dating of rocks is time—the
time since the rock began to retain the
daughter element. In most cases being
discussed here, this is the time since
the rock solidified from an earlier

molten material, such as a lava flow. Once the rock solidifies, any daughter atoms are trapped in the solid mineral structure.

Earth's Age

To measure earth's age, you might expect that we would merely search

◄ *Figure 3·6* **A. The wrinkled Alps, seen here from space, were once thought to be caused by a shriveling and contraction of the earth. This view covers portions of Italy (foreground, dark), Switzerland (middle), Germany (top), and France (upper left). (NASA photo by Skylab 4 astronauts) B. A ground view of the French Alps, shown in the upper left part of A, reveals rugged mountains, ridges, valleys, and contorted strata unlike features known on other planets. They are caused by a combination of massive crumpling during collisions of tectonic plates, and erosion by flowing water; among planets of this solar system both processes are probably unique to the earth. (See pages 63–64.)**
▼

for the oldest known rock on earth and assume that it dates from earth's creation. In reality it is not so simple. The earth is geologically so active that rocks dating from earth's formation have long since eroded away. For instance, many rocks in the Rocky Mountains of the United States formed about 60 million years ago or less; many rocks in the eastern United States formed a few hundred million years ago. Older, more stable parts of North America exist around Hudson's Bay, where rocks a few billion years old have been found. Such stable, ancient regions yield the oldest known rocks. These regions have been found in several continents, and are called **continental shields,** after their flat, circular shapes. The oldest known earth rocks, formed about 3.9 billion years ago, were found in Greenland, the eastern extension of the Canadian shield. Rocks that have survived so long are extremely rare.

These results mean that the earth must have formed before 3.9 billion years ago. Without additional information, we would not be able to fix the age exactly. However, as we will see in the next few chapters, more information comes from outside the earth. Lunar rocks and meteorites show that *all the planets formed within a relatively short interval of about 50 to 90 million years, about 4.6 billion years ago* (Pepin, 1976). The **age of the earth** is therefore believed to be 4.6 billion years.

Earth's Internal Structure

Studying the earth's interior can give us ideas about the evolution of the planet as a whole. We have already seen that the amount of oblateness in earth's shape, caused by rotation, is one clue to the state of the interior material. Another clue is contained in waves that pass through the earth's material. If you snap your finger on the surface of a calm swimming pool, waves radiate across the surface from the disturbance. Observers at other edges of the pool could gain information about the disturbance by observing the waves; for example, they could locate the disturbance by comparing the directions from which the waves come. In addition, less obvious waves travel through the water toward the bottom of the pool, and they would also carry information.

In the same way, waves are generated in the earth by earthquakes, and these carry information, not only about the earthquake, but also about the earth's material. Waves traveling through the earth (or other planets) are called **seismic waves.** Some travel along the surface and others penetrate through the earth. The waves have

B

different velocities and characteristics, depending on the type of rock or molten material that they traverse. For example, as shown in Figure 3•7, waves that travel through the earth from an earthquake on the far side have traversed deep interior regions. Such waves differ from waves that have traveled a shorter distance. Waves from adjacent earthquakes reveal distinct differences in rock layers at different depths. Through such studies, three major regions have been mapped in the earth: the core, mantle, and crust (see Figure 3•8).

The Core

The **core,** the innermost region of the earth, reaches an estimated density of 10 to 12 g/cm^3 at its center (ordinary rocks are about 3 g/cm^3 and water is 1 g/cm^3). The depth to earth's center is 6,378 km (about 4,000 mi), while the depth to the core's surface is about 2,900 km. The core is believed to consist mostly of iron and nickel, with a possible admixture of up to 20 percent sulfur or silicon.[2] Such a mixture probably originated very early in the earth's formation, when much of the interior was molten. Heavy metals sank to the center, while light silica minerals floated to the crust. Evidence for a nickel–iron core includes not only the high density but also nickel–iron nodules, or lumps, brought up from great depths in volcanic vents. More important, some meteorites are nickel–iron,

---

[2]Nonetheless, we have no *direct* proof of the core composition. As the geophysicist F. Birch remarked in a moment of cynicism, when a geophysicist says "pure iron," he really means "uncertain mixture of all the elements."

proving that *their* parent planet(s) (see Chapter 10) did melt and form regions of nearly pure metal.

Seismic studies show that the core has a solid center but a molten outer layer. High internal temperatures keep the outer part melted, but pressures near the center are so great that they compress the central material, which raises its melting point and thereby makes it solid. Liquid metal in the outer part is believed to circulate at 0.02 cm/sec, setting up electric currents that create the earth's **magnetic field,** the force field that aligns magnetic compasses. This theory could help to explain the origins of similar magnetic fields found on some, but not all, other planets.

The Mantle

Although the **mantle** lies only a few tens of kilometers beneath our feet, it is almost as remote from scientific investigation as the core. No drill hole has penetrated it. An American effort, the Mohole Project, was launched in the 1960s to drill to the mantle through the ocean crust, but it provoked political as well as technical controversy, and was finally canceled. The continuing energy crisis may lead to new attempts at deep drilling to tap the mantle for geothermal energy. Meanwhile, our only *direct* evidence about the mantle comes from material blown out of deep volcanic vents. These vents have brought up rocks from the uppermost mantle. Such rocks are denser and richer in iron compounds than the familiar crustal rocks. They indicate that the most common elements in the mantle are oxygen, silicon, magnesium, iron, and calcium. These elements form crystalline compounds called **minerals,** which in turn combine in different groups to

form various kinds of **rocks.** The mantle, unlike the core, is thus rock.

Earthquakes occur as deep within the mantle as 700 km, but no deeper. This shows that rocks in the upper part of the mantle are brittle enough to fracture, whereas the deeper material is so hot that it flows rather than build up stresses. How do stresses build up again after earthquakes relieve them? Though solid, the upper mantle can slowly deform (like cool tar or asphalt), and flows in circulation patterns driven by the escape of heat energy from the earth's interior. Rocks are thus stretched until new fractures occur, sensed by us as new earthquakes.

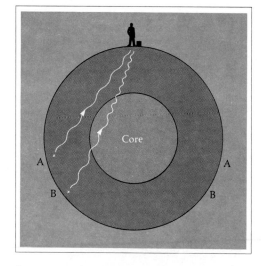

▲

Figure 3•7 **Differences in the nature of seismic waves from earthquakes in nearby regions A and B provide evidence of a distinct core region inside the earth. Waves that pass through the core are altered in character. The wave paths are curved because rock layers at different depths have different properties.**

Figure 3•8 **A simplified diagram of the earth's interior structure, as revealed by modern seismic studies. Dots indicate typical earthquake positions.**

The Crust

The **crust** is the outermost shell of the earth—the thin stage for all human activity. It is a layer of rock with lower density, more silicate, and less iron than the mantle. It is as thin as 5 km under the oceans but as thick as 30 km under the continents, reaching even deeper levels under major mountain ranges. Much of it is composed of a dark grey volcanic rock known as **basaltic rock.** Continental regions tend to contain more of a lower-density, silicate-rich rock type known as **granitic rock.**

If the outer mantle is in motion, then the thin surface crust can hardly be stable. Indeed it is not, but because the time-scale for mantle flow and crustal disruption is so much longer than a human lifetime, it took geologists until recent decades to recognize the great disruptions that are constantly, but slowly, occurring in the crust. Even the solid rock beneath our feet can flow like a fluid if given enough time, as shown by highway roadcuts that reveal folded strata resembling layers of distorted putty. To understand these contortions—and to compare earth with other planets—we must investigate *processes,* as well as present-day *structure,* inside the earth.

Processes Affecting Earth's Evolution

Both internal and external processes with different time scales have interacted to shape the surface of our planet. One group of processes, called **differentiation,** has separated some groups of chemical elements from other groups. Iron exists in nearly pure metallic form in the earth's core. Silicate rocks of different types occur in the mantle and crust. Certain gases exist in the atmosphere, but other types of gases, common throughout the universe, are absent. Again and again as we study other planets we will encounter similar situations where different elements are segregated, not always in the same proportions as on earth. Much can be learned about the history of the earth and other planets by studying how different element groups differentiated in different places—on the earth, on the moon, on other planets, on meteorites, and even in dust grains drifting between the stars. One of the most effective agents of differentiation is heat, since it may melt at least some rock materials and allow free flowing of chemical substances.

Differentiation by Complete Melting

Many geologists believe that the main differentiation of earth into an iron-rich core and silicate mantle occurred by melting of the earth's interior. One heat source may have been radioactivity. Since heat is just another name for motions of atoms and molecules, the motions of particles shot out of disintegrating radioactive atoms cause heat. Deep in the earth's interior, this heat cannot readily escape because of the insulating effect of overlying rock layers; so the tempera-

Crust (granites and basalts)

Continent

Continent

30

5

700

Mantle (basic rocks)

Depth 2,900 km

Outer core (liquid nickel-iron)

5,100 km

Inner core (solid nickel-iron)

Midocean ridge

Mohorovičić Discontinuity (base of crust)

Zone of earthquake activity

ture must have slowly increased until melting occurred. Just as in a smelter vat, melting allowed heavy metal-rich minerals such as nickel-iron to sink, while lighter silicate-rich mixtures rose to the surface. Because of a chemical affinity of uranium for silicate minerals, much of the radioactive uranium was concentrated in the silicate-rich crust, thereby reducing the heat sources in the deep interior. Today the interior has cooled until only the outer core is still molten. The temperature still increases at greater depth.

Differentiation by Partial Melting

In some regions inside the earth, even today, radioactive minerals or hot material from below may accumulate. These sources may provide enough heat to melt some minerals but not enough to completely melt all the minerals. In that case, the rock becomes partly fluid, with a matrix of still solid material, rather like slush. The fluid part may flow away, removing a portion of the chemicals to a different area, while the solid minerals stay behind. Although other processes of differentiation exist, the processes of complete and partial melting show that the earth, even if it started out as uniform as a billiard ball, would soon become nonuniform.

Outgassing

Any heating, partial melting, or complete melting of underground material is likely to drive off light gaseous substances, such as hydrogen, water, and carbon dioxide. This release of gas is called **outgassing.** The gases have been added to the atmosphere, changing the original atmosphere into what we breathe today. Even after-

wards further processes occurred. Much hydrogen, being light, rose and escaped into space; water collected in oceans; much carbon dioxide dissolved in the oceans and reacted with silicates, eventually forming carbonate rocks.

Convection and Continent Formation

How does heat get out of the interior of an astronomical body like the earth? Heat always flows from the hottest regions to the coolest regions. There are three important mechanisms. **Conduction** is the familiar flow of heat through material, as when the protruding end of a fireplace poker gets hot after it is left in the fire too long; the heat is transported by mechanical jostling of adjacent molecules in the material. **Radiation** occurs when energy is carried by electromagnetic waves, such as light or infrared radia-

tion. Just as radiation can travel from a fireplace through air or glass, some radiation can travel through rock materials in the earth. However, difference in temperature between the earth's core and the outer regions is so great that conduction and radiation are probably inadequate to carry away the heat.

What other mechanism is there? Suppose we model the situation with a pan of water or cooking oil, to which heat is added from below. If heat is added at the bottom at a great enough rate, blobs of material become so hot that they expand and become less dense than their surroundings,

Figure 3·9 **Outgassing can be seen at volcanic sites, as in this view of Kilauea crater, Hawaii, as steam and other volcanic gases rise from the temporarily dormant crater. Such volcanic gases have slowly altered the earth's atmospheric composition.**
▼

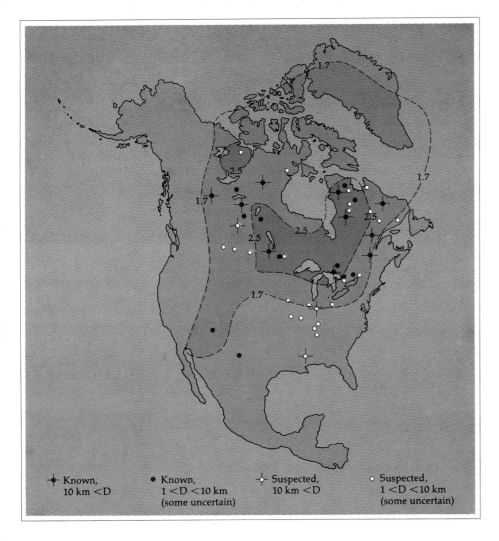

- Known,
  10 km < D

- Known,
  1 < D < 10 km
  (some uncertain)

- Suspected,
  10 km < D

- Suspected,
  1 < D < 10 km
  (some uncertain)

▲
Figure 3·10 **In the oldest parts of North America, especially in the two-billion-year-old Canadian shield, eroded impact craters testify that the earth has been bombarded by sizable interplanetary bodies. Representative rock ages (billions of years) are shown in shaded areas by dashed lines; craters are more common in older regions. Key at bottom of figure gives approximate diameter, D, of crater.**

and hence rise, just as a hot air balloon rises through cooler air. Ascending currents, or cells, form in some areas, and descending currents form in other areas. This process is called **convection,** or heat transport by motions of material. The convection cells can actually be seen if a strong light

is used to observe the highly heated cooking oil. The same phenomenon is familiar in the atmosphere; cumulus clouds often mark ascending convection currents, and airplane passengers will recall choppy riding near such clouds as the plane encounters ascending and descending currents.

Ascending and descending currents probably occur in the upper parts of the mantle and perhaps in other regions. We usually call these regions "solid rock," but there is no inconsistency. Many materials which appear solid and brittle during short periods of time can actually flow if subjected to stress over longer periods. A familiar example is pitch or asphalt, which

may shatter if struck with a hammer but will flow if subjected to mild, steady pressure.

Convection currents in earth not only transport heat, they aid in differentiation. They recycle material. Suppose some silicate minerals get caught in such a current. Being lower density than most rocks, they would normally remain floating at the surface once they got there. If they happen to get caught in a descending current, they may descend, return to the surface, and remain the next time around. In this way, silicate-rich rock masses may have accumulated at the surface, even if the surface had no such rocks to begin with. Many geologists believe that the silicate-rich, granitic continental masses accumulated this way over periods of hundreds of millions of years (Hargraves, 1976).

Impact Cratering

Interplanetary debris ranging from millimeters to kilometers in size occasionally crash into the earth and other planets, creating explosions that blast away crustal rock and suddenly create circular depressions called **impact craters.** Comparisons of terrestrial and lunar data show that this process was much more common in the first billion years of geologic history than it is today. However, the process still continues, and ancient continental shields show eroded impact craters that have accumulated over the last billion years or so (Figure 3·10). In the earliest geologic time, large impact craters may have been important in blowing away crust from some regions and piling it up in other regions, perhaps creating primitive ocean basins and continents. Cratering is an example of a direct extraterrestrial or astronomical process in landscape formation, as opposed to most geologic processes, which are generated inside the earth (see Figure 3•11).

Volcanism

Volcanism is the eruption of molten materials from a planet's interior onto its surface. On earth, a zone about 70 to 200 km below the surface contains pockets of partly melted materials, as indicated by seismic wave analysis. This underground molten rock, called **magma**, is often charged with gas such as steam, is less dense than surround-

ing rock, and is highly corrosive. Therefore it tends to work its way to the surface, especially in regions where fractures provide access. When it reaches the surface it erupts and is called **lava.** It may shoot, foam-like, into the air or ooze out and flow for many kilometers across the ground. If

enough lava is erupted, it may accumulate into volcanic mountains. During intervals ranging from years to millions of years, volcanism thus creates landforms ranging from flat lava flows to clusters of volcanic peaks. Also, the crust may collapse into magma chambers, creating craters that mimic impact craters (see Figure 3•12).

Erosion and Deposition

If these were the only processes, we might expect our planet's landscape to consist of craters, lava flows, and vol-

Figure 3•11 **A circular lake is all that remains of the highly eroded Manicouagan meteorite impact crater in Canada. Formed about 210 million years ago, the original crater was about 65 km in diameter.** (NASA photo by Skylab 3 astronauts)
▼

▲
Figure 3•12   **Crater Lake, Oregon, is a volcanic crater formed by collapse of a volcanic mountain summit during explosive eruptions about 7,000 years ago. The lake is nearly 10 km across.** (Dale P. Cruikshank)

canic mountains. However, the earth undergoes very active processes which break down these landforms. **Erosion** includes all processes by which rock materials are broken down and transported across a planet's surface; such processes include water flow, chemical weathering, and windblown transport of dust. **Deposition** includes all processes by which the materials are deposited and accumulated; such processes include deposition of sediments in lake and ocean bottoms and dropping of windblown dust in dune deposits.

On earth, erosion and deposition are by far the most important landscape forming processes, but this is not true on all other planets. The extreme activity of erosion and deposition on earth is responsible for wearing away or covering up the most ancient rocks, and for wearing away ancient impact craters so rapidly that they were not even recognized on the earth until the last few decades. Similarly, most familiar mountain ranges do not display their initial forms, which may have been caused by volcanism, fracturing, and folding; instead, most mountain ranges are

resistant rock units left after erosion of softer units.

Plate Tectonics

Tectonics is the name given to all processes of fracturing and movement of the earth's crust. The movement occurs as a result of the sluggish currents in the upper mantle, described earlier. These currents drag at the crustal rocks and create slowly increasing stresses. Just as a rubber band can slowly stretch until it suddenly snaps, rocks in the earth can slowly deform, flow, and stretch until the stress is too great or too sudden, and then they crack. Any such sudden cracking produces an **earthquake.** If the rock units actually move with respect to each other, as well as simply cracking, the result is called a **fault.** Faulting near the surface disrupts the ground; many California earthquakes, for example, cause offsets along the famous San Andreas fault. Until recent decades, geologists knew that some areas were tectonically more active than others, but no planet-wide pattern had been discovered.

However, as early as 1620, English naturalist Francis Bacon noticed that the east and west shores of the Atlantic Ocean could fit together like pieces of a jigsaw puzzle. He speculated that the Americas had once touched Europe and Africa, but had been broken by giant faults. By 1922, German geophysicist Alfred L. Wegener and others championed the idea that the present continents were merely drifting fragments of one or more large, primeval proto-continents (Hallam, 1975).

Until 1960, most geologists rejected this theory of the earth's major features, believing that there was insufficient energy available to "push the continents around." But during the

1960s, evidence from new geological exploration confirmed it. Mapping of the ocean floor, especially in the mid-Atlantic and Caribbean, revealed regions where the crust was being pushed apart and occasional eruptions of lava formed new ocean floor. The fault along which America had been split from Europe and Africa was identified as the mid-Atlantic ridge. Geologic rock units in the eastern Americas were identified as fitting onto units of the same age and composition in western Europe and Africa. In short, discoveries during the 1960s and 1970s showed that the ascending and descending convection currents in the mantle have broken the crust into distinct units, called **plates,** whose edges are defined by fractures, earthquakes, and volcanoes caused by the currents themselves (Figure 3•13). **Plate tectonics** is the process of crustal disruption by flow of materials in these plates.

Rock ages show that the tectonic breakup that produced the modern continents began surprisingly recently in earth's history. Modern continents are fragments of two major land masses (Figure 3•14). These began to break apart some 300 million years ago, significantly separating by 100 million years ago.[3] In other words, the crust's modern continental configuration formed only in the last few percent of geologic time. Still earlier configurations have been mapped from rock units dating back 2.5 billion years, and a single massive continent (sometimes called Pangea) may have

---

[3]Geologists have named the northern and southern proto-continents Laurasia and Gondwanaland, respectively, after modern geological provinces where their rocks are traceable. Geologists with a political sense of humor have been seen with bumper stickers reading "Re-unite Gondwanaland."

existed around 3 billion years ago (Engel et al., 1974; Hargraves, 1976).

The convection currents that ascend on one side of a plate, push material across the plate, and descend on the other side have pushed continental masses toward regions of descending currents, a process called **continental drift.** In regions such as the Gulf of California, the Red Sea, and East Africa, crustal blocks are even now being split apart by motions away from sites of ascending currents. As

a result of the movement of plates, continental masses have collided, as shown in Figure 3·15. These collisions cause faulting and massive crumpling of crustal rock strata. This process is now believed to be the cause of the major mountain masses of earth, such as the Himalayas, Alps, and Andes—the mountains which geologists once mistakenly attributed to a simple shrivelling process like that of a drying apple. The evolution of the earth, we now see, is determined by forces

much more dynamic than those shrinking a drying apple.

Figure 3·13 **A map of the distribution of earthquakes and volcanoes reveals lines of features that define the boundaries of the major tectonic plates. For example, the line of shallow earthquakes down the middle Atlantic lies along the mid-Atlantic ridge, the fracture that allowed the Americas to break off from Europe and Africa.**
▼

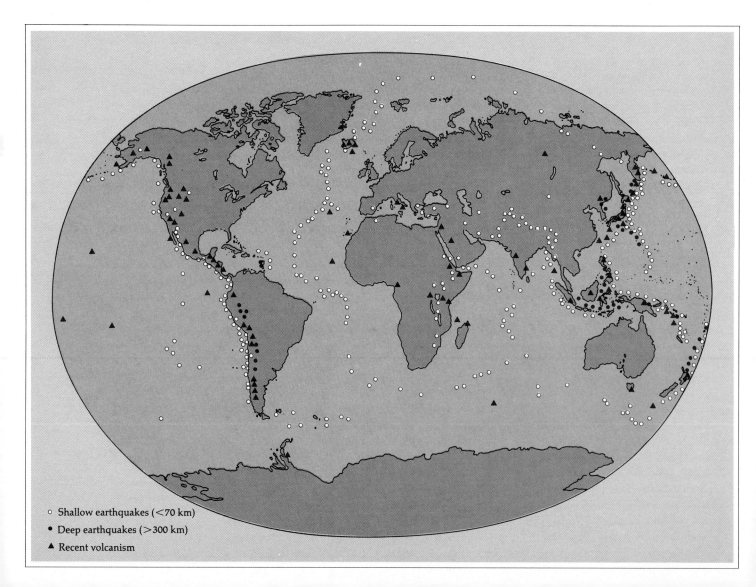

○ Shallow earthquakes (<70 km)

● Deep earthquakes (>300 km)

▲ Recent volcanism

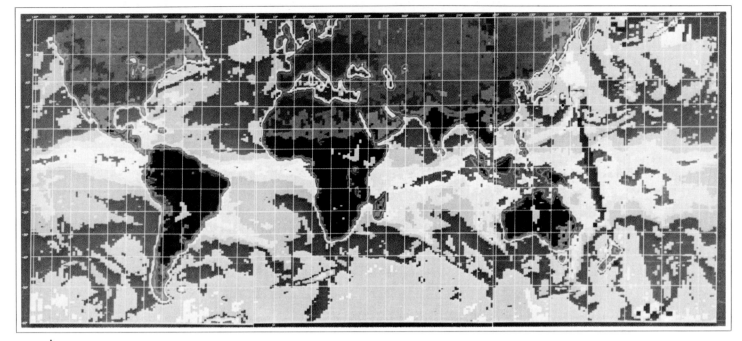

A

Figure 3·14 A. A map of the earth compiled from microwave radio scanning from the Nimbus-5 satellite clearly reveals the present outlines of the continents indicated by temperature difference from the oceans. Light-colored equatorial oceans are warmer than darker, high-latitude oceans. B. A schematic map of the earth as it may have appeared roughly 200 million years ago shows how the present continents were joined together into one or two large landmasses. *Laurasia* and *Gondwanaland* are names given by geologists to two major areas. Relatively recent splittings of land have produced the Atlantic Ocean, the Red Sea, and the Gulf of California.

B

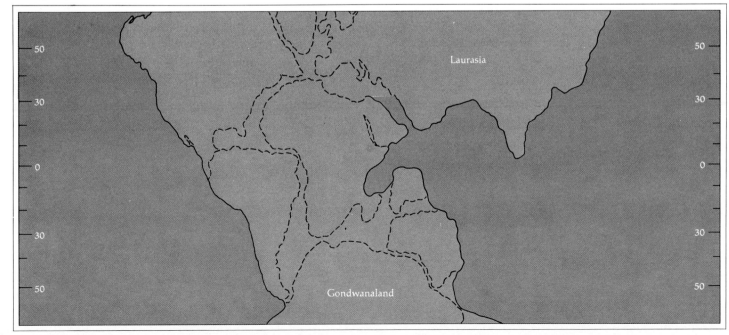

3

Summary:
The Earth's History

Geologic studies of strata and fossils have established a chronology of events in the history of the planet earth, and radioisotopic dating of rocks has established the actual ages of these events. The sequence, known as the **geologic time scale,** is shown in Table 3·1. It is broadly divided into intervals called *eras;* only the last 14 percent of earth's history has yielded enough rock evidence, such as fossils and dates, to provide finer divisions which geologists call *periods.*

Drawing from material presented in this chapter, we can now construct a thumbnail sketch of the earth's history.

Some four and a half eons ago the earth formed. Giant meteorites struck much oftener than today, scarring the primal landscape with great impact craters. The surface was lifeless. Since hydrogen was the dominant gas in the early solar system, the earth's original atmosphere was probably composed of hydrogen and hydrogen-rich com-

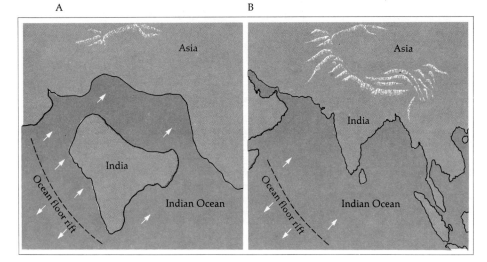

A.   B.

◄ *Figure 3·15* **A. Evolution of southern Asia illustrates continental drift and its effects. According to geological evidence, currents (arrows) drove the subcontinent of India into the Asian plate. B. This activity caused the region of the collision to crumple and form the Himalayan mountain range.**

Figure 3·16 The parallel ridges of the Appalachians, photographed from about 30,000 feet, mark resistant folded strata revealed by erosion. Major mountain belts of this type are produced by slow collisions of major shifting crustal plates.
▼

Table 3·1 The Geologic Time Scale

Era	Age (millions of years)	Period	Life Forms	Events
	0			
		Quaternary	Man	Technological environment modification, Ice Ages
	3			
Cenozoic		Neocene		
	22			
		Paleocene	Mammals	Building of Rocky Mountains
	62			
		Cretaceous		Recent continental studies
	130			
Mesozoic		Jurassic	Dinosaurs	
	180			
		Triassic	Reptiles	
	230			
		Permian	Conifers	Building of Appalachian Mountains
	280			
		Pennsylvanian	Ferns	
	310			
Paleozoic		Mississippian		
	340			
		Devonian	Fishes	
	405			
		Silurian	Early land plants	
	450			
		Ordovician		
	500			
		Cambrian	Trilobites	Earliest good fossil records
	570			
		Ediacarian	Small soft forms	
	640			
	1000		First macroscopic life-forms, sexual reproduction	Growth of proto-continents
Proterozoic				
	2000		Oxygen-producing microbes	Oxygen increasing in atmosphere
	2600			
				Crustal and atmospheric evolution
Archeozoic	3000		Microscopic life	
	3600			
				Oldest rocks—crustal formation?
	4000			
				Heavy meteoritic cratering
	4500			
	4600			Formation of sun and planets
	12,000?			Evolution of galaxy
	16,000?			Origin of universe

Source: Data in part from Schopf (1975).

pounds such as ammonia (NH_3), methane (CH_4), and water (H_2O).

Although an entire day and night at first lasted only about 5 hours, the day rapidly grew to nearly its present length. Initial heat or radioactively added heat caused major differentiation as metals drained to the center to form the core. Breakup of NH_3 gas yielded nitrogen (N_2). Outgassing released large amounts of steam (H_2O) and carbon dioxide (CO_2) gas as well as lava. As these gases accumulated, the atmosphere began to evolve toward its present composition. Some of the steam also condensed and accumulated as surface water. Much of the CO_2 was used up in forming carbonate rocks. Life probably originated within a few hundred million years of the formation of large pools of liquid water. This must have happened within about 1.3 eons of the earth's formation, since the earliest microscopic fossils date back about 3.3 eons (Schopf, 1975).

The heating of the earth's interior promoted convection currents that stirred crustal materials. Lighter masses of sedimentary rock tended to drift on the surface and coagulate, whereas heavier masses were apt to be pulled down into the mantle by the earliest plate-tectonic activity. Continents formed and were repeatedly split and rejoined as rift zones broke landmasses, and plate motions caused them to drift into each other.

Mountainous landscapes thus replaced the original cratered crust, though vestiges of the old surfaces still appear in the cratered continental shields. With the coming of plants and animals, the landscape became familiar to our eyes.

In the scale of planetary evolution, recent events such as the Ice Ages some thousands of years ago, or the peculiar warming around A.D. 1000 that allowed the Vikings to colonize Greenland, seem the merest of details. Recent work suggests that the major Ice Ages are caused by slight changes in the earth's orbit around the sun, caused by gravitational attractions of other planets. But we still do not know whether other climate changes result from purely terrestrial events such as the blocking of sunlight by volcanic dust, or from astronomical events such as the orbital changes or changes in the sun's radiation. Yet it is important to understand such cosmic "details." Small shifts in climate during recent decades may have produced the droughts, and could affect world food production.

Let us conclude our summary of the earth's history by compressing events into a single day. Life evolves sometime in the early morning, but the fossil-producing trilobites that begin the traditional geologic time scale in the Cambrian period do not appear until about 9:30 in the evening. By 10 P.M. there are fishes in the sea, and by 11 P.M., dinosaurs on the land. Mammals do not appear until about 11.40. Human beings, who have been here, depending on our definition of *human*, perhaps two million years, make their appearance only 30 seconds before midnight. The last few thousand years occur in a tenth of a second —represented by the pop of a single flashbulb at midnight. The question is, what will be here a tenth of a second after midnight?

Figure 3•17 **An imaginary landscape on the primitive earth. From the rim of an impact crater, one sees the crater's lifeless interior, a glimpse of the ocean beyond, and an unusually large moon (studies of the moon's orbit show it was once much closer to the earth than it is today).** (Painting by author)

Figure 3•18 **A unique scene in the solar system, but the most common scene on earth—an ocean of liquid water. When water degassed from the earth's interior, it encountered surface conditions warm enough to prevent it all from freezing, but not warm enough to evaporate it all. Earth is the only planet where liquid water is known to exist.**

Concepts

oblate spheroid

coriolis drift

Foucault pendulum

principle of relativity

declination

right ascension

radioactive atom

parent isotope

daughter isotope

half-life

radioisotopic dating

continental shields

age of the earth

seismic waves

core

magnetic field

mantle

minerals

rocks

crust

basaltic rock

granitic rock

differentiation

outgassing

conduction

radiation

convection

impact craters

volcanism

magma

lava

erosion

deposition

tectonics

earthquake

fault

plates

plate tectonics

continental drift

geologic time scale

Problems

1. When were the last major earthquakes in your region? Is your region seismically active or quieter than average? How is it located with respect to the boundaries of tectonic plates?

2. If a rock sample can be shown to contain one-eighth of its original amount of radioactive U-235, how old is it?

3. In view of preservation of ancient craters and lack of folded mountain ranges on the moon, would you predict the moon to have more or less earthquake activity than the earth?

4. Fossils of apelike predecessors of the genus *Homo* (for example *Australopithecus*), found in Africa, are believed to date back at least 2 to 3 million years before the present. What percent of the earth's age is this? Can you accept, philosophically, that events happened during most of the history of the earth before there was anyone around to see, hear, or record them?

5. If the earth is 4.6 eons old, about how much more radioactive U-238 did it have when it formed? Would the heat-production rate from radioactivity when the earth formed have been more, less, or the same as it is now?

6. The composition of the earth's atmosphere is probably much changed from what it first was.

a. *What happened to the abundant hydrogen atoms initially present or produced by breakup of molecules such as* CH_4?

b. *Given that much ammonia* (NH_3) *was initially present, and that ammonia molecules break apart when struck by solar particles in the atmosphere, account for one source of the earth's now-abundant nitrogen.*

c. *What two gases were added abundantly by volcanoes, and where did these two gases finally end up?*

7. Describe how you would expect conditions on the earth to be if earth were so close to the sun that the mean surface temperature were above 373 K (100°C). What if the earth were far enough from the sun so that the mean surface temperature were below 273 K (0°C)?

Advanced Problems

8. The average near-surface temperature gradient in the earth is 25°C/km.

a. *Assuming the surface temperature is about 20°C, how many km would one have to drill to reach a depth where water would boil? (Temperature of boiling water is 100°C.) Compare this depth with that of the deepest mines, roughly 3.5 km.*

b. *If such depths could be reached economically, dual pipes could be lowered, with water pumped down one pipe, converted to steam, and blown up the other pipe. Steam-powered plants could thus tap the planetary energy source in any part of the world. Describe possible economic and political consequences of such a project.*

9. As seen from Mars, the sun subtends an angle of about $\frac{1}{3}°$. Suppose that the earth passes exactly between Mars and the sun.

   a. *Is the earth big enough to cause a total eclipse of the sun as seen from Mars?*

   b. *Assuming that the human eye can resolve a disk as small as 2 minutes of arc (120 seconds of arc), could the earth be detected by the unaided eye as it crossed the sun, as seen from Mars? (Hints: Use the small angle equation. Earth's diameter = 12,756 km; earth's distance from Mars = 60,000,000 km).*

Project

---

1. Place cooking oil an inch or two deep in a flat pan over low heat, under a single strong light. Because the oil does not readily boil, heat is soon transmitted to the surface in visible convection currents. Note how the currents divide the surface into cells—or regions of ascent, lateral flow, and descent. These cells are analogous to tectonic plates in the earth's surface layers. Sprinkle a slight skin of flour on the oil's surface to simulate floating continental rocks and watch for examples of "continental drift" and plate collisions. Experiment with different depths of oil and different temperature gradients (by changing the heat setting), and record the results.

The Conquest of Gravity and Space

The events that led to our leaving the cradle and visiting the nearest planetary body were events of science, engineering, and sociology. In this chapter we will begin with early dreams of space travel, move to Newton's analysis of the gravitation that binds us to the earth, and then describe how physical principles were applied to allow travel to the moon and beyond.

The Dream of a Flight to the Moon

Flights to the moon appear in literature as far back as Graeco-Roman times and recur throughout modern

European history. The Greek satirist Lucian (ca. A.D. 190) had one of his characters put on vulture and eagle wings, take off from Mt. Olympus, and fly to the moon to learn how the stars came to be "scattered up and down the heaven carelessly." Around 1500, Leonardo da Vinci sketched devices for manned flight, and in 1528 the Spaniard Eugenio Torralba confessed to the Inquisition that he had flown near the moon under the guidance of a demon. Lucian's stories of lunar flight were translated into English in 1634, and from then to our own time literary trips to the moon occurred often. Among them are

opportunity! The dream is probably as old as mankind.

Isaac Newton and Gravity

Newton was a father not only of physics and astronomy, but of space travel as well. Between the ages of 23 and 25, while attending Cambridge, he almost single-handedly invented

Figure 4·2 **The rocket was still not recognized as the best mechanism for space travel by the late 1800s, when French artist Gustave Doré made this illustration of a lunar voyage.**
▼

calculus, discovered the principle of gravitational attraction and certain properties of light, and invented the reflecting telescope (in which a curved mirror replaces the lens). Newton once said that he made his discoveries "by always thinking about them," a trait that no doubt contributed to his reputation for absentmindedness.

At age 41, Newton began writing his famous *Principia*, a collection of his results, and published it three years later, in 1687. He became president of the Royal Society at 60, died at 84, in 1727, and was buried in Westminster Abbey. Of Newton, Alexander Pope wrote:

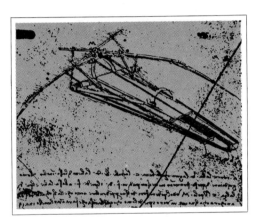

▲
Figure 4·1 **Leonardo da Vinci drew this engineering sketch for a flying machine in about 1500.**

Cyrano de Bergerac's 1656 account of "Empires of the Moon," and Daniel Defoe's "Journey to the World in the Moon" in 1705. Over a hundred European accounts of voyages to the moon were published between 1493 and the first balloon ascensions in 1783 (Nicholson, 1948). Clearly, flight to the moon was not an idea conjured by twentieth-century engineers upon hearing the knock of technological

Nature and Nature's laws lay hid in night:
God said, "Let Newton be!" and all was light.

In 1726, an acquaintance of Newton's gave this account of his early thoughts on gravity, based on Newton's own conversations (Ball, 1893):

The first thoughts [were those] he had, when he retired from Cambridge in 1666 on account of the plague. As he sat alone in a garden, he fell into a speculation on the power of gravity: that as this power is not found sensibly diminished at the remotest [height] to which we can rise, neither at the tops of the loftiest buildings, nor even on the summits of the highest mountains, it appeared to him reasonable to conclude that this power must extend much farther than was usually thought; why not as high as the moon, said he to himself?

The moon must be attracted to the earth by some force, Newton thought, because it does not travel in a straight line as it would if no force were pull-ing on it, but rather follows a path curved around the earth. Newton, supposedly inspired by a falling apple, realized that he could calculate from the moon's known orbital motion how fast it "fell away" from that hypothetical straight line. He would then compare this acceleration rate, away from a straight line, to the rate of a falling body at the earth's surface. After gathering accurate data, Newton showed that whereas the moon was 60 times farther from the earth's center than the earth's surface, its gravitational acceleration was about 1/3600, or $1/(60 \times 60)$, of the acceleration experienced at the earth's surface. The force diminished as the *inverse square of the distance.* Gravitational force is thus said to follow an **inverse square law.**

This result is not surprising. If anything spreads from a point in straight lines in all directions, it must become less concentrated as it gets farther from that point. Light, radio waves, and water spray from a fast-rotating sprinkler are examples. The inverse square relation is illustrated by the following thought-experiment. Imagine light from a candle shining into a cone, as shown in Figure 4.3. If the cone is cut at the point where its base has an area of 1 square centimeter, then all the candle light that reaches the tip of the cone passes through this square centimeter. At twice the distance of the 1 cm² base, the base of the cone is twice as wide and so has four times the area—but receives no more light. The original amount of radiation is now dispersed over 4 square centimeters, or is one-fourth as much per unit area at twice the distance. Thus the inverse square law is in effect with respect to light radiation.

Newton now knew how distance affected gravity, but he did not know what else affected it. He eventually

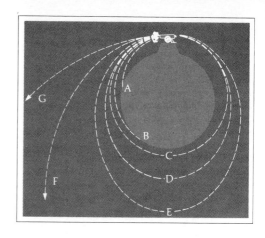

Figure 4.4 **Newton realized that a vehicle launched parallel to the surface of the earth from a mountaintop could travel varying distances, depending on its launch velocity. At low speed, it would hit at nearby point A. At higher speed, it could travel halfway around the earth, hitting at B. Slightly higher speed would put it in an elliptical orbit with a low point (or perigee) at C. Higher speed, called circular velocity, would put it in a special type of elliptical orbit—the circular orbit D. Slightly higher speed would create an elliptical orbit with farthest point (or apogee) at E. Escape velocity would create a parabolic orbit, F, that never returns to earth. Still higher speed creates a hyperbolic orbit, G, which also never returns.**

showed that the gravitational attraction between two bodies is proportional to the amount of material in each body—that is, to its **mass.** The more mass, the more attraction. Once both factors, distance and mass, were known, **Newton's law of gravitation** took the form:

$$F \text{ is proportional to } \frac{Mm}{r^2}$$

Where:

F = gravitational force of attraction between two bodies

M = mass of larger body

m = mass of smaller body

r = distance between centers of M and m.

Figure 4.3 **The inverse square law. At twice the distance from the source, the light is spread over four times the area.**
▼

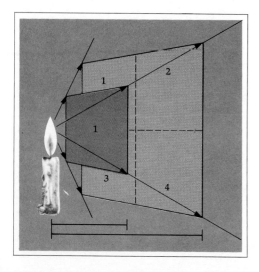

Thus if either mass doubles, the force between them doubles; but if the distance between their centers doubles, the force drops by a factor of four.

Mass should not be confused with *weight.* The weight of an object is merely the gravitational force by which the earth pulls the object against the earth's surface. "Weight" is merely a convenient name for gravitational force. The weight of an object would be different on different planets because the gravity of different planets differs. The mass, or amount of material in a body, however, always remains the same (except during nuclear reactions).

Circular Velocity:
How to Launch a Satellite

Newton realized that an object could be launched into orbit around the earth. His *Principia* contains a diagram of an earth-satellite fired from a cannon on a mountaintop, with the barrel pointed parallel to the ground (Figure 4·4). He merely applied his *first law of motion:* A projectile keeps moving forward unless a contrary or modifying force acts on it. In this case, of course, the force of gravity pulls the cannonball toward the ground. If its launch speed is too slow, it falls to earth near the cannon. At a higher speed it travels farther. At a high enough speed, it curves toward the ground, but the earth, being round, curves away at the same rate. Thus the

projectile never reaches the ground, but travels all the way around the earth and returns to the mountaintop.[1] This speed—the speed at which an object must move parallel to the surface of a body in order to stay in circular orbit around it—is called **circular velocity.**

The farther from the earth or other central body, the less the force of gravity that must be overcome, and therefore the lower the necessary circular velocity. At earth's surface, 6,378 km from its center, circular

---

[1]This neglects two realities: In the length of a cannon barrel, no shell could be accelerated to 18,000 mph without shattering; and the earth's atmosphere would retard the satellite, making it fall back to the ground. Thus, rockets must be used, and the projectile shot above the atmosphere.

Optional Math Equation II
Calculating Circular and Escape Velocities

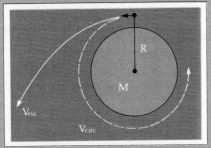

Figure 4·5 The equation for circular velocity gives the velocity, V_{circ}, required to place a small object in circular orbit (dashed line) around any large mass, M, from an initial position at any distance R. The velocity for escape, V_{esc}, from this same position, is always $\sqrt{2}\ V_{circ}$.

One of Newton's achievements in systematizing physics was that his results, once derived for a specific case such as the earth, can be generalized. Thus, all of these principles apply as well to bodies orbiting around other planets or around stars. For example, for any small satellite orbiting around any large mass M, the circular velocity is

$$V_{circ} = \sqrt{\frac{GM}{R}}$$

G = Newton's gravitational constant = 6.67×10^{-8} in cgs units

M = Mass of central body

R = Distance of orbiter from center of central body

This is the second of five simple but important equations we will be presenting in this text (the first was the small angle equation described in Chapter 2, page 38). This simple

result shows that the greater the mass M, the greater the circular velocity; and the greater the distance R, the less the circular velocity. As an example, we can compute the velocity required to achieve circular orbit not far above earth's surface. Expressing units in the cgs system, we start with the mass and radius of the earth:

$M = 5.98 \times 10^{27}$ g

$R = 6.38 \times 10^8$ cm

Substituting these into the equation, we find that a vehicle must reach a speed of 7.91×10^5 cm/sec, or about 8 km/sec, to stay in circular orbit, confirming the value quoted above.

Another useful fact is that the escape velocity of a small satellite is *always* $\sqrt{2}$ times the circular velocity. Thus the vehicle above would have to reach about 11 km/sec in order to escape earth on a parabolic orbit.

velocity is 8 km/sec, or nearly 18,000 mph. The moon, 384,000 km from that same center, moves at only about 1 km/sec in its circular orbit.

In short, launching a satellite into earth orbit is a seventeenth-century idea! All that is needed is to get an object above the atmosphere (so that air resistance isn't a problem), point it in a direction parallel to the ground, and accelerate it to 8 km/sec.

The orbiting body's point of closest approach to the earth is called its **perigee;** its point of furthest departure, its **apogee.**

Escape Velocity

As a body is launched at higher and higher speeds, the apogee point is farther and farther away, making the orbit an **ellipse**—a type of curve describing the closed orbit of one body around a second body. High enough speed would send it from perigee near the earth to apogee outside the solar system. At slightly higher speed it would travel infinitely far from earth, following a curve the shape of a **parabola,** and it would never come back. This speed, which allows the object to escape from earth forever, is called **escape velocity,** or sometimes *parabolic velocity.* At a point near the surface of the earth, escape velocity is about 11 km/sec; it is less at more distant points. Launched at still higher speeds, a body travels a similar curve called a **hyperbola,** and these speeds are called *hyperbolic velocities.*

Rockets and Spaceships

How could a body launched from earth reach circular velocity or escape velocity? Jules Verne, in *From the Earth to the Moon* (1865), imagined using a 900-foot-long cannon and 400,000 pounds of explosive. But as noted above, a cannon is impractical. H. G. Wells, in *The First Men in the Moon* (1901), avoided the technological problem by having his hero invent an antigravity substance that simply floated a craft to the moon.

A more realistic technology was needed. The spacecraft must (1) carry its own means of propulsion, and (2) operate in the vacuum of space. Item (2) ruled out propeller or jet aircraft. Around the turn of the century, several experimenters and visionaries realized that rockets were ideal. Their use first recorded in China and Europe in the 1200s, rockets work essentially by **Newton's third law of motion:**

For every action, there is an equal and opposite reaction.

For example, if you sit on a wagon and throw a large mass (like a cinder block) out the back, the wagon coasts forward; the force needed to expel the mass causes an opposite force on the vehicle from which the mass is expelled. In the same way, the force used to expel high-velocity gases out the back of a rocket nozzle pushes the rocket forward with equal force. This force exerted by exhaust gases is called **thrust.**

The Russian experimenter Konstantin Tsiolkovsky, beginning in 1898, and the American Robert Goddard in the 1920s studied and fired rockets, but both were mavericks and their work was not widely used. In the 1920s in Germany, Hermann Oberth, who remarked that Verne's book was an inspiration, published several books on rocket-powered space travel. Oberth's work attracted a group of enthusiasts, including Wernher von Braun, whose astronautical experiments were converted into the V-2 guided missile program under the Nazis. At war's end, about 125 German rocket experts, including von Braun, moved to the United States and continued the chain of space-travel development that stretched back to Lucian, Newton, and Verne. A remarkable footnote to our times is Oberth's survival through war and political vicissitudes to attend as a NASA guest the launch of the first successful moon flight in 1969.

The First Satellites

After several secret postwar studies of satellites had been made, President Eisenhower announced in 1955 that the United States would launch a

Figure 4·6 **The first artificial satellite. Russian artist A. Sokolov and cosmonaut A. Leonov collaborated on this painting of Sputnik I.**
▼

Figure 4·7 **This time exposure of Sputnik II shows the trail left by the satellite (lower right) as it passed a few degrees from the moon.** (Photo by Donald L. Strittmatter, January 1958)

satellite as part of the International Geophysical Year (1957–58). This program was to be a civilian program using a nonmilitary rocket called Vanguard. Within days, Soviet scientists announced their plan to launch satellites larger than the American ones. These statements were not taken seriously in the West, Americans holding an image of the Soviets that the Soviets themselves romanticized in their poster art: that of the unsophisticated, shirt-sleeved tractor driver.

On October 4, 1957 the Soviet Union astonished the world by launching the first artificial satellite, a 184-pound instrumented sphere named Sputnik I (Russian for "traveling companion"). In November the half-ton Sputnik II went up, carrying a dog as a biological test. In December, under hasty orders, American technicians tried to launch a small satellite in one of the Vanguard test rockets, but it blew up on the launch pad as millions watched by live television.

These three months in late 1957 produced a crisis in Western confidence and soul-searching changes in American education. After years of

chafing at the bit, the Army team under von Braun was given the go-ahead in November, and, 84 days later, on January 31, 1958, orbited the first American satellite, Explorer I. Vanguard I, a three-pound sphere, went into orbit in March, while Sputnik III went up in May. At 3,000 pounds, Sputnik III was 56 times as massive as the three American satellites combined, and reinforced American anxiety during that spring.

The first satellites were designed primarily to probe the nearby environment of space. Among their discoveries were the **Van Allen belts,** doughnut-shaped zones of energetic atomic particles surrounding the earth, and the earth's "pear shape," or slight southern-hemisphere bulge.

The First Manned Space Flights

American engineers had concentrated on miniaturizing precision instruments, while Soviet engineers had concentrated on rocket power. Though either approach would work for orbiting the earth, the Soviets' large, powerful rockets gave them an edge. After putting the first probe on the moon and photographing the moon's far side for the first time in 1959, the Russians went on to test the biological possibilities of space flight with dogs, some of which they recovered from orbit. On April 12, 1961, in a 5-ton craft, the 27-year-old Russian Yuri Gagarin became the first person to

Figure 4·8 **The evolution of public consciousness of space exploration is reflected in the number of related articles published each year in major periodicals.**

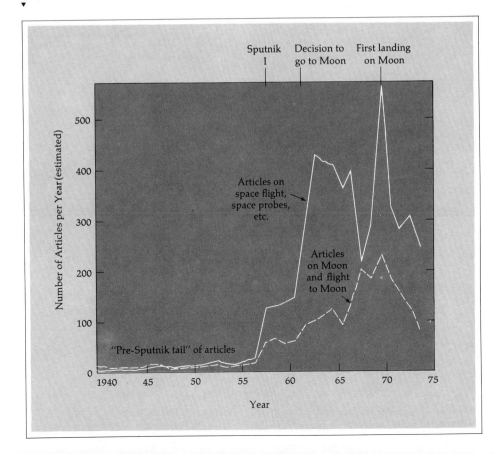

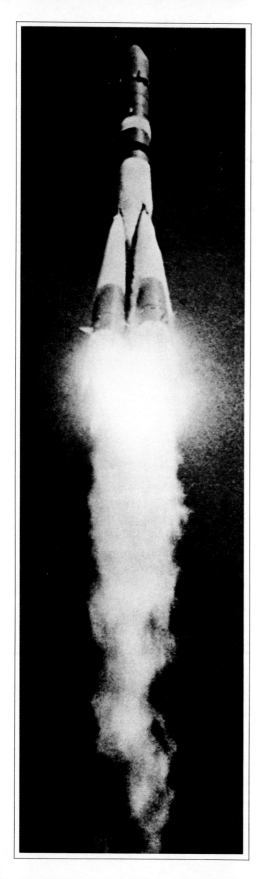

◄ *Figure 4•9* The launch of Vostok-1 and Yuri Gagarin, the first human to successfully circle the earth, in April 1961.

▲
Figure 4•10 Astronaut Alan Shepard became the first American in space during a 15-minute suborbital flight, May 5, 1961. In February 1971 he commanded the Apollo 14 flight to a successful landing on the moon. (NASA painting by Ted Wilbur)

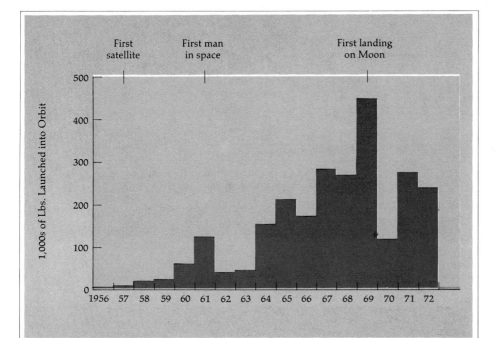

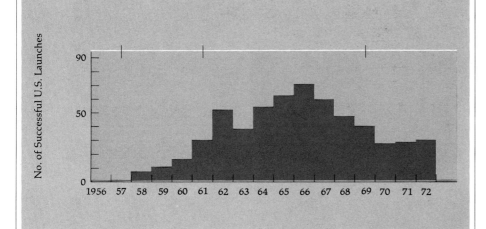

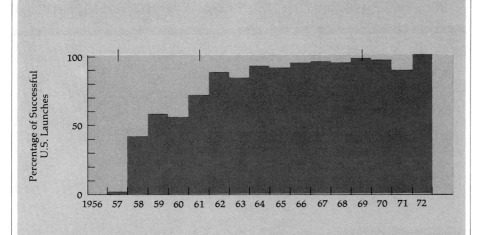

circle the earth, taking 108 minutes to complete the flight.

The first American manned flight was Alan Shepard's 15-minute suborbital flight on May 5, 1961. On August 7, Russian cosmonaut Gherman Titov made 17 orbits in a full-day flight. America's first single-orbit flight came six months later, February 20, 1962, as a Mercury capsule was piloted by John Glenn. One scientific result of these flights was to allay fears that the Van Allen radiation belts or meteoroids might prevent manned space flight.

The Decision to Explore the Moon: Politics and Science

Space exploration obviously had not been a purely scientific effort divorced from politics or national goals. The whole enterprise was paced by social judgments and funding decisions. In this it is rather unlike science in past centuries, and more like many medical, environmental, and energy programs today. The technical resources needed to attack a major scientific problem have become so complex that they demand, not a single genius in a hand-built laboratory, but a coordinated, well-managed program with heavy financial, political, and industrial support. Even a wealthy Isaac Newton could hardly be expected to manufacture the heavy castings, special glass,

◄ *Figure 4·11* **The growth of space technology, shown by the total mass launched into space, the number of U.S. launches, and the percent of successful U.S. launches over a 16-year period.**

and solid-state electronics necessary for sophisticated astronomical observations today, not to mention a fueled, 364-foot Saturn rocket! Just as Newton's work might have been lost in a society that did not favor inquiry, disseminate opinions, or preserve results, exploring the moon required a particular set of social conditions.

Verne imagined a voyage to the moon undertaken by a group of shrewd Yankees, funded by international subscription to build their cannon. In the late 1930s Robert Heinlein imagined it as a commercial venture taking place in 1978, funded by an industrial tycoon. In reality it was to be a $20 billion government-inspired enterprise, conducted by an agency whose greatest problem may have been not the technology but the coordinating of widespread resources necessary for the undertaking.

Between 1957 and 1961, planners had roughed out technical requirements and timetables for a lunar voyage. Though President Kennedy, who took office in January, 1961, sought national goals that would spur creative effort, he was at first skeptical about a lunar program, particularly faced with a Russian lead. (Gagarin had made the first manned orbital flight in April 1961.) Kennedy's science advisor recalls Kennedy remarking (Logsdon, 1970):

> 'If you had a scientific spectacular on this earth that would be more useful . . . or something that is just as dramatic and convincing as space, then we would do it.' We talked about a lot of things . . . and the answer was that you couldn't make another choice.

Shepard's May 5 suborbital flight gave some reason for optimism. On May 25, 1961, the goal was set in a presidential speech before Congress:

> . . . the dramatic achievements in space which occurred in recent weeks should have made clear to us all, as did the Sputnik in 1957, the impact of this adventure on the minds of men everywhere. . . .
>
> I believe that this nation should commit itself to achieving the goal, before this decade is out, of landing a man on the moon and returning him safely to earth.

Kennedy's assassination in November 1963 invested the NASA program with the sense of memorial to a vision, making serious cuts or curtailment unlikely. The leap into space thus came about by a special combination of political, technological, and social factors.

Figure 4·12 **A night view of the Apollo 17 spacecraft, the 364-foot Saturn-5 rocket.** (NASA)
▼

Space Exploration: Costs and Results

The results of space exploration will be clearer to our grandchildren than to us, but we can at least compare some costs and benefits.

Figure 4·13 **A Skylab-2 astronaut climbs outside the space station to retrieve film packs from a solar telescope.** (NASA)
▼

Costs

For the decade 1959 to 1969, leading up to the first lunar landing, the total budget of the United States was $1,400 billion, whereas the NASA budget (including traditional aircraft research as well as spacecraft development) was $35 billion or about 2.5 percent of the total. In the early 1970s funding of all scientific research (civilian and military) put together was about 6.5 percent of the total U.S. Budget. In contrast, the Departments of Defense

and Health, Education, and Welfare, each spent about one-third of the total budget.

There has been a real question whether society might be improved by canceling space exploration and spending the money on programs directly designed to alleviate such social problems as poverty, illness, malnutrition, or the energy crisis. But as the above figures show, the science budget is too small to have much impact in these areas. As NASA has been quick to point out, in 1969 the

government spent 15 times as much on social programs (income security, health, veterans, education, housing, and so on) as on the NASA budget; and 31 times as much in 1972. In other words, canceling all space exploration or even all basic research could increase the budgets of direct social programs by only a few percent. Moreover, there is no guarantee that any funds cut by Congress from the space exploration budget would be reappropriated to social spending.

Practical Results

Seven centuries elapsed between the first probable use of rockets (in a Mongol battle in 1232) and their first scientific application (in an atmospheric research flight conducted by Goddard in 1929). Constructive benefits have come only in recent years. The first weather satellite launched in 1960 led to nearly continuous monitoring of weather conditions. Most evening TV weather reports now display satellite photographs, used by the Weather Bureau to predict storms both at sea and on land. Thousands of lives and millions of dollars have been saved through these programs.

Another important practical consequence of space flight is the communications satellite. Events ranging from Chinese ballet and Olympic games to outbreaks of war, when viewed internationally, clarify relationships among peoples. Marshall McLuhan, Buckminster Fuller, and others predict that this improved communication will strengthen human community on this planet, producing a "global village"—just as the bickering American colonies eventually came to accept a common identity.

Intercontinental telephone communications are made possible by a group of satellites orbiting about 26,000 miles out from the earth's center. With an orbital period of 24 hours, these satellites stay fixed in relation to any particular transmitting station. From an earthbound antenna, a signal beamed at the satellite is retransmitted to another part of the earth.[2]

Another use of satellites is photography for mapping. Amazingly enough, in some remote areas of earth—such as the Amazon jungles of Brazil and certain parts of Ethiopia—satellite photos are better than any existing map. Ultraviolet and infrared satellite sensors are being used in agriculture, forestry, and prospecting. Satellites can spot certain crop blights by subtle color changes before they are detected on the ground; they can detect and track pollutants in rivers, coastal waters, and the atmosphere; and they can locate mineral deposits by "reading" geologic features and subtle soil colorations. These applications are likely to expand in the next decade.

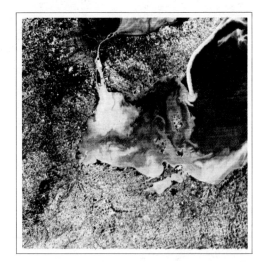

▲
Figure 4·15 **Photo made on March 27, 1973 by Earth Resources Technology Satellite (ERTS) shows polluted waters entering the southwest corner of Lake Erie, from highly urbanized areas along the Detroit (top) and Maumee (lower left) rivers. Lighter tones, brought out by contrast enhancement, reveal currents of polluted water. Similar satellite photos can be used to aid in enforcing pollution-control standards. (NASA)**

Figure 4·14 **Use of space photography in weather forecasting has become familiar on the evening news. This photo, made from the Applications Technology Satellite (ATS) on September 27, 1971, shows U.S. cloud cover and two tropical storms: Ginger (right) and Olivia (lower left). Advance storm warnings reduce property damage in the United States and other countries. (NASA)**
▼

[2]This system was first proposed in the late 1940s by science fiction writer Arthur C. Clarke. In practice today, it causes curious pauses in transcontinental phone calls. If your call goes by normal ground links from east coast to west coast, the gap between your question and your friend's answer is about 0.2 seconds, the normal response time for the brain to frame an answer. However, if your call is beamed up to a satellite about 36,000 km above the surface, then these radio waves must complete two round trips (144,000 km) at the speed of light (300,000 km/sec) between question and answer. This takes roughly 0.5 seconds. Added to the 0.2 seconds mentioned above, it gives a typical pause of 0.7 seconds between dialog elements in a space-routed telephone call.

we carry in us a stubborn disinclination to treat all men as brothers. On the other hand, we live on a shrinking and vulnerable planet which our lack of foresight is rapidly turning into a slum. Never again on this planet will there be unoccupied land, cultural isolation, freedom from bureaucracy, freedom for people to get lost and be on their own. Never again on this planet. But how about somewhere else?

Wernher von Braun commented shortly after the first lunar landing (Lewis, 1969):

> *. . . the ability for man to walk and actually live on other worlds has virtually assured mankind of immortality.*

▲
Figure 4·16 **The manned shuttle (winged, left center) is planned for extensive use in the 1980s, and will allow the launch, construction, and repair of many astronomical observatories and other scientific satellites in orbits around the earth.** (NASA painting by Robert McCall)

NASA supporters point also to the many valuable spinoffs from the space program—new fabrics, new food-processing techniques, and miniaturization methods now used for heart-pacers, surgical equipment, and the like. Some critics have asked if these benefits could have come more cheaply from direct research in those specific areas. But scientific and technological advance most often seem to happen in indirect and even informal ways. A broad research and technological effort is more effective in some ways than highly directed doses of research funding.

Intangible Results

There are two important intangible results of space exploration. First, it demonstrates a successful management technique to accomplish an extremely broad technological objective.

Second, space flight engenders a cosmic perspective. It is probably no coincidence that the ecological movement erupted just at the time of the first lunar flights, when photographs of the earth from space became poster art, and astronaut Lovell, on Christmas Eve, 1968, radioed from lunar orbit that the "vast loneliness . . . of the moon . . . makes you realize just what you have back there on earth."

The cosmic perspective also opens the possibility of human survival even in the event of a natural or human-caused catastrophe on the earth (Dyson, 1969):

> *We are historically attuned to living in small exclusive groups, and*

Future Plans

In considering space exploration, our perspective is broadening to include not just flights to the moon and planets, but exploration of the whole universe as well. The moon and nearby planets are, of course, the easiest extraterrestrial places to reach. All planets and satellites in the inner solar system have now been studied from close range on more than one occasion by spacecraft. Current flights are planned to visit more distant planets in the solar system.

In the meantime, the space shuttle will provide a reusable craft to carry people and supplies into space near the earth. This will allow physical experiments monitored by astronauts in long-term space stations. The space shuttle has important consequences for telescopic astronomy as well. The earth's atmosphere distorts astronomical images and absorbs certain colors, or wavelengths, of light. Telescopes above the atmosphere can and do obtain new information about stars and galaxies far beyond the solar system. The space shuttle will also provide new opportunities to deliver, construct, and repair orbiting telescopes above the atmosphere. An orbiting device called the Space Telescope will be one of the first of this new generation of astronomical instruments for probing the universe.

Finally, various new devices are being planned for communication, energy generation, weather prediction, mapping, and other purposes. Long range studies are under way to consider commercial and industrial utilization of materials and energy in space.

▲
Figure 4·17 **The scene of earth rising over the lunar landscape provided humanity with a new view of our planet.** (NASA photo by Apollo astronauts in lunar orbit)

4

Summary

The idea of carrying people or instruments into space existed long before the technological possibility. Newton's theory of gravitational attraction made it possible to calculate how fast objects would have to go in order to orbit around the earth or escape from the earth. Development of rocket technology early in this century provided the means to reach these speeds. Political and social factors, including strong public support for space exploration, allowed the implementation of these means.

The technology of space travel has been used in three areas: (1) improving conditions such as communications and weather prediction on earth; (2) exploration of other bodies in the solar system; (3) improving telescopic observations of stars and galaxies far beyond the solar system. Space flight has provided a sense of exploration unmatched since the Renaissance voyages to the New World. Plausibly, space travel's greatest value could come in ensuring human survival in other environments in the event that warfare or environmental changes make this planet uninhabitable.

Concepts

inverse square law

mass

Newton's law of gravitation

circular velocity

perigee

apogee

ellipse

parabola

escape velocity

hyperbola

Newton's third law of motion

thrust

Van Allen belts

Problems

1. If an astronaut is flying in circular orbit just above the atmosphere and wants to escape from the earth, how much additional velocity does he need?

2. Rocket engineers typically speak of the difference between one orbit and another in terms of velocity difference: for example, so-and-so many feet per second. Why is this? (Consult Problem 1 if necessary.)

3. Because the earth rotates, the equatorial regions move at about 1,000 mph. Why is it easiest to launch a satellite in a west-to-east orbit over the equator?

4. You have just rowed a rowboat to a point at rest about a foot from the dock. You step off the rowboat toward the dock.
   a. *Why does the rowboat move away from the dock?*
   b. *In this instance, how are you analogous to the exhaust from a rocket?*

5. Do you think a proposal to spend $20 billion to land a man on the moon for the first time would be endorsed by Congress or the public if it happened this year instead of 1961? Explain your answer, accounting for similarities or differences in public attitudes between these two periods of history.

6. Do you believe that in the next century manned flights to other planets, satellites, or asteroids will be common? Do you think this would be a healthy or unhealthy enterprise in respect to social progress, intellectual stimulation, cost, sources of raw materials, etc.?

Advanced Problems

7. If the moon were moved three times as far away, how much weaker would be the force that attracts it to the earth?

8. If the earth were half as massive, how much weaker would be its attraction for the moon?

9. Suppose an astronaut is orbiting the earth at the same distance as the moon, in a circular orbit, but not near the moon. How much faster would he or she have to move to escape the earth altogether?

10. Compute the velocity of an artificial satellite in a circular orbit 42,500 km from the earth's center. Show that it would be in a synchronous orbit—that is, it would revolve with a period of 24 hours and thus stay above the same point on the earth's surface.

Projects

1. Observe the moon as it passes above or below a bright star. Remembering that the moon is about 3,480 km in diameter, observe how many of its own diameters it moves in one hour, and calculate its orbital velocity.

2. Observe an artificial satellite. What illuminates it? Why are artificial satellites commonly seen at dusk or dawn, and not at midnight? Observe whether the satellite passes into the earth's shadow, and whether it reddens slightly as it does so.

84

The Moon

The same moon that we see, with almost the same features (give or take a few craters), shone down on the breakup of the continents, gleamed in the eyes of the last dinosaurs, and illuminated the antics of the first proto-humans. The same moon, with exactly the features we see, was seen by all the historical figures we have mentioned: Stonehenge builders, Chaldean astrologers, Akhenaten, Mayan eclipse observers, Aristarchus, al-Battani, Isaac Newton, Jules Verne. All those people asked what it is, how it is composed, where it came from. Today it is a little different; it has footprints on it and we know a few of the answers.

The moon is at once familiar and unfamiliar. Ours is the first generation to have seen a few of its stark landscapes and to know that it is at an opposite extreme from the earth's environment, with no air, no water, no weather, no blue sky, no clouds, no life. Yet many people are still confused about its simple phases. Can you recall which way the horns of the crescent point in the evening sky? Cartoonists often draw the horns pointing down toward the horizon. Not so. Since the fully illuminated edge of the crescent must face the sun, which has just set below the

Figure 5·1 **The phases of the moon as seen by a northern-hemisphere observer at sunset on the indicated days of the cycle.**
▼

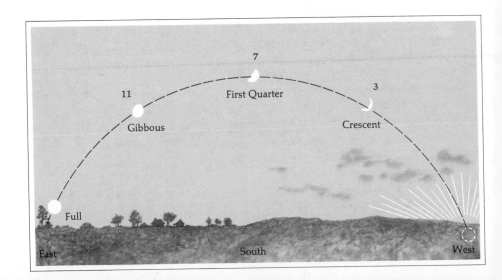

horizon, the horns must point upward, away from the sun. Scoff, too, at the novelist who describes the full moon rising at midnight! The fully illuminated moon is always opposite the sun, and hence must rise as the sun sets.

The whole subject of the moon's motions around the earth, though it might at first seem mundane, provides interesting clues about the ancient history of the earth–moon planetary pair. We will thus begin with considerations of these motions, move on to a description of the moon's surface

Figure 5·2 **The moon's motion around the earth. Synchronous rotation is shown by a mountain (indicated by a triangle) that always faces the earth. The moon completes one rotation during one complete revolution around the earth. Phases seen by an earthbound observer are indicated at different points in the orbit.**
▼

and the astronauts' discoveries, and end with the puzzling problem of the moon's origin.

Synchronous Rotation of the Moon

It took many centuries for people to accept the idea that the earth is a spinning, round planet. Perhaps this idea would have come centuries earlier if people could have watched the moon spin in the sky, with mountainous regions and grey plains carried around the lunar face and out of sight onto the back side, only to reappear again a few days later. Instead, we do not seem to see the moon spin. From day to day we see only the same familiar features of "the man in the moon"—a pattern of grey lava patches —only lit from different directions by

the sun as the moon goes through its phases. As early as 1680, the French astronomer G. D. Cassini correctly explained this in a statement that is sometimes hard to grasp:

The moon rotates on its axis with a period equal to its orbital revolution period around the earth, so that it keeps the same side facing earth at all times.

The rotation of any satellite around any other body in this way is called **synchronous rotation** since it is synchronized with its own revolution. How can the moon rotate at all, you might ask, if it always keeps the same side toward the earth? The best answer is an experiment. Put a chair in the middle of a room. The chair is earth; the walls, the distant stars. To represent lunar orbital motion, walk around the chair. If you walk around the chair always facing it—so that an earthly observer in the chair never sees your back—you will find that you have faced all sides of the room during one circuit. In other words, you have rotated once on your axis and made one revolution around the chair.

Even writers who should know better sometimes speak of the moon's "eternally dark side." This mistake comes from a popular belief that the side eternally hidden from us must always be dark. But the far side, just like the near side, has day and night— periods of sunlight and periods of darkness—each of which lasts about two weeks. The moon takes about four weeks to make a complete rotation. There *is* always a dark side, but it isn't necessarily the *far* side.

To be more precise, the moon takes 27.32 days to complete one revolution around the earth, relative to the stars—

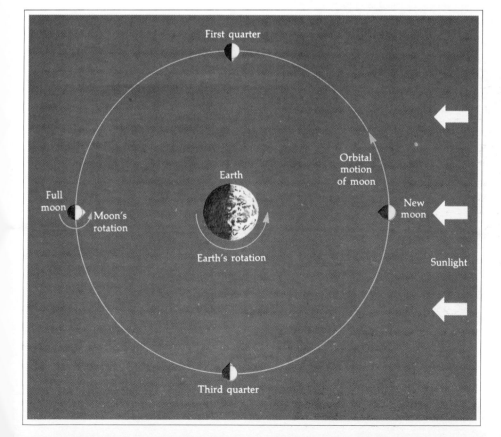

an interval called the moon's **sidereal period.** During this period, of course, the earth has moved roughly 27°

Figure 5·3 **The moon's motion is studied by electronic techniques such as radar and laser ranging. Here a laser beam from the Kitt Peak solar telescope (used at night) is projected toward the moon (overexposed in upper right). Reflective devices placed on the moon during space missions reflect the beam, allowing accurate tracking.** (Kitt Peak National Observatory)

▼

around the sun, so that the moon has to move through this additional angle to complete its cycle of phases relative to the sun. Therefore, the cycle of lunar phases takes 29.53 days—an interval called the **synodical month.**

Is there a reason why the moon keeps one side toward earth? When asking for a reason, a scientist normally means to ask, "Could the observation be explained by some more fundamental properties of nature, so

that it becomes a special case of a more general phenomenon?" Here the answer is yes. In the 1780s and 1790s, Joseph Louis Lagrange and Pierre Simon Laplace used Newton's laws to show that if the moon were slightly egg-shaped or football-shaped, gravitational forces would make the longest axis point toward the earth at all times. And indeed, one axis of the moon is about 2–3 km longer than the others and points steadily toward earth.

If the moon always kept exactly the same side toward earth, earthbound observers could never see more than exactly 50 percent of it. Careful mapping, however, has shown that the moon "wobbles," and that during a period of years, 59 percent of the moon can be seen, with first one limb and then another being turned slightly farther toward earth than its average position. These apparent irregularities of the moon's rotation are called **librations.**

Tidal Evolution of the Earth–Moon System

Gravitational forces in the earth–moon–sun system cause **tides,** or bulges in the shape of the earth and moon. These tides have intriguing consequences. When two bodies are near each other each exerts a gravitational force on the other. Thus, the side of the moon facing the earth has a stronger force on it than the far side, because the facing side is closer, as shown in Figure 5·4. This means that the moon stretches slightly along this line, limited by the small elasticity of its rock interior. This stretching is called a **body tide.** Similar forces acting on the earth produce not only a body tide, but an **ocean tide** in the much more flexible layer

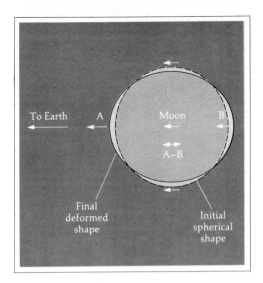

Figure 5•4 **The gravitational attraction of the earth on the moon is greater on the near side (A) than on the far side (B). The difference (A − B) acts as a stretching force deforming the moon from its unstressed spherical shape to a flattened shape with tidal bulges on the near and far side.**

of water on the earth's surface. These tides take the form of bulges on the front and back side of both the earth and moon, since each body comes into equilibrium with the gravitational forces by stretching along the earth–moon line.

The earth's body tides are hard to detect, but its ocean tides are obvious to anyone who visits a beach for more than an hour. They range in height from about 2 feet to over 50 feet. One might suppose that high tide would always occur when the moon is over-head, with highest tides at new moon or full moon, because the tidal forces act along the earth–moon line or, in the strongest case, along the earth–moon–sun line. However, ocean tides do not behave this way. Although ocean tides at new moon or full moon (so-called **spring tides**) tend to be stronger than those at quarter phases (so-called **neap tides**), they are not smoothly related to the position of the moon and sun, because of complicated

effects involving motions of water around the irregularly shaped oceans and seas, and because of earth's rotation. Coastline geometry in some places can produce remarkable wave effects associated with tides, but so-called "tidal waves" are related not to tides but to earthquakes or volcanic activity at sea; they are properly called by their Japanese name, **tsunamis.**

Tidal bulges raised on earth and moon have four major effects, first described by George Darwin (1898), son of the famous naturalist.

The Moon's Synchronous Rotation

The effect of tides explains why the moon keeps one side facing the earth. As mentioned above, any elongation in the moon's shape would tend to

Figure 5•5 **The cause of the moon's tidal recession. Because of the earth's relatively rapid rotation, the earth's tidal bulge (MM') gets dragged off the earth-moon line by an angle (θ). Thus, in addition to the normal gravitational force of the earth (F_E), there is a net forward force (F_M) caused by the difference in attractions between the nearer bulge (M) and the farther bulge (M'). This force (F_M) pushes the moon ahead in its orbit, causing it to spiral slowly outward.**
▼

make one side always face the earth. Tides guarantee such an elongation, since they create bulges. If the moon were initially spherical or rotating at a non-synchronous rate, gravitational forces acting on it would create tidal bulges and slow the rotation until it became synchronized and one side faced earth at all times.

Recession of the Moon

The second effect is a slow **tidal recession** of the moon away from the earth because of gravitational forces on tidal bulges. As shown in Figure 5•5, the earth's rapid rotation drags its tidal bulge slightly ahead of the earth–moon line by some angle θ. The gravity of the bulge M, closest to the moon, then tends to pull the moon ahead in its orbit. This produces the effect of a small rocket attached to the trailing side of the moon, accelerating the moon forward and making it spiral out from the earth. This effect also indicates that the moon was much closer to the earth several billion years ago. Analyses do not indicate the exact date, but most researchers believe the moon was closest to earth about 4.6 billion years ago during formation of the two bodies.

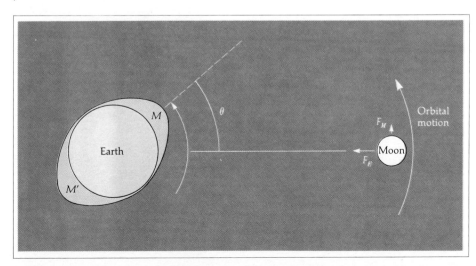

Slowing the Earth's Rotation

The third effect of tides is that the earth's rotation slows down, just as the earth has slowed the moon to synchronous rotation. The moon's weaker gravity has not been able to slow the earth to full synchronous rotation, however. The effect can be seen in Figure 5·5, where the moon will pull backward on the tidal bulge M, acting like a brake on the earth. Studies indicate that billions of years ago, when the moon was closer, the earth's day was only about 5 or 6 hours long, instead of the current 24 hours. These theoretical results have been confirmed by an unexpected source. Certain marine creatures create daily and monthly banded structures in their shells or other hard parts, allowing biologists to count the number of day-bands in a monthly cycle. Results give good evidence that whereas the present synodic month is 29.5 days long, it was only about 29.1 days 45 million years ago. Older fossils show that tides existed as long as 2.8 eons ago, and that the synodic month was as little as 17 days long (Kaula and Harris, 1975). We cannot extrapolate such data further back to determine exactly when the moon was closest to earth, because the tidal effects depend on the configuration of terrestrial oceans and continents, and these configurations were different by unknown amounts in the past.

The moon could eventually recede so far from the earth that the earth's gravitational pull would be too weak to keep it in orbit. It could then escape into an orbit around the sun. However, before that point is reached, small tides raised on the earth by the *sun* will slow the earth's rotation so that the length of the day and the month become equal, stopping the tidal recession. The earth and moon would then *both* be in synchronous rotation. Eventually, because of this small solar effect, the earth's rotation will slow so much that the day will exceed the month, and the tidal process will reverse; the moon will begin to approach the earth again. It may eventually break into fragments when it arrives inside a distance called Roche's limit.

Roche's Limit

The fourth application of tidal theory shows that a small body, if close enough to a large body, can be torn apart by tides. Calculated around 1850 by French mathematician Edward Roche, **Roche's limit** is the distance between any two bodies within which the tide-raising force exerted on the smaller body is sufficient to disrupt it. The effect occurs because the *difference* between gravitational attraction of the near and far sides of a satellite increases as the satellite moves closer to the primary body, as shown in Figure 5·6. The position of Roche's limit depends on the size, density, and strength of the satellite. For instance, a small metal spacecraft is not torn apart while orbiting near the earth. But a body as big as the moon, if similarly placed for a long period, would develop internal stresses and fractures, and eventually disintegrate into a cloud of orbiting particles, like Saturn's rings. The moon could not remain an intact body much closer to earth than about 18,000 km (11,000 mi) (its present distance is 384,000 km or 240,000 mi).

Tidal effects obviously have a bearing on the origin of the moon. Was it ever inside Roche's limit? We will return to such questions at the end of the chapter, after examining other kinds of lunar data.

Figure 5·6 **Roche's limit. If a satellite is located at A, a small tidal bulge develops. At closer distance B, a larger bulge develops. Within a critical distance, called Roche's limit, the differential stretching force between the near and far sides is so great that the satellite is torn apart.**

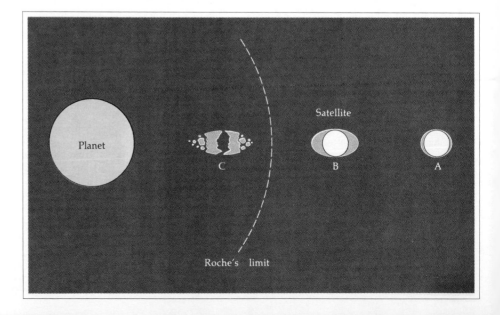

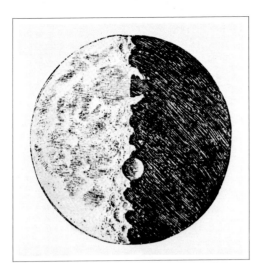

Figure 5·7 **The first telescopic observations of the moon were published in 1610 by Galileo. This appears to be his view of the whole moon at third quarter, showing the grey plain, Mare Imbrium (top), a prominent crater, and the irregular terminator.**

Figure 5·8 **This is one of the first known photographs of the moon, an 1851 daguerreotype, probably by the American astronomer J. W. Draper or his son Henry Draper. Compare with Figure 5·9, a modern photo at same phase. The smallest details are about 35 km across. (NASA)**
▼

Features of the Moon

Until the telescope was invented around 1608, no one had much idea of the lunar surface features, except for the grey patches that make up the

Figure 5·9 **A modern view of the moon at a phase similar to that in Figure 5·8, illustrating a century's advance in astronomical photography. The smallest details are about 1 km across. (Lunar and Planetary Laboratory, University of Arizona)**
▼

face of "the man in the moon." Some thought the moon was a polished sphere. Galileo Galilei (1564–1642), the first person known to have seen the moon's features through a telescope, reported his observations in 1610. He saw that shadows were pronounced along the **terminator,** the line dividing lunar day from lunar night. Details are less well seen under high lighting (full moon) or at the edge of the disk, called the **limb.** The shadows indicated the presence of mountains

and other rugged relief. After brief inspection, Galileo was able to report:

> The moon certainly does not possess a smooth surface, but one rough and uneven, and just like the face of the earth itself, is everywhere full of vast protuberances, deep chasms, and sinuosities.

Most of the roughness was caused by thousands of **impact craters** of all sizes, ranging up to 1200 km diameter. The dark grey patches that are visible to the naked eye and form "the man in the moon" Galileo found to be much smoother than the brighter, cratered areas. He mistook them for seas. Using Latin, as scientists did in the 1600s, he called them **maria** (MAR-ee-ah; the singular is **mare** [MAH-ray]). These "seas," as Galileo himself probably eventually realized, are actually vast plains. (We now know they are covered with dark lava. Mare surfaces cover much of the front side of the moon, but only 15 percent of the whole moon.) Astronomers named the bright regions **terrae,** or "lands" (singular, **terra**). They are heavily cratered, rugged upland areas.

Galileo estimated the height of certain lunar peaks and discussed the possibility of an atmosphere. His work greatly strengthened the conception that the moon is an earthlike, planetary body having familiar features such as mountains. He also noted that the moon shines by reflected sunlight, concluded that the earth must do the same, and used this as a proof against those who argued that the earth was not to be included among the planets.

The first reasonably accurate lunar map was produced by the German Johannes Hevelius, in 1647. Hevelius correctly estimated that peaks in the lunar Apennines (near the Apollo 15 landing site) were about 5,000 m high.

In 1651 an Italian priest, Riccioli, started the present practice of naming craters after well-known scientists and philosophers, such as Copernicus, Tycho, and Plato. The "seas," or lava plains, were given poetic and fanciful names, such as Mare Imbrium (Sea of Rains) and Mare Nectaris (Sea of Nectars). Lunar mountains were named for prominent terrestrial ranges, such as the Apennines, the Alps, and the Carpathians. These mountains, however, are unlike terrestrial folded ranges; instead they are the rims of

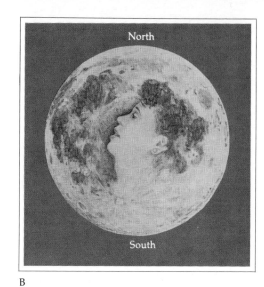

B

▲

as the bright rays emanating from the crater Tycho (bottom). B. Several folklore figures, such as "the woman in the moon" are formed by the pattern of dark lava plains.

Figure 5·10 A. At full moon, the rugged terrain is minimized by the absence of shadows. Compare with Figure 5·9. Full lighting emphasizes different features, such

▼

A

vast craters, called **basins,** which in turn contain the mare plains. Bright streaks, called **rays,** radiating from various craters, but showing no relief, are fine debris blasted out of the craters. Most maria are merely seas of lava that have flooded ancient basins. Meandering canyonlike valleys, probably eroded by hot lavas, are called **rilles.**

Astronomers of the 1700s and 1800s undertook a tantalizing endeavor, searching with ever-larger telescopes at ever-greater magnification for key diagnostic details that might show how the moon formed. Were there any signs of changes, of new craters forming, of volcanic activity? Were there any artificial structures that might have been built by the inhabitants that popular writers continued to place on the moon?[1]

Despite years of search, no changes or artificial structures were found. Lunar photographs, first made in 1849, showed the same structures year after year. Because there seemed nothing new to discover, interest in the moon dwindled, and many astronomers saw it as a sterile nuisance that lit up the night sky and blotted out the faint objects they wanted to observe.

This early work proved that the moon is much less geologically active than the earth. However, evidence for minor geologic activity on the present-day moon came in 1963, when on two occasions observers at Lowell Obser-

◄ *Figure 5•11* **The Orientale basin is the most famous multi-ring complex on the moon. The outermost ring of cliffs is nearly 1,000 km in diameter; the central, dark lava plain is the size of New York State. Lava appears to have erupted from fractures along the bases of some of the outer cliffs.** (NASA Orbiter photo)

Figure 5•12 **A schematic cross-section of a major lunar basin. The original impact crater is in the center. Faulting and subsidence may have created the outer ring of cliffs. Vertical hatching indicates later intrusions of lava (vertical scale exaggerated).**
▼

---

[1]Or—an idea conceived only recently—by alien interstellar travelers who might have visited the solar system in past eras? See, for example, Arthur C. Clarke's book *2001.*

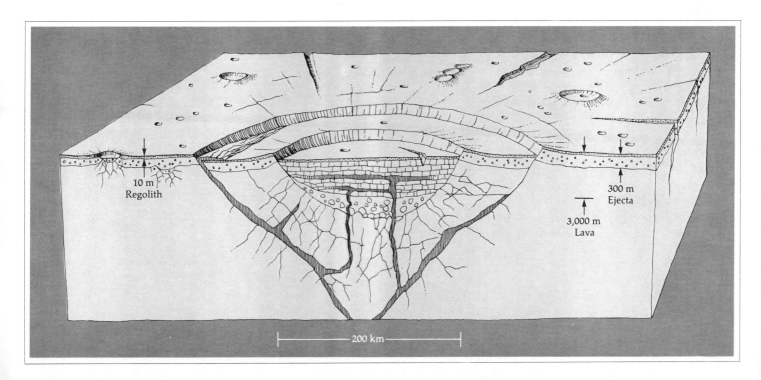

10 m
Regolith

300 m
Ejecta

3,000 m
Lava

200 km

vatory saw a red glow near the crater Aristarchus. This glow may have been a volcanic eruption or gas discharge; later data from Apollo flights indicated occasional gas emission in this region.

Why is the moon's geologic character different from the earth's? Why do craters dominate? Where did the moon come from? What would it be like to stand on the moon? Such questions motivated actual journeys to the moon in the last decade.

The Flights to the Moon

The first human device to reach the moon was a Russian spacecraft that carried little scientific equipment and crashed into the moon in 1959. The first close-up photos of the surface, from the American probe Ranger VII in 1964, showed that the surface was not craggy, but covered by a gently rolling layer of powdery soil, scattered rocks, and shallow craters of various sizes. This type of soil cover, present in nearly all parts of the moon, came to be called **regolith** (rocky layer). The lunar regolith is made primarily of debris blasted out of lunar craters as the craters formed. Each well-preserved lunar crater is surrounded by a sheet of such material, called an **ejecta blanket.** The regolith is therefore said to be composed of overlapping ejecta blankets.

Table 5•1 lists the six Apollo lunar landings and earlier test flights. The first two landings were cautious tests of the spacecraft, touching down on flat smooth plains. This allowed sampling of the rocks in the lava flows that make up the maria. These lavas were $3\frac{1}{2}$ billion years old, about as old as the oldest known earth rocks.

Following the unsuccessful flight of Apollo 13, which returned to earth,

Apollo 14 landed on the ejecta blanket from the vast Imbrium basin. This structure, about 1,200 km in diameter from rim to rim, is the largest visible crater on the moon. Apollo 14 samples from a region outside the rim were about 3.9 billion years old and confirmed that fractured rocky material and debris had been thrown hundreds of kilometers out of the Imbrium

Figure 5•13 **The Apollo 11 launch, July 16, 1969. The following figures through 5•23 portray the stages of an Apollo voyage to the moon and back. (See also Color Photos 5–7.)**
▼

region to form the basin. Probably the basin was formed when a meteorite about 150 km in diameter crashed into the moon about four billion years ago. (Typical basin structure is shown in Figure 5•12.)

The next flight landed at the edge of the lava plains inside the Imbrium basin. Rock samples from the rim of the basin, the Apennine Mountains, confirmed that the rim rocks of the basin were older than the lavas that later covered the basin floor to form Mare Imbrium (Figure 5•17).

Apollo 16 was sent to an upland, or *terra,* site near the crater Descartes.

From rock samples it seemed that this terra region was composed of overlapping layers of fractured debris ejected from various craters, roughly 4 billion years old.

The last flight, Apollo 17, went to a still more complex site near some dark deposits that scientists had interpreted as recent volcanic strata. Instead of finding young rocks, however, the astronauts found more ancient lavas about 3.8 billion years old. A fragment of fractured rock was dated at 4.5 billion years, making it the oldest rock found on the moon. (For further summaries of Apollo flights see Taylor, 1975; El-Baz, 1975.)

◄ *Figure 5·14* **The first man to set foot on the moon. Neil Armstrong on the morning of the Apollo 11 launch, July 16, 1969.** (NASA sketch by Paul Calle)

Figure 5·15 **The lunar landing module preparing to descend to the surface of the moon.** (NASA)
▼

Mission	Date	Landing Site	Results
		**Table 5·1 Manned Apollo Explorations of the Moon**	
Orbital Missions			
Apollo 8	Dec 24 68[a]	—	First lunar orbit. Orbital mapping. 115 km minimum altitude.
Apollo 10	May 21 69[a]	—	Test of approach to approx. 17 km minimum altitude.
Landing Missions			
Apollo 11	Jul 20 69	Mare Tranquillitatis	First landing. Samples of mare material.
Apollo 12	Nov 18 69	Oceanus Procellarum	Samples of mare material.
Apollo 14	Feb 5 71	Fra Mauro (ejecta from Imbrium basin)	Samples of ejecta from Imbrium basin.
Apollo 15	Jul 30 71	Edge of Mare Imbrium at foot of Apennine Mts.	Samples of material from Apennine Mts., forming rim of Imbrium basin. Samples of mare material. First use of roving vehicle.
Apollo 16	Apr 20 72	Lunar uplands near crater Descartes	First landing in uplands. Samples of upland materials.
Apollo 17	Dec 11 72	Taurus Mts.	Samples from a region suspected of recent volcanism.

[a]Date lunar orbit began.

◄ *Figure 5·16* The landing module on the moon, at the Apollo 15 site at the base of the Apennine Mountains showing the tracks of the lunar roving vehicle. (NASA)

Figure 5·17 An astronaut with a scientific package on the lunar surface. Note the partially buried, rounded boulder, and footprints in regolith powder in the foreground. (NASA)
▼

What Do Lunar Rocks Say about the Earth and Moon?

Rocks are cosmically significant. Traced back far enough in time, they can be seen to be solid materials that crystallized from earlier melted solids, which in turn crystallized from still earlier melted solids; the original material once existed as solid particles condensed in space. Rocks tell the histories of their parent planets.

Rocks can be analyzed in various ways:

1. in terms of their elements;

2. in terms of their minerals, which are chemical compounds made out of elements;

3. in terms of their structures, which reflect their origin and environment;

4. in terms of their ages, determined from isotopes.

We will describe various findings and then summarize their meanings at the end of the chapter.

The elements and minerals in lunar rocks are quite similar to those of the earth. No new elements are found on the moon, and the most common lunar minerals are also common on earth. Most lunar rocks are *igneous* or modified **igneous rocks;** that is, rocks that have solidified from molten materials. On the earth, these rocks are common only in the volcanic regions; covering only about a quarter of the earth's crust. The rest of the earth is covered mostly by **sedimentary rocks,** rocks made out of particles deposited from the sea or air. Absence of sedimentary rocks on the moon's

◄ *Figure 5•18* **The powderlike properties of the lunar regolith, or soil, are seen in this view of the imprints made by the footpad of a Surveyor unmanned spacecraft.** (NASA)

▲
Figure 5•19 **Most lunar regions visited by astronauts have been relatively smooth and powder-covered, chosen for safety. This view of the interior of a fresh, rock-strewn crater (foreground) was made by Apollo 15 astronauts near the Apennine Mountains (background). It hints at the rugged** landscapes in still unvisited regions of the moon. (NASA)

Figure 5•20 **Geologist Harrison Schmitt was the first trained geologist on the moon. Here he examines a huge boulder found during the Apollo 17 explorations.** (NASA)
▼

Figure 5·21 Blastoff from the moon was televised to the earth by a remote camera set up by Apollo 16 astronauts. Debris flying away from the spacecraft includes material blown off the descent stage (lower) by ignition of the ascent engine. (NASA)

Figure 5·22 The lunar module, minus the landing platform and legs (see Figure 5·16), climbs from the lunar surface to meet the orbiting Apollo 16 command ship, from which this photo was taken. (NASA)

◄ *Figure 5·23* **Return to earth by parachute, following the reentry of the command module (bottom). The final Apollo mission, Apollo 17, landed December 19, 1972, near American Samoa. (NASA)**

rich in aluminum and potassium. These probably represent lava flows around four billion years old. A third type of lunar rock includes a few fragments of **ultrabasic rocks,** or dense rocks rich in iron minerals. These may represent rocks from the lunar mantle, below the anorthosite-basalt crust, and include a 4.5 billion-year-old fragment found by Apollo 17 astronauts.

Some differences between lunar and terrestrial rocks do indicate important differences in the fundamental composition of earth and moon. First, if we consider the moon as a single giant rock, we find that its **bulk density** (total mass divided by total volume) is much less than the earth's—3.3 g/cm^3 for the moon; 5.5 g/cm^3 for the earth. This proves that the moon is made of a different *distribution* of elements and minerals than found on earth. The main difference is that the moon does not have as much iron or other heavy metals as earth has, especially in its dense core.

Second, **volatile elements** (elements that would evaporate if the rocks were strongly heated) such as hydrogen, helium, lead, and mercury are less plentiful on the moon than on the earth. **Refractory elements** (elements with high boiling temperatures, such as aluminum and titanium, which would be left behind if the rocks had been strongly heated) are more common on the moon's surface than on earth. These findings suggest that the moon's material was once heated to high temperature, and perhaps fragmented so that the volatiles escaped,

surface shows that the moon has never had water or atmosphere in its environment, but has been subject to processes of rock melting, resolidification, and fragmentation.

The igneous rocks in the lunar maria, or plains, are primarily lava flows of **basalt,** a type of lava very common in volcanic areas of earth. These basalts are three to four billion years old. The lunar terrae, or uplands, are rich in a lighter-colored type of rock called **anorthosite,** or anorthositic

gabbro. On earth, anorthositic rocks are often found in the ancient cores of continents, such as the Canadian shield. Lunar anorthosites are believed to represent the original lunar crust, probably formed 4.4 to 4.6 billion years ago, when much of the earliest lunar crust melted because of unknown heat sources. This anorthositic crust has since been modified by events such as meteorite bombardment. Also found in the terrae, or uplands, are certain types of basalts

while silicates and refractories were left behind. This would imply that the earth and moon, in some poorly understood way, had different histories.

The amount of **lunar differentiation** is uncertain. The relatively small proportions of the lunar iron core, if it exists at all, do not prove a lack of differentiation, since the moon had very little iron to drain to the center in the first place. The total amount of melting that occurred in the moon's

deep interior is uncertain. The chemical difference between the moon's anorthositic crust and its more ultrabasic mantle indicates some melting and differentiation, but not as much as occurred in the earth to produce

earth's granitic crust and ultrabasic mantle. This lunar differentiation, along with moonquakes and lunar lava flows, prove that the moon has not been geologically dead.

A prominent age difference between surface rocks of earth and moon shows up at once. Lunar rocks are old, generally dating from three to four eons ago (see Table 5.2); terrestrial rocks are young, generally less than an eon

Figure 5.24 **Apollo 15 astronauts David Scott (left) and James Irwin (right) join NASA geologists in examining lunar rock samples in the sample receiving laboratory in Houston.** (NASA)

▼

Figure 5.25 **A lunar rock sample. This lava rock is a fragment of a 4 by 5-foot boulder found by Apollo 16 astronauts in the lunar uplands. It contains numerous vesicles (cavities) and inclusions (fragments of foreign rock).** (NASA)

▼

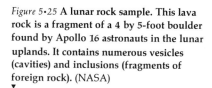

Figure 5.26 **A microscopic view of lunar soil particles, showing angular fragments and spherical beads of glass. The glass fragments give this sample an unusual orange color.** (NASA)

▼

Table 5·2 Summary of Dated Lunar Rocks

Mission	Rock Type	Ages[a] (Eons = 10⁹yr.)	Notes
Apollo 11	Mare basalts	3.48–3.72	7 rocks
Apollo 12	Mare basalts	3.15–3.37	9 rocks
Apollo 14	Fractured rocks ejected from Imbrium basin	3.85–3.96	6 rocks; date of Imbrium impact?
Apollo 15	Mare basalts	3.28–3.44	6 rocks from 2 sample sites.
	Upland anorthosite	4.09	1 rock
Apollo 16	High-Al basalt	3.84	
	Upland breccia	3.92	
Apollo 17	Mare basalt	3.77	1 rock
	Upland breccia	3.86	1 rock
	Fractured dunite	4.48	1 rock
Luna 16	Mare basalt	3.45	1 rock
Luna 20	Data not yet available on returned upland sample		

Comparison Ages for Other Rocks

Meteorites	4.6	Typical formation date
"Whole moon"	4.6	Formation of moon
Oldest terrestrial rocks	3.9	Very rare specimens
Typical terrestrial rocks	0.01 to 0.5	

[a]Date of solidification of crystalline rocks, or formation of breccias, based primarily on Rb-Sr results of Wasserburg, Papanasstasiou, Tera, and colleagues at the California Institute of Technology.

in age. This difference confirms that the moon, though not geologically dead, is much less geologically active than the earth; the moon's ancient rocks have been preserved whereas the earth's have been dragged down into the mantle, crumpled into mountain ranges by plate-tectonic activity, or buried by layers of sediment. The moon's less disturbed rocks tell us much more about the early history of the earth–moon system than earth's rocks do.

The Interior of the Moon

Two other types of Apollo data shed further light on the interior of the moon, in addition to the chemical data suggesting that the moon lacks a large iron core. First, the moon has virtually no magnetic field. This again suggests lack of a molten iron core, because scientists believe planets' magnetic fields originate in currents in such cores. Nonetheless, magnetic measurements on lunar rocks indicate that when they solidified billions of years ago, there was a lunar magnetic field roughly 4 percent the strength of the earth's field. Because no such field exists today, the implication would be that the moon might have had a small, molten iron core in the past, when its interior temperatures were higher. Such a core might still exist but reach no farther than about 350

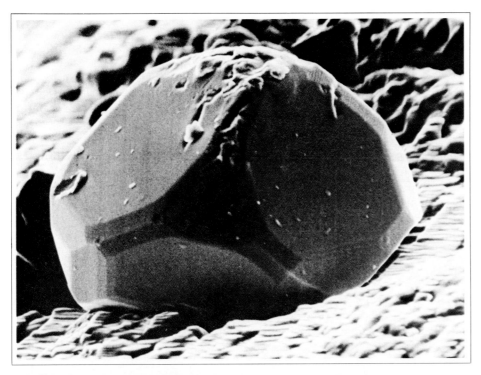

◄ Figure 5·27 Electron microscope photography reveals the detailed structure of crystals from this lunar rock sample. This is an iron crystal that grew on one face of a larger crystal of pyroxene (calcium-magnesium-iron silicate). Its structure indicates that it formed from a hot vapor as the parent material cooled, perhaps after an explosion caused by a high-speed impact. (NASA, Apollo 15 sample)

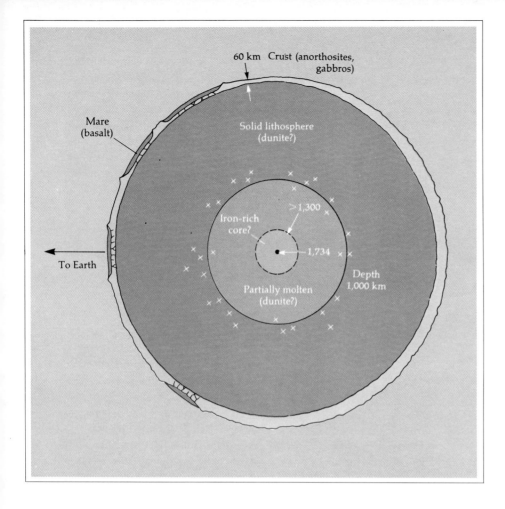

60 km Crust (anorthosites, gabbros)

Mare (basalt)

Solid lithosphere (dunite?)

Iron-rich core?

>1,300

To Earth

1,734

Depth 1,000 km

Partially molten (dunite?)

▲
Figure 5·28 **Cross-section of the moon as revealed by Apollo and other data. The crust is thinner on the front side than on the far side. Fractures under large impact craters on the front side have allowed lava to reach the surface and create more mare lava flows than on the far side. An earthquake zone marks the bottom of the lithosphere. The existence of an iron-rich core is uncertain.**

km from the moon's center, or about 20 percent of the radius (Lammlein et al., 1974).

The second type of data, recorded by automated devices left by lunar astronauts, includes seismic measurements of moonquakes. The moon shows much less quake activity than the earth. Large lunar quakes rank only 0.5 to 1.3 on the Richter scale,

compared to 5 to 8 for occasional major earthquakes. In a year, the earth expends about 100 million million times as much seismic energy as the moon. Nonetheless, moonquakes tell us something about the moon's interior. They occur mostly at depths of 700 to 1,200 km. Many occur in monthly cycles associated with tidal flexing as the moon's slightly elliptical orbit brings it toward, then away from the earth. Seismic waves from these quakes indicate that the quake zone is at the bottom of a solid rock layer, or **lithosphere,** and at the top of a partly melted zone (Toksoz et al., 1974). This again indicates that parts of the moon became hot enough to melt at some time in its past. Some quakes detected by Apollo instruments had a different

source—the impacts of modest-sized meteorites (too small to make craters visible from earth).

Cratering of the Moon and Earth

Lunar rock analysis also reveals the history of cratering of the moon, and, by extension, of the earth. Comparison of crater statistics and lunar rock ages shows that over four billion years ago, in its earliest history, the moon underwent an **early intense bombardment,** with a cratering rate thousands of times higher than today's rate. From three to four billion years ago, the rate declined to one that has been nearly constant for the last three billion years. The earth has had roughly the same cratering rate as the moon in the last three billion years, and scientists assume that both earth and moon experienced similar histories.

The meaning of this discovery seems to be that the moon records the final stages of formation of the planets. The early intense bombardment of the moon represents the final stages of sweep-up of debris left over after planet formation. Probably planets formed by accumulating various bits of floating rocky material in the early solar system. The now-visible lunar craters may represent only the last interplanetary bodies to be swept up, with still earlier craters obliterated by the ones we now see. Calculations reveal that the bodies that formed the craters ranged from abundant microscopic particles through numerous kilometer-sized chunks, to objects over 100 km in diameter. The latter made craters about 1,000 km across.

Repeated cratering by a rain of meteorites has had a notable effect on lunar topography. In the lava plains, there have been only enough craters

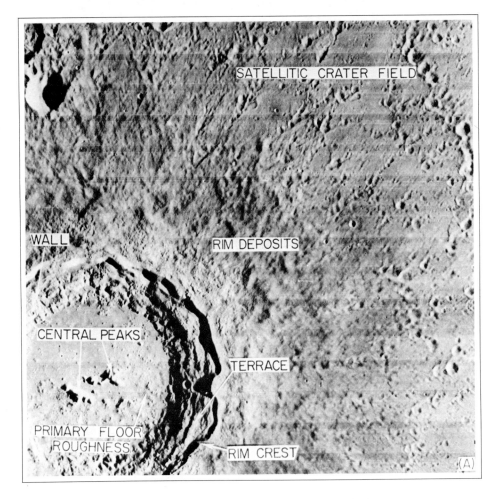

in the last three eons to cover a few percent of the surface with multikilometer craters, and to grind up a layer of lava into the regolith some tens of meters thick. But in the uplands, the 4- to 4½-eon-old surface is virtually saturated with multikilometer craters. Small impacts and overlapping ejecta blankets have smoothed the once-craggy landscape, producing the gently rolling hills photographed by astronauts. Repeated cratering may have pulverized ancient layers to as much as 2 km deep, producing a layer of rocky soil sometimes called a **mega-regolith.** Cratering effects are illustrated in Figures 5·29 and 5·30.

Cratering, then, has badly damaged the record of the moon's earliest history by pulverizing most of the earliest rocks. For this reason, although the moon has yielded the secrets of a period earlier than that revealed on earth, it has not yielded ultimate secrets of planetary origin, as had been hoped before the Apollo flights.

Figure 5·29 **The typical structures in a large lunar crater are exemplified by this Orbiter photograph of the crater Copernicus. The main bowl was excavated by the impact of a meteorite a few kilometers in diameter. Central peaks and terraces are rebound and slump features. Rim deposits and satellite craters are caused by debris blasted out of the crater during its formation.** (Courtesy James W. Head, III; Brown University)

Figure 5·30 **Only the youngest lunar terrains ▶ are not blanketed by a softly rolling layer of debris. In this craggy landscape on the floor of the fresh-looking crater Tycho, the smallest structures are roughly 100 m across. The surface may be a few hundred million years old.** (NASA Orbiter photo)

Theories of the Moon's Origin

The question of the moon's origin has long frustrated theorists. Scientists have argued for years about three theoretical scenarios, each of which has its counter-evidence. The first was the **fission theory,** whereby a "daughter" broke off from the earth. George Darwin, carrying his tidal analysis to its logical conclusion, proposed not only that the moon was once near the earth, but that it actually was part of the earth. He hypothesized that rotational and solar tidal forces combined to raise a bulge that broke away in a swarm of particles to form the moon. Later theorists added that the split probably happened after the earth's iron had drained into the central core, explaining why the moon got little iron and has a low density. The fission theory was attacked by geophysicists because tidal forces seemed insufficient to raise such a bulge or to account for the material's escape beyond Roche's limit. Furthermore, because abundances of isotopes and trace elements in lunar rocks differ from those in rocks on the earth, lunar origin from earth-material is questionable. In the second theory, the **sister-planet theory,** the moon was formed from nearby particles left over after the earth formed. Such particles might have accumulated into the moon. One objection was that Venus and Mars lacked the comparable sister-satellites that might have been expected if the process was one of normal planetary evolution. A second objection was that the earth's debris would not be likely to form a moon with little iron and a mean density of only 3.3 g/cm^3, in view of the earth's large iron core and its density of 5.5 g/cm^3.

A third theory was the **capture theory,** in which the moon originated elsewhere in the solar system, inter-acted with some other planet, was thrown into an orbit that approached earth, and then was captured into orbit around earth. This theory explained the moon's difference in composition, because a moon formed elsewhere might differ from earth in composition, as other planets do. But the theory's chief weakness is that such a capture would involve very special and unlikely orbital conditions during the moon's approach to earth.

Because each of these three theories[2] seemed to have major defects, some pre-Apollo theorists facetiously proposed that the moon must not exist at all!

Apollo results have driven scientists to combine some of the theories. One new theory is a **collision-condensation theory,** in which crustal silicates were blasted off earth by a collision with a large piece of interplanetary debris, forming a massive encircling ring or cloud of fine matter. Iron-rich material, having already been concentrated in the earth's core, was absent. Being hot and fragmented into fine dust, this matter now lost its volatiles. The silicates and refractories condensed into particles that formed the moon as a sister planet.

A **co-accretion theory** proposes that one of many small bodies in the early solar system was captured into orbit around the earth. It then gradually accumulated the last debris of earth-like (mantle-like) material left over near the earth, and grew into the moon (Kaula and Harris, 1975).

None of these theories—neither the old nor the hybrid—has been proved. Much lunar material remains in the Houston archives, and more material from other lunar sites will likely be returned by automated spacecraft,

such as the Russian Luna vehicles which brought back samples from three sites through 1976. Such material may give us new clues about the origin of the moon. The firmest information from rocks studied so far is about not the mode but the time of lunar origin. The moon formed about 4.6 eons ago, and the process—whatever it was—apparently took not more than about 90 million years. This is the same time interval believed to mark the formation of the entire solar system.

▲
Figure 5·31 **The human legacy on the moon.** (NASA)

<hr>

[2]Geologist Bevan French has called them the daughter, sister, and pick-up theories.

5

Summary: History of the Earth–Moon System

The moon is an ancient planetary body, little disturbed since the formative days of the solar system. Much of the information in this chapter yields a chronological history of the moon. Because the earth shared much, if not all, of this history—although erosion obliterated most of the earth's early records—the information from the moon is especially useful in reconstructing major events in the early part of earth–moon history.

Origin Lunar rocks and meteorites reveal that the moon and planets were formed 4.6 eons ago. Probably earth and moon were associated at this time or very shortly thereafter, but the exact sequence of events remains unknown.

Duration of Formative Process Differences in ages among lunar and meteorite specimens indicate that the moon and planets reached approximately their present sizes a few million to 90 million years after the formative process began.

Early Heating and Differentiation At least one lunar rock, 4.5 billion years old, together with analyses of other lunar rocks, indicates that heating, partial melting, differentiation, and cooling of lunar material began that long ago. Whether the moon was entirely molten or only partly molten is uncertain. By 4.4 billion years ago, the moon was probably already accumulating its anorthositic crust. The earth also heated and differentiated, and its relatively greater iron content led to a much larger iron core than in the moon. Massive anorthositic accumulations in the cores of ancient continental shields may be remnants of a primitive, moon-like crust on earth.

Early Intense Bombardment Nearly all lunar rocks that formed before 4.1 billion years ago have been pulverized by repeated bombardment. According to Apollo data, at about that time the cratering rate was 1,000 times the present rate, and before that it was still higher. Because this high ancient rate probably represents the sweep-up of interplanetary debris by early planets, the earth was presumably also bombarded at that time, which, along with erosion, may account for the absence of earth-rocks older than four billion years. Many large craters formed on the earth and moon during this period.

Tidal Movement of the Moon The moon was once much closer to earth than it is now. Tidal analysis proves that from its moment of closest approach, it moved out quickly, reaching about half its present distance in only 100 million years, or 2 percent of the time since earth's formation. If the moon formed alongside earth as earth formed, then it probably approached its present orbit between 4.5 and 4 billion years ago, and earth's day approached its present length at that time.

Further Internal Evolution Further heat in the moon and earth may have come from meteoritic bombardment, magnetic effects, or a long-term buildup of heat from radioactive elements such as uranium. As a result

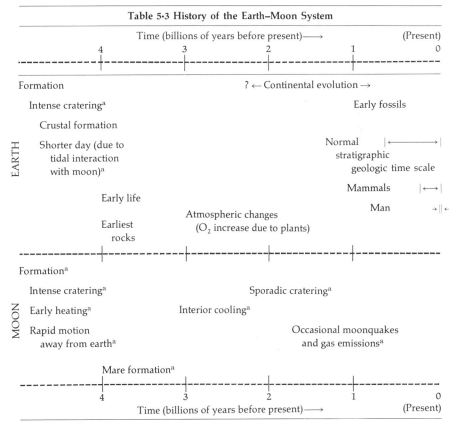

Table 5·3 History of the Earth–Moon System				

Time (billions of years before present)⟶ (Present)

EARTH

Formation ? ← Continental evolution →

Intense cratering[a] — Early fossils

Crustal formation

Shorter day (due to tidal interaction with moon)[a] — Normal stratigraphic geologic time scale

Mammals

Early life — Man

Earliest rocks — Atmospheric changes (O_2 increase due to plants)

MOON

Formation[a]

Intense cratering[a] — Sporadic cratering[a]

Early heating[a] — Interior cooling[a]

Rapid motion away from earth[a] — Occasional moonquakes and gas emissions[a]

Mare formation[a]

Time (billions of years before present)⟶ (Present)

[a]Information discovered or improved through Apollo-related lunar research.

of the moon's size and configuration, this heating apparently peaked in the outer parts of the moon about 3.8 to 3.2 billion years ago, producing an underground reservoir of liquid mush containing magma and crystals of unmelted minerals. On earth such heat, sustained for a longer time, continues to cause volcanism today.

Mare Lava Flows The liquid (basaltic) material erupted in many places on the front side of the moon, where the crust was apparently thinnest because of large-scale impacts. Lava broke through the fractured floors of most large basins, flooding them with successive lava flows that formed the mare plains 3.8 to 3.2 billion years ago. Because the moon is smaller than earth, it cooled more quickly and most volcanism died out around 3 eons ago.

Sporadic Recent Cratering By the end of mare formation, about 3 billion years ago, the meteorite impact rate had declined to its present low value. The ancient uplands retained multitudes of old craters, but the newly formed maria were relatively uncratered. Occasional large impacts, involving meteorites a few kilometers across, excavated major craters, throwing out bright rays of pulverized ejecta across the dark maria and old uplands. Smaller meteorite impacts churned the upper soil, producing the regolith. On the earth some of the large craters formed by meteorites during the last eon have also been preserved (as described in Chapter 3).

Terrestrial Erosion As shown in Table 5·3, the familiar parts of terrestrial history studied by most geologists are only the tail-end of earth–moon history. Because of plate tectonics and atmospheric erosion (as emphasized

in Chapter 3), the best-dated surface rocks of earth represent only about 12 percent of earth–moon history, whereas the lunar surface structures date back through 75 percent of it. The moon shows more clearly than earth the middle parts of geologic history and the combined results of internal geologic processes (for example, partial melting) and external astronomical processes (impacts). The moon has given us new understanding of processes that shaped our own world.

Concepts

synchronous rotation

sidereal period

synodical month

librations

tides

body tide

ocean tide

spring tides

neap tides

tsunamis

tidal recession

Roche's limit

terminator

limb

impact craters

mare (maria)

terra (terrae)

basins

rays

rilles

regolith

ejecta blanket

igneous rocks

sedimentary rocks

basalt

anorthosite

ultrabasic rocks

bulk density

volatile elements

refractory elements

lunar differentiation

lithosphere

early intense bombardment

megaregolith

fission theory

sister-planet theory

capture theory

collision-condensation theory

co-accretion theory

Problems

1. At what time of day (or night) does:
   a. *The first quarter moon rise?*
   b. *The full moon?*
   c. *The last quarter moon?*
   d. *The new moon?*

2. When a terrestrial observer is recording a new moon, what phase would the *earth* appear to have to an observer on the moon?

3. Explain why the moon keeps one side toward the earth.

4. Why might one naively expect the highest tides to occur at noon or midnight on the date of new moon or full moon? Why don't the highest tides always occur at these times?

5. If astronauts orbiting just above the atmosphere released, into an orbit of their own, two ping-pong balls just in contact with each other, would you expect them to stay in contact with each other indefinitely? Why or why not?

6. In selecting a lunar landing site,

a. *What type of feature might offer fresh bed rock, where regolith layers have been stripped away?*

b. *Would you expect the landscape inside a young 100-km-diameter crater to be rougher or smoother than inside an old crater of the same size?*

c. *Where might astronauts land to seek evidence of recent volcanic activity?*

7. Imagine you are an astronaut exploring the moon.

a. *Would an ordinary compass work on the moon? Why or why not?*

b. *What celestial object could serve as a navigational beacon for astronauts on the moon's front side?*

c. *What property of this object's apparent motion would make it especially useful as a navigational aid during a lunar stay of several months?*

8. If the ages of the earth and moon are identical, as believed, why are typical rocks found on the moon so much older than typical earth rocks?

9. Suppose sedimentary rocks had been discovered on the moon. How would this affect our beliefs about the moon's history?

10. By comparing pictures of lunar maria (3 to 4 billion years old) and uplands (4 to 4.5 billion years old) prove that the meteoritic cratering rate was much higher during the first few hundred million years of lunar history than during the last 3 billion years.

Advanced Problems

11. The moon has about 0.012 of the earth's mass and 0.27 of the earth's radius.

a. *Using Newton's law of gravity, show that the moon's surface gravity is about one-sixth that of the earth.*

b. *How much would a 180-lb. person weigh on the moon?*

12. If you have a telescope that resolves details as small as 1 second of arc, what would be the diameter of the smallest craters and mountains you could see on the moon?

13. How fast must a projectile move to escape from the moon? (Mass of moon $= 7.35 \times 10^{25}$ g; radius $= 1.74 \times 10^{8}$ cm.) Compare this with escape velocity from the earth.

Projects

1. Observe the moon at different phases with a telescope of at least 2 inches' aperture. Locate and compare the texture of terrae (upland regions) with maria (dark plains). Compare visibility of detail near the terminator and away from the terminator, and explain the difference. Sketch examples of craters, ray systems, mountains, and rilles.

2. With a telescope, find an example of a bright-ray crater, such as Tycho or Copernicus, and compare its appearance at full moon (high lighting) with its appearance near the terminator (low lighting). Why do the rays disappear under low lighting? Make a simulation of this effect by scattering a thin dusting of white flour or powder on a slightly darker, textured surface in a raylike pattern. Illuminate with a bright lightbulb from above (full-moon) and from the side at a low angle (low lighting), and compare the appearance.

3. Prepare a box with white flour several inches deep and a light dusting of darker surface powder (flush with top edge of box). Drop different-size pebbles into the box to make craters. Illuminate with a bright light bulb from various angles and compare with the appearance of the lunar surface. Compare the number of craters needed to simulate a mare region and a terra region; can the surface be saturated with craters if enough stones are dropped? What physical differences exist between this experiment and lunar reality? (Example: These stones hit at a few meters/sec, whereas meteorites hit the moon at several km/sec and cause violent explosions.)

The Solar System

In questions of science the authority of a thousand is not worth the humble reasoning of a single individual.

Galileo

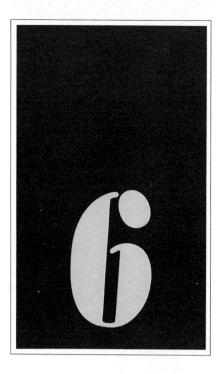

Discovering the Solar System

If we journey far beyond the moon and look back at the solar system, we discover that the earth is only the fifth-largest of many worlds that orbit around the average-size star we call the sun. This is a far cry from the conception of twenty generations ago, when the earth was viewed as a kind of imperial capital of the universe—a unique stationary scene of human activities around which sun, moon, planets, and stars moved. The exciting transition from the older idea to the modern conception began what astronomer Carl Sagan has called "the cosmic connection," the realization that we are only one part of a larger system of worlds—an idea still growing in our consciousness even today. The transition began during the Renais-

sance when the concept of a system of earth-like worlds seemed such an attack on traditional knowledge that it became an issue of life and death for some of the participants, as we will see. Today, the idea that we live in a **solar system**—the sun and its surrounding worlds—suggests a new conception of a large theater in which we may travel, investigate many new examples of geological, meteorological, and biological processes, and perhaps exploit sources of energy and materials.

A Survey of the Planets

The solar system is defined as the sun, its nine orbiting planets, their own satellites, and a host of small interplanetary bodies such as asteroids and comets. Starting in the center of the solar system, the major bodies and their symbols are:

⊙ **Sun**
☿ **Mercury**
♀ **Venus**
♃ **Earth**
♂ **Mars**
♃ **Jupiter**
♄ **Saturn**
♅ **Uranus**
♆ **Neptune**
♇ **Pluto**

The symbols, mostly derived from ancient astrology, are sometimes used as convenient abbreviations today. A traditional memory aid for this outward sequence is *Men Very Early Made Jars Stand Upright Nicely, Period.* To avoid confusion about the positions of Saturn, Uranus, and Neptune, remember that the (*S, U, N*) is a member of the system, too.

Some simple facts about the bodies of the solar system are useful to remember. For example, the sun is about 10 times the diameter of Jupiter, and Jupiter is about 10 times the diameter of the earth. Whereas the sun is a **star,** composed of gas and emitting radiation by its own internal energy sources, **planets** are bodies at least partly solid, orbiting the sun, and known to us primarily by reflected sunlight. **Satellites,** in turn, are solid bodies orbiting around the planets.

The planets divide into two different groups. **Terrestrial planets** are the inner four planets, Mercury through Mars. They are nearest to earth and roughly resemble it in size and in rocky composition. The **giant planets** are the four large planets of the outer solar system, Jupiter through Neptune. Much bigger than the terrestrial planets, they also have a different composition, being rich in icy or gaseous hydrogen compounds such as methane (CH_4), ammonia (NH_3), and water (H_2O). Pluto falls into neither category, being a special, mysterious case, as we will see.

Table 6·1 (p. 112) presents a comprehensive list of data. One little-known fact is that some satellites are bigger

Figure 6·1 **The arrangement of the solar system as it is now known. Orbits of the nine main planets and a typical comet are shown, plus positions of typical asteroids. Note eccentricities of the orbits of Pluto, Mercury, and Mars.**
▼

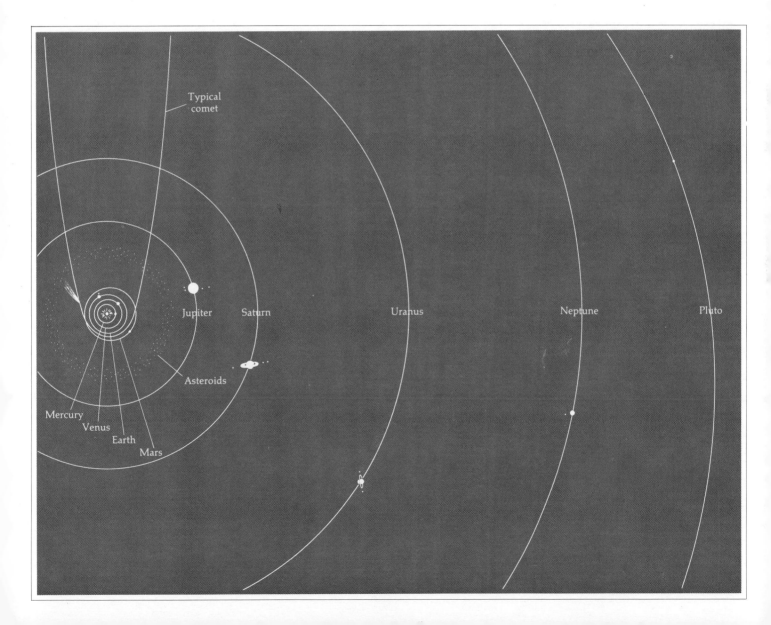

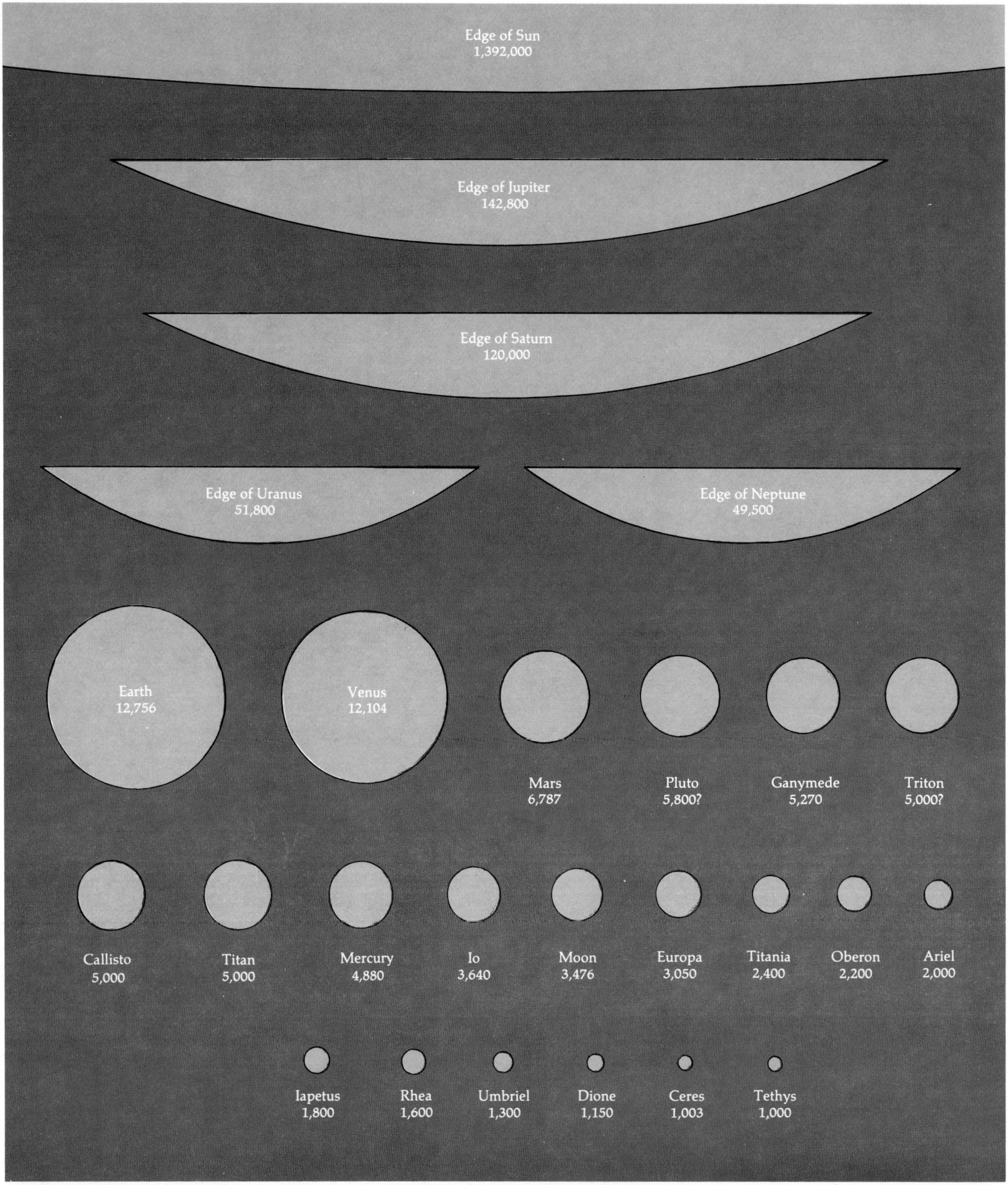

Edge of Sun
1,392,000

Edge of Jupiter
142,800

Edge of Saturn
120,000

Edge of Uranus
51,800

Edge of Neptune
49,500

Earth
12,756

Venus
12,104

Mars
6,787

Pluto
5,800?

Ganymede
5,270

Triton
5,000?

Callisto
5,000

Titan
5,000

Mercury
4,880

Io
3,640

Moon
3,476

Europa
3,050

Titania
2,400

Oberon
2,200

Ariel
2,000

Iapetus
1,800

Rhea
1,600

Umbriel
1,300

Dione
1,150

Ceres
1,003

Tethys
1,000

than some planets! The diameters of the remotest satellites are uncertain because they are hard to observe, but the largest satellite in the solar system may be Jupiter's moon, Ganymede, with diameter about 5,270 km. Other large moons include Jupiter's moon Callisto, Saturn's moon Titan, and Uranus' moon Triton, all with diameters around 5,000 km. Each of these satellites is probably bigger than the planet Mercury. Similarly, the large asteroid Ceres (diameter 1,003 km), which orbits the sun in a planet-like orbit between Mars and Jupiter, is bigger than half the known satellites of the solar system. As Figure 6·2 shows, there are probably 26 worlds in the solar system larger than 1,000 km in diameter.

The Solar System's Configuration

The most important step in mapping the solar system was a careful study of planets' motions as seen from earth (Figure 6·3). Two important results were the discovery that two planets have orbits inside the earth's orbit, and the discovery of peculiar motions of Mars.

Inferior and Superior Planets

An important clue about the arrangement of the solar system is that, to an earthly observer, Mercury never strays more than 28° from the sun, and Venus never more than 47° (see Figure 6·4). All other planets can appear at any angular distance from the sun along the zodiac. This obser-

◄ Figure 6·2 **The 26 largest bodies in the solar system, shown to true relative scale. Diameters are given in kilometers.**

▲ *Figure 6·3* **The motion of Jupiter in the constellation Pisces during one week in 1975. Vertical field is about 6°. A few days' observation quickly distinguishes the moving planets from the stationary background stars. Similar observations allowed early astronomers to measure planetary motions long before the invention of the telescope. (Exposure with 35 mm camera, 135 mm lens, 5 minutes, f2.8, 2475 Recording film; magnitude of Jupiter is** −2.4; magnitude of bright pair of stars immediate right of Jupiter is +6.)

Figure 6·4 A key observation shows that Mercury and Venus lie between the earth and the sun. Their maximum angular separations from the sun are 28° and 47°, respectively, and they can sometimes be seen transiting in front of the sun. Exterior planets, however, can appear at opposition, 180° from the sun.
▼

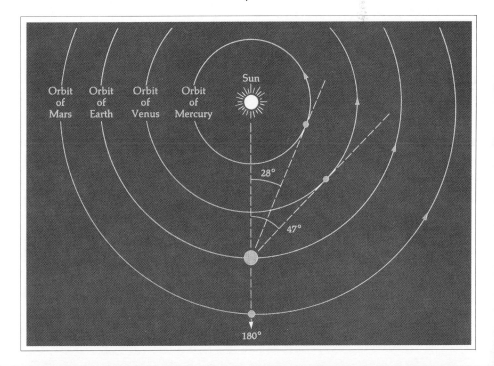

vation indicates that Mercury and Venus lie closer to the sun than the earth does, whereas the other six planets are farther away from the sun, outside earth's orbit. For this reason Mercury and Venus came to be called **inferior planets,** while the more distant planets were called **superior planets.**

Final proof of this arrangement may have come when ancient astronomers observed actual **transits,** the passage of an inferior planet directly between earth and sun. Only a planet closer to the sun than earth could pass between it and the sun. During a transit, a sharp observer looking through fog or a smoked glass can see Venus (but not Mercury, which is too small) as a tiny black spot moving across the sun. Transits of Mercury are the more common of the two. The next two transits of each are:

Mercury	Venus
12 Nov. 1986	8 June 2004
14 Nov. 1999	6 June 2012

Retrograde Motion of Mars

Superior planets slowly shift position from night to night, usually west-to-east relative to the background stars. Mars, however, performs a maneuver that puzzled all early analysts of the solar system. Observers who plotted its position from night to night found that as Mars approaches a point opposite the sun in the midnight sky, it slows down, reverses itself, and drifts westward in so-called **retrograde motion** for some days before resuming its normal eastward (prograde) motion. We now know that Mars does not really reverse its motion. The appearance is an illusion caused by the motion of Mars relative to earthly observers. The earth moves faster

around the sun than Mars does, following an "inside track" closer to the sun. Mars seems to go backward whenever we overtake and pass it. Similarly, if you ride in a fast racing car on the inside of a circular track and overtake a car on an outside track, you would see its initial motion [eastward], watch it slip behind you as you pass [westward drift], and then from a distance see it resume its initial motion relative to background structures of the racetrack.

Problems with the Ptolemaic Model

In Chapter 2, we showed how ancient scientists arrived at a solar system model with planets moving in deferent circles and epicycles around the earth, as shown in Figure 6·5. But the relative motions of inferior and superior planets, especially Mars, were destined to cause trouble. Inferior planet motions were accounted for by putting Mercury's and Venus' orbits

Table 6·1 Objects in the Solar System
(Including the sun, all planets, satellites known through 1976, and the five largest asteroids)

Object	Equatorial Diameter (km)	Mass (g)	Rotation Period (days)	Distance from Primary (10³ km unless marked)	Orbit Inclination[c] (degrees)	Orbit Eccentricity
Sun	1,392,000	1.99 (33)	25.4	0	—	—
Mercury	4,880	3.30 (26)	58.6	.387 AU	7°.0	.206
Venus	12,104	4.87 (27)	243 R[a]	.723 AU	3.4	.007
Earth	12,756	5.98 (27)	1.00	1.00 AU	0.0	.017
Moon	3,476	7.35 (25)	27.3	384	18–29	.055
Mars	6,787	6.44 (26)	1.02	1.52 AU	1.8	.093
Phobos	22	?	0.32	9	1.1	.021
Deimos	12	?	1.26	23	1.6	.003
Asteroids						
1[b] Ceres	1,003	1.2 (24)?	0.38	2.77 AU	10.6	.08
2 Pallas	608	?	0.4?	2.77 AU	34.8	.24
4 Vesta	538	2.4 (23)?	0.22	2.36 AU	7.1	.09
10 Hygiea	450	?	?	3.15 AU	3.8	.10
31 Euphrosyne	370	?	?	3.16 AU	26.3	.22
Jupiter	142,800	1.90 (30)	0.41	5.20 AU	1.3	.048
5 Amalthea	240	?	0.4?	181	0.4	.003
1 Io	3,640	8.91 (25)	1.77	422	0.0	.000
2 Europa	3,050	4.87 (25)	3.55	671	0.5	.000
3 Ganymede	5,270	1.49 (26)	7.16	1,070	0.2	.001
4 Callisto	5,000	1.06 (26)	16.69	1,880	0.2	.01
13 Leda	8?	?	?	11,110	26.7	.146
6 Himalia	170?	?	?	11,470	27.6	.158
10 Lysithea	19?	?	?	11,710	29.0	.130
7 Elara	60?	?	?	11,740	24.8	.207
12 Ananke	17?	?	?	20,700	147 R[a]	.17
11 Carme	24?	?	?	22,350	164 R[a]	.21
8 Pasiphae	27?	?	?	23,300	145 R[a]	.38
9 Sinope	21?	?	?	23,700	153 R[a]	.28

between earth and sun, and adjusting their deferent and epicycle rates of motion so they would never get more than 47° from the sun. Mars' motions were explained by adjusting its epicycle motion so that it would sometimes appear to be moving east-to-west.

The main goal of these schemes was to be able to predict the positions of planets, often for astrological purposes. The basis of all the popular solar system models was to break down the seemingly irregular motions into combinations of circular motions or uniform cycles. (The same method is used today to represent seemingly irregular curves as combinations of regular sinusoidal curves, or harmonics, in a technique called Fourier analysis.) Even before the **Ptolemaic model** was perfected, Eudoxus (ca. 360 B.C.) reportedly represented planetary motions by a combination of motions of 27 rotating concentric spheres carrying the planets; Aristotle (ca. 360 B.C.) reportedly used 55 spheres; Apollonius of Perga (ca 220 B.C.) reportedly multiplied the number of possible combinations of cycles by introducing epicycles; and Ptolemy perfected the **epicycle model** around A.D. 140.

In the *Almagest,* Ptolemy showed how a properly chosen set of epicycles could account for all observations of planetary positions accumulated by his time, and he made relatively accurate predictions of planetary positions. Arab astronomers introduced this scheme into Europe in the early Middle Ages. However, by the 1200s and 1300s, Arab astronomers in Damascus, Syria, realized that further adjustments were needed to get accurate predictions, and discussed adding small epicycles to the main epicycles.

In 1252, King Alfonso X of Castile, in a medieval version of the National Science Foundation, supported Arab and Jewish astronomers for 10 years to calculate the extensive Alfonsine Tables, an almanac of predicted planetary positions based on Ptolemaic theory. These tables became the basis of planetary predictions for the next three centuries. But the solar system now seemed very complicated, with all its overlapping cycles. (On seeing the complexity of the Ptolemaic calculations, Alfonso is said to have remarked that had he been present at the creation he could have suggested a simpler arrangement!) New epicycles had to be added to make positions agree with observations, and the calculations thus became ever more complicated.

Around 1340, the English scholar William of Occam enunciated his famous principle, called **Occam's razor,** applicable to all branches of science and philosophy:

Table 6·1 (cont.) Objects in the Solar System
(Including the sun, all planets, satellites known through 1976, and the five largest asteroids)

Object	Equatorial Diameter (km)	Mass (g)	Rotation Period (days)	Distance from Primary (10^3 km unless marked)	Orbit Inclination[c] (degrees)	Orbit Eccentricity
Saturn	120,000	5.69 (29)	0.43	9.54 AU	2.49	.056
11	300?	?	?	151?	0?	.0?
10 Janus	220?	?	?	160	0	.0
1 Mimas	400?	3.7 (22)	?	186	1.5	.02
2 Enceladus	500?	8.5 (22)	1.37	238	0.0	.00
3 Tethys	1,000?	6.3 (23)	?	295	1.1	.00
4 Dione	1,150	1.2 (24)	2.7	377	0.0	.00
5 Rhea	1,600	1.5 (24)	4.4	527	0.4	.00
6 Titan	5,000	1.4 (26)	15.95	1,222	0.3	.03
7 Hyperion	500?	?	?	1,481	0.4	.10
8 Iapetus	1,800	?	79.33	3,560	14.7	.03
9 Phoebe	200?	?	?	12,930	150 R[a]	.16
Uranus	51,800	8.76 (28)	0.96? R[a]	19.18 AU	0.8	.05
5 Miranda	800?	4 (23)?	?	130	3.4	.02
1 Ariel	2,000?	4 (24)	?	192	0	.00
2 Umbriel	1,300?	2 (24)	?	267	0	.00
3 Titania	2,400?	1.1 (25)?	?	438	0	.00
4 Oberon	2,200?	9 (24)	?	586	0	.00
Neptune	49,000	1.03 (29)	0.92?	30.07 AU	1.8	.01
1 Triton	5,000?	3.4 (26)	5.9	354	160.0 R[a]	.00
2 Nereid	700	?	?	5,570	27.5	.76
Pluto	5,800?	6.6 (26)	6.4	39.44 AU	17.2	.25

[a]R in column 4 indicates retrograde rotation; in column 6, retrograde revolution.
[b]Numbers assigned to asteroids and outer planets' satellites indicate order of discovery, except for largest satellites.
[c]To ecliptic for planets; to planets' equator for satellites.

Data from D. Morrison and D. P. Cruikshank (1974), *Space Science Review 15,* 641; D. Morrison, private comm. (1975); S. Larson, private communication (1977). (Numbers in parentheses are powers of ten.)

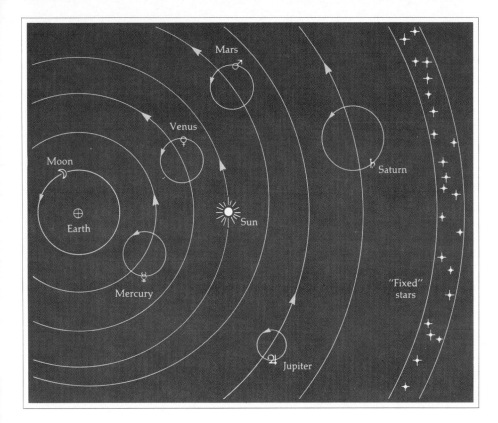

country. He associated with several astronomers and mathematicians and made his first astronomical observations at age 24. A few years later, a cathedral post gave him the economic security to continue his observations. At age 31, he observed a rare conjunction that brought all five known planets as well as the sun and moon into the constellation of Cancer. He found that their positions departed by several degrees from predictions in his set of Alfonsine Tables.

Familiar with several classical alternatives to Ptolemy's system, Copernicus analyzed planetary motions by various methods, including use of small "epicyclets." He soon began to realize that the solar system would be simpler and the task of prediction easier if the sun were placed at the center and the earth viewed as one of

▲
Figure 6·5 **The solar system as it might have been conceived by a Ptolemaic astronomer between** A.D. **100 and 1500.**

> **Multiplicity ought not be posited without necessity.**

In other words, among competing theories, *the best is the simplest, the one requiring the fewest assumptions and modifications in order to fit observations.* This idea may have contributed to suspicion of the tacked-on epicycles. Some European scholars, concerned about how the planets moved and remained suspended, regressed to Aristotle's more simplistic idea of spherical shells (see Figure 2·12). Purbach (ca. 1460) thought the spheres were made of transparent crystal with special hollows for the epicycles; Fracastorius (ca.

1550) proposed a scheme with 79 spheres!

According to astronomical historian Owen Gingerich (1973a,b), the growing dissatisfaction with the Ptolemaic system was aesthetic (a result of its increasing unwieldiness) as much as intellectual (a dissatisfaction with its results). Obvious errors of a few degrees had accumulated in predicted positions of planets. One observer remarked that a new model would be more "pleasing to the mind." This observer was Nicolaus Copernicus.

The Copernican Revolution

Copernicus' Theory

Nicolaus Copernicus was born February 14, 1473, the son of a Polish merchant. During his university education in Italy he became excited by the surging scientific thought of that

▲
Figure 6·6 **Nicolaus Copernicus (1473–1543). Copernicus is holding a model with a central sun circled by the earth, and the earth by the moon.**

its orbiting planets. In 1512, he circulated a short comment (*Commentariolus*) containing the essence of the new thesis: the sun was in the center, with the planets moving around it and the stars immeasurably more distant. This comment was not widely distributed, however, and few of Copernicus' acquaintances at that time realized that the work he was pursuing would scandalize and revolutionize the medieval world.

He continued his studies but, fearing controversy, delayed publication for many years. Finally, encouraged by visiting colleagues, including some in the clergy, he allowed the *Commentariolus* to be more widely circulated. News of Copernicus' work spread rapidly. Late in life, Copernicus prepared a synthesis of all his work, *De Revolutionibus* (*On Revolutions*, 1543). In this book he laid out the evidence about the solar system's arrangement:

> . . . *Venus and Mercury revolve around the sun and cannot go farther away from it than the circles of their orbits permit* [*the 47° figure mentioned earlier*]. . . .
>
> *According to this theory, then, Mercury's orbit should be included inside the orbit of Venus.* . . .
>
> *If, acting upon this supposition, we connect Saturn, Jupiter, and Mars with the same center, keeping in mind the greater extent of their orbits . . . we cannot fail to see the explanation of the regular order of their motions.*
>
> *This proves sufficiently that their center belongs to the sun.* . . .

Having thus laid out the correct nature of the solar system, Copernicus tackled the solar system's relation to the stars:

> *The extent of the universe, however, is so great that, whereas the distance of the earth from the sun is consider-*

able in comparison with the other planetary orbits, it disappears when compared to the sphere of the fixed stars. I hold this to be more easily comprehensible than when the mind is confused by an almost endless number of circles, which is necessarily the case with those who keep the earth in the middle of the universe.

Although this may appear incomprehensible and contrary to the opinion of many, I shall, if God wills, make it clearer than the sun, at least to those who are not ignorant of mathematics. . . .

Now the stage was set for turmoil. The printer of *De Revolutionibus*, a Lutheran minister, had tried to defuse the situation by extending its title to *On the Revolutions of Celestial Orbs*, as if to imply that the earth was not necessarily included. He had also inserted a preface stating that the new theory need not be accepted as physical reality but could be seen merely as a convenient model for calculating planetary positions. From a philosophical viewpoint, we might accept this (see Einstein's thought experiments, Chapter 3), but it did not deter medieval critics. Already Copernicus himself had come under fire from Protestant fundamentalists: In 1539 Martin Luther had called him "that fool [who would] reverse the entire art of astronomy. . . . Joshua bade

the sun and not the earth to stand still."

In a world of strong dogmas, tampering with established ideas is dangerous. The Reformation era of ideological clashes was no exception. In the 1530s Michael Servetus had been criticized for certain writings on astrology and astronomy; in 1553 he was burned at the stake as a heretic for a mysterious theology that offended both Protestants and Catholics.[1]

Copernicus himself missed the height of the violent debate. He was ill in his last year and the first copies of his book were reportedly delivered to him on the day of his death, in 1543, at age 70. But the **Copernican revolution,** which put the sun instead of the earth at the center of the solar system, was under way.

About 1584, a 36-year-old Italian theologian and naturalist, Giordano Bruno, came into intellectual circles in London and Paris, where he reportedly found Oxford faculty members being fined for criticizing Aristotle. Bruno became known for tracts vigor-

---

[1]Both Protestants and Catholics were involved in outrageous suppression. It was John Calvin himself who masterminded Servetus' execution, although, in a fit of moderation, he recommended beheading instead of burning. Servetus, a man of wide learning and varied interests, had improved geographic data on the Holy Land and had discovered blood circulation in the lungs.

Table 6·2 Five Key Figures in the Copernican Revolution		
Nicolaus Copernicus	1473–1543	Proposed circular motions of planets around sun
Tycho Brahe	1546–1601	Recorded planets' positions
Johannes Kepler	1571–1630	Analyzed Tycho's records; deduced elliptical orbits and laws of planetary motion
Galileo Galilei	1564–1642	Made telescopic discoveries supporting Copernican model
Isaac Newton	1642–1727	Formulated laws of gravity and used them to explain elliptical planetary orbits

ously defending the Copernican view against academicians. His message included between-the-lines exhortations that warring Protestant and Catholic factions in Europe could be brought together by a better appreciation of the new scientific view of earth's and humanity's place in the universe (Lerner and Gosselin, 1973).

Bruno expanded on the Copernican cosmology. The stars, he said, were all worlds like the sun. Many planets might orbit around them, offering abodes for other races. Bruno traveled in Europe, lecturing on the theological implications of astronomy and science, often with remarkable imagery that was at once poetic and scientific. For example, he discussed optics using "light" in both the physical sense and as "divine light." When he returned to Italy in 1592 he was arrested by the Inquisition, a church court established to detect and punish heresy. In 1600, after eight years of investigation of his philosophical and political views, he was burned at the stake.

In 1575, a correspondent wrote to the astronomer Tycho Brahe, "No attack on Christianity is more dangerous than the infinite size and depth of the heavens." Was the earth to be taken as merely a minor province of the universe? Where was heaven? In 1616, the Catholic church banned reading of *De Revolutionibus* "until corrected." It was corrected in 1620 by removing nine sentences asserting that it contained actual fact, not just theory. Nevertheless, as Will Durant summarized the situation:

> . . . The heliocentric astronomy compelled men to reconceive God in less provincial, less anthropomorphic terms; it gave theology the strongest challenge in the history of religion. Hence the Copernican revolution was far profounder than the Reformation; it made

the differences between Catholic and Protestant dogmas seem trivial. . . .

Tycho Brahe's Sky Castle

Tycho Brahe was a flamboyant naturalist who wore a silver nose to cover a dueling mutilation. With funds from the king of Denmark, he built the first modern European observatory, named Uraniborg (Sky Castle), at his island home near Copenhagen. Tycho's observations, all made by the naked eye (the telescope was not yet invented), were catalogs of star and planet positions. By demonstrating that stars and other bodies showed no angular shift in position as our position shifts with rotation of the earth, Tycho proved that stars and planets were many times farther away than the moon, for which he *could* detect a

▲
Figure 6•7 **Tycho Brahe (1546–1601), as shown in an old print. Silver plate on his nose covers a dueling scar.**

shift (called *parallax;* see Chapter 2, p. 36).

At age 16, Tycho had noticed errors in predicted planetary positions in the same Alfonsine Tables that Copernicus had used. At age 25, in 1572, Tycho saw a **nova** (newly-bright star; Tycho's was a type now called *supernova*). These are now known to be exploding stars, but Tycho and his contemporaries knew nothing of their cause. However, by demonstrating that it had no parallactic shift, he disproved the popular belief that it was an object in the earth's atmosphere, or in the region of the earth–moon system. In 1577 he observed a bright comet and showed that it was also a remote object, far beyond the moon. These discoveries were critical in overturning pre-Copernican theories. They meant that new objects could appear in the supposedly unchangeable heavens, and planets could not be attached to crystalline spheres, because such spheres would be smashed by the comets.

These observations inspired Tycho to catalog the precise positions of stars and planets, which he did between 1576 and 1596. Not being able to convince himself that the earth could move, Tycho invented a compromise solar system in which the earth was central and stationary, but other planets were in correct sequence in orbit around the sun. The inferior planets Mercury and Venus were correctly placed in orbits between earth and sun, and the whole configuration moved around the earth.

His pension withdrawn by the King of Denmark, Tycho moved to Prague in 1599, where he was joined in 1600 by a 30-year-old assistant named Johannes Kepler. When Tycho died, in 1601, Kepler inherited the great compendium of Tycho's observations, with all its potential for fruitful analysis.

▲
Figure 6•8 **Johannes Kepler (1571–1630).**

Kepler's Laws

Devoutly religious and a believer in astrology, **Johannes Kepler** was sure that planetary motions must be governed by hidden regularities—"the harmony of the spheres." With Tycho's material, Kepler first went to work on the orbit of Mars, which presented the most notable case of retrograde motion—the backtracking that had plagued astronomical theorists since Ptolemy. He found something astonishing: after all the centuries of debate over the arrangement of circular orbits, the orbit that fitted Mars' motion best was not a circle at all, but an **ellipse,** a roughly egg-shaped figure symmetric around two interior points called foci (singular, **focus**). Kepler found that the sun lay exactly at one focus of Mars' orbit. Kepler went on to discover two other related principles, and these *three laws of planetary motion* were published in two books, *New Astronomy* (1609) and *The Harmony of the Worlds* (1619). **Kepler's laws** describe how the planets move (without attributing this motion to any more general physical laws); show

that the sun is the central body; and allow accurate prediction of planetary positions:

1.

Each planet moves in an ellipse with the sun at one focus.

2.

The line between sun and planet sweeps over equal areas in equal intervals of time.

3.

The ratio of the cube of the semi-major axis to the square of the period [of revolution] is the same for each planet.
(This is sometimes called the *Harmonic Law.*)[2]

Although the planetary orbits are ellipses, they are only slightly eccentric—that is, they are nearly circular. This is why the Ptolemaic system of circles worked as well as it did.

In the case of the earth, the semi-major axis, or average distance from the sun, is one **Astronomical Unit** (AU), and the other planets' distances are multiples of this unit. Working in AUs and years, we can confirm the third law numerically, using a as the semi-major axis and P as the period

$$\text{Earth:} \frac{a^3}{P^2} = \frac{(1 \text{ AU})^3}{(1 \text{ yr.})^2} = \frac{1}{1} = 1.00$$

---

[2]The major axis of an ellipse is its longest diameter; the semi-major axis is half that. Since most planets in our solar system have nearly circular orbits, the semi-major axes of their orbits are essentially the orbital radii.

The same formulation works for other planets.

Kepler's laws solved the mystery of the apparent retrograde motion of Mars. Kepler's laws say that because the earth is closer than Mars to the sun, the earth moves faster in its orbit than Mars. Therefore, the earth catches up to Mars, like the driver on the inside track in the racing car analogy. As earth passes Mars, Mars' motion appears to be retrograde.

Galileo's Observations

Kepler's laws might not have gone so far in establishing the Copernican model of the solar system had it not been for the contemporary invention of the telescope and extensive observations by an Italian scientist, **Galileo Galilei.** Unlike Kepler, Galileo had a

▲
Figure 6•9 **Galileo Galilei (1564–1642).**

superbly practical turn of mind. For example, after reportedly watching the regular swing of a lamp in the Pisa cathedral, he applied the periodic motion of the pendulum to regulate clocks. As early as 1597, Galileo wrote to Kepler:

> Like you, I accepted the Copernican position several years ago. . . . I have not dared until now to bring [my writings on the subject] into the open. . . .

Galileo perfected the telescope and began astronomical observations with it in late 1609. By 1610 he had made some of the most important observations in the discovery of the solar system. For example, he found four satellites revolving around Jupiter—*proof at last that some bodies do not revolve around the earth.* These discoveries electrified European intellectuals.

Because Galileo wrote in Italian rather than Latin, he built a popular following outside the universities as well. Academics and churchmen saw in him a threat, and Galileo soon found himself being attacked from local pulpits. His invitations to reactionary academics and churchmen to look through his telescope and see for themselves led nowhere. Some looked and said they saw nothing; some refused to look; some said that if the telescope were worth anything, the Greeks would have invented it.

From 1613 to 1633 Galileo was in frequent contact with church authorities, even in Rome. In 1616 a cardinal read him an order that he must not "hold or defend" Copernican theory, though he could discuss it as a "mathematical supposition." In 1632, his great book *Dialogue of the Two Chief World Systems* appeared. It featured a fictionalized debate between Copernican and Ptolemaic advocates. In 1633, 69-year-old Galileo was ordered

to Rome to stand trial before the Inquisition, where a curious episode occurred. The court produced a purported copy of the cardinal's order of 1616 telling Galileo not to "hold or defend" Copernicanism. The Inquisition's copy, still in the files today, also ordered him not to "teach" or "discuss" it in speech or in writing (contrary to the cardinal's 1616 order). The document in the Inquisition files is not the original and lacks the names of the alleged witnesses. Historians now suspect this document was a fraud created to frame Galileo.

The Inquisition jurors were inclined to be lenient if Galileo would only repudiate his work. The elderly Galileo saw no point in getting himself killed; his book was already published and he had faith that intelligent people could see plain truth, through telescopes or in print. So he recited a prepared recantation and was sentenced to prison, a sentence commuted by the Pope to house arrest on Galileo's own estate, where he died in 1642.

Newton's Synthesis

In spite of the efforts of the Inquisition, the evidence now accumulated so rapidly that the Copernican revolution was almost complete. The main element still lacking was an overall theoretical scheme that would draw together the empirical Copernican rules into a concise physical explanation of the behavior of the solar system. To be intellectually satisfying, this theory ought to start with a few universal principles and show that the Keplerian orbits and the Galileian satellite motions *had* to exist as a consequence of these principles. The man who achieved this synthesis—the man usually deemed the greatest physicist who ever lived—was **Isaac**

▲
Figure 6·10 **Isaac Newton (1642–1727).**

Newton (whose spectacular career was sketched in Chapter 4).

Earlier scientists had been hindered by a misconception that no object could move unless a force acted on it. (This often *seems* true in everyday experience, because of friction, which early scientists did not understand.) One of the main arguments against Copernicanism had been that no one could imagine a force strong enough to keep the earth moving in an orbit around the sun. But no such force was needed; only a force of attraction to the sun was needed to keep the earth near the sun, so it wouldn't drift off into interstellar space. It was Newton who realized this, corrected the misconception, and solved the riddle. But his solution led to another riddle: How could the sun influence the planets in their Keplerian orbits if it never touched them, and always stayed at such a great distance from them? And how could the planets stay in the sky without spheres to hold them up?

Newton answered all these questions of "action at a distance" with his laws of motion and gravity, the basis of most modern physics except for the corrections made necessary by work on relativity during the present century. These laws were enunciated in Newton's book, *Principia,* in 1687. *They are quite unlike Kepler's three laws* in that they are not merely empirical rules based on observation; rather, *they are fundamental postulates,* from which a host of other phenomena can be correctly predicted and observed. **Newton's laws of motion** are:

1.

A body at rest stays at rest and a body in motion moves at constant speed in a straight line unless a net force acts on it.

2.

For every force acting on a body, there is a corresponding acceleration proportional to and in the direction of the force, and inversely proportional to the mass of the body.
In other words,
Force = Mass × Acceleration

3.

For every force (sometimes called action) on one body, there is an equal and opposite force (called reaction) acting on another body.

Newton's law of gravitation is:

Every particle in the universe attracts every other particle with a force proportional to the product of their masses, and inversely proportional to the square of the distance between them.

The properties of Keplerian orbits follow from Newton's laws. It is an important exercise in advanced astronomy courses to *derive all three of Kepler's laws from Newton's laws.* This exercise shows that if Newton's laws are true, the Copernican theory and Kepler's laws also have to be true.

Newton's laws thus tidied up the miscellaneous observations of preceding centuries and completed the Copernican revolution. If the force of gravitation did *not* obey the inverse-square law, orbits could be nonellipti-

cal. The law of gravitation, when combined with the first and second laws, shows why planets do not move in straight lines, but are always deflected toward the sun. The third law explains why rockets work, and led ultimately to man-made satellites and spacecraft orbiting moons and planets in Keplerian orbits.

By the time of Newton's death, at age 84 in 1727, the solar system was conceptualized essentially as we see it today, lacking only the discovery of the three outer planets, Uranus, Neptune, and Pluto.

Bode's Rule

A curious relationship discovered by the German astronomer Titius and popularized by his colleague Johann Bode, in 1772, is helpful in memorizing the distances of the planets from the sun. **Bode's rule** is: Write down a row of 4s, and add the sequence 0, 3, 6, 12, 24, and so on, doubling each time, as shown in Table 6·3. By putting in the appropriate decimal point, you get the number of Astronomical Units between each planet and the sun.

Because Bode's rule, unlike Kepler's laws, does not necessarily follow from Newton's laws, it is considered more descriptive than explanatory, and not a law of physics. Nonetheless, it seems to imply that the process that formed the planets—whatever it was—

Table 6·3 Bode's Rule—Distances of Planets from the Sun										
	Mercury	Venus	Earth	Mars	Asteroids	Jupiter	Saturn	Uranus	Neptune	Pluto
	4	4	4	4	4	4	4	4	4	4
	0	3	6	12	24	48	96	192		384
Estimated distance	0.4	0.7	1.0	1.6	2.8	5.2	10.0	19.6	—	38.8
Actual distance	0.4	0.7	1.0	1.5	2.8	5.2	9.5	19.2	30.0	39.4

Note: All distances are in terms of the Astronomical Unit, the average distance of the earth from the sun.

partitioned the solar system so that each planet formed nearly twice as far from the sun as the next inner planet. No one has fully explained why this happened.

In 1781 Bode's rule was strengthened with the discovery of Uranus at its predicted position, about twice as far from the sun as Saturn. Astronomers then noted that the rule predicted a planet between Mars and Jupiter. Telescopic searchers set out to find this missing planet, and on January 1, 1801, they found the first and largest asteroid, Ceres, just at the right distance! Ceres might have been called an ordinary but small planet, except that many other small asteroids have since been discovered between Mars and Jupiter. The discovery of Neptune in 1846 somewhat reduced the credibility of Bode's rule by putting a planet outside a predicted position, though the still later discovery of Pluto filled a predicted position.

6

Summary

The planets are tiny specks circling the sun. If you backed off far enough to see the system as a whole, the outer giants would hardly be noticeable and the inner planets would be lost in the glare of the sun. This conception of the solar system was accepted only after one of the major intellectual upheavals in human history, only about four centuries ago. The key to this Copernican revolution was the work of the five scientists listed in Table 6·2. Of special importance were Kepler's three laws, which described how planets moved, and Newton's laws of motion and gravity, which revealed the underlying forces that explain Kepler's laws.

Concepts

solar system

star

planets (names and order)

satellites

terrestrial planets

giant planets

inferior planets

superior planets

transits

retrograde motion of Mars

Ptolemaic model

epicycle model

Occam's razor

Copernican revolution

Tycho Brahe

nova

Johannes Kepler

ellipse

focus

Kepler's laws

Astronomical Unit

Galileo Galilei

Isaac Newton

Newton's laws of motion

Newton's law of gravitation

Bode's rule

Problems

1. To an observer north of the plane of the solar system, do the planets appear to revolve around the sun clockwise or counterclockwise? Which way to an observer south of the plane? (Hint: All planets revolve in the same direction as the earth rotates, from west to east.)

2. Which planet can come closest to the earth? (See Table 6·1.)

3. Which planet moves the largest number of degrees per day in its orbit around the sun? Which the least?

4. How many times bigger in diameter than the earth is the largest planet? How many times more massive than the earth is the most massive planet? (See Table 6·1.)

5. Can the full moon ever occult Venus (pass between earth and Venus)? Draw a sketch to show why or why not.

6. One of Galileo's telescopic discoveries was that Venus, like the moon, goes through a complete cycle of phases from narrow crescent to full. How did this disprove the Ptolemaic model, which restricts Venus to positions between sun and earth?

7. Which planet can come closer to Jupiter, earth or Uranus?

8. Which planet can come closer to Uranus, earth or Pluto?

9. Does Mars' actual orbital motion around the sun change in any special way during the time it appears to execute retrograde motion to a terrestrial observer?

10. If Venus, earth, Mars, and Jupiter are in a straight line on the same side of the sun, what phenomenon does an observer on earth see? What would an observer on Mars see? An observer on (or near) Jupiter?

11. List systematic differences in physical properties between inferior planets and giant planets. Include size, temperature, density, satellite numbers, and orbital properties (see Table 6·1).

12. Do you think Copernicus and other major figures in the Copernican revolution would have viewed themselves as revolutionaries? Why? Contrast the causes of scientific revolutions and political revolutions. What roles do factual discoveries, strong personalities, controversy, and publicity play in each?

13. Why is Bode's rule not a general law of nature in the sense that Newton's law of gravity is? In how many instances can it be tested by experiment or observations?

Advanced Problems

14. A small telescope will show the disks of Mars and Jupiter when they are closest to earth. Assume that Mars approaches within 60 million km and Jupiter within 630 million km. Use the small angle equation and data from Table 6·1 to calculate the angular sizes of the two planets in seconds of arc. Which appears larger in the telescope in angular size?

15. Calculate and compare the orbital velocities of earth (1.5×10^{13} cm from the sun) and Pluto (about 5.9×10^{14} cm from the sun). The sun's mass is 2.0×10^{33} g.

16. If a comet on a parabolic orbit made its closest approach to the sun at a point very near the earth, and moved in the same direction as the earth, how fast would it appear to an observer on earth to be going by?

Projects

1. On a piece of typewriter paper try to make a scale drawing of the orbits of the planets, based on orbital radii listed in Table 6·1. Which orbits are hard to show clearly? What size dots could represent planets at this scale?

2. Make a large wall chart showing the orbits of the planets out to Saturn in scale. Mark the motion of each planet in one-day or one-week intervals, as appropriate; using the *American Ephemeris and Nautical Almanac,* an astronomy magazine, or similar source, locate each of the planets' relative positions for the current date. Update the chart during the semester and watch for alignments that represent conjunctions, elongations, and so on. Confirm these in the night sky.

3. If a planetarium is convenient, arrange for a demonstration of planetary motions from night to night. Demonstrate the apparent retrograde motion of Mars.

4. With a telescope of at least 2 inches' aperture examine Jupiter and confirm that it is attended by four prominent satellites, as discovered by Galileo. (On any given night, one or more satellites may be obscured by Jupiter or its shadow.) By following Jupiter from night to night, confirm that the satellites move around Jupiter, proving the Copernican dictum that not all celestial objects move in earth-centered orbits.

Mercury and Venus

In February 1974, the American spacecraft Mariner 10 sailed less than 6,000 km from Venus' cloudy surface, taking photos and scientific measurements, and then proceeded on toward the planet Mercury. The same spacecraft passed within 6,000 km of Mercury in March, sending back the first close-up photos of that planet. Then it sailed on around the sun, made a second pass by Mercury in September 1974, repeated the whole maneuver, and made a third pass only 327 km above the surface of Mercury again in March 1975. In October 1975, two Russian spacecraft, Venera 9 and Venera 10, parachuted through the clouds of Venus and made soft landings, sending back photos of that planet's surface.

Prior to these flights, Mercury was known only as a blurry disk in earth-bound telescopes, and Venus was known only as a featureless world covered with enigmatic clouds. Those flights culminated a long effort to learn the properties of the innermost planets of the solar system.

The Planet Mercury

Mercury is a moon-like body, only about 40 percent the size of the earth, and about 40 percent bigger than the moon. It has no atmosphere. A telescope with an aperture larger than about six inches reveals that it has phases and very faint dusky markings on its illuminated side. These markings are probably lava plains similar to the patchy dark maria of the moon. Mercury's surface has a slight pinkish cast. No satellite has been found.

From the nineteenth century until the 1960s, observers drew maps and tried to use the pattern of markings to clock the planet's rotation. Some observers concluded that Mercury always kept one side toward the sun, just as the moon keeps one side toward earth. But in the 1960s radar showed that Mercury does *not* keep one side to the sun; it rotates very slowly, so that a Mercurian day is about 59 earth days. Some of the earlier observers' reputations were salvaged when it was found that the 59-day period causes Mercury to present similar sides to the earth and sun during every third *apparition,* or close approach to earth (Cruikshank and Chapman, 1967).

The Puzzle of Mercury's Orbit

Mercury's orbit has long been puzzling in several respects. Excluding remote Pluto, its orbit is the most highly inclined to the ecliptic plane (by 7°) and the least circular. More importantly, its orbit has motions that seem, at first glance, to contradict the laws of Kepler and Newton. The *perihelion*—the point nearest to the sun—shifts in position slowly around the sun from year to year. This is called **orbital precession.** Some precession had been predicted from Newton's laws, but observers found an excess shift of 43 seconds of arc each century—a tiny amount, but enough to consternate orbital theorists. Analysts such as U. J. Leverrier, around 1860, thought that the excess shift might be caused by the gravity of a small, undiscovered planet close to the sun, inside Mercury's orbit, and even gave the planet a name—Vulcan (Moore, 1954). Leverrier had already successfully predicted Neptune's existence from similar gravitational disturbances observed in the motion of Uranus, but he was not correct in the case of Mercury. Twentieth-century observations show that no planet exists inside Mercury's orbit.

But how, then, can Mercury's orbital precession be explained? The solution came in 1915, when Albert Einstein showed that the great mass of the sun disturbs the orbits of the planets closest to it. Einstein's theory of relativity predicted almost exactly the excess precession observed—43.03 arc seconds/century. Einstein predicted

smaller excesses for Venus and earth, and these, too, were confirmed by observation. Thus, solving the puzzle of Mercury's orbit played a major role in the acceptance of the theory of relativity.

Mercury's Temperature and Surface Features

We want to go beyond merely describing the appearance of Mercury, or of the other planets as well. We want to describe the properties of their surfaces, such as temperature, composition, roughness, and so on. Measurement of each quality requires special techniques, and most such techniques are more complicated than

we can discuss here. However, as an example, we introduce the third of five major physical relations to be used in this book. This is **Wien's law,** discovered in 1898 by German physicist Wilhelm Wien (pronounced *Veen*):

The hotter an object, the bluer the radiation it emits.

This effect is familiar in everyday life. Ordinarily, a nail emits no visible radiation. If you heat it over a stove, it begins to radiate a dull red light. Higher temperature leads to a bright orange-red glow. At a still higher

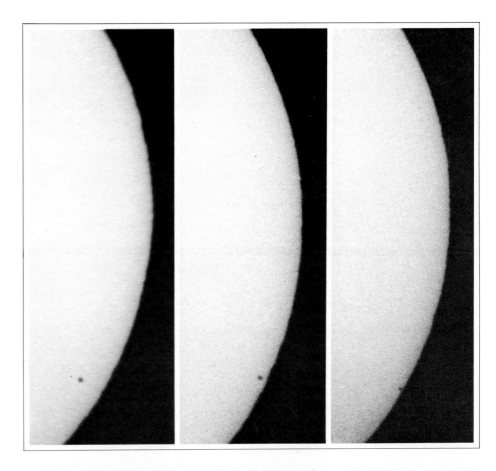

Figure 7·1 **Three frames made during a transit of Mercury on November 10, 1973, show Mercury (the small black disk at bottom) moving off the disk of the sun. Because these frames show the size of Mercury's disk during its closest approach to earth, they illustrate how poor an earthbound observer's view of the planet is.** (Photos with a Questar telescope by W. A. Feibelman)

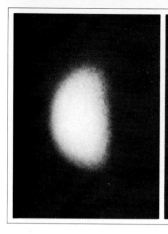

◀ *Figure 7-2* **Three photographs of Mercury in 1934 (left to right, June 7, 11, 12) showing the changing phases of the planet. Virtually no surface detail can be seen; very few photos from the earth show reliable detail.** (Lowell Observatory)

temperature, the dominant color is "white-hot," or yellowish white. If the nail could be heated enough, its radiation would become distinctly bluish. This is the sequence of colors in the **spectrum**—or array of all colors of light (see Color Photo 30). Each of these colors is radiation that can be assigned a certain wavelength, and the spectral sequence described leads smoothly from longer-wavelength (redder) colors to shorter-wavelength (bluer) colors.

Despite appearances, the nail is radiating regardless of its temperature. When it is at room temperature the light it radiates is so red (of such long wavelength) that we cannot see it because our eyes are not sensitive to it; radiation of such long wavelength

is called **infrared light.** Similarly, a hot enough object radiates such blue light that we cannot see it; such short-wave radiation is called **ultraviolet light.** Though we cannot see these forms of radiation, we can build instruments to detect them; radios, for example, detect especially long wavelengths of infrared radiation (usually called radio waves). The eye detects radiation with wavelength about 4×10^{-5} to 6.5×10^{-5} cm, corresponding to deep blue and deep red, respectively.

Instruments show that solid bodies radiate mixtures of all colors all at once; but for any temperature, there is one dominant color, and this color is bluer as the object gets hotter. Wien's law tells which wavelength corresponds to the maximum amount

of radiation for each temperature. Thus, measurement of the radiation allows estimation of temperature.

Because Mercury is so close to the sun, its daytime surface is much hotter than the earth's. In accord with Wien's law, infrared detectors can measure radiation from both the day and night side of Mercury. These measures show that temperatures in the upper few millimeters of Mercurian soil range well above 500 K (441°F) in the "early afternoon" near perihelion to lows of about 100 K (−279°F) at night. In some areas, depending on soil type, the temperature might exceed 600 K (621°F). Though science fiction of a few years ago described pools of molten metal on the daylight side, this now seems unlikely, for several reasons. First, the melting point of lead is about 600 K; of aluminum, 832 K. Second, nothing in geology suggests that such metals are likely to drain from Mercury's surface rocks and remain in pools on the surface. Third,

---

Optional Math Equation III
•
Measuring Temperatures of Astronomical Bodies: Wien's Law

Wien's law tells which wavelength, W, corresponds to the maximum amount of radiation at each tempera-

ture, T. Using cgs units, the wavelength is given in centimeters and the temperature in degrees Kelvin (K). The law is:

$$W = \frac{0.290}{T}$$

The number 0.290 is a constant of proportionality and remains the same in all applications of the law. Thus, as T increases, W decreases, giving shorter wavelengths and hence bluer light.

An example is instructive. The sun has a temperature of about 5,700 K. Therefore, we can calculate the wavelength of the strongest solar radiation. Using cgs units and powers of ten,

$$W = \frac{2.9 \times 10^{-1}}{5.7 \times 10^3} = 0.509 \times 10^{-4}$$
$$= 5.09 \times 10^{-5} \text{ cm}$$

This, of course, is exactly in the middle of the range to which the eye is sensitive. (Otherwise we could not see sunlight!) Light of this wavelength corresponds to a yellow color.

◄ *Figure 7·3* A mosaic of Mariner 10 photos of Mercury shows a heavily cratered planet that resembles the moon. (NASA)

Figure 7·4 The earth's moon shown at the same scale as Mercury, emphasizing their similar features.
▼

because of the insulating effect of the overlying soil, the temperature slightly below the surface is only 314 K to 446 K.

Mariner 10 produced the best available data about Mercury, including photographs of the surface, and magnetic measurements. The surface resembles the moon's in its abundance of rugged craters, and in its huge, multiringed craters known as basins. The most prominent basin, shown in Figure 7·6, is named the Caloris Basin (in keeping with the high Mercurian temperatures). Its concentric ring structure, more than 1,200 km in diameter,

A

Figure 7·5 Compare these Mariner 10 photos of Mercury with pre-Mariner views from earth (Figure 7·2). The left view shows a Mariner 10 picture made at a distance; next is the same with contrast enhanced; next is the same blurred to simulate terrestrial observing conditions; at right is a drawing of the same side of Mercury, at the same phase, revealing approximate agreement of the major bright features. Not until Mariner's 1974 fly-by was it known that these bright features are large impact craters. (Sequence courtesy of A. Dollfus)
▼

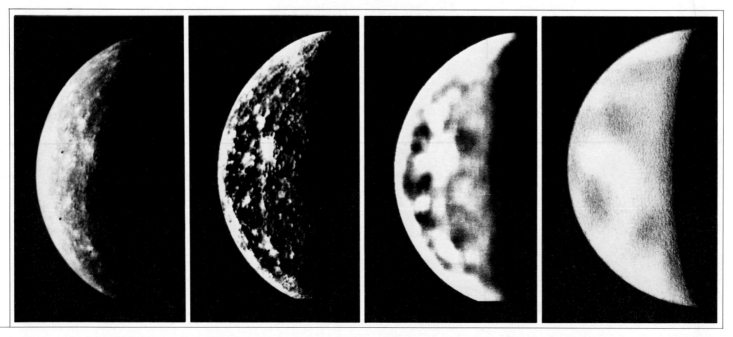

Figure 7·6 Comparison of Mercury and moon photos shows striking similarity of surface features. **A.** The region of the Caloris basin on Mercury. The center of the basin lies in shadow out of the frame to the left. The left half of the frame is dominated by curved cliffs and fractures surrounding the impact site. The photo from top to bottom covers an area of about 1,300 km; the cliffs are believed to be about 2 km (6,000 feet) high. (NASA) **B.** A portion of the central face of the moon, showing smooth lava plains and heavily cratered uplands. The large crater Alphonsus, about 110 km across, lies near center of photo. (Lunar and Planetary Laboratory, University of Arizona)

strongly resembles the Orientale ringed basin on the moon (see Figure 5·11). Evidently the same impact and lava-flow processes occurred on Mercury as on the moon some $3\frac{1}{2}$ to $4\frac{1}{2}$ eons ago. From this discovery, most scientists believe all planets suffered an intense bombardment by interplanetary debris at the close of the planet-forming process.

Cliffs such as those in Figure 7·7 appear to mark huge faults, suggesting that Mercury may have contracted slightly because of cooling, or from tidal forces caused by the nearby sun. Parts of the ancient cratered surface

appear to have been leveled by a more recent melting process.

The magnetic field of Mercury is only about one percent that of the earth. Nonetheless, Mercury's magnetic field has the form calculated for an internally generated field like the earth's, with magnetic axis tilted about 7° to the planet's rotation axis. These results suggest that it is caused by weak currents in an iron core like earth's core. A substantial iron core of this kind could also account for Mercury's mean density (5.5 g/cm³), which is surprisingly higher than the moon's (3.3 g/cm³).

B

Figure 7·7 A heavily cratered portion of Mercury resembles the lunar uplands. Arrows mark what appear to be major faulted cliffs. (NASA)

Landscapes on Mercury

Mercury's landscape probably superficially resembles the lunar landscape: Many areas would show rolling, dust-covered hills and plains, eroded by eons of bombardment by small meteorites, and covered by regolith. Fresh impact craters or faults might display rugged boulders and outcrops of craggy rocks. The sky would be black, dominated during the nearly 30-day period of sunlight by a sun looking about $2\frac{1}{2}$ times bigger in angular size than the sun in earth's sky. Besides the sun, two extremely bright planets, much brighter than any stars, would occasionally dominate the sky. These would be yellowish-white Venus and the bluish earth.

Figure 7•8 **Landscapes on Mercury are believed to resemble those on the moon. In this imaginary view, Venus (the brighter, upper object) and earth (lower) are seen in conjunction low over Mercury's horizon. (Painting by author)**

▼

The Planet Venus

Entirely different from Mercury is the planet Venus. Sometimes called "earth's sister" because of its similarities to earth, Venus has about 95 percent of the earth's diameter and about 82 percent of its mass. Although it approaches within 42,000,000 km (26,000,000 miles) of us, Venus is mysterious because it is completely obscured by clouds. This brilliant yellowish-white, nearly blank cloud surface is the first feature to impress a telescopic observer.

When Venus passes between the earth and sun, Venus appears as a very thin, back-lighted crescent. At this time the "horns," or **cusps,** of the crescent often extend part way around the dark side (see Figure. 7•9.) This is because of sunlight scattered in Venus' atmosphere, which apparently has a foggy or hazy quality. This was discovered by the Russian scientist, M. Lomonosov, in 1761; it was the first proof that Venus has an atmosphere.

Telescopic observers see only the vaguest markings in the Venusian atmosphere.[2] Such markings include dusky patches, streaks, faint bands, and bright polar regions. In 1928, the American astronomer Frank Ross photographed much more clearly defined bands and other dusky patterns by using plates and film sensitive to ultraviolet light (see Figure 7•10). These patterns are formed by cloud layers differing in composition, particle size, or altitude.

Telescopic observers occasionally report a faint, coppery glow, known as **ashen light,** on the dark side of Venus. Russian probes in 1975 also detected such a glow; their results suggest it is similar to earth's auroral glows, caused by radiation from molecules that have interacted with solar radiation. No satellite has been discovered near Venus.

---

[2]Adjectives for planets are controversial. Some writers, claiming that "Venusian" is an ugly word, use "Cytherean" (from a name of Aphrodite), which is merely confusing. "Venereal" is already preempted by other areas of human endeavor. While "Venusian" has the sanctity of science fiction tradition, many astronomers use "Venerian" or the noun "Venus" as an adjective.

Venus' Retrograde Rotation

Since clouds hide all surface features of Venus, telescopic viewers were unable to determine the planet's rotation rate, though most astronomers suspected that it had **prograde rotation**—turned from west to east like the earth, moon, Mercury, and Mars—with a period of from one to several days. In 1962 American radio astronomers bounced radar signals off the planet and discovered the surprising result that Venus takes 243 days to rotate and that it has an east-to-west or **retrograde rotation.** The axis of rotation is within a degree of being perpendicular to Venus' orbital plane. The cause of this unusual reverse spin may involve tidal forces between earth and Venus or an ancient collision with a body larger than the moon, of which there may have been many in the early solar system (Goldreich and Peale, 1970; Singer, 1970). (See Figure 7·11, next page.)

Venus' Dense Atmosphere

Many nineteenth-century scientists assumed that **Venus' atmosphere** was like earth's. However, in 1932 Mount Wilson astronomers observing the spectrum of Venus detected extraordinary amounts of carbon dioxide (CO_2, the same gas dissolved in carbonated soft drinks). Later data showed the atmosphere is about 97 percent CO_2.

Still undiscovered was the composition of the opaque cloud layer and the air below it. In the 1940s and 1950s some writers imagined stormy clouds of water droplets, a surface swept by torrential rains, and vegetation like that of a Brazilian rain forest. Some supposed that the highly reflective clouds would shade the surface and moderate the climate in spite

◄ *Figure 7·9* **A telescopic view of Venus while passing nearly between the earth and the sun. Backlighting by the sun illuminates the atmosphere of Venus all the way around the disk. This effect, first recorded in 1761, not only proves the existence of an atmosphere but gives information on its structure.** (New Mexico State University)

Figure 7·10 **Part of a sequence of ultraviolet photographs of Venus made with earth-based telescopes. These pictures are among the best made with ground-based equipment, and show faint cloud patterns.** (Lunar and Planetary Laboratory, University of Arizona) ▼

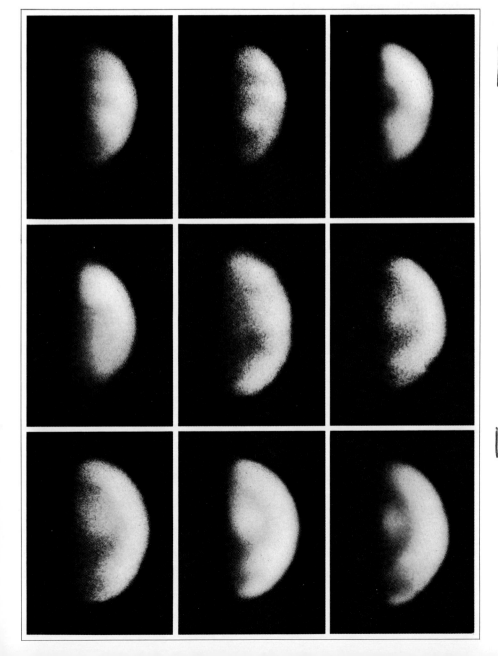

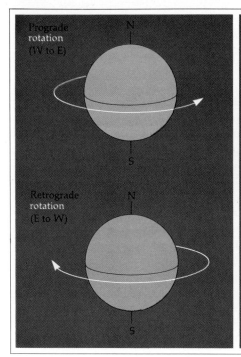

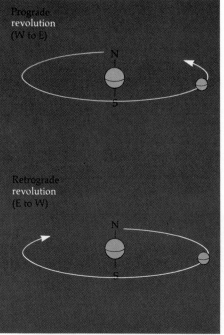

of Venus' closeness to the sun. However, in the 1960s, when the planet's thermal radiation was measured at far-infrared and radio wavelengths, Wien's law revealed that the lower atmosphere has a temperature of about 750 K (891°F)—hardly conducive to liquid water or life as we know it!

In 1970 a Soviet probe, Venera 7, the first spacecraft to successfully land on another planet, transmitted data from the Venusian surface for 23 minutes. The data confirmed the high temperature and revealed an atmospheric pressure about 90 times as great as earth's. Instead of our familiar 14.7 pounds per square inch, the pressure on Venus is about 1,320 pounds per square inch, equivalent

▲
Figure 7•11 **The difference between rotation (motion around an internal axis) and revolution (motion around an external body) is illustrated by the right and left drawings. Prograde motion (top) is west to east. Most planets, including the earth, have prograde rotation and revolution. The rotation of Venus is retrograde, as diagramed in the lower left figure. Remember the difference between rotation and revolution by noting that the common handgun should be called a rotator, not a revolver.**

Figure 7•12 **Contrast-enhanced image of ▶ Venus in ultraviolet light shows details of the structures of clouds, which are arranged in bands roughly paralleling the Venusian equator.** (NASA from Mariner 10)

to that endured by a diver about 264 feet below the terrestrial ocean surface. Later Soviet spacecraft landings confirmed these results. Venus is a stranger world than most humans had imagined!

Still other unexpected discoveries came from earth-based spectrometers (which identify gases by their characteristic absorption of certain colors of light). In addition to CO_2, Venus' atmosphere contains minor traces of other gases, including hydrogen chloride (HCl) and hydrogen fluoride (HF), as shown in Table 7·1. The principal clouds turned out to be not water droplets, but tiny droplets of sulfuric acid (H_2SO_4) about 0.0002 cm in diameter.

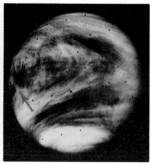

Figure 7·14 **Three ultraviolet images of Venus made by cameras on the Mariner 10 spacecraft as it flew by Venus on February 7, 1974. The arrow pointing to a 1,000-km-wide dark cloud-feature shows motions in the atmosphere during a 14-hour period. (NASA)**

Figure 7·13 **Earth, photographed by Apollo 10 astronauts from 58,000 km is reproduced here at the same scale as Venus in Figure 7·12. Earth is 5 percent bigger, and shows some cloud bands similar to those on Venus, but has more cyclonic spiral cloud patterns, caused by coriolis forces produced by earth's faster rotation (described in Chapter 3). (NASA)**

Figure 7·15 **Haze layers extending above the featureless cloud deck of Venus, from an enlarged telephoto view by Mariner 10. (NASA)**

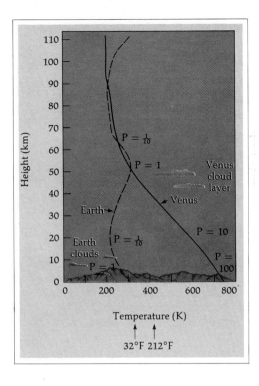

◄ Figure 7·16 **Physical data on Venus' atmosphere (solid curve) compared with earth's atmosphere (dashed). The greenhouse effect greatly heats the lower atmosphere of Venus. Pressures are marked (in units of earth's surface pressure).**

These discoveries raise many questions. If Venus is so similar to earth in size and distance from the sun, why did it evolve to such a different state, as shown by the comparison in Table 7·1? How do these highly reactive acids affect Venus' rocks? Why isn't there more water on Venus, as on the earth? Studying these mysteries of Venus has given geochemists valuable insight into the nature of earth, as described in the next two sections.

The Greenhouse Effect

Planets absorb sunlight and, following Wien's law, radiate infrared light. A planet's surface temperature is determined by the balance between the amount of visible sunlight it absorbs and the amount of infrared radiation it emits. If the solar energy absorbed each second is greater than the infrared energy radiated each second, the planet heats up. If the incoming amount is less than the outgoing amount, the planet cools. The mean, or equilibrium, temperature is reached when the two rates are equal.

The incoming rate is easily calculated as the total sunlight striking the planet minus the amount reflected off the planet. The total energy striking the planet per square centimeter in one second is called the **solar constant** for that planet. The outgoing radiation increases as the planet's surface temperature increases

Suppose the planet has no atmosphere. Then the situation is easy to predict. The surface rocks heat up until their temperature is so high that outgoing infrared radiation equals the incoming sunlight rate. But if the planet has an atmosphere that absorbs some sunlight, the atmosphere and the surface both warm up and radiate energy in the infrared. Because of the nature of atmospheric gases, the outgoing infrared energy may not escape into space. In fact, the gases CO_2 and H_2O absorb a great deal of the outgoing infrared, thus adding energy to the atmosphere and warming it even more. The warming continues until the amount of infrared escaping at the top equals the amount of incoming sunlight. On Venus, the lower atmosphere has to reach about 750 K before this condition is met.

This heating is called the **greenhouse effect** because of its resemblance

Table 7·1 Comparison of Atmospheres of Venus and Earth[a]			
Venus		**Earth**	
Gas	Percent by Volume	Gas	Percent by Volume
CO_2	97	N_2	78.1
N_2	<2	O_2	20.9
CO	0.002	H_2O	0.05 to 2 (variable)
H_2O	0.0001?	Ar	0.9
HCl	0.00002	CO_2	0.03
HF	0.0000001	Ne	0.0018
		He	0.0005
		CH_4	0.0002
		Kr	0.0001
		H_2	0.00005
		N_2O	0.00005
		Xe	0.000009

[a]Compositions are for near-surface conditions, with terrestrial data other than H_2O tabulated for dry conditions. CO_2 on the earth is probably increasing by 2 to 3 percent of the listed amount in each decade, because we are burning fossil fuel so voraciously. This may be modifying the earth's climate. The symbol < means "less than"; the symbol > means "greater than."

Source: data from Marov (1972).

to the physics of a greenhouse. The glass panes of a greenhouse admit sunlight but block the escape of the infrared. (They also keep the warm air from escaping—an important function that, in the case of planets, is performed by gravity.) Hence the inside becomes warmer than the outside. The greenhouse effect explains why Venus is so hot: Its massive CO_2 atmosphere blocks outgoing infrared radiation. The greenhouse effect also explains why a cloudy night often stays warmer than a very clear night on the earth, since the water vapor in the cloud layer blocks outgoing infrared radiation from the cooling earth.

Venus' Surface Conditions and Chemistry

The high temperature and atmospheric pressure of Venus help explain its unearthly geochemistry. The surface rocks are at such a high temperature that volatile substances, such as Cl and F, are driven into the atmosphere. This helps to account for atmospheric compounds such as HCl HF.

Why should Venus' atmosphere be mostly CO_2 instead of N_2 and O_2, like the earth's? The answer is clearer if we rephrase the question: Why does earth *not* have a massive CO_2 atmosphere? The two planets actually have similar amounts of CO_2, but earth's is mostly dissolved in ocean water and partly trapped in carbonate rocks, such as limestone. This can be understood through an important geochemical reaction, sometimes called the **Urey reaction,** that occurs in the presence of water:

$$MgSiO_3 + CO_2 \rightleftharpoons MgCO_3 + SiO_2.$$

On the water-rich earth the CO_2 dissolves in ocean water, making weak carbonic acid that attacks the magnesium silicates ($MgSiO_3$) of the silicate rocks, converting many of them to carbonates ($MgCO_3$) and silica-rich sediments. Both Venus and the earth probably produced large amounts of gaseous H_2O and CO_2 through volcanic eruptions, but Venus apparently retained less H_2O than earth (probably because the light gas

▲
Figure 7•18 **Russian artist's impression of one of the Venera space probes parachuted to the surface of Venus.** (From a painting by Soviet artist A. Sokolov)

H_2 floated to the top of the hot Venusian atmosphere and escaped into space). Thus, the reaction in the equation above was driven to the left on Venus, but to the right on earth. The earth's "missing" CO_2, therefore, is hidden in its sediments and rocks, explaining why the earth's atmosphere is not nearly as massive as Venus'.

The surface structure of Venus is poorly known. However, radar signals bounced off Venus indicate a relatively flat terrain, lacking massive mountain ranges but containing craters just as on the moon, Mercury, Mars, and earth (Figure 7•17). One radar image suggests a major canyon, perhaps created by limited tectonic activity. Some form of erosion appears to have erased smaller craters and filled in floors of large craters, making them quite shallow, compared to lunar or Mercurian craters. In general, the surface appears older and less broken by tectonic activity than the earth's surface.

Figure 7•17 **Craters are revealed in a high-resolution image (left) of Venus constructed from radar signals bounced off Venus from terrestrial transmitters. At right the view of the whole planet shows bright regions (α, β, δ) of high radar reflectivity, of unknown cause.** (NASA)
▼

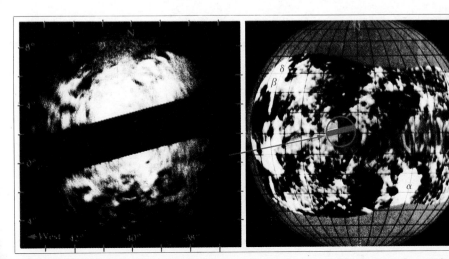

Landscapes on Venus

By the 1970s, scientists had discarded the old image of the Venusian surface as a temperate rain forest. Many thought the surface might be very dusty, hazy, and dimly lit because of the clouds. On October 22 and 25, 1975, Soviet spacecraft Venera 9 and 10 landed at two sites on Venus and returned the first photos of the surface. The photos were surprising in revealing brighter scenes than had been expected, with clear views to the horizon at least 100 meters away (Figure 7•19).

The scenes were not of dusty dunes but of angular boulders at one site and smoother rocks and soil at the other. The seemingly varied states of erosion at the two sites, with relatively

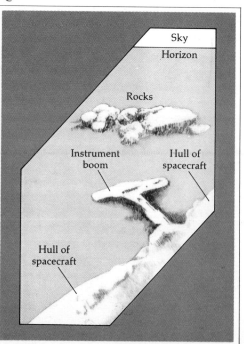

▲

Figure 7•20 **Angular rocks and soil near Venera 9 spacecraft on the surface of Venus. The white arc at bottom is part of the spacecraft. The fact that massive rock units** are visible and are neither eroded smooth nor weathered into fine soil indicates recent rock-forming or rock-exposing geological activity, such as lava eruption or faulting.

Figure 7•19 **The surface of Venus, photographed from two Venera soft-landed spacecraft, October 1975. Each image is a portion of a wide-angle panorama, oriented** ▼ with horizon at top. A. Angular boulders at landing site of Venera 9. B. Light-toned bedrock outcrops at landing site of Venera 10. Interrock regions are apparently covered by dark soil, scattered in places on the rock surfaces. Other frames nearby indicate rock is jointed with nearly parallel fractures. C. Key to features.

A

B

C

young-looking rocks at the first, suggested different degrees of geological activity in different regions. There were no signs of life. These Soviet spacecraft and five earlier ones that parachuted into the atmosphere confirmed the surface temperature of 750 K. They found only mild winds at the surface, but "jet-stream" winds (up to 185 kph) at altitudes around 40 km.

Why Do Some Planets Lack Atmospheres?

Our discussion of the earth in Chapter 3 showed that planets probably formed with some initial gas concentration called a *primitive atmosphere,* which changes to a *secondary atmosphere* as new gases are added by the process of outgassing. Why, then, do some planets lack atmospheres while other planets have dense ones? The explanation comes from three principles that govern motions of gas molecules in atmospheres:

1. The higher the temperature, the higher the average speeds of the molecules.

2. The lighter the molecules, the higher their average speeds. Light gases like hydrogen and helium have faster average speeds than heavier gases such as oxygen, nitrogen, carbon dioxide, or water vapor.

3. The bigger the planet, the higher the speed needed for a molecule to escape into space.

If you could heat a planet's atmosphere, more and more molecules would be moving faster than escape velocity, and fast-moving molecules moving upward near the top of the atmosphere would shoot out into space, never to return. First hydrogen, then helium, then heavier gases would leak away into space. Thus, cold, massive planets are the ones most likely to retain all the gases of their primitive and secondary atmospheres; hot, small planets (with weak gravity and low escape velocity) are most likely to lose all their gases. Calculations based on these principles show that planets as small as Mercury and the moon have lost virtually all of their gases. Venus and earth have lost most of their hydrogen and helium, but have kept heavier gases.

7

Summary: Comparative Planetology

The planets can be compared to help explain phenomena on all of them, and particularly phenomena on earth. Study of planets, their origins, and their development has come to be called **planetology. Comparative planetology** is the comparison of different planets to understand what makes them alike or unlike.

Only two decades ago, virtually all knowledge of major phenomena on the earth, such as rock chemistry, weather circulation, or mountain building, had to be derived from knowledge found on the earth alone. Today we can gain insights from other planets, treating them as special "laboratory examples" of other conditions. Venus, for instance, is an earth-sized planet nearer the sun with a different rotation rate—from which we learn that a hotter earth probably would not have developed so many carbonate rocks, but would have retained its volcanically released CO_2 as a thick atmospheric gas.

From Venus we also learn that a more slowly rotating earth would have more linear weather circulation and less well-developed cyclonic spiral systems. From the moon and Mercury we learn that smaller planets apparently do not develop enough internal energy to drive the plate tectonic activity that has broken and re-formed the earth's original, cratered crust.

From our studies of the earth, moon, Mercury, and Venus we can derive three general principles which will clarify phenomena of other planets as well.

Geological Activity The largest planets are most likely to have internal geological activity. Internal heat is the energy source that drives geological activity such as tectonic faulting, earthquakes, and volcanism; the larger a planet, the more radioactive minerals it contains and the more radioactivity there is to release heat. Also, the larger a planet, the better insulated the interior and the harder it is for the heat to escape. Small planets, on the other hand, cool rapidly and lose whatever heat they may have generated. The earth has enough internal energy to drive plate tectonics, which Mercury and the moon lack.

Surface Age The larger a planet, the younger its surface features are likely to be. This principle follows from the one above. The more internal heat, the more the surface is likely to be broken by recent geological activity. Small planets that cooled long ago retain very ancient surface features undisturbed by the passage of time. The earth and probably Venus retain fewer ancient craters than Mercury and the moon.

Atmospheres The larger and cooler a planet, the more likely it is to have an atmosphere, and the more likely this atmosphere will retain its original gases.

In summary, Mercury is a moon-like world with an ancient cratered surface little disturbed by internal geological activity except from some lava flows and large faults. Venus is a more earth-like world, but its lack of water has allowed carbon dioxide to form a dense atmosphere, causing a greenhouse effect that produces extremely high temperatures by earth standards. No other major planets or satellites exist in the region inside earth's orbit.

Concepts

orbital precession of Mercury

Wien's law

spectrum

infrared light

ultraviolet light

cusps

ashen light

prograde rotation

retrograde rotation of Venus

Venus' atmosphere

solar constant

greenhouse effect

Urey reaction

planetology

comparative planetology

Problems

1. List processes of planetary evolution that have occurred on both Mercury and the moon. List processes that are unique to each, if any. Is there any indication of plate tectonic activity on Mercury? What might this indicate about heat flow and mantle convection inside Mercury?

2. Some of the best telescopic observations of Mercury and Venus are made during midday instead of after sunset or before sunrise. Why? Why are none made at midnight?

3. Which is hotter, Mercury or Venus? Why?

4. Venus and earth are about the same size and mass, and degassing volcanoes on each probably produced both CO_2 and H_2O gas. Why is CO_2 a major constituent of the atmosphere only on Venus, and H_2O a *major* constituent of the atmosphere on neither planet?

5. If astronauts were going to walk on both Mercury and Venus, what differences in space suit design might be needed for the two ventures? Could a suit for Venus be made entirely of flexible material? Why or why not?

6. Compare the cloud patterns on earth and Venus (Figures 7·13 and 7·14).
a. *What types of features are similar?*
b. *These two planets have nearly the same radius, mass, and surface properties, but different rotations. What might be learned about circulation (wind patterns) of planetary atmospheres by comparing patterns of Venus and earth?*
c. *Would coriolus forces be greater or smaller on Venus? How would this affect air flow and cloud patterns?*

7. Clouds of water vapor tend to absorb infrared radiation. Use this fact and the theory of the greenhouse effect to explain why desert climates on the earth have greater temperature extremes, day to night, than moist climates.

8. State which of the following characteristics of Venus suggest a primitive surface (little disturbed since planet formation), and which ones an evolved surface (affected by geologic processes such as erosion, differentiation, and plate tectonics).
a. *craters*
b. *a large, rifted canyon*
c. *lack of high mountain ranges*
d. *granitic surface rocks*
How would you compare the state of Venus' geologic evolution with that of the moon, Mercury, and earth?

9. What are the chances that life as we know it exists on Venus? Why?
a. *If Venus were to have a surface temperature of about 300 K and an abundance of H_2O in its clouds, how would you rate the chances for life? Why?*
b. *If Venus were exactly like earth, would life necessarily exist there?*

10. How close is Venus during its nearest approach to earth? (See Table 6·1). How many times farther is this than the distance to the moon?

11. If a telescope shows Venus to be a thin crescent, where is Venus relative to earth and sun?

Advanced Problems

12. When earthbound observers can see Mercury partly illuminated, it is at a distance of about 0.9 AU, or 135 million km.
a. *If you used a telescope that revealed details of angular size 1″, what would be the smallest features you could see on Mercury? (Hint: Use the small angle equation.)*
b. *Why can virtually no detail be seen on the surface of Mercury when it is at its closest point to earth, about 0.61 AU away?*
c. *With the telescope in part a, what would be the thinnest cloud layers visible*

on Venus at its closest approach to earth?

13. If Venus has a surface temperature of 750 K, at what wavelength is its strongest radiation emitted? Is this ultraviolet, visible, or infrared radiation?

Projects

1. Try to see Mercury. This is likely to be harder than it sounds, because Mercury is prominent only during its elongation period, which lasts a few days, and then only for an hour or less just after sundown or before sunrise. Determine from an almanac or astronomy magazine when Mercury comes to evening elongation and find a site with a very clear western horizon. It is best to start a few days before the elongation so you can become familiar with the background stars in the appropriate region of the sky.

2. Determine whether Venus will be prominent in the evening or morning sky during the course, and observe its motions and brightness from day to day. Observe on which date it is farthest from the sun and estimate this angle.

3. Observe Venus in a telescope of at least 2 inches' aperture on several dates a few weeks apart. Observe and sketch the changes in phase, and explain them in terms of Venus' motion relative to the earth and sun.

4. If a planetarium is convenient, arrange a demonstration of the motions of Mercury and Venus.

Mars

On a Martian summer morning in 1976, the sun came up as usual on a rock-strewn plain. The dawn temperature was around −120°F, but by afternoon the air warmed to about −20°F. The wind was slight and the rust-colored rocks lay about as they had for the last few million years. The first sign of something unusual came at about 4:11 in the afternoon when a tiny, white, starlike object appeared high in the dusky red Martian sky. It was the 53-foot-diameter white parachute of the Viking 1 landing craft.

In a few moments, the contraption would have been clearly visible from the surface. At an altitude of 4,000 feet the lander's three engines fired with a pale, transparent flame. Within seconds the parachute cut loose and drifted away. Slowed by its rocket engines, the spacecraft dropped for another 40 seconds or so. As it dropped the last few feet, reddish dust swirled

into the air. The first of three lander legs hit the ground about as hard as you would if you jumped off a chair, and the jolt automatically switched off the engines. The ungainly spacecraft bumped to rest and Viking 1 became the first human-built machine to gather data on the surface of Mars.

The motivation for this feat of exploration is that Mars is such an earth-like planet that its climate, geology, and history may tell us something about earth's climate, geology, history. More importantly, the presence or absence of organisms on Mars may tell us something about life and its origin on earth. As might be hoped for any successful voyage of exploration, Viking's voyage confirmed some pre-existing ideas, but caused surprising revisions of some other ideas.

Mars as Seen in Earth's Telescopes

When Mars is on the far side of the sun from the earth, it is roughly 380 million km away, but at its closest it comes within about 56 million km (35 million mi)—closer than any other planet but Venus. When it is that close, a telescope of only a few inches' aperture will show features on its reddish surface, including polar ice fields, clouds, and dust deposits— features quite like those of earth.

The Markings of Mars

First to see Mars through a telescope was probably Galileo, who wrote in 1610 that he could see its disk and phases, indicating that it was a spherical world illuminated by the sun. When telescopes improved, observers began mapping dusky and bright patches that we can still recognize today. Dutch physicist Christian Huygens first clearly sketched these markings in 1659, and French-Italian observer Giovanni Domenico Cassini tracked them a few years later to determine that **Mars' rotation period** is 24^h 37^m, only a bit longer than earth's (Figures 8·1—8·3).

Seasonal Changes on Mars

Mars has seasons just like earth, though each season lasts about twice as long, since the Martian year is nearly twice ours. Telescopic observers in the 1700s and 1800s discovered **seasonal changes in the features of Mars.** In the Martian hemisphere that is experiencing summer, the bright, white polar cap shrinks away and may disappear from view, while the dusky markings darken and grow more prominent. Some early observers mistakenly thought the dark areas were oceans, and called them **maria,** just as on the moon. Brighter, orange areas came to be called **deserts,** which turned out to be more appropriate.

While retaining roughly constant shapes, the markings also change slightly from year to year, as shown in Figure 8·5. Once these changes were established, many observers thought that the dark areas were regions of vegetation, perhaps losing their leaves in winter and turning dark and lush in summer, as on earth. To understand the more modern studies of Mars, it is important to realize that many observers in the late 1800s erroneously believed that Mars had climate and vegetation like the earth. This opinion evolved even further as a result of the celebrated affair of the Martian canals.

Canals on Mars?

In 1869 Father Angelo Secchi in Rome mapped streak-like markings he called *canali,* maintaining the convention of naming dark areas after bodies of water. In 1877, Giovanni Schiaparelli, director of an observatory in Milan, popularized the term and drew the streaks much narrower and more linear than earlier observers had. He showed them as forming a network of lines on Mars. These features came to be called **canals.** Other observers did not see such features. Said the famous American astronomer E. E. Barnard, of Lick Observatory, in 1894:

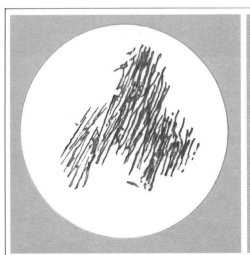

Figure 8·1 **One of the earliest known sketches of Mars was drawn by Christian Huygens on November 28, 1659. The northward-extending triangle is believed to be a dark region still visible today and known as Syrtis Major.** (After Huygens)
▼

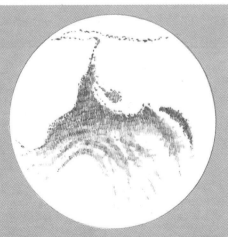

Figure 8·2 **Mars as drawn by English observer W. R. Dawes during the 1864–65 opposition. As in Figure 8·1, Syrtis Major extends northward but Dawes recorded a streaky extension of a type later called a "canal." Clouds covering the north polar ice cap are outlined by dotted line at the top.**
▼

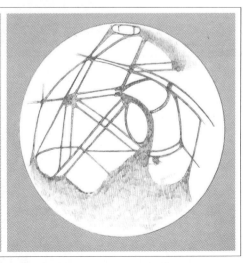

Figure 8·3 **Mars as drawn by Italian observer Giovanni Schiaparelli on June 4, 1888. As in preceding figures, Syrtis Major is the north-extending streaky dark area. Schiaparelli first popularized the "canals," shown by him here as nearly straight lines and line-pairs covering Mars. Spacecraft show they do not exist in this form. The north polar ice cap is at the top.**
▼

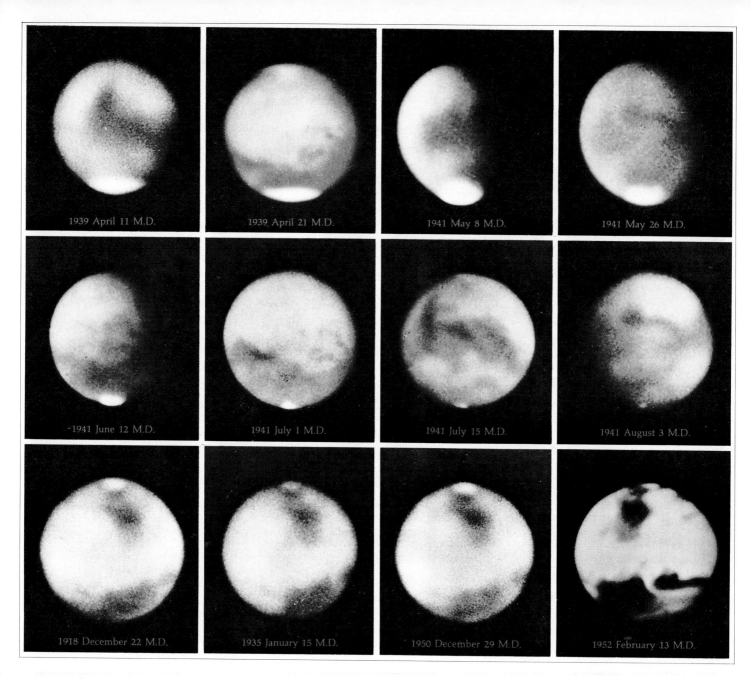

1939 April 11 M.D.	1939 April 21 M.D.	1941 May 8 M.D.	1941 May 26 M.D.
1941 June 12 M.D.	1941 July 1 M.D.	1941 July 15 M.D.	1941 August 3 M.D.
1918 December 22 M.D.	1935 January 15 M.D.	1950 December 29 M.D.	1952 February 13 M.D.

▲
Figure 8•4 **A sequence of photos from different years, arranged according to Martian seasonal date (M.D.) in the southern hemisphere. Note the melting (or more correctly, subliming) of the south polar cap from Martian April to Martian July.** (Lowell Observatory; last photo, 200-inch Palomar telescope)

To save my soul I can't believe in the canals as Schiaparelli draws them. . . . I see details where some of his canals are, but they are not straight lines at all. When best seen, these details are very irregular and broken up. . . .

The controversy grew hotter in 1895 with a vivid description of the canals in a book, *Mars,* by Percival Lowell.

Lowell was a wealthy Bostonian who founded his own observatory in the exceptionally clear air of Flagstaff, Arizona. Lowell said the canals were very sharp lines:

. . . it is the systematic network of the whole that is most amazing. Each line not only goes with wonderful directness from one point to another,

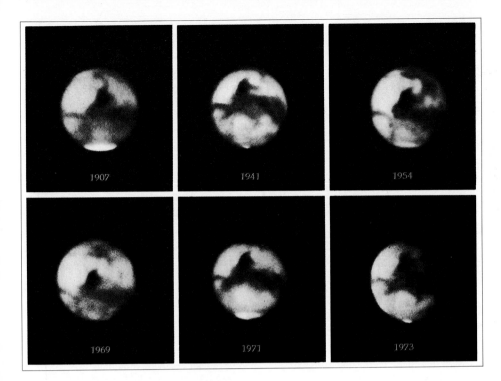

Figure 8·5 **Syrtis Major region of Mars (see Figures 8·1 to 8·3), as photographed from 1907 to 1973. Variations in the patterns of markings can be seen. A dark "wing" to the right of Syrtis Major, and a brightening of the near-circular Hellas bright region south of Syrtis Major, are evident in the 1941 photo.** (Lowell Observatory)

but at this latter spot it contrives to meet, exactly, another line which has come with like directness from quite another direction.

Lowell concluded that the features *really were canals*—artificial ditches built by intelligent creatures to carry water. He pointed out that spectroscopic measurements had revealed Mars to be a dry place. He hypothesized that a once-moist climate was becoming desert-like as water evaporated from the thin Martian atmosphere into space, and that a Martian civilization had turned to massive irrigation canals to carry water from their polar snow fields to the dry, but warmer, equator.

This exciting hypothesis attracted raging debate for several decades. Some observers confirmed canal-like streaks while many others could see none. The argument ranged from objective scientific measurements of low temperatures and thin air on Mars, to claims of whose observatory had better observing conditions. Skeptics said that if the great nation of France had failed in its attempt to build the 42-km Panama canal in the 1880s, then the heathen Martians certainly could not have succeeded in building much larger canals!

Today, after spacecraft visits to Mars, we know that the canal network does not exist as Lowell and some others drew it. What went wrong with the observations and the hypothesis of Martian civilization? Astronomer Carl Sagan has commented that drawings

of geometric networks on Mars certainly do imply that intelligence is present, but drawings alone don't tell which end of the telescope it is on. Modern evidence indicates two reasons why canals existed only in observers' minds. First, streaky markings including faulted canyons and dust deposits do exist on Mars; seen through the shimmery atmosphere of earth they may resemble patterns of lines. Second, and more importantly, some people are more likely than others to perceive streaky patches as straight lines, especially if they already believe the lines are there. Lowell was one of these; he even drew lines on Venus.

Names of Martian Features

Schiaparelli, a classical scholar as well as astronomer, named the larger Martian dark and light regions (as well as the illusory canals) after historical, mythological, and geographical features of his native Mediterranean area. These names, such as Hades, Arabia, and Libya, are still used for

Figure 8·6 **Percival Lowell, observing with** ▶ **the Lowell Observatory telescope which he used to report elaborate systems of canals on Mars, around 1900.** (Lowell Observatory)

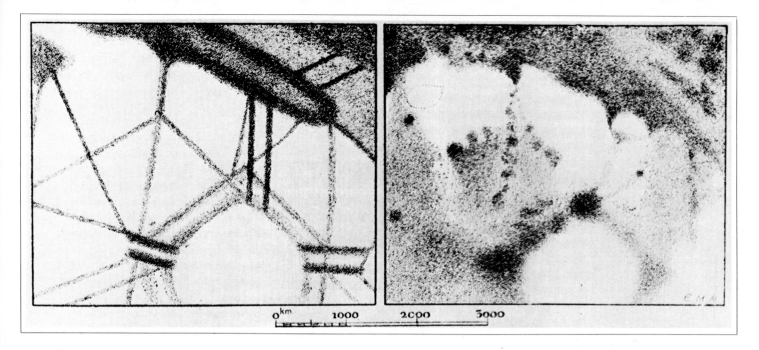

▲
Figure 8•7 An explanation of the canals, proposed by the French observer E.M. Antoniadi, around 1930. Under mediocre observing conditions, canal-like streaks may be seen in some regions of Mars, as Lowell described them (left). During moments of excellent atmospheric stability, observers with large telescopes find that these streaks break apart into a complex pattern of mottling and dark patches. The canals are a product of poor observing conditions combined with the physiological tendency of the eye to connect dots into lines.

the major dark and light areas. When the Mariner 9 spacecraft in 1971 revealed actual geologic structures such as craters, mountains, and canyons—all too small to be seen from earth—these were assigned additional names. As on the moon, craters were named after scientists. The largest canyon complex, big enough to stretch across the United States from coast to coast, was named Valles Marineris (the Valleys of Mariner).

The Lure of Mars

On October 30, 1938 Orson Welles broadcast a realistic radio play in which listeners heard "newsmen" reporting that Martians had invaded and were laying waste to New Jersey. Thousands believed it and the resulting panic caused a national scandal.

In July 1965, Mariner 4 sent back the first close-up pictures of Mars, which were widely published on front pages of newspapers; the reporting rate of UFOs shot up by a factor of six for several weeks. Why has Mars, of all the planets, held such a strong fascination for the public?

The answer lies in the hypotheses we have just discussed. If Copernicus knocked the earth out of the center of the universe, Lowell and the Victorians made people realize that earth's civilization might not be the only one around. Giordano Bruno and others in the 1600s and 1700s had suggested that other worlds might be inhabited—an idea known as the *plurality of worlds;* and Darwin's theory of evolution, published in 1859, made more plausible the idea that other planets might produce totally alien species specially adapted to their environments.

This idea electrified leading thinkers. Tennyson and other poets wrote about it. In 1898, H. G. Wells published *The War of the Worlds* in which Martians invade earth to escape their dying planet. Wells even pointed out that aliens would not necessarily be friendly:

> *The Tasmanians, in spite of their human likeness, were entirely swept out of existence in a war of extermination waged by European immigrants. . . . Are we such apostles of mercy as to complain if the Martians warred in the same spirit?*

There was even a UFO scare in the 1890s that produced reports of Martian spaceships (described as looking like the first dirigibles, which were then flying).

Between the days of Lowell and the first space flights to Mars, several generations of readers grew up on stories by Edgar Rice Burroughs, Ray Bradbury, and others, which pictured a Lowellian Mars with remnants of a dying civilization holding out in nearly deserted cities on a dying, drying planet. While incorrect in many details, these blends of theory and fancy fostered a valid excitement about whether life might really exist on Mars, and helped to prepare public opinion for actual voyages to the red planet. Yet, by the time of the first Martian voyages, earth-based measurements had already revealed that Mars had colder, thinner, dryer air than had been thought. Could such a planet support advanced life-forms? Microbes? Or no life at all?

Voyages to the Surface of Mars

The first three human-made devices to reach the surface of Mars were unsuccessful Russian probes. Mars 2 crashed in November 1971; Mars 3 landed in December 1971, but failed after 20 seconds on the surface; another probe sent back data while parachuting through the atmosphere in 1974, but failed moments before

Figure 8·8 **A mosaic of Mariner 9 photographs shows most of the Martian disk and its major features. At top is the north polar cap, composed of CO_2 and H_2O snowfields. Most of this face of Mars consists of desertlike lava plains and faults** ▼

surrounding Olympus Mons and associated volcanoes. Olympus Mons is the mountain near the bottom left edge of the lower segment of the mosaic; other volcanoes are to the right of it. (NASA)

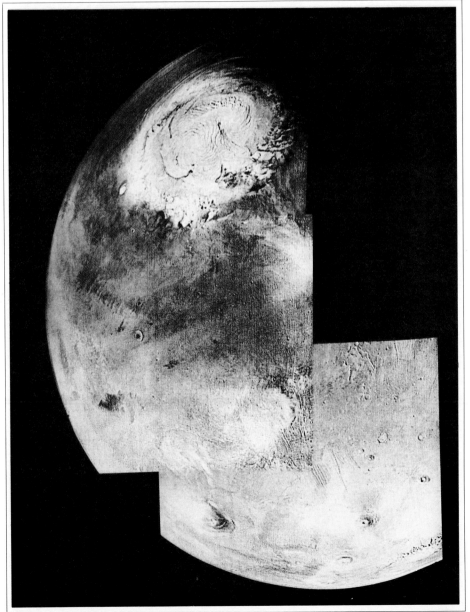

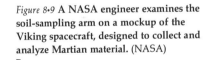

Figure 8·9 **A NASA engineer examines the soil-sampling arm on a mockup of the Viking spacecraft, designed to collect and analyze Martian material.** (NASA) ▼

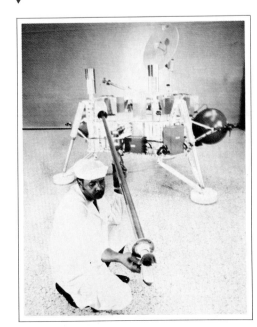

▲
Figure 8·10 The first panorama of the Martian suface was sent from the Viking 1 lander on July 20, 1976. The view, looking south in the late afternoon, shows a rocky desert. Some hills on the horizon may be distant crater rims. Part of the Viking spacecraft is seen in the foreground. (See also Color Photos 8–11.) (NASA)

Figure 8·11 A plain of strewn boulders greeted the Viking 2 lander in the Martian region known as Utopia. The largest rock in center is about 1 by 2 feet. The horizon is estimated to be about 3 km (2 miles) away. Vesicular, or bubbly, texture of rocks is believed to indicate that they are blocks of lava, perhaps eroded or broken out of a nearby (or dust-buried?) lava flow. (NASA)
▼

Figure 8·12 First close-up of Martian soil was sent from Viking 1 lander on July 20, 1976. Fine dust and rock chips are shown. Landing leg is at right. The landing dislodged soil, which settled in the center of the concave pad at the end of the leg. The large central rock is about 5 cm (2 inches) across. The soil shows no signs of life. (NASA)
▼

Figure 8·13 The largest boulder ▶ photographed by Viking landers lies only 8 m (26 feet) from Viking 1. (Landing on this rock would probably have destroyed the spacecraft.) The boulder is about 1 by 3 m (3 by 10 feet), dark grey, probably volcanic, and covered on its top surface by a thin mantle of reddish dust, which may have filtered down out of the atmosphere after the last dust storm. Sand dunes surround the boulder. The same boulder is shown in Color Photo 9. (NASA)

touchdown. Causes of the failures may have been design problems or hostile Martian conditions; a dust storm was raging during the 1971 landings.

The first successful landing on Mars was that of the **Viking 1** spacecraft, which touched down July 20, 1976 (seven years to the day after the first human landing on the moon). It was followed on September 3, 1976 by a duplicate spacecraft, **Viking 2.** Both landings were in plains that looked relatively smooth from orbit, but turned out to be rock-strewn. Boulders as wide as three meters were photographed near the landers. Sand dunes were prominent at the first site. Missing were the Martians, deserted cities, canals, or strange vegetation claimed by early writers.

Mars turned out to be a desolate yet beautiful desert. The *reddish color* of Mars was vividly shown by color photos from Viking landers. Though some rocks appeared dark grey, like terrestrial lavas, most rocks and soil particles were covered with a coating of rust-like, reddish iron oxide minerals. Similar iron minerals are what give terrestrial deserts their familiar red-to-yellow coloration, especially when moisture is present only occasionally. Viking scientists were surprised to find the sky of Mars reddish-tan instead of blue. The sky color is caused by much fine red dust stirred from the surface into the air by winds (see Color Photo 9).

The Martian Climate

During the weeks after landing (Martian summer), **air temperatures** at the two Viking sites ranged from nighttime lows around 187 K ($-123°$F) to afternoon highs around 244 K ($-20°$F). Temperatures of the soil, which absorbs more sunlight than the

▲
Figure 8·14 **Martian sand dunes and rocks appear in this Viking photo. The white boom holds meteorology instruments that measure air temperature and other properties.** (NASA)

air, exceed freezing (273 K or 32°F) in some summer afternoons, so that any frost formed at night near the surface could melt and produce moisture or water vapor. Winds at the two sites were usually less than 17 kph, with gusts exceeding 50 kph. However, much higher winds are believed to occur at certain seasons, raising clouds of dust that can be observed from earth. **Air pressure** at each site was only about 0.7 percent that on earth.

Rock Types

Martian rocks at the two sites appear to be fragments of lava flows, as judged by their textures and colors, and by the chemistry of the associated soil. Abundances of elements measured in the soil resemble those for soils de-

rived from basaltic lavas on earth and moon. For example, Martian soils show silicon contents of 15–30 percent by weight; iron, 14 percent; calcium, 3–8 percent; aluminum, 2–7 percent; titanium, 0–1 percent. These figures are similar to those of terrestrial and lunar basalts. A minor component in the soil is water, about 1 percent by weight—not, however, in the form of liquid, but as H_2O molecules in the crystal structure of the rock particles. It was released when the rocks were heated to temperatures around 800°F during automatic experiments inside the Viking spacecraft. It may be a remnant of more abundant water in the past, and raises the possibility that future explorers could get water by heating Martian soil. Many investigators believe that ice may be present among soil particles a few meters below the surface. This would form a layer of **permafrost,** or ice-rich soil similar to that in arctic tundra regions of the earth.

Atmospheric Composition

The composition of the **Martian atmosphere,** shown in Table 8·1, gives several clues about the planet's history. Like Venus, Mars has an atmosphere that is mostly CO_2, probably generated chiefly by planetary degassing through volcanic activity. As noted in our discussions of earth and Venus, volcanic gases, generated from melting of interior rocky matter, are rich in CO_2. The argon in Martian air is especially informative. Because it is chemically inert, none that was released from the planet would have combined chemically into the rocks. And because it is also heavy, it could not have floated to the top of the atmosphere and escaped into space. The present amount of argon is therefore a measurement of the total amount ever degassed from the interior. Because the composition of volcanic gases on Mars can be estimated from experience with earth's volcanoes, the amount of argon can be used to estimate the amount of other accompanying gases. These results show that a *massive* atmosphere must have been degassed in the past, perhaps enough to create a surface pressure a quarter or half that of the earth's, compared to the 0.7 percent pressure now prevailing. Much of this gas should have been nitrogen, carbon dioxide, oxygen, and water vapor.

In other words, Mars probably had a more extensive, earthlike atmosphere in the past. Where did the missing gases go? Meteorologists have shown that atoms and molecules of some gases, such as nitrogen, can react in the high atmosphere in such a way that atoms are accelerated out of the atmosphere into space. Other gases, such as H_2O, have apparently frozen out of the atmosphere or combined with the soil.

The Martian polar caps give vivid evidence that important gases freeze out of the Martian atmosphere. On the hemisphere that is having summer

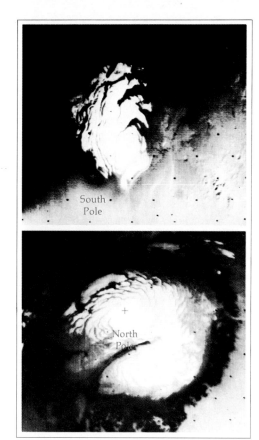

Figure 8·15 One of the most Marslike regions on earth is this arid coastal desert of Peru. Here, rainfall is very rare, so that landforms are controlled mostly by windblown dust. As on Mars, erosion and transport of some materials may have occurred in the past from episodes of water flow.

Figure 8·16 The north and south polar ice caps of Mars and their relatively featureless surrounding plains, as shown by Mariner 9. Shown here near their minimal extent (500–1,000 km across), they grow in size during the Martian winter. Snow deposits may be less than 1 km thick. (NASA)

Table 8·1
Composition of Martian Atmosphere

Gas	Percent by Volume[a]
CO_2 (Carbon dioxide)	95
N_2 (Nitrogen)	2.7
Ar (Argon)	1.6
CO (Carbon monoxide)	0.6
O_2 (Oxygen)	0.15
H_2O (Water vapor)	0.03 (varies)
Kr (Krypton)	trace
Xe (Xenon)	trace
O_3 (Ozone)	0.000003

[a]Amounts of these gases vary slightly with season and time of day. Water vapor is especially variable. Some CO_2 condenses out of the atmosphere into the winter polar cap; changing cap sizes cause small changes in the total Martian atmospheric pressure.

there is a small, permanent cap of frozen water, because the temperature is below the freezing point of water. It is not known how thick this ice deposit is, or how much Martian water it contains. During Martian winter, temperatures plunge below the 146 K ($-197°F$) level at which carbon dioxide freezes. Polar fogs of CO_2 clouds form, and CO_2 snow, or "dry ice," accumulates on the ground to form a large polar cap that eventually shrinks away during Martian spring.

The Major Martian Geological Structures

While data and pictures from the Viking landers have given vivid ideas of the Martian landscape, the mapping of large-scale geological structures from orbiting spacecraft has yielded information just as intriguing. This information has been gained primarily by the first Martian orbiter, Mariner 9, in 1971–1972, and by two Viking orbiters in 1976–1977.

Martian Dust: The Markings Demystified

The dark markings, once thought to be vegetation, are probably caused mostly by deposits of dust. Evidence for this theory includes vast dune fields, patches, and streaks associated with the leeward sides of crater rims and hills. These patches have been simulated by dust blowing over model craters and hills in wind tunnels. Both dunes and leeward deposits of dust show up at small scale in the Viking landers' photos. The shapes of the markings change during dust storms. The seasonal changes long noted from earth arise because Martian seasonal

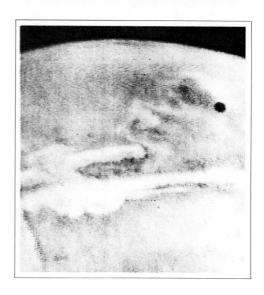

Figure 8·17 One of the early pictures sent by ► Mariner 9 showed the planet obscured by a raging dust storm. This view shows the curved horizon and a vast straight canyon system, Valles Marineris, filled with bright clouds of blowing dust. Black spot at right is a photographic defect. (NASA)

Figure 8·18 Telephoto view reveals that the dark patch in the floor of one Martian crater is a large complex of dunes about 1–2 km apart. This and other pictures indicate that many of the major dark markings may be windblown dust deposits. (NASA)
▼

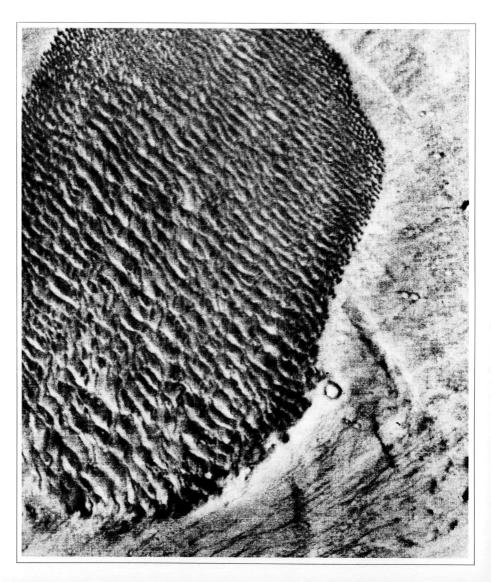

winds and summer dust storms often redistribute the dust (Figure 8·18).

Martian Craters: Clues to Surface Processes

Martian craters are interesting for three reasons. First, they are abundant and show that impacts have been common on Mars as well as the other planets. Second, they show varied states of degradation, which indicate that erosive processes have been much more active on Mars than on the moon or Mercury. In some regions they have been obliterated by young-looking lava flows; in other regions, by massive accumulations of sediments or wind-blown dust hundreds of meters deep. Third, the craters give a way to estimate the ages of Martian surface features. By estimating the rate at which interplanetary debris would hit Mars and create the craters, analysts have estimated that the older, heavily cratered regions are probably a few billion years old, like the cratered surfaces of the moon; but the youngest, sparsely cratered volcanoes, lava flows, and eroded surfaces may be as young as 100 million years or less.

Martian Volcanoes: Clues to Crustal Stability

The largest known **volcanoes** in the solar system were discovered on Mars by Mariner 9 in 1971. The highest volcano, *Olympus Mons* (Figure 8·20), rises about 24 km above a reference level in the lower Martian deserts. Mt. Everest, on earth, is only 9 km above sea level, 13 km above the mean ocean floor, and 20 km above the greatest ocean depths. The base of Olympus Mons is about 500 km across and would nearly cover the state of Missouri. The caldera, or volcanic crater, at the summit is about 65 km across. Olympus Mons is thus several times bigger than any similar type of volcanic cone on earth. For example, Mauna Loa, in Hawaii, reaches about 9 km above the sea floor, is about 120 km in diameter at its sea-floor base, and has a summit caldera a few kilometers wide.

Nearly a dozen other large, distinct volcanic mountains dot Mars. Most are a few hundred kilometers across at the base, and one has a summit caldera 140 km across. They are clustered in two major volcanic regions that form broad domes or swellings on the face of Mars, about 1800 to 2500 km across. They are surrounded by swarms of radial fractures that extend

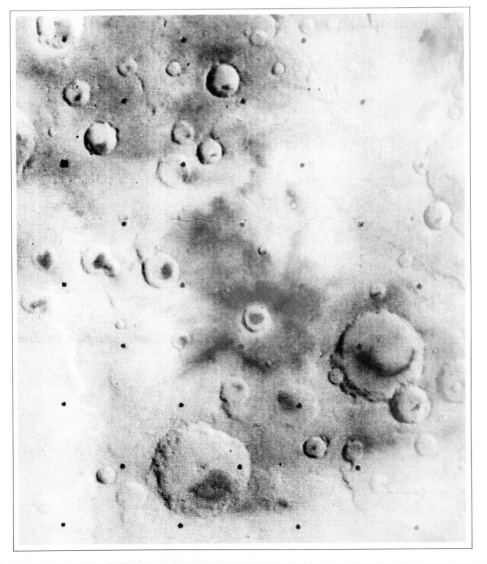

Figure 8·19 **Some regions of Mars are fairly heavily cratered. Blowing dust often collects in dark patches in the bottoms of craters. Craters here are some tens of kilometers across. Geometric array of black dots is artificial, used for camera calibration. (NASA)**

thousands of kilometers from their centers. One of these fracture swarms is a vast canyon system, the Valles Marineris.

The large volcanoes, broad domes, and fracture systems all suggest that the Martian crust in these areas has been massively disturbed by lava upwelling from below. Uplift and volcanism may have been caused by rising hot currents similar to those suspected inside the earth. The Martian volcanoes have probably reached such enormous size because moving tectonic plates have not formed on the surface of Mars. A given spot on Mars therefore stays directly over a given ascending magma current, and the lava eruptions during many years pile up into single huge mountains. In contrast, Mauna Loa and many other terrestrial volcanoes—in fact, the whole string of Hawaiian islands—are believed to be separate lava accumulations that occurred as plates in the earth's crust drifted over a single region by intermittent mantle activity under the Pacific. A Martian volcano shows what might happen if all the Hawaiian islands were piled into one eruptive unit.

Figure 8·20 **This mosaic of pictures shows the volcanic mountain Olympus Mons, surrounding desert, and fractures. The mountain's base is nearly 500 km across.**
▼

Figure 8·21 **An intensely fractured Martian region shows interlacing faulted valleys a few kilometers wide. These faults are associated with the volcanism that produced Olympus Mons and other major volcanoes.** (NASA)
▼

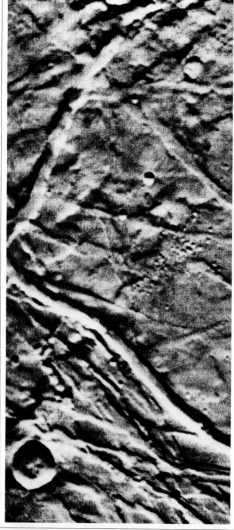

Martian Riverbeds: The
Mystery of Ancient Martian Climate

So far Mars might sound like the moon or Mercury with a few extra volcanoes and a little air to blow the dust around. Mariner 9 shattered this conception by returning with photographs of **channels** that look like dry **riverbeds.** These channels meander in sinuous curves and often have tributaries. They get wider and deeper in a downslope direction. They have sedimentary deposits on their floors. In short, they have all the features of riverbeds. Other theories of their origin, such as that they might be lava-flow channels, do not explain all their features.[1]

The greatest surprise of Martian exploration was to send cameras to this very arid planet and discover what appear to be riverbeds. Evidently, Mars once had flowing rivers of liquid water. The number of impact craters interrupting the channels (and thus of

---

Figure 8•22 **Viking orbiter photos were combined into mosaic of a region about 250 by 200 km (155 by 120 miles) where water has apparently flowed across the surface. Streamlined erosion channels from left to right mark the course of flow. Background terrain is believed to be lava flows, dotted by impact craters. At right, water has apparently cut through ridges formed by the lava. (NASA)**

[1]Although a few of the major riverbeds lie near some reported positions of the once-popular "canals," there is little correspondence in general. The channels do not explain the reports of canals.

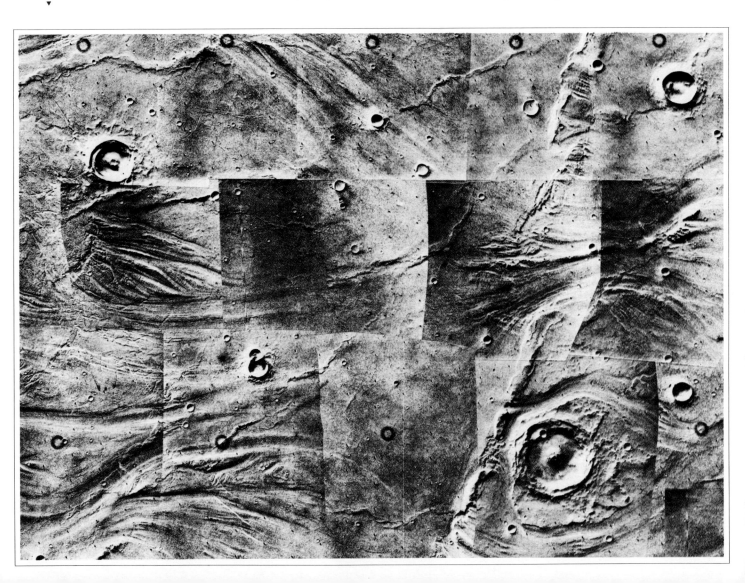

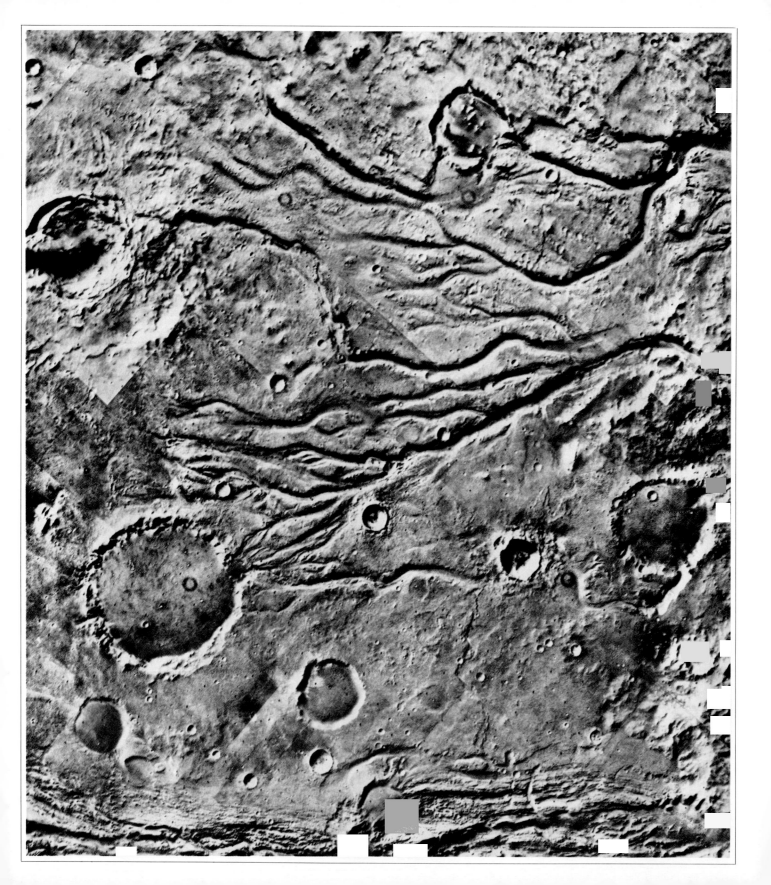

more recent date) indicate that the channels are perhaps hundreds of millions of years old. They are younger than the most ancient cratered regions, but older than most volcanoes. Some channels emanate from chaotic collapsed areas believed to have formed when ice deposits melted and released water onto the surface. Others —a network of fine channels near the equator and large channels with tributaries—suggest that some water came from other sources, possibly rainfall from a once-denser atmosphere. Other signs of ancient erosion, such as degradation of craters and buildup of sedimentary layered deposits, also suggest much more atmospheric and erosive activity in the past.

These observations lead to exciting conclusions and new questions. Although Mars' present conditions are inhospitable to life as we know it, Mars' *ancient* conditions may have been more clement.

But where has the water gone? Is it frozen beneath the surface and in the polar ice, or has most of it escaped into space? Many Mars analysts believe much of the water is still on Mars. We know that much water is frozen in the permanent polar caps and that about a percent of the surface soil is chemically bound water; there may be still more water frozen as permafrost below the surface.

More provocatively, what could have caused the transition from the earlier conditions—with more liquid

water, more air, and higher temperature—to the present arid, freezing conditions? This question has led astronomers and meteorologists to combine forces to try to understand what factors could change planetary climates within periods of a hundred million or a billion years.

Under present conditions, water is very unlikely to flow on Mars. It is usually frozen, and even if it warmed enough to melt, it would very rapidly evaporate into the thin Martian air (in many regions the air pressure is so low that the water would spontaneously boil away into the air). But calculations indicate that if Mars were warmer, water might be able to exist more readily in liquid form.

Cornell scientists have found that if solar radiation striking the Martian polar caps increased by only 10 to 15 percent, much of the frozen CO_2 in the cap would *sublime* (go from solid to gas), passing into the atmosphere. This denser atmosphere would make Mars warmer, carry more heat to the poles, and cause still more polar dry ice to sublime. Thus, there would be a striking feedback effect: a small initial change in polar climate could result in a large change in global climate (Sagan, Toon, and Gierasch, 1973).

These results led scientists to look for ways in which more sunlight might have reached the Martian poles in the past. One such process was revealed by calculations of changes in the rotation of Mars caused by the pile-up of lavas in the volcanoes in the domed region around Olympus Mons. These calculations indicated that before that volcanism occurred, the poles of Mars might have been inclined toward the sun by as much as 45°, instead of the present 25°. This dip would mean that each summer pole would receive enough warmth to melt all the water

in the permanent ice cap. The water would circulate in the atmosphere on its way toward freezing at the winter pole. Rain might occur and create the observed riverbeds.

An alternate effect, possibly occurring at the same time, is that the radiation from the sun itself might vary by small amounts from one era to another, thus causing climates on *all* planets to vary.

These Martian discoveries give another example of how astronomical exploration of other planets can illuminate our condition on earth. Everyone is familiar with the Ice Ages that have occurred on earth in the last hundred thousand years, and geological records indicate even greater climate changes in the last few hundred million years on earth. The Martian evidence may suggest that planetary climate changes are common. In fact, calculations of the same rotational effects mentioned above have been made for earth since the discovery of Martian channels, and some scientists have concluded that similar changes in the axial tilt of the earth, caused by known gravitational effects, fully explain the earth's Ice Ages.

Where Are the Martians?

Much of the motivation and excitement in exploring Mars has been in the search for extraterrestrial life. Discovery of life on Mars would be of major cultural importance. Just as the discoveries of the Copernican revolution showed the earth was not the center of the solar system, **life on Mars** would show that humans are not necessarily the lords of creation. On the other hand, proof that life never evolved on Mars in spite of favorable conditions would present

◄ *Figure 8·23* **Dramatic evidence of ancient riverbeds on Mars is found in these channels and tributary systems. The area is about 180 km (110 miles) wide and drops about 3 km in the direction of flow, from the west edge (left) to the east. Water apparently cut into some old craters, but predated others. (NASA Viking photo from orbit)**

an exciting challenge to our present concepts of the nature of life, since biological experiments suggest (but do not prove) that life should evolve wherever conditions are suitable.

The Viking mission was specifically designed to look for life[2] on Mars. Of the five experiments involved, two gave negative results, but three gave what might be called ambiguously positive results. First, the cameras showed no signs of life. But that was not a strong test, since theorists had always supposed Martian life might exist only as primitive microscopic organisms. The second experiment, a soil analysis, revealed no organic molecules in the soil at either Viking site, at a sensitivity of a few parts per billion. Since **organic molecules,** or massive molecules containing carbon, are essential building blocks of life as we know it, this test is a strong indication that living organisms do not now exist in Martian soil, and have not existed in the recent past.

The other three experiments were designed to look for ongoing biological processes, such as metabolism and photosynthesis, by taking soil samples, putting them in special chambers (some with nutrients), and watching for chemical changes that would indicate living organisms processing the material in the chamber. All three of these tests found the type of changes that indicated life in terrestrial samples!

After preliminary analysis, a few investigators believe Viking may have actually detected microbial life on Mars, but in a quantity of less than one organic molecule per billion parts of soil. However, the magnitude and

[2]By this scientists generally mean "life as we know it," the ability of carbon atoms to combine with other atoms and form very complex molecules, which in turn form organisms that grow and reproduce.

rate of the changes were quite unlike those for terrestrial microbes. Furthermore, because most investigators believe microbes would be unlikely, if not impossible, at the measured low abundance of organic molecules, the Viking results seem to reveal unexpected but ordinary chemical reactions in the soil, *not life.* One explanation of the reactions is that solar ultraviolet radiation reaching the surface of Mars may cause different mineral properties than those familiar on earth; in their altered states, Martian minerals may react in unanticipated ways in the test chambers.

Thus, Viking has shown that Mars "is not teeming with life from pole to pole" (to use the memorable understatement by astronomer Carl Sagan). It probably has no present or recent microbial life. Why? Ultraviolet light would break down organic molecules exposed for long periods; therefore, life might have existed in the past, but its traces might have been wiped out in the surface soil. Alternatively, life might have existed or exist now in subsurface layers or other protected places not sampled by Viking. Perhaps we need to seek life on moist, sunfacing slopes near the water-rich polar cap. Another theory is that life may never have evolved on Mars. Liquid water may never have existed long enough to allow complex organisms to evolve. Future exploration may

◄*Figure 8•24* **Sampler arms on both Viking 1 and 2 landers scooped up soil samples for chemical analysis inside the spacecraft. Here a Viking 2 sampler arm pushes aside a frothy-textured lava rock to gather a soil sample from under it. Protected soil in such locations would have less irradiation by solar ultraviolet light, expected to be harmful to any possible Martian life-forms. This experiment occurred on October 9, 1976. (NASA)**

be aimed toward clarifying whether ancient life existed, as well as toward detecting present life. The dusty plains of Mars have not given up their last secrets about the origins of life in the solar system.

Martian Satellites: Phobos and Deimos

Not all of the mysteries of Mars are on its surface. In 1877, the American astronomer Asaph Hall became the first human to see a satellite of Mars. Shortly afterward, he charted the positions of two Martian moons, and named the inner satellite Phobos ("fear") and the outer one Deimos ("terror") after the chariot horses of Mars in Greek mythology. Mariner 9 obtained the first close-up photographs of these potato-shaped, cratered chunks of rock (Figures 8·25, 8·26). Measurements of these photographs reveal the dimensions given in Table 8·2, p. 157.

An interesting bit of history accompanies the discovery of these satellites. Both Jonathan Swift, in *Gulliver's Travels* (1726), and Voltaire, in *Micromegas* (1752), gave Mars two moons over 100 years before their discovery. Some pseudo-scientific writers have claimed Swift and Voltaire had access to occult knowledge or records of mysterious ancient observers. But the answer lies in history. Still earlier, around 1610, Kepler had speculated about Martian moons. Influenced by medieval numerology, Kepler proposed that because Venus, earth, and Jupiter (in order outward from the sun) had 0, 1, and 4 moons,[3] respectively, Mars might have 2 to fit into the sequence between earth and Jupiter. Swift was undoubtedly repeating

[3]Galileo had discovered four large satellites of Jupiter.

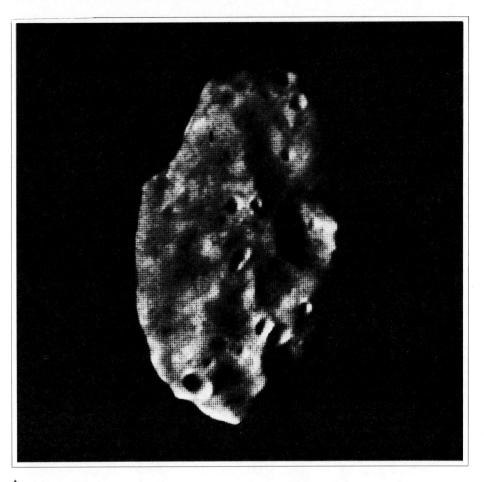

Figure 8·25 **The inner satellite of Mars, Phobos, appears to be a cratered, floating rocky chunk about 28 km long.** (NASA photo by Mariner 9 orbiter)

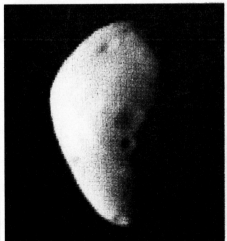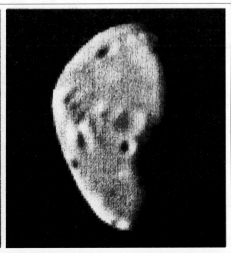

Figure 8·26 **Mariner 9 images of the 12-km satellite, Deimos. Left image shows the picture as originally received; right image shows the picture after computer-processing to enhance contrast and small details. Most spacecraft photography of planets and satellites has been similarly enhanced by computer.** (NASA)

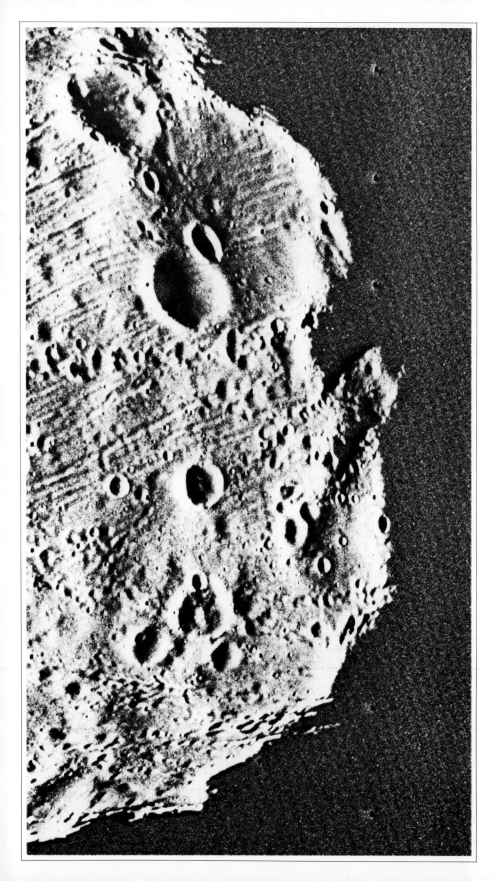

Kepler's idea, which was based on faulty information (we know of at least 14 satellites of Jupiter) and a faulty premise, but was nevertheless confirmed.

Phobos and Deimos turned out to be highly cratered bits of cosmic debris, probably resembling asteroids. The craters were caused by collisions with smaller bits of meteoritic debris. The largest crater, on Phobos, is 8 km (5 mi) across. It is named Stickney, the maiden name of Mrs. Hall, who encouraged her husband's successful search for the satellite. A collision violent enough to create such a crater would release as much energy as 100,000 atom bombs of the Hiroshima size, or about 1,000 hydrogen bombs of megaton size.

A peculiar feature of Phobos was shown when Viking orbiters made close-up photos. Networks of grooves exist on the surface, reaching widths around 100 m and lengths as much as 10 km (Figures 8·27 and 8·28). Some are rows of adjoining craters. Perhaps they are fractures from past collisions between Phobos and large meteorites. Or perhaps they involve fractures on some larger parent body, of which irregularly-shaped Phobos is only a fragment.

The origins of Phobos and Deimos are also unknown. Their surfaces are dark in color and apparently of a composition resembling a carbon-rich type of meteorite (called carbonaceous

◄ *Figure 8·27* **Close-up of Mars' satellite, Phobos, by Viking orbiter revealed surprising grooves covering the asteroid-sized moon. The photo, made from 880 km (545 miles) away, shows objects as small as 40 m (130 feet) across. Grooves might be traces of fractures occurring on Phobos (or its parent body, if Phobos is a fragment) during impacts that made the larger craters. (NASA)**

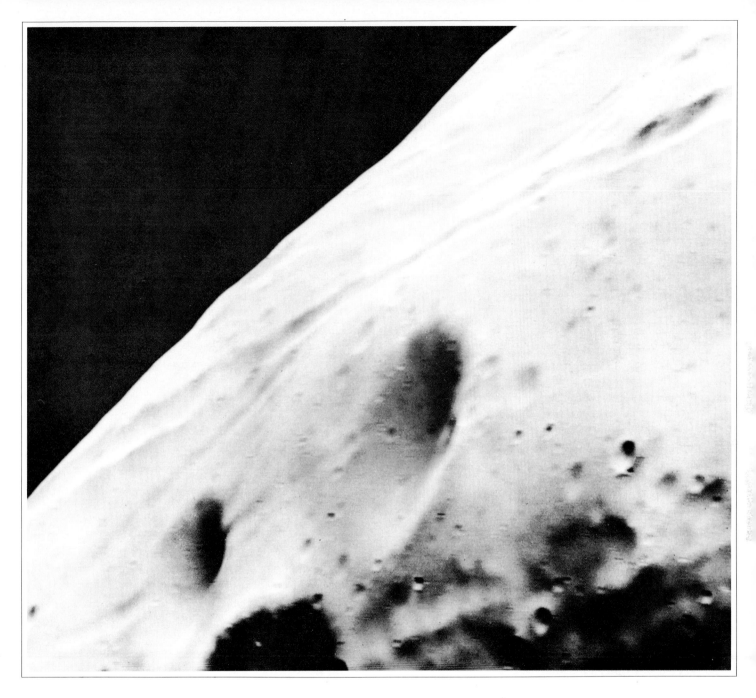

Table 8·2 Dimensions of Phobos and Deimos		
	Phobos	Deimos
Longest diameter	28 km (17 mi)	16 km (10 mi)
Intermediate diameter	23 (14)	12 (7½)
Shortest diameter	20 (12)	10 (6)

Figure 8·28 **A landscape of Phobos. This photo, made during a very close fly-by between the Viking orbiter and the Martian satellite, is slightly blurred by the spacecraft's motion. Distance was only 120 km (75 miles). The area shown is about 3 km wide, and includes craters and one of the grooves, whose origin is unknown. Smallest details are about 10 m across.** (NASA)

chondrite) known to be common in the nearby asteroid belt. Are Phobos and Deimos merely captured asteroids? Their orbits seem too circular and too close to Mars' equator to permit this explanation. Were they formed in orbit around Mars? Are they fragments of a much larger satellite blown apart in a collision with a large meteorite? No one is certain. Further exploration of the red planet may provide us with some of the answers to our questions.

8

Summary

Mars, the planet once thought to have fields of vegetation or even a dying civilization, has been revealed by spacecraft to be mostly barren desert lacking any advanced life-forms. Biological experiments aboard Viking landers indicate that interesting chemical reactions take place in Martian soil. These may be nonbiological reactions, but if they are biological they would indicate that microbial life has evolved on Mars. Further experimentation and possibly further missions to Mars may be required to answer this question.

Evidence about Mars' ancient climatic history seems to contradict its present-day barrenness. Nearly all water on Mars today is locked in polar ice, frozen in the soil, or chemically bound in the soil; virtually no liquid water exists on Mars today. But liquid water apparently once flowed and eroded the surface, indicating different past climates. Current research indicates that the climates of both Mars and earth may have varied over hundreds of millions of

years, possibly due to dynamical wobbles of the rotation axis.

Concepts

Mars' rotation period

seasonal changes in the features of Mars

maria

deserts

canals

Viking missions

Martian air temperatures

Martian air pressure

permafrost

Martian atmosphere

volcanoes

channels (riverbeds)

life on Mars

organic molecules

Problems

1. Describe a day on Mars as a future astronaut might experience it. Include: length of day, appearance of landscape, possible clouds and winds, possible hazards, and objects visible in the sky.

2. What scientific opportunities for long-term exploration does Mars offer, as compared to the moon? What qualities of the Martian environment might make operating a long-term base or colony easier on Mars than on the moon, once initial materials were delivered to the site?

3. Compare photos of craters on Mars and on the moon.

a. *Assuming that all craters had similar sharp rims when fresh, which craters have suffered most from erosion?*
b. *What does this say about lunar versus Martian environments?*

4. Give examples of how the Martian environment and geology are midway between those of the smaller planet Mercury and the larger planet earth. Include comments on atmosphere, craters, volcanism, and plate tectonics.

5. What scientific knowledge might be gained from close-up investigation of Phobos and Deimos? What measurements would be of interest if rocks from Phobos and Deimos were available for study?

6. Imagine you are a visitor from outer space exploring the solar system.
a. *If two Viking-type spacecraft landed at random places on earth, took photos, measured the climate, and took soil samples, what might they reveal about earth?*
b. *How many landings might be needed to adequately characterize earth?*
c. *To characterize Mars to the same degree, how many might be needed?*

Advanced Problems

7. In a gravitational field, objects fall through a height h in a time given by $t = \sqrt{2h/a}$, where a is the acceleration of gravity. Suppose you are an astronaut exploring Deimos, where a is about 0.42 cm/sec². Suppose you drop a tool from eye level.
a. *How long does it take to reach the ground?*
b. *How long would it take on earth, where a is about 1,000 cm/sec². (Hint: Because we are using the cgs system, express* h *in centimeters).*

8. With a telescope that can resolve details as small as one second of arc, what is the smallest detail you can see on Mars during its closest approach to earth, at a distance of about 56 million km?

9. Calculate the velocity needed to launch an object into circular orbit around Deimos from a point on the highest "mountain" on Deimos, assumed to be 8 km from the center. Assume the mass of Deimos is 6 $\times$ 10^{18} g. Many people can throw an object at about 30 m/sec. Would you need a rocket to launch a satellite from Deimos?

Projects

---

1. Observe Mars with a telescope, preferably within a few weeks of opposition date and with a telescope of aperture at least 6 to 8 inches. Magnification around 250 to 300 is useful. Sketch the planet. Can you see any surface details? Usually the most prominent detail is one or the other of the polar caps, a small, brilliant white area at the north or south limb, contrasting with the orangish disk. Can you see any dark regions? Compare the view on different nights, and at different times of night. (Because Mars turns about once in 24 hours, the same side of Mars will be turned toward earth on successive evenings at about the same hour.) If no markings can be seen, three explanations are possible: (1) observing conditions are too poor; (2) the hemisphere of Mars with very few markings may be turned toward earth; (3) a major dust storm may be raging on Mars, obscuring the markings.

2. For the previous observations, determine which side of Mars you were looking at. (Your instructor may need to assist you.) First determine the date and Universal Time of your observations (UT = EST + 5 hr, = PST + 8 hr. Thus 10 P.M. EST on April 2 = 03^{h}00^m on April 3,

Universal Time). In *The American Ephemeris and Nautical Almanac,* the table "Mars: Ephemeris for Physical Observations" gives the central meridian (or longitude on Mars of the center of the side facing earth) at 0^{h}00^m Universal Time on each date. From these tables you can find the Martian central meridian for the time of your observation. (Mars turns about 14.7°/hour.) Compare your observations with a map of Mars, locating the part of Mars that you observed.

The Outer Planets

In March 1972, the Pioneer 10 space-craft left earth with the highest velocity achieved by any space vehicle up to that time. It took only 11 hours to reach the moon's orbit, compared to the three days taken by Apollo space-ships. Twenty-one months later, on December 4, 1973, Pioneer 10 sailed past the planet Jupiter at only 2.8 Jupiter radii from the planet's center. While monitoring Jupiter's environ-ment, this first probe into the outer solar system was deflected at high speed by Jupiter's strong gravitational pull. It then sailed "up," out of the ecliptic plane, thereby initiating an-other cosmic journey by being the first human-made device to leave the solar system and enter interstellar space.

Pioneer 10 is now on its way out of the solar system. It will measure the environment at greater distances from the sun, but will take 80,000 years to reach a distance as far as the nearest star. Although it will be an inoperative derelict long before that time, it carries a plaque describing its human builders—a "message in a bottle" floating in uncharted seas of space.

Features of the Giant Planets

The outer solar system contains three other **giant planets** besides Jupiter—Saturn, Uranus, and Neptune—and finally the small planet, Pluto. The name of the monarch of the Roman gods is fitting for Jupiter. The biggest planet, it contains 71 percent of the total planetary mass—nearly two and a half times as much as all other planets combined. The four giant planets together contain $99\frac{1}{2}$ percent of the total planetary mass, and harbor about 91 percent of the known satel-lites. Clearly the outer planets contain important information about the nature of the solar system; yet they are so remote that we lack the detailed infor-mation available for the nearer planets.

The four giant planets have much lower mean densities than the terres-trial planets—0.7 to 1.6 g/cm³ as com-pared to 3.9 to 5.5 g/cm³ for the rocky metallic terrestrials. Saturn, at 0.7, would float like an ice cube if we could find a big enough ocean (water's density is 1.0 g/cm³). The image is significant. The giant planets evidently *are* made largely of ices.

Jupiter measures a little over ten times earth's diameter, and Saturn just under ten times. Uranus and Neptune have diameters about four times earth's. Placed on the face of

Jupiter, earth would look like a dime on a dinner plate. (See Color Photos 14 and 17.)

The four giant planets have similar massive atmospheres—rich in hydrogen, methane, and ammonia retained from earliest times by the planets' strong gravities. Such atmospheres, though poisonous to current terrestrial life, are believed to resemble earth's original atmosphere, before volcanic emissions and photosynthesis supplied it with carbon dioxide and oxygen. Clouds and haze in these atmospheres hide the true surfaces of the giant planets. The atmospheres of Jupiter and Saturn, the giant planets that

we can see best, have east–west bands of dark and bright clouds. Spotty and streaky clouds form, evolve for several weeks or longer, and finally dissipate.

Figure 9·1 **Artist's conception of cloud-belted Jupiter and its four large satellites. Limb darkening of Jupiter—a faintness of the disk's edge caused by absorption of sunlight—is emphasized in this view. (NASA)**
▼

Satellites of the Giant Planets

Unlike the terrestrial planets, each giant planet has several large satellites. Earth has only one large satellite, Mars has two small ones, and Venus and Mercury have none.

These groups of satellites circling the giant planets seem to be small-scale analogs of the solar system. Study of them may reveal information on the origin and history of the solar system as a whole.

In the 1970s new types of equipment were trained on satellites in the outer solar system, which had previously

been mere "points of light." New measurements revealed that many of these satellites are worlds comparable to terrestrial planets. Some are bigger than Mercury. Many have snow and rock. Some have atmospheres; at least one has clouds. Being

Figure 9·2 **Semipermanent cloud formations** ▸ **of Jupiter, visible from earth in small telescopes.**

Figure 9·3 **Photographs of Jupiter spanning 81 years, showing the changing array of Jupiter's belts and zones. The dark North Temperate Belt is relatively permanent, but the Equatorial Zone changes from bright (1891) to dark (1972). Many of these images show the Great Red Spot (1891, 1937, 1964, 1966, 1968, 1972). (Lowell Observatory)**
▼

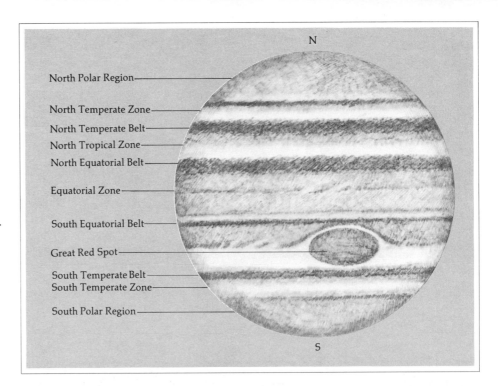

N

North Polar Region

North Temperate Zone

North Temperate Belt

North Tropical Zone

North Equatorial Belt

Equatorial Zone

South Equatorial Belt

Great Red Spot

South Temperate Belt
South Temperate Zone

South Polar Region

S

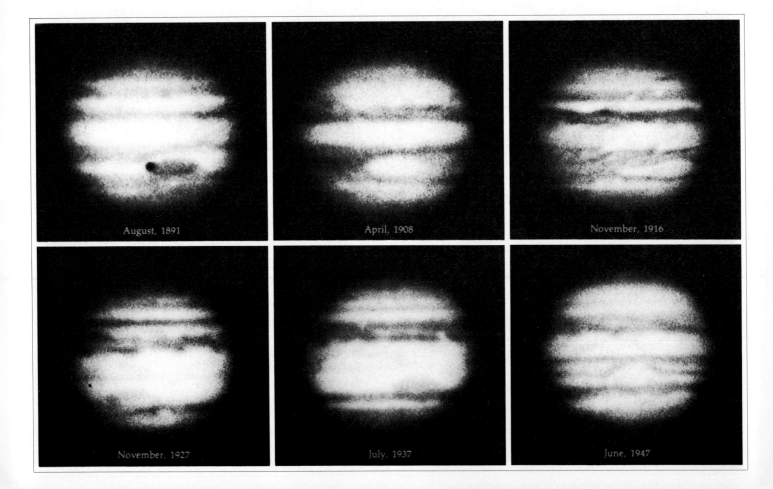

August, 1891

April, 1908

November, 1916

November, 1927

July, 1937

June, 1947

composed largely of ices rather than silicate rocks, they offer us a chance to understand new geological processes, as well as new principles of meteorology and perhaps even biochemistry.

The Planet Jupiter

Jupiter's Atmosphere

Even a week's observations with a backyard telescope reveal the changing cloud patterns of Jupiter. The most obvious pattern is the system of dark and light cloud bands parallel to Jupiter's equator. The dark ones are called **belts,** and the bright ones, **zones.** Within these bands, wispy spots and streaks arise, develop, and die out. These features look small, but some of them are larger than the earth! Though the smaller ones evolve in days, the larger ones may last for months or years. Dark clouds may grow and darken an entire bright zone for months at a time (Figure 9•3).

The belts and zones have distinct colors. Usually, belts are brown, reddish, or even greenish, whereas zones are light tan or yellowish. These colors are attributed to photochemical reactions among minor constituents of the clouds, perhaps involving hydrogen sulfide, organic molecules, or metallic sodium particles.

Jupiter's atmospheric composition is mostly hydrogen and helium, as indicated in Table 9•1. Water, ammo-

Table 9•1 Composition of Jupiter's Atmosphere

Gas	Estimated Percent by Mass[a]
H_2	60%
He	36
Ne	2
H_2O	0.9
NH_3	0.5
Ar	0.3
CH_4	0.2

[a]Of these gases, only H_2, CH_4 (methane), and NH_3 (ammonia) are directly observed from earth. The other gases are inferred from theory and from measured abundances in stars.

Source: Data from Owen (1970).

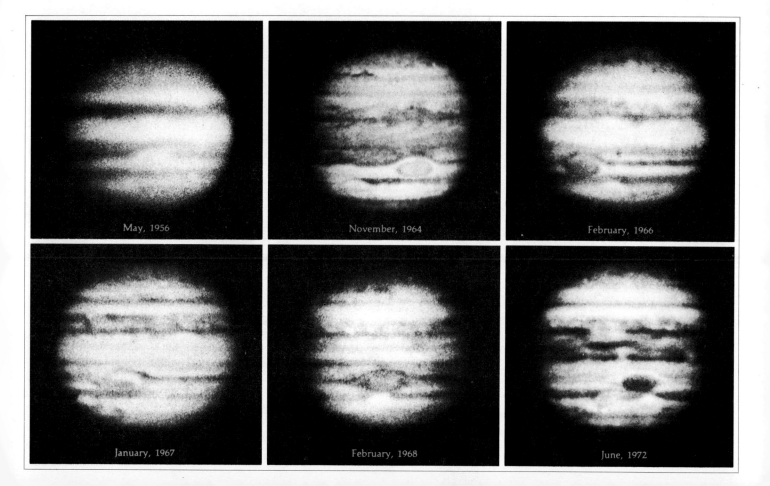

May, 1956   November, 1964   February, 1966

January, 1967   February, 1968   June, 1972

nia, and methane are also present, along with the minor compounds responsible for the colors. The clouds are believed to be primarily ice crystals of ammonia, ammonium hydrosulfide, and frozen water.

Infrared radiation from the upper 20 km of the atmosphere reveals **Jupiter's temperature** to be about 133 K (−220°F), both on the sunlit and nighttime sides of the planet, according to Pioneer 10 data (Figure 9·5). The poisonous upper clouds of Jupiter are cold and forbidding by human standards.

The lower atmosphere of Jupiter may be more interesting. Infrared observations through occasional "holes" amid the highest clouds have revealed lower haze layers with temperatures around 250 K (−9°F). Still lower regions may be even warmer, resembling the primeval atmosphere of earth where terrestrial life originated. A few scientists speculate that some level may even be warm enough for complex organic molecules to evolve into simple organisms that could float in the atmosphere.

A recent model of Jupiter's atmosphere, for example, calls for temperatures similar to those at the earth's surface at a level about 60 km below the Jovian cloud tops, where the pressure would be about ten times the earth's surface pressure (Lewis and Prinn, 1970). Such conditions might be hospitable to primitive life, but strong updrafts would probably draw hypothetical organisms away from the hospitable level. Thus, most scientists doubt that any advanced life-forms exist on Jupiter.

Figure 9·4 **Close-up view of Jupiter from above the Red Spot, taken by Pioneer 11 at a distance of 1,100,000 km. (NASA)**

Figure 9·5 **A heat map of Jupiter. Infrared sensors on Pioneer 10 detected infrared radiation (wavelength 40 microns) dependent on temperatures in Jupiter's atmosphere about 30 km above the cloud tops. The warmest areas (bright in this image) were the dark belts; the coolest areas (dark) were found to correspond to bright zones (compare with Figure 9·2). (NASA)**

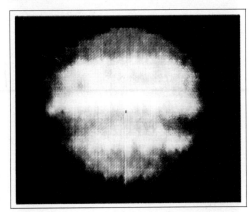

Jupiter's Red Spot

For at least 300 years a storm three times bigger than the whole earth has been raging on Jupiter. In the telescope it appears as an enormous reddish oval in the south tropical zone, probably first studied by G.D. Cassini in 1665. It became very prominent in 1887, when it was rediscovered and called the Great Red Spot. It has

reached diameters of 40,000 km. The **Red Spot** and other, smaller transient spots are probably vast, hurricanelike storm systems in Jupiter's atmosphere. Small clouds approaching them get caught in a counterclockwise circulation like leaves in a great whirlpool.

Jupiter's Rotation

Cloud belts and zones have a rotation period of about 9ʰ50ᵐ near the equator, but several belts and zones at higher latitudes average 9ʰ56ᵐ. The Red Spot has its own rate, sometimes lagging behind or drifting ahead of nearby clouds. Radio radiations from Jupiter give a fourth period, perhaps reflecting deep-atmosphere electrical

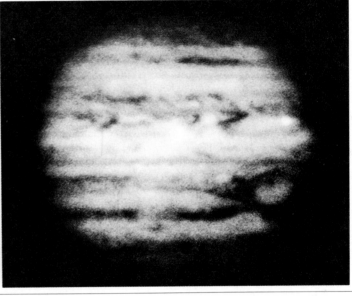

▲ *Figure 9·6* **Examples of the best earth-based photographs of Jupiter, through blue and red filters. In the blue view (left), little light from the Red Spot (low right) passes through the blue filter, making it appear dark. Most of its light passes through the red filter, making it appear bright in the red photo. Black-and-white photos, with filters used in this way, can thus give information on the color of astronomical objects.** (Photo with 61-inch telescope, Lunar and Planetary Laboratory, University of Arizona)

◄ *Figure 9·7* **Artist's conception of a Pioneer spacecraft passing close by Jupiter, near the Red Spot.** (NASA)

Figure 9·8 **A close-up view of swirls and other cloud features of Jupiter, photographed by the Pioneer 10 spacecraft from a distance of about 992,000 km. The limb, or curved horizon, of Jupiter is dimly visible at the top.** (NASA)

storms. No one is sure if any of these periods marks the "true" rotation of a well-defined solid or liquid surface beneath the clouds—or whether such a surface exists. But the general rotation of the planetary globe is probably nearly 10 hours.

Jupiter's Radiation and Magnetism

Besides reflected sunlight, Jupiter emits three other types of radiation. In order of increasing wavelength they are: infrared thermal radiation, shortwave radio radiation, and long-wave radio radiation.

Infrared Thermal Radiation The first is called **Jupiter's infrared thermal radiation** because it is due to the heat of the planet itself. Wien's law reveals a temperature of about 125 K ($-234°$F) for the average radiating layers of the upper atmosphere at both equator and poles. The same measurements allow scientists to calculate the total amount of energy being radiated by the planet. Surprisingly, this turns out to be about twice as much energy as it absorbs from the sun! Extra energy must be coming from somewhere. But from where?

The interior of Jupiter must be emitting much heat of its own. Theorists believe Jupiter is slowly contracting, releasing gravitational energy as heat and radiation. Just as a brick gains energy in falling toward the center of the earth, and releases it as vibration, noise, and heat as it strikes the ground, Jovian material would release heat as it settled toward the planet's center.

Although Jupiter is radiating its own energy, it is not usually considered to be a true star, because its energy is not produced by thermonuclear fusion. Jupiter's mass is not enough to create the central pressure and heat necessary for star-like fusion reactions in its interior.

Shortwave Radio Radiation The second kind of radiation emitted by Jupiter is shortwave radio radiation, often called **Jupiter's decimeter radiation** because its wavelength is around $\frac{1}{10}$ meter ($= 1$ decimeter). The distribution of this radiation with respect to wavelength shows that it is **synchrotron radiation,** a type of radio wave emitted when very fast electrons move through a magnetic field. (These electrons move at nearly the maximum possible speed for any objects: the speed of light. The same radiation is found in synchrotron atom-smashers on earth.)

Synchrotron radiation implies that Jupiter must have a magnetic field. The field was confirmed in 1973 and 1974 when Pioneers 10 and 11 flew through it and measured it. As shown in Table 9·2, it is the strongest planetary field known. Like earth's field, it is not perfectly aligned with the rotation axis but is about 15° off. The polarity is opposite earth's, with the north *magnetic* pole on the south side of the planet; a compass would point south on Jupiter. Because such fields are believed to originate in the circulation of conductive material deep inside the planets, Jupiter's field indicates such a core.

Another similarity to earth is that charged particles emitted by the sun have been trapped by the magnetic field in donut-shaped rings around the planet, like earth's Van Allen belts. The particles are 100,000 to a million times more concentrated than near earth, making a zone of quite hazardous radiation near Jupiter's inner satellites.

Long-wave Radio Radiation The third kind of radiation, the strong, erratic **decameter radio radiation** (wavelength $\simeq$ 10 meters), was discovered in the mid-1950s. Its sources are located on Jupiter itself rather than in nearby space. They may be localized storm areas. The strangest property was discovered by radio astronomer E.K. Bigg in 1964: We receive the bursts of radio noise primarily when Jupiter's satellite Io is in certain positions *with respect to the earth*. Why should radiation emitted from Jupiter depend on the orbital position of *one* of its 12 moons—particularly its position relative to earth? Apparently, Io moves through the magnetic field of Jupiter, disturbing this field and its particle motions in such a way as to produce beams of radiation that, like searchlight beams, sweep through space as Io moves along. Intermittently they point toward earth.

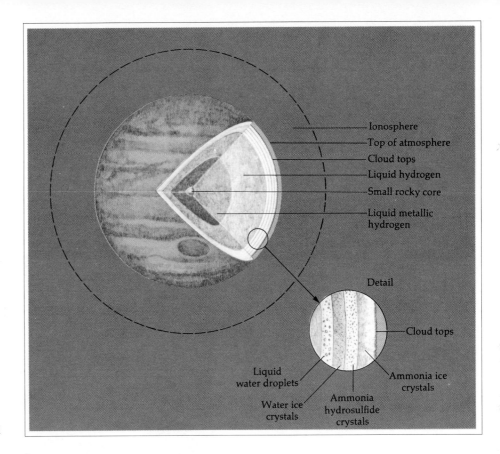

Figure 9·9 **A model of the interior and atmosphere of Jupiter, based on telescopic and spacecraft data. The atmosphere depth to the liquid zone is 1,000 km. (NASA)**

Table 9·2 Magnetic Fields in the Solar System		
Location	Approximate Field Strength (gauss[a])	Source of Data
Sun (average surface)	2	spectra
Sun (sun spot)	1000	spectra
Interplanetary space	0.00004	various spacecraft
Mercury	0.006	Mariner 10
Venus	low[b]	Mariner 10
Earth (surface)	0.3 to 0.8	field measures
Mars	low[b]	Mariner 9
Jupiter	3 to 14	Pioneer 10, 11
Saturn	low	radio data

[a]The gauss is a unit of magnetic field strength. Think of it as measuring the strength of response of a compass placed in the field.

[b]Spacecraft near Venus and Mars detected little or no increases over the interplanetary field strengths near these planets.

Jupiter's Internal Structure and Surface

What is Jupiter like under its clouds? Its density is too low for it to be an iron-silicate body like earth. Instead, **Jupiter's interior** is believed to consist of some 60 percent hydrogen, the rest being helium with small amounts of silicates and other "impurities." The heavier elements may have sunk to the center, so that a core of silicate or silicate-iron material—like rocks of terrestrial planets—may exist near the center. According to some theoretical models, the core may be earth-sized and have a temperature around 30,000 K.

Under the high pressure within Jupiter's interior, hydrogen takes on an unfamiliar form called **metallic**

hydrogen, which can conduct electric currents. As shown in Figure 9•9, much of Jupiter's interior may be taken up by a mantle of liquid metallic hydrogen, in which convection currents carry heat to the surface and electric currents generate Jupiter's magnetic field. Much of the outer half may be a sea of ordinary liquid molecular hydrogen, H_2. There may be no solid surface at all, only a slushy mixture of liquid hydrogen and ice crystals.

The gravity at this "surface" is stronger than that of any other planet. A person weighing 150 pounds on earth would weigh about 400 pounds on Jupiter! Similarly, the pressure of the thick atmosphere is roughly 100 times the air pressure on earth. Combining the various kinds of data, we conclude that the surface region may be a vast, cloud-obscured, high-pressure ocean about 120 km below the cloud tops.

Jupiter's Satellites

Jupiter has 4 large and at least 10 small moons. Using the newly-invented telescope, Galileo and the German astronomer Marius independently discovered the four large satellites on two consecutive nights in 1610. Named in outward order Io, Europa, Ganymede, and Callisto for friends and lovers of Jupiter, these four are called the **Galilean satellites.** Another system for naming the satellites assigns Roman numerals in order of discovery.

Ganymede (III) is one of the largest known satellites, about 5,270 km (3,300 mi) in diameter, exceeding the size of the planet Mercury. The other three Galilean satellites range from about 3,050 to 5,000 km in diameter. The fifth largest satellite, Amalthea (V), only about 240 km across, is in a

tight orbit inside Io's. These five satellites are closest to Jupiter, all having very circular orbits with **prograde** (west-to-east) motion in the plane of Jupiter's equator.

The outer moons divide into two groups. Moons VI, VII, X, and XIII are prograde and have orbits around 12,000,000 km from Jupiter, inclined about 27° to Jupiter's equator. The outermost moons, VIII, IX, XI, and XII, all move in **retrograde** (east-to-west) orbits about 23,000,000 km from Jupiter. Moons XIII and XIV were discovered in 1974 and 1975 by Charles T. Kowal.[1] These two groups of small

---

[1] XIV's orbit and direction of motion were uncertain at the time this book went to press.

moons range from about 8 to 130 km in diameter. Each group may consist of fragments of a larger moon smashed by an asteroid, comet, or another satellite. Or they may be captured asteroids. Other small moons may yet be found.

Because the four large moons resemble our moon and the terrestrial planets, their surface conditions are of special interest. Infrared observations indicate daytime temperatures from 127 to 155 K ($-231°$ to $-180°$F). Temperatures at night and during eclipse, as the moons pass through Jupiter's shadow, are even colder. The rapid drops in temperature during eclipse indicate that Ganymede is covered by a powdery material that serves as a good insulator, preventing heat from being stored in subsurface layers.

What is this material? Because frozen water is detectable on the surfaces of Galilean satellites, it may be powdery snow or a powdery mixture of snow crystals, ice fragments, and soil. The snow cover is only partial, ranging from perhaps 10 percent on Callisto to about 75 percent on Ganymede. Darker regions are perhaps richer in silicate soils (Figures 9·12, 9·13).

Io would be an especially strange world to visit. Pioneer 10 measurements indicate that Io has a thin atmosphere, with only about a billionth the surface pressure of the earth's. The atmosphere or frost deposits may be related to a strange effect: when Io emerges from eclipse, after passing through Jupiter's shadow, it is sometimes a few percent brighter than usual! Europa may show the same effect. The brightness lasts 10 or 15 minutes, and might be caused either by temporary condensation of ammonia frost from the atmosphere during the cold period of eclipse, or by brightening of certain minerals when they are cooled during eclipse.

◄ *Figure 9·10* **The crescent Jupiter, a view never seen from earth.** The photo was made from beyond Jupiter by Pioneer 10, after passing through the planet's radiation belts. Pioneer 10, on an orbit that would take it out of the solar system, obtained this view after the sun had emerged from behind the planet, moving one NASA experimenter to write in response to this picture,

We went regardless, in nineteen-seventy-two.
Some died on Earth, but on we went, unable to stop even the final plunge through hell, emerging as a child of man seeking audience in the stars.
Released forever, we twist back to view the past and witness our first sunrise on Jupiter.
(NASA)

▲ *Figure 9·11* **A spectacular view of Jupiter could be seen from its closest satellite, the small moon Amalthea. Jupiter dominates the sky, subtending an angle of about 41°.** (Painting by James Hervat)

Figure 9.·12 **A view of Ganymede from Pioneer 10 as it flew by the satellite at a distance of 751,000 km (about twice the distance from earth to moon).** Dusky markings reported by earth-based observers are confirmed, but their nature is not clarified. (NASA) ►

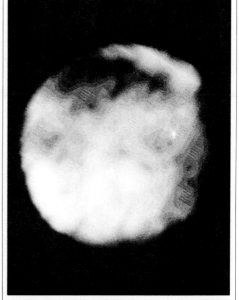

Io has colored regions, including dusky reddish polar caps and a brighter yellowish equatorial region. Their composition is uncertain. During Io's winter, a thin atmosphere of ammonia would freeze to form polar deposits of ammonia snow, evaporating during the "warmer" seasons (still nearly 200°F below zero). Sulfur deposits may also be involved. Trying to understand such surfaces, geochemist Fraser Fanale predicted that if an icy-rocky satellite were heated and partly melted by internal energy sources, the surfaces could become coated with salts (such as sodium chloride) or salt-rich snows, just as evaporation creates salt flats in earth's dry lake beds.

This hypothesis was confirmed in 1973 when observers discovered that Io is sometimes surrounded by a yellow glow. Sodium atoms from the salty snows are probably being knocked off the satellite by the high-energy atoms in Jupiter's radiation belt. Struck by sunlight, the sodium atoms emit a yellow glow (the same glow produced by sodium atoms in candle flames). The glow has been detected as far as 10 Io-radii from Io,

and might be visible to a future astronaut on Io as a faint auroral sky glow at certain times during Io's 1.8-day trip around Jupiter.

With its reddish and yellowish surface deposits, its sporadic post-eclipse brightenings, its faintly yellowish sky, and its peculiar relation to Jupiter's radio emissions, Io hints at the bizarre environments that may be discovered as other distant satellites are explored.

The Planet Saturn

Saturn's Globe and Atmosphere

The next planet beyond Jupiter is Saturn, famous for its ring system. The globe itself is less interesting than Jupiter's, lacking such striking features as Jupiter's Red Spot. Yellowish and tan cloud zones and belts hide the surface. Occasional bright disturbances erupt and last for weeks. (See Color Photo 18.)

As on Jupiter, atmospheric circulation varies with latitude. Rotation periods near the equator are about

Figure 9·14 **Varying aspects of Saturn during its 29-year orbit around the sun, as photographed with earth-based telescopes. Because the rings maintain a fixed relation to the ecliptic, the earth-based observer sometimes sees from "above" and sometimes**

◀ *Figure 9·13* **An imaginary view of Jupiter as seen from its satellite Callisto. The inner satellite, Io, is seen left of Jupiter, surrounded by its cloud of glowing sodium. (Painting by author)**

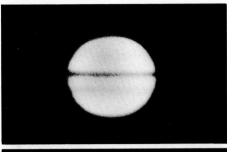

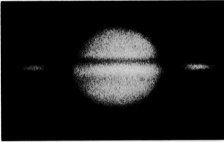

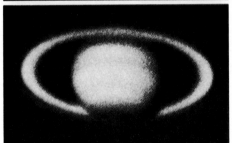

▲

Figure 9•14 (cont.) **"below" the ring plane. This series shows half the 29-year cycle, including two views that show the "disappearance" of the rings as the earth passes through the ring plane.** (Lowell Observatory)

Figure 9•15 **Features of Saturn. Telescopes as ▶ small as 5 cm (2 in.) aperture will show the rings; telescopes larger than about 25 cm (10 in.) will sometimes show all these features.**

10^h14^m, but cloud formations nearer the poles show longer periods, such as 10^h40^m.

That the mean density of Saturn is the lowest of any planet, less than that of water, indicates Saturn has only a small core of silicates and ices. Its layer of metallic hydrogen is smaller than Jupiter's. Most of its bulk is molecular hydrogen in either liquid or highly compressed gaseous form.

Saturn's atmosphere resembles Jupiter's (Table 9•1), although ammonia is less prominent, probably being frozen out in the lower atmosphere by Saturn's colder temperatures. Methane may be considerably more abundant than on Jupiter, perhaps approaching the abundance of helium or hydrogen.

Saturn's Rings

Galileo's first sight of Saturn, in 1610, was a strange one: a blurry disk with blurry objects on either side. He drew it as a triple planet (Figure 9•17). Not until 1655 did Christian Huygens discover that a **ring system** encircled the planet.[2] A small modern telescope easily shows the astonishingly beautiful rings, largest in the solar system. The rings' proportions are surprising. They stretch about 274,000 km (171,000 mi) tip to tip, but they are only about 3 km (2 mi) thick! The system is so thin that terrestrial observers lose sight of it altogether when it appears edge-on for a few days once every 15 years as the earth passes through the ring system's plane.

The rings are divided by several gaps. In 1675 G.D. Cassini discovered

[2]Huygens announced this discovery in the form of a famous anagram (succession of apparently meaningless letters). Huygens later explained that the anagram was to represent the words "Annulo cingitur, tenui, plano, nusquam cohaerente, ad eclipticam inclinato" (it is surrounded by a thin flat ring, nowhere touching, inclined to the ecliptic). In the days before copyrights, such anagrams were a common form by which scientists published initial results, thus establishing priority while giving themselves time for more observations. Not until about 1661 were most observers convinced of the accuracy of Huygens' discovery (Alexander, 1962).

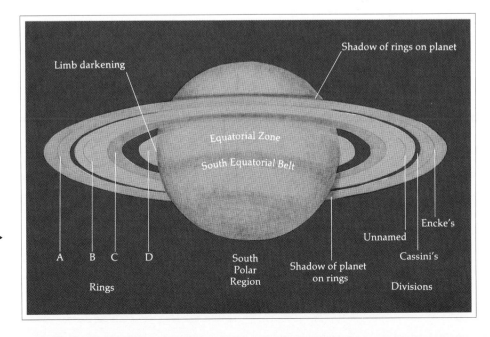

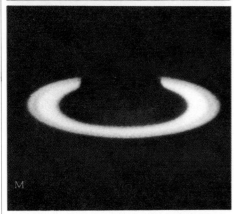

Figure 9·16 **Photographs of Saturn with films and filters sensitive to different colors of light. B = blue (4,500 A), the most common view (see Figure 9·14). M = wavelength absorbed by methane gas (8,975 A); abundant methane in the atmosphere absorbs most of the light, while the rings remain bright, being covered with ice that reflects sunlight at this wavelength.** (Lunar and Planetary Laboratory, University of Arizona)

a prominent gap between the dusky outer ring (called ring A) and the brighter inner ring (B); it is called **Cassini's division.** Other, narrower gaps have been found, further subdividing the rings. Ring C is a faint ring inside and bordering ring B. A still fainter ring has been found on some photos, well inside ring C and separated from it by a large gap; it apparently extends nearly to the atmosphere of Saturn.

Several discoveries clarify the nature of the rings. First, in 1859, James Clerk Maxwell made a crucial finding: the rings are not solid disks, because they are inside Roche's limit. As described in Chapter 5, a disk inside Roche's limit would be pulled apart into myriads of particles. In 1895 American astronomer James Keeler confirmed spectroscopically that the rings consist of small particles, each following its own near-circular orbit around Saturn, obeying Kepler's laws.

Second, once the rings are known to consist of small particles, the gaps in the rings can be explained. Cassini's division, for example, is probably caused by gravitational forces of the satellite Mimas acting on the ring particles. Mimas has an orbital period twice that of particles near the division, causing repeated disturbances believed to accelerate the particles out of the gap. Other gaps are believed associated with similar forces.

Third, the ring particles are composed of (or covered by) frozen water. (Lebofsky, Johnson, and McCord, 1970). Most of the particles are roughly a few centimeters in diameter. Visitors to the rings might find themselves surrounded by a swarm of floating hailstones (Figure 9·19).

Fourth, brightness asymmetries indicate that the ring particles all keep the same face toward Saturn and are probably rarely disturbed by collisions with other particles. Apparently, their paths are nearly parallel, like paths of racers on a track.

How did such a remarkable ring system form? The first clue is that any particles in the location of Saturn's rings would have to form a flat ring system. Mutual collisions plus gravitational forces from Saturn's oblate mass force the particles into a ring in Saturn's equatorial plane. Since they are inside Roche's limit, tidal

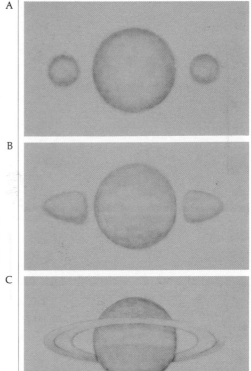

Figure 9·17 **When Galileo first observed Saturn, he thought it was a disk with two satellites (as in A). Later telescopes showed a view like B. A few decades later, better telescopes revealed the true situation to be C.**

forces prevent them from coalescing into a satellite.

There remains the problem of where the particles came from in the first place. They may have been:

1. particles condensed from gas as Saturn formed;

2. part of a satellite blown apart by collision with a comet or asteroid;

3. part of a comet or asteroid broken apart by tidal forces after approaching too close to Saturn.

Close-up study of the composition and form of the individual particles

Figure 9·18 Imaginary view of Saturn from a high angle north of the ring plane. The Milky Way, the Coalsack nebula, and the constellation of the Southern Cross are shown in the background. (Painting by Chesley Bonestell)

Figure 9·19 An imaginary view within the rings of Saturn, showing the multitude of cobble-size ice or ice-covered chunks that compose the rings. The shadow of the rings is seen on Saturn, lower left. (Painting by author)

might reveal whether one of these theories is correct.

Saturn's Satellites

Like Jupiter's satellite system, **Saturn's satellites** include some exotic worlds that may display hitherto unknown phenomena.

Saturn has 11 known moons (see Table 6·1). The innermost two satellites are small moons just outside the ring. Photographic studies reported in 1977 suggest other small moons may exist in this region. They may be related in origin to the rings.

The next five moons outward range from about 320 to 1,080 km in diameter and are spaced about 60,000 to 150,000 km apart. If their estimated low densities (about 1 to $2\frac{1}{2}$ g/cm^3) are correct, they are probably rich in ices (recall that the density of frozen water is 1 g/cm^3).

The seventh and most interesting moon is the giant Titan, discovered in 1655 by Huygens. At nearly 5,000 km in diameter, it is one of the largest known satellites. Titan's atmosphere, still the densest known for any satellite, is chiefly gaseous methane, but molecular hydrogen is also abundant.

As wearers of Polaroid sunglasses know, sunlight reflected by different surfaces is polarized in different ways. Titan's polarization shows that it is covered by a layer of clouds. Their reddish color may be due to certain complex molecules produced when sunlight strikes atmospheres rich in hydrogen and hydrogen compounds. Probably because of a *Venuslike* greenhouse effect (Chapter 7), Titan's temperatures of about 150 K ($-189°$F) are nearly double the predicted 87 K. Theoretical models of Titan's atmosphere indicate a surface pressure near 0.4 of the earth's.

These data raise the exciting possibility that Titan's surface environment resembles what has been postulated for the primitive earth, except for the low temperature. And even this difference might be erased in localized warm environments, if there were any volcanoes or other heat sources on Titan. Could liquid water temporarily form from melting ice? Could primitive biological activity occur in such microenvironments? Could "ice volcanoes" erupt water instead of lava? These questions make Titan a fascinating target for future exploration.

The ninth satellite out from Saturn is also large and interesting. Iapetus, about 1,800 km in diameter, was discovered in 1671 by G.D. Cassini, who announced a mystery: Iapetus could be seen easily when it was on the west side of the planet, but when it swung around to the east side, it vanished! Modern telescopes reveal that it does not truly vanish, but diminishes greatly in brightness. Iapetus always keeps the same face toward Saturn, so that one hemisphere leads in the orbit and the other hemisphere trails. The solution to Cassini's mystery is that the trailing side is fully six times brighter than the leading side.

But why? Probably the trailing hemisphere is covered with snow or ice (which reflects 28 to 60 percent of the sunlight), whereas the leading side is covered with dark, loose soil (reflecting only 4 to 9 percent). Perhaps one face of Iapetus has lost snow or gravel through blasting by a few large meteorite impacts, or has accumulated much dark meteoritic debris.

The anomaly may involve the next and outermost moon, Phoebe. A small moon about 260 km in diameter, Phoebe is the only Saturn satellite in retrograde orbit. Debris knocked off Phoebe by ordinary meteorite impacts would strike the leading side of Iapetus

Figure 9·20 **The discovery photograph of Saturn's tenth satellite, Janus, made in 1966 by the French observer A. Dollfus. Diagram shows the identification of the rings and other satellites.** (Courtesy A. Dollfus)
▼

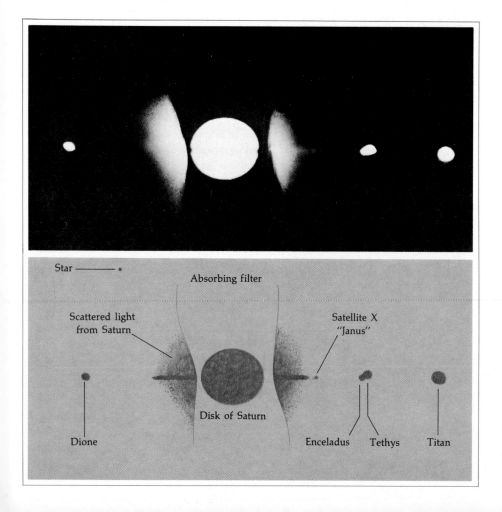

Star ——————— *

Absorbing filter

Scattered light from Saturn

Satellite X "Janus"

Disk of Saturn

Dione

Enceladus   Tethys   Titan

▲
Figure 9·21 **An imaginary view of Saturn rising over a rocky plain on its satellite Rhea. Because the satellite's orbit lies in the plane of the rings, the rings would be visible only edge-on, as a narrow line.** (Painting by James Hervat)

 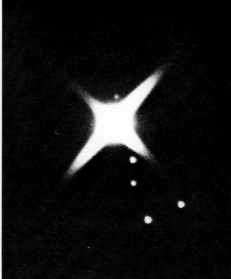 

Figure 9·22 **Uranus (overexposed) and its five satellites, photographed with the McDonald Observatory 2-meter telescope.** (NASA photo, Lunar and Planetary Laboratory, University of Arizona)

at unusually high speed. This may have blasted away the snow cover on the leading side, causing a concentration of dark soil, partly originating on Phoebe (Soter, 1974).

The Planet Uranus

In 1781 William Herschel became the first known person to discover a new planet when he detected Uranus during an ambitious star-mapping project. The planet's name was suggested by J. Bode (of Bode's law) because in mythology Uranus was the father of Saturn, who in turn was the father of Jupiter. Because of Uranus' long, 84-year period of revolution around the sun, it has made only two orbital circuits since its discovery.

Herschel found several satellites as well. The orbits of these satellites are steeply inclined to their planet's

orbital plane by 98°. Laplace noted in 1829:

If the various satellites of a planet move in a plane greatly inclined to that of its orbit. . . they are kept in that plane by the [gravitation] of the planet's [equatorial bulge].[3]

Thus, Uranus itself must have an equatorial bulge, and its equator must be inclined 98° to its orbital plane. This **obliquity,** or deviation from being parallel or perpendicular to the plane of orbit, being greater than 90°, means that the **rotation of Uranus** is *retrograde,* or east to west, and that the poles of Uranus lie nearly in the plane of the orbit.

Thus Uranus must have a unique seasonal sequence. When the north pole points almost directly toward the sun, the southern hemisphere is plunged into a long dark winter, lasting for about a quarter of the planet's 84-year revolution period. After this

[3]The same force that would bring a cloud of orbiting particles into a ring over a planet's equator would keep the satellites in their plane, as Laplace noted in 1829 (see earlier discussion of Saturn's rings).

21-year south polar winter and north polar summer, the planet has moved a quarter of the way around its orbit and the sun is now shining on the equatorial regions. Each point on the planet now goes from day to night during the planet's 23-hour rotation period. After 21 more years the south

Figure 9·23 **A photograph of Uranus from an altitude of 24.7 km above the earth with a 3-ton, 0.9-meter telescope carried aloft by an unmanned balloon, Stratoscope II, in 1970.** (NASA)
▼

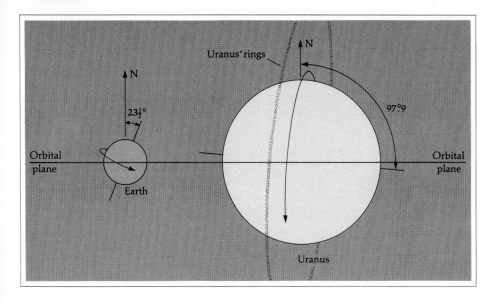

Figure 9·24 **Comparison of the sizes and rotations of earth and Uranus. Earth has an obliquity (or axial tilt) of 23½° and a prograde (west-to-east) rotation. Uranus has much steeper obliquity and retrograde rotation.**

pole points approximately toward the sun and the southern hemisphere experiences a long summer.

Infrared observations in the 1930s revealed the gas methane in the atmosphere of Uranus and observations in the 1960s revealed an even greater abundance of hydrogen. Apparently Uranus' atmosphere is clearer than Jupiter's. Instead of distinct clouds, the atmosphere may be more like a thin mist or fog in which our line of sight simply disappears into the haze. About half the interior diameter is believed occupied by a core of rocky and icy material. The outer layers are probably mostly molecular hydrogen in liquid or high-pressure gas form.

Satellites and Rings of Uranus

Uranus has five known satellites, three of them more than 2,000 km (860 mi) in diameter. All five orbit accurately in the plane of Uranus'

Figure 9·25 **What Uranus and its rings may look like as seen from its nearest satellite, Miranda. Uranus would subtend an angle of 22°. Because Miranda orbits in the plane of the rings, they could be seen only as a thin line. The cloudy disk of Uranus is believed to resemble those of Saturn and Jupiter.** (Painting by author)

▼

equator rather than in the plane of the solar system. Thus, the origin of satellite systems is seen to be closely involved with the history of the parent planet.

In March, 1977 Uranus passed in front of a relatively bright star, as seen from Africa and the Indian Ocean. A number of astronomers set up observing equipment, expecting to watch the star dim as it passed behind Uranus' upper atmosphere, thus to learn about the atmospheric cloud or haze structure. Unexpectedly, but more significantly, they saw the star dim and brighten several times at some distance from the Uranian disk, but symmetrically on each side of the disk.

The observations indicate a ring or rings of small bodies around Uranus inside the satellite system, somewhat similar to the Saturn system. However, preliminary analysis indicates that instead of Saturn's wide rings separated by narrow gaps, **Uranus' rings** are narrow and separated by wide gaps. The particles' nature is uncertain; they could be anything from centimeter-scale ice grains to small moons a few hundred kilometers across, or (as may be true for Saturn's system) a mixture of many small

grains and a few big ones. In telescopes, the rings are invisible, but future spacecraft flights may reveal their nature.

The Planet Neptune

The discovery of Uranus led to the discovery of Neptune. After 1800, theorists tried unsuccessfully to fit observations of Uranus' position into Kepler's laws of planetary motion. Uranus seemed to have its own somewhat irregular motions. A few scientists thought this might signal a breakdown of Newton's law of gravity at large distances from the sun. Others correctly suggested that Uranus was being attracted by a still more distant planet. In the 1840s an English astronomer and a French mathematician set out independently to predict where the new planet should be.

Both men tried to shorten their laborious calculations by following Bode's rule and guessing that the new planet was 38 AU from the sun.[4] Unfortunately, Neptune happens to be the one serious exception to Bode's rule. The English astronomer J.C. Adams finished his prediction in 1845 after two years of calculations, but had trouble getting his senior professors at Cambridge interested in searching for the new planet with a telescope. In July and August of 1846 English astronomers began a desultory search. Several times they actually charted the new planet but failed to realize what they had seen.

Meanwhile, the French mathematician Leverrier, who had finished his work soon after Adams, interested two young German astronomers in

---

[4]Today the necessary calculations could be done quickly on computers, but in the 1800s they took months or years of effort.

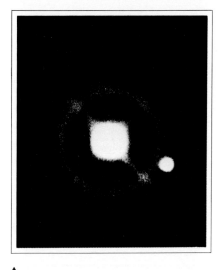

Figure 9·26 **Neptune and its largest satellite, Triton. The overexposed image of Neptune is surrounded by a halation ring, an optical effect produced in the telescope-film system.** (McDonald Observatory)

the search. Armed with Leverrier's predictions, they located the new planet within half an hour of starting their search on September 23, 1846. Although Adams and Leverrier are now both credited with the discovery, the incident became an international scandal at the time because of the Britons' failure to grasp their opportunity.

Neptune is so far away that features are very difficult to observe.

Methane has been found in its atmosphere, and the planet is believed to resemble Uranus in general. It apparently rotates in about 22 hours.

Neptune has a peculiar satellite system. One very large satellite, Triton, seems to have a diameter approaching 5,000 km (3,000 mi), although this figure is very uncertain. Conceivably, remote Triton might turn out to be the largest satellite in the solar system. Triton revolves in a highly inclined retrograde orbit, unlike most large satellites. Nereid, a small satellite, revolves in an elliptical prograde orbit.

The Planet Pluto

Because the motions of Uranus and Neptune were still unaccountably irregular, Percival Lowell began a search for a ninth planet. Finally detected at Lowell Observatory in 1930 by Clyde Tombaugh, it lies close to the position predicted by Bode's rule

Figure 9·27 **Portions of the photographs on which Pluto (arrow) was discovered. The planet was distinguished from stars by its motion.** (Lowell Observatory)

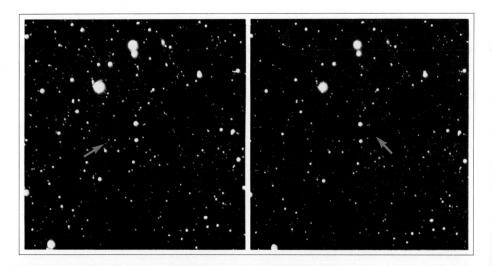

for a planet, about 39 AU from the sun. It was given the name Pluto, and a new planetary symbol was created from the first two letters, P̶L, which were also Lowell's initials.

Pluto is so far away that little is known about its physical nature. Its light variations indicate that it rotates in 6.39 days and is probably tilted more than 50° to its orbit. Its estimated density is about 2 g/cm³, like that of other outer planets and satellites. In 1976, spectroscopy revealed that Pluto is covered by a frost of frozen methane at estimated temperatures around 40 K. Pluto is about the size of the moon.

Pluto's orbit is inclined 17° to the plane of the solar system—more than twice that of the next most inclined planetary orbit. The orbit is so elliptical that Pluto occasionally approaches closer to the sun than does Neptune. It is the only planet to cross another planet's orbit. Several astronomers have suggested that Pluto might not be a true planet at all, but an escaped satellite of Neptune. If Pluto started out as a Neptune satellite, it could have been ejected from Neptune's system by a near-miss with the large satellite, Triton. This hypothesis about **Pluto's origin** could explain the peculiar orbit of Triton, which may have been disturbed by a gravitational interaction with Pluto, and Pluto's crossing of Neptune's orbit. Pluto crosses Neptune's orbit out of the plane of the solar system, so there is no chance of a collison between the two in the foreseeable future.

Planet X

Several astronomers have sought dynamical or photographic evidence of a planet beyond Pluto, sometimes called **Planet X.** Clyde Tombaugh, the discoverer of Pluto, conducted a long search of the region beyond Pluto, in the plane of the solar system, and ruled out any planet as large as Neptune out to a distance of around 100 AU. Occasional Soviet reports of such a planet through 1974 were later discounted. It thus *appears* unlikely that any more *large* planets will be discovered in this region. However, comets traverse the region, showing that at least some smaller material exists there.

9

Summary

The four giant planets are the dominant planets in the solar system in terms of diameter, mass, and numbers of satellites. They have such strong gravity that they have retained light gases such as hydrogen, helium, and hydrogen compounds that were once common in the solar system but have escaped from terrestrial planets due to their low gravity. The giant planets thus have larger proportions of these gases, and have high-density cores of solid or liquid materials that may contain silicate minerals and ices. Their atmospheres are rich in molecular hydrogen, methane, and ammonia gases.

Each of the four giant planets has at least one large satellite (ranging from lunar size to the size of Mercury) and smaller satellites a few hundred kilometers across. Since the giant planets and their satellites form miniature solar systems resembling the sun–planet family, they may contain clues about the origin of the solar system as a whole. The larger satellites of the outer solar system, neglected until recent years, include exotic worlds with atmospheres, where ices may play the role that rocks play on the terrestrial planets. Pluto is anomalously small for the outer solar system, resembling one of the larger satellites; it may have had a different origin than the four giants.

Concepts

giant planets

belts

zones

Jupiter's atmospheric composition

Jupiter's temperature

Red Spot

Jupiter's infrared thermal radiation

Jupiter's decimeter radiation

synchrotron radiation

Jupiter's decameter radio radiation

Jupiter's interior

metallic hydrogen

Galilean satellites

prograde and retrograde satellite orbits

Saturn's atmosphere

Saturn's ring system

Cassini's division

obliquity

Saturn's satellites

Uranus' rotation

Uranus' rings

Pluto's origin

Planet X

Problems

1. Answer the following problems:
a. *How might studying cloud patterns on Jupiter, Saturn, Mars, and Venus help us understand terrestrial meteorological theory?*
b. *What planetary or environmental characteristics might figure in such a theory (example: rotation period)?*
c. *How do these factors vary from one of these planets to another?*
d. *Why are other planets not included on the list?*

2. Which planet has:
a. *The highest surface gravity?*
b. *The lowest surface temperature?*
c. *The highest atmospheric pressure?*
d. *The most atmospheric hydrogen?*
e. *The highest percentage of atmospheric oxygen?*

3. List places in the solar system where you might expect to find droplets, pools, or oceans of liquid water.

4. Describe the appearance of the sky to an observer flying above the clouds in the upper atmosphere of Saturn at dusk at low latitudes. Do the same for the polar regions.

5. How would you expect gravity on the Galilean satellites to compare with gravity on the moon? (See data in Table 6·1.) Which three bodies in the solar system would you expect to have general environments most like the moon's?

6. Describe a landscape on Io.

7. What difficulties might be met in sending a spacecraft through the rings of Saturn to examine the ring environment? Consider a pass *perpendicular* to the ring plane versus a pass *in the* ring plane. (Note that without elaborate retrorockets, such a spacecraft would probably travel about 20 to 30 km/sec relative to the rings.)

8. Describe the seasons and other effects that would occur if the earth had the same obliquity as Uranus.

9. Do orbits of planets ever cross, as seen from far north or south of the plane of the solar system?

10. Where would Bode's rule, if extended, predict a tenth planet? Would it be easier to detect if covered with snow or rock? Why?

11. Two large planets have the same size and mass, but different orbits.
a. *Which would you expect to have more hydrogen? Why?*
b. *If they have had different amounts of volcanism, which would you expect to have more carbon dioxide? Why?*

Advanced Problems

12. Jupiter's four Galilean satellites revolve in orbits about 400,000 to 2 million km from the planet.
a. *What would be their maximum angular separation from the planet when Jupiter is at its closest distance of about 4 AU from the earth?*
b. *These satellites have a brightness of about 6th magnitude, equal to the faintest stars visible to the unaided eye. The eye can normally distinguish details as little as 2 minutes of arc apart. Why, then, are the Galilean satellites not normally visible to the unaided eye?*

13. If you had a telescope that could reveal details as small as $\frac{1}{2}$ second of arc, would you be able to see dark markings on Ganymede (diameter, 5,270 km)?

14. If a spacecraft with an infrared sensing device flew by Jupiter or Saturn and measured a cloud formation whose strongest radiation came at a wavelength of 2×10^{-3} cm (a wavelength sometimes called 20 microns), what would be the temperature of the material in the cloud?

15. Assume the mass of Saturn is 5.7×10^{29} g.
a. *What is the orbital velocity of a particle orbiting around Saturn in Saturn's rings, about 250,000 km from the center of Saturn?*
b. *Suppose this particle hits another one which orbits 1 km farther away from Saturn. Estimate how fast they come together, using whatever mathematical techniques you know.*

Projects

1. Observe Jupiter with a telescope of at least 3 inches' aperture. Sketch the pattern of belts and zones. Which are the most prominent belts? Which zones are brightest? Compare these results with photos in this book. Is the Red Spot or other dark or bright spots visible on the side of Jupiter being observed?

2. Observe Jupiter with large binoculars or a telescope of at least 1 inch aperture. How many satellites are visible? Observe the satellite system at different hours over a period of several days and try to identify the satellites. (This could be done as a class project, with different students making sketches at different hours.)

3. With a telescope of at least 6 inches' aperture, determine the rotation period of Jupiter by recording the time when the Red Spot is centered on the disk. Note that intervals between appearances must be an incremental number (1, 2, 3, etc.) of rotation periods.

4. With a telescope of at least 3 inches' aperture, observe Saturn. Sketch the rings. Estimate the angle by which the rings are tilted toward earth during your observation. Does this angle change much from day to day? From year to year? Can you see any belts or zones? (Usually they are less prominent than on Jupiter.) Can you see Cassini's division? Sketch nearby star-like objects that may be satellites and track them from night to night. Identify Titan, the brightest satellite.

5. With a telescope of at least 6 inches' aperture, observe Uranus and Neptune. What color are they? Can you see any detail? Any satellites?

Comets, Meteors, Asteroids, and Meteorites

On June 30, 1908, a mysterious explosion occurred in Siberia. English observatories, 3,600 km (2,200 mi) away, noted unusual air-pressure waves. Seismic vibrations were recorded 1,000 km (600 mi) away. At 500 km, observers reported "deafening bangs" and a fiery cloud. The explosion was caused as an unknown object struck the earth's atmosphere from space.

Some 200 km (120 mi) from the explosion, the object was seen as "an irregularly-shaped, brilliantly white, somewhat elongated mass . . . with [angular] diameter far greater than the moon's." Carpenters were thrown from a building and crockery knocked off shelves. An eyewitness 110 km from the blast reported:

Suddenly . . . the sky was split in two and [about 50°] above the forest the whole northern part of the sky appeared to be covered with fire. . . . I felt great heat as if my shirt had caught fire . . . there was a . . . mighty crash. . . . I was thrown onto the ground about [7 m] from the porch. . . . A hot wind, as from a cannon, blew past the huts from the north. . . . Many panes in the windows [were] blown out, and the iron hasp in the door of the barn [was] broken.

Probably the closest observers were some reindeer herders asleep in their tents about 80 km (49 mi) from the site. They and their tent were blown into the air and several of them lost conciousness momentarily. "Everything around was shrouded in smoke and fog from the burning fallen trees."

The cause of this remarkable event was a collision between the earth and a relatively modest bit of interplanetary debris, of which there are many examples circling the sun. The Siberian impact is the largest-scale event of this type known in well-recorded history; probably it is the largest impact to have occurred on land in the last few centuries.

Smaller impacts have been recorded more often; interplanetary stones fall from the sky in various locations every year (sometimes remote and sometimes inhabited—several houses have been hit in recent decades). Tiny dust grains that burn up in the atmosphere before hitting the ground can be seen every night—they are sometimes called "shooting stars."

Two conclusions are apparent. First, events ranging from small meteorite impacts to large, rare impact-explosions may have occurred in the ancient past, perhaps influencing the beliefs of ancient people about forces in the sky. For example, the well-preserved Arizona impact crater was created roughly 20,000 years ago by an interplanetary body (even bigger than the Siberian one) that exploded with hundreds of times the energy of the Hiroshima atomic bomb.

The second conclusion is that the solar system contains not just the planets and satellites we have already discussed. Floating between the planets are many different sizes of cosmic debris, collectively known as the **meteoritic complex.** They have been divided into different types depending on different properties that can be observed or inferred.

Comets are icy worlds a few kilometers across. **Asteroids** are rocky-metallic worlds ranging from a few hundred meters or less up to 1,000 km in diameter. **Meteoroids** are still smaller bits of debris, probably dislodged from comets and asteroids, floating in their own orbits in space.

Some of these bodies have planet-like orbits that are nearly circular, prograde, and lying in the plane of the solar system. Others have very different orbits, highly elliptical, perhaps retrograde, and perhaps inclined at steep angles to the solar system's plane.

The ones whose orbits cross the orbits of the earth and other planets can collide with the planets. When they collide with the earth, they are heated by friction with the atmosphere (just as returning space vehicles are). **Meteors** are the smaller ones (usually up to a few centimeters) that burn out completely before striking the ground. **Meteorites** are the larger ones that survive and fall to the ground: they are found to have rocky or metallic composition.

Since most of these bodies are much smaller than planets and satellites, you might suppose that they are much less important in our attempt to understand the origin and history of the solar system. Not so. The meteorites, especially, are free samples of other planetary worlds. Their properties testify that they broke off of larger bodies, called **parent bodies,** in the remote past.

All of these bodies—comets, asteroids, and meteoroids of all sizes—are debris left over from the origin of the solar system, 4.6 billion years ago, just before the present planets formed. At that time the solar system was filled with innumerable small, preplanetary bodies, ranging up to 1,000 km across, known as **planetesimals.**

These bodies even suggest the conditions in which the sun itself formed, and by implication, the formative conditions for the other stars. Thus, surprisingly, these smallest bodies of the solar system provide a link with the most distant stars, which will be a topic of much of the rest of this book.

Comets

Comets are the most spectacular of the small bodies in the solar system. They drift slowly from night to night among the stars, often with remarkable appearance. They have several parts. The brightest part is the **comet head,** which consists of a starlike point, called the **comet nucleus,** surrounded by a bright, diffuse glow, called the **comet coma.** The **comet tail** is a fainter glow extending out of the coma, usually pointing away from the sun. Although the tail of a typical comet can be traced by the naked eye for a few degrees, binoculars or long-exposure photos may reveal fainter

▲ *Figure 10·1* **Comets are a common spectacle in the sky. This view of Comet Ikeya-Seki was made in 1965 with a stationary 35 mm camera, 20-second exposure at f1.9 on Tri-X film.** (Steven M. Larson)

▲ *Figure 10·2* **Comet Kohoutek, January 1974. Photo made with 42-cm Schmidt telescope; 10-minute exposure.** (NASA)

parts tens of degrees long, or even extending clear across the night sky.

In 1704 the English astronomer Edmund Halley discovered that comets travel on *long, elliptical orbits* around the sun and that certain comets re-appear. Calculating the orbits of 24 well-recorded comets by Newton's methods, Halley found that four comets (seen in 1456, 1531, 1607, and 1682) had the same orbit and a peri-odicity near 75 years. The orbit passed far beyond Jupiter and Saturn, but approached the sun nearer than the earth does.

Halley correctly inferred that he was dealing with a single comet, and that its slight irregularities in peri-odicity were caused by gravitational disturbances of the planets, especially Jupiter. Halley predicted that this comet would return about 1758. It did so on Christmas night, 1758. It was named Halley's comet[1] and is due again about 1986.

---

[1]Comets discovered today are first named for their discoverers but are later given scientific names in order of their passage around the sun, for example, Comet 1978 I.

Comets as Omens

Halley's discoveries helped to dispel earlier fears that comets were evil, fearsome omens. For example, the appearance of Halley's comet in A.D. 66 was said to have heralded the destruction of Jerusalem in A.D. 70. Five circuits later it was said to mark the defeat of Attila the Hun in 451. In 1066 it presided over the Norman conquest of England. In 1456, its ap-pearance coincided with a threatened invasion of Europe by the Turks, who had already taken Constantinople three years before. Pope Calixtus III ordered prayers for deliverance "from the devil, the Turk, and the comet." Of course, we now recognize that noteworthy events are likely by chance in most cultures every decade or so, so that comets are no longer regarded as portents.

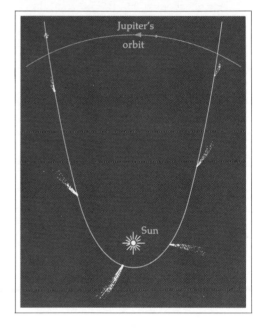

▲ *Figure 10·3* **The orbit of a typical comet brings it from the outer solar system into the inner solar system where it loops rapidly around the sun and moves outward again. A tail structure develops only in the inner solar system as the nucleus is heated by the sun. The diagram shows how the tail trails during approach to the sun, but leads as comet recedes.**

Figure 10·4 **Comet West, photographed in March 1976. Note the complex tail structure, including a straight segment (lower) and a broad curving segment (upper). Trees are silhouetted in lower left; part of the Milky Way can be seen in upper right.** (Steven M. Larson)

The erroneous belief in omens probably delayed a methodical investigation of comets' nature and behavior. Connecting each comet with a specific earthly event kept observers from realizing that some comets, like Halley's, return periodically.

Comets' Orbits

In ancient times no one knew how far away comets were, and many thought they were phenomena in our own atmosphere. Seneca, the Roman contemporary of Jesus, wrote:

> *Some day there will arise a man who will demonstrate in what regions of the heavens the comets take their way. . . .*

That man was Tycho Brahe, who in 1577 arranged observations of a bright comet from two different locations. Finding no parallactic shift in the comet's angular position relative to the stars (see Chapter 2), Tycho correctly concluded that it was more distant than the moon, and thus not terrestrial.

Studies of cometary orbits by Halley and others revealed several important characteristics. First, orbits are in-

Figure 10·5 **Comet Humason, showing rayed coma structure and filamentary tail. This 60-minute exposure tracked the comet's motion, causing background star images to blur.** (Naval Observatory, Official U.S. Navy photograph)

clined at nearly random angles to the plane of the solar system. Thus, comets populate a roughly spherical volume of space around the sun, rather than being confined to a disk-like shape as the planets are.

From statistics of comets' orbits, Dutch astronomer Jan Oort discovered around 1950 that this spherical swarm of comets is very extensive, and that most comets spend most of their time about 50,000 AU from the sun. This swarm, containing millions of comet nuclei, is called the **Oort cloud.** Here, comets are so far from the sun that gravitational attractions of nearby

stars, as well as the sun, are important. Following Kepler's and Newton's laws, comet nuclei drift slowly at these great distances, being disturbed by stellar forces, until they eventually fall back into the inner solar system.

As Kepler's laws predict, comets swing very rapidly around the sun, usually passing among the planets in only a few years. Only then can they be seen by earthbound observers. Then they swing slowly back toward the Oort cloud, where they may spend hundreds or thousands of years. Occasionally some of these comets pass near Jupiter or other planets and are redirected by gravitational forces into orbits that remain in the inner solar system. Many of the best-known comets, which return every few decades, are of this type.

No comets have been observed to reach or leave the solar system with speeds much greater than escape velocity. Therefore, they are all believed to be truly members of the sun's family (probably formed along with the planets) and not to come from or escape to regions as distant as the stars.

Studying Comets by Spectroscope

Although possible spacecraft missions to comets are being planned, no instruments or humans have visited a comet yet. How, then, can we tell the physical nature of a comet? The answer is to use the most important instrument in telescopic astronomy: the spectroscope.

Different kinds of gases, liquids, and solids emit and reflect different amounts of light in various colors. The intensity of light of each color depends on the composition of the object. The **spectroscope** breaks the light into its constituent colors and measures the amount of light in each color interval. (Since each color corresponds to a certain wavelength of light, astronomers usually use the term *wavelength* instead of color.) The array of all colors, in order of wave-

length, is called the **spectrum** of the object (see Color Photo 30).

If the material of comets were made up of solid particles, the comet would produce no light of its own and the spectrum would be merely that of reflected sunlight. But if diffuse material were glowing gas, then the spectrum would be more complex, emitted in certain colors (called **emission bands**) that depend on the composition of the gas.

This test was first carried out on a comet's coma in 1868, by the English astronomer William Huggins (quoted by Shapley and Howarth, 1929):

Figure 10·6 **Three views of the head of Comet Bennett (1970 II) at the same scale illustrate some problems of astronomical photography. The left photo shows a typical long exposure, recording faint details but overexposing the coma. The central and right views are two prints of an exposure on special low-contrast film, revealing the great intensity of the nucleus relative to the coma. The true nucleus would be a microscopic dot.** (Alan Stockton, University of Hawaii)
▼

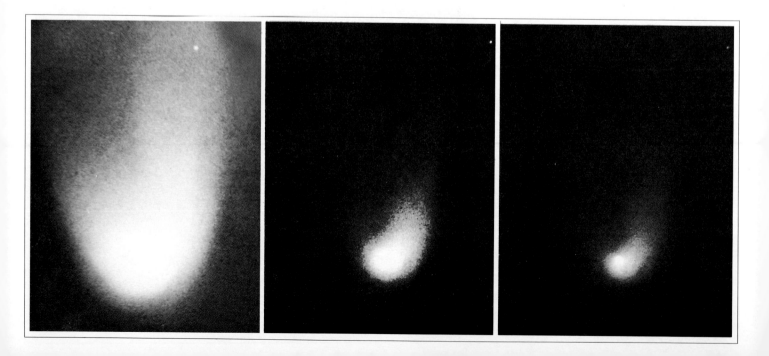

The comet's light was resolved into a spectrum of three bright bands [which] agreed in position with three [bands] in the brightest part of the spectrum of carbon. . . . The comet, though 'subtle as the Sphinx,' had at last yielded up its secret. The principal part of its light was emitted by luminous vapor of carbon.

Most of the light of comas, then, is emitted by *gas*—molecules and ions (charged molecules or atoms) originating in the nucleus and traveling out into the coma at velocities of 0.1 to 3 km/sec. Once the atoms or molecules break loose from the solid nucleus, they are exposed to solar radiation, which may alter the electron structure of the atoms or knock electrons out of them. These processes

Figure 10·7 **Linear gaseous streamers in the head are shown in this telescopic photo. The nucleus and central coma are much brighter than the other structure, and are consequently overexposed.** (University of Michigan Observatory/Cerro Tololo) ▼

lead to the radiation from the atoms or ions of gas. The carbon Huggins found was only one bright constituent, not necessarily the most abundant. Modern spectroscopes have identified neutral molecules and atoms such as C (carbon), CN, CH, and Na; and ions such as CH^+, OH^+, N_2^+, CO^+, CO_2^+.

Spectroscopic study of comet tails reveals two types. Type I are relatively straight and composed of gas, mostly ionized. Type II are curved and composed of solid dust particles. Comet tails cover enormous distances, and may stretch all the way from one planet's orbit to another's. The tail's gas and dust are very thin. Caught in the "wind" of outrushing solar gas and radiation, the tail streams out behind the comet as the comet approaches the sun, but leads as the comet recedes from the sun. It is like the hair of a girl walking in a strong wind, streaming out behind as she walks into the wind, but in front if she reverses direction (Figure 10·3).

Comets' Nuclei

All the spectacular gas and dust of comets are released from the nucleus. The secret of a comet's nature, then, must be hidden in the nucleus, which is too small to see with the naked eye from earth. Evidence suggests the **composition of comet nuclei** follows what astronomer Fred Whipple calls the **dirty iceberg model**—chunks of frozen gases several kilometers across, laced with bits of silicate dust.

The first clue in this direction is that the gases released from the nucleus are rich in carbon (C), hydrogen (H), oxygen (O), and nitrogen (N). In the cold, outer parts of the solar system, from which comets come, these gases would most likely exist

as ices such as frozen methane (CH_4), ammonia (NH_3), and water (H_2O).

A second, consistent, clue is that comets remain inactive—frozen—in the outer solar system, where it is too cold for the gases to be released. Calculations indicate that only within about 3 AU from the sun would ices become warm enough to release appreciable amounts of gas; observations show this is just the distance where most comets begin to develop gaseous comas and tails.

A third clue is that comet nuclei are fragile. Several have been observed to break into pieces as they approached the sun, apparently due to the stretching tidal forces of solar gravity. This might be expected for icy bodies weakened by gas loss, but seems unlikely for solid rocks. Irregular motions of comets have also been observed and attributed to jetlike outbursts of gas produced as ice chunks sublime or break off. The evidence for solid particles embedded in the nucleus includes the dust particles observed in Type II tails, and the fact that some groups of meteors apparently originated in comets.

The Origin of Comets

Jupiter's massive atmosphere, trapped since the solar system's origin by its gravity, shows that gases such as C, H, O, and N were common in the early solar system. Calculations indicate that in the cold outer regions of the early solar system, these gases probably condensed into icy planetesimals made of frozen CH_4, NH_3, and H_2O—just the materials believed to occur in comet nuclei. Chemical calculations indicate that small silicate particles would also form and be included in the icy structures, thus explaining the "dirt" in the "dirty icebergs."

Comets, then, are probably survivors of this early population of small bodies formed in the outer solar system. How were these objects preserved? Study of the orbits shows that as comets passed near Jupiter and the other giants, their gravity would have flung many of the comets far beyond Pluto, thus forming the Oort cloud. Here, their motions would be affected not only by the sun but also by the gravity of passing stars still farther away. After billions of years, the ones that fall back into the inner solar system are being discovered and observed.

Meteors and Meteor Showers

Meteors that flash momentarily across the sky might seem totally unrelated to comets. However, a study of their frequency on different nights reveals a direct connection. On an average night you may see about 3 meteors/hour before midnight and about 15 meteors/hour after midnight.[2] But on certain dates each year you may see **meteor showers** of 60

---

[2]You see more meteors after midnight because you are then located on the leading edge of the earth as it moves forward in its orbit, sweeping up interplanetary debris.

Table 10·1 Properties of Selected Small Bodies in the Solar System

Class	Example	Diameter (km)	Orbit Semi-major Axis (AU)	Orbit Eccentricity	Orbit Inclination (degrees)	Remarks
Comets						
Short-period	Encke	−2	2.2	0.85	12	Probably icy
	Halley	few?	18	0.97	162	
Long-period	Kohoutek	few?	very large	1.0	14	Probably icy
Meteors						
Shower	Perseid	10^{-6}	40	0.97	114	Cometary debris
	Taurid	10^{-6}	2.2	0.80	2	
Fireballs	July 31, 1966	?	32	0.98	42	May be related to comets or to asteroids
	May 31, 1966	?	3.0	0.80	9	
Asteroids						
Belt	1 Ceres	1003	2.8	0.08	11	
	2 Pallas	608	2.8	0.23	35	Probably rocky
	3 Juno	247	2.7	0.26	13	
	4 Vesta	538	2.4	0.09	7	
	14 Irene	170	2.6	0.16	9	
Trojan	624 Hektor	100 × 300	5.1	0.02	18	Unusual shape
Apollo	433 Eros	7 × 19 × 30	1.5	0.22	11	Elongated
	Apollo	?	1.5	0.56	6	Lost before orbit precisely measured
Meteorites (chondrites)	Pribram	10^{-4}	2.5	0.68	10	
	Lost City	10^{-4}	1.7	0.42	12	
	Leutkirch[a]	10^{-4}?	1.6	0.40	2.5	

[a]An object photographed over Europe in 1974. No fragments recovered as of late 1974, but believed to be a stone meteorite.

Sources: Data from Chapman and Morrison (1974); Hartmann (1972, 1975); Gehrels (1972).

meteors/hour or more, all radiating from one direction in the sky (Table 10·2).

The best-known example is the Perseid shower, which occurs every year around August 12, when bright meteors streak across the sky every few minutes from the direction of the constellation Perseus. (A shower is named for the constellation most prominent in the area of the sky from which the shower radiates, called the radiant.) Occasionally the showers are so intense that meteors fall too fast to count. During the Leonid shower of November 17, 1966, meteors fell like snowflakes in a blizzard for

Figure 10·8 **A portion of the solar system showing orbits of a few selected interplanetary bodies including comets, main-belt asteroids, Trojan asteroids, Apollo asteroids, meteorites, and large meteors, or fireballs.**
▼

about half an hour, at a rate estimated to be more than 2,000 meteors per minute (see Figure 10·9).

What is the connection between the showers and the comets? In 1866 G.V. Schiaparelli (of Martian canal fame) discovered that the Perseid meteor shower occurred whenever the earth crossed the orbit of Comet 1862 III. The Perseids, then, must be spread out along the orbit of that comet. Other relationships were soon found

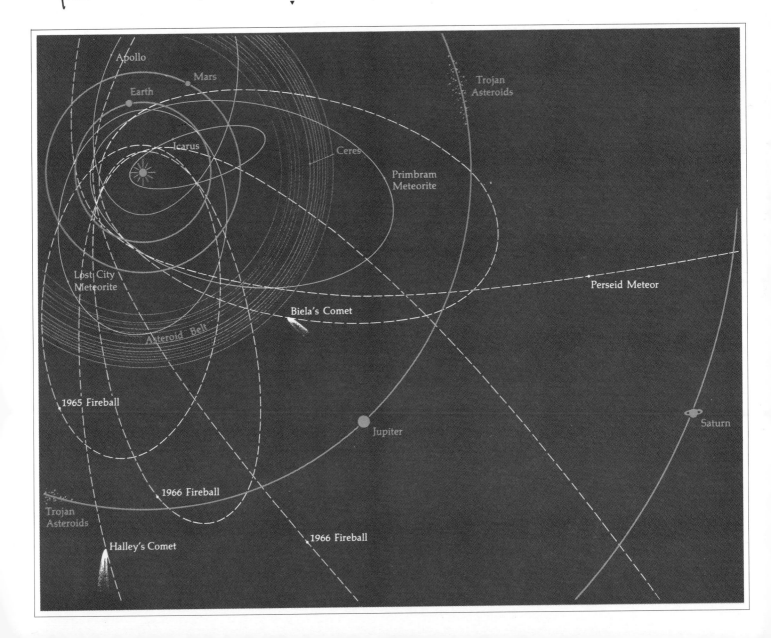

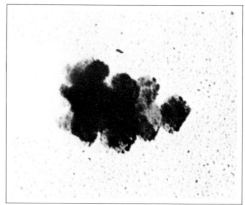

◀ *Figure 10•9* **A rare meteor shower: the Leonids of November 17, 1966. The rate of meteors visible to the naked eye was estimated to exceed 2,000 per minute. The brightest star (upper left) is Rigel, in the constellation Orion. This exposure of a few minutes' duration was made with a 35 mm camera.** (D.R. McLean)

Figure 10•10 **An example of a microscopic meteoroid particle (diameter 0.3 micron) collected by rockets at altitudes of 75–150 km. Most such particles are nonmetallic and are believed to be debris from comets.** (Dudley Observatory)
▼

between specific meteor showers and specific comets, as Table 10.2 shows. Even the orbits of individual sporadic meteors, tracked photographically, often resemble long- or short-period cometary orbits. Therefore, *most meteors must be small bits of debris scattered from comets.*

Since the 1960s, rockets and balloons have collected microscopic fragments believed to be meteoroids. As shown in Figure 10•10, they are irregular glassy silicate and metallic particles. If they are of cometary origin, they support the theory of comet nuclei as "dirty icebergs" with bits of entrapped

Table 10•2 Dates of Prominent Meteor Showers

Shower Name (Constellation)	Date of Maximum Activity Each Year[a]	Associated Comet
Lyrid	April 21, morning	1861 I
Perseid	August 12, morning	1862 III
Draconid	October 10, evening[b]	Giacobini-Zinner
Orionid	October 21, morning	Halley
Taurid	November 7, midnight	Encke
Leonid	November 16, morning	1866 I
Geminid	December 12, morning	?

[a]Showers can last several days before and after the peak activity on the listed date. Observations are best when the constellation in question is high above the horizon, usually just before dawn.

[b]The Draconids are now weak because their orbits have been disturbed by gravity of planets, but further disturbances may again strengthen the shower in the future.

grit. They are too small to reach the ground, but occasional large ones, called **fireballs**, are very bright and spectacular.

Asteroids

Asteroids are the largest of the interplanetary bodies, ranging from the largest, Ceres (about 1,000 km in diameter), down to a few hundred meters across and probably less. Spectral studies show they are rocky, often with metal embedded in them. They appear in various parts of the solar system, but are most abundant in the **asteroid belt**—a region between Mars and Jupiter.

Discovery of Asteroids

Asteroids have played an interesting role in the mapping of the solar system. Too small normally to be seen by the naked eye, they were unknown before 1800. Bode's rule, confirmed by the discovery of Uranus in 1781, called for a planet at 2.8 AU from the sun, in the large space between Mars and Jupiter. Therefore, astronomers set out to find the "missing planet" in 1800; success came on the first night of 1801 when Ceres was discovered, just at 2.8 AU[3].

Between 1802 and 1807, three more small, planet-like bodies turned up between 2.3 and 2.8 AU from the sun. Because of their small size, they came to be called minor planets, or *asteroids*,

---

[3]This discovery reportedly came just at the time when the philosopher Hegel had "proved" philosophically that there could be no more than the seven then-known planetary bodies, Mercury through Uranus. Which suggests that going out and looking is worth more than sitting at home speculating.

a name which now applies to any interplanetary body that is not a comet (not known to emit gas). By 1890, 300 asteroids were known. In 1891, German astronomer Max Wolf began searching for them photographically by time exposures, detecting many new asteroids by their tell-tale motion among the stars, as shown in Figure 10·11.

Asteroids are known by numbers (assigned in order of discovery, but only after the orbit has been accurately identified) and a name (chosen by the discoverer); examples are 1 Ceres, 2 Pallas. The names cover a wide range of human interests, from mythology (1915 Quetzalcoatl) and politicians (1932 Hooveria!) to spouses and lovers. The 1977 catalog of asteroids contained 1,940 with known orbits, but perhaps 100,000 observable ones remain uncharted.

Asteroids outside the Belt

Several subgroups of asteroids have orbits outside the main asteroid belt. **Apollo asteroids,** for example, cross the earth's orbit. They are named after Apollo, first of the group to be discovered. About two dozen are known, including one as small as 800 meters across. Many approach rather close to the earth. In 1968, Icarus passed within about 6 million km (4 million mi),[4] and was observed to be a ball about 1 km across, rotating in 2^h15^m. Others are irregular in shape.

Apollos generally last only a few hundred million years before crashing

---

[4]Icarus' approach touched off scare articles in the tabloids. Reporters claimed it might crash into earth even though astronomers had applied Kepler's laws and predicted a miss after observing the orbit. Icarus obeyed Kepler's laws and missed the earth.

into earth or other planets and creating impact craters. Yet some still exist, eons after the formation of the solar system, so they must be replenished. And indeed, they are almost certainly asteroids that have been thrown out of the belt by gravitational forces of Jupiter. New ones will probably be ejected in the future.

Mars-crossing asteroids are similar to Apollos but cross only Mars' orbit, not earth's. Collisions with such asteroids may have caused many of the craters on Mars.

Trojan asteroids[5] lie in two swarms in Jupiter's orbit, 60° ahead of and

---

[5]The name *Trojan* comes from the tradition of naming asteroids after heroes in Homeric epics.

Figure 10·11 **Two images of asteroid 887 Alinda (center). A. The telescope was guided on asteroid; star images trailed. B. The telescope was guided on stars, showing movement of asteroid during exposure.** (New Mexico State University Observatory)
▼

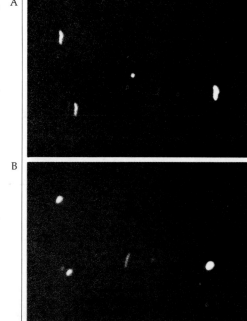

A

B

60° behind the planet, called **Lagrangian points.** The astronomer Joseph Louis Lagrange discovered that particles can be held in the two swarms by Jovian and solar gravitational forces. At least 45 Trojans have been found at one Lagrangian point, and observers predict a total of about 700 observable Trojans. Because they are darker in color than most asteroids, they probably have a different composition (possibly more carbon and internal ices?), and may have formed near Jupiter, not in the belt. One unusually large Trojan is Hektor, estimated to be 100 km wide and 300 km long, tumbling end over end in Jupiter's orbit.

Physical Nature of Asteroids

Spectroscopic observations show that asteroids are chunks of rocky material, often with embedded metals. Spectra show that the rocky material is basalt-like lava in some cases (resembling lunar, terrestrial, and Martian

Figure 10·12 **No one has observed an asteroid from close range, but most astronomers suspect asteroids resemble the 22-km Martian satellite Phobos. Different asteroids** have different compositions and histories, but many have irregular shapes and perhaps cratered surfaces, as indicated in this close-up view of Phobos, made by cameras in the Viking spacecraft. The smallest craters shown here are house-size. (NASA)

lavas). In other cases the rock is more water-rich material that has apparently never been melted. Metallic material is probably nearly pure nickel–iron alloy, similar to metal meteorites. The commonly irregular shapes of asteroids suggest that many may be fragments of larger bodies. Their possible close-up appearance is suggested by Figure 10·12.

Mines in the Sky

Because some asteroids probably have iron-rich metal portions (similar to the nearly pure nickel–iron meteorites discussed later), and because some come very near the earth, asteroids may someday be economically exploited. Studies are already underway to examine the feasibility of flying to Apollo asteroids for scientific as well as economic exploration. Massive blocks of nickel-iron might be mined in space, brought to space stations for processing, or shaped into crude entry vehicles and landed on earth in remote areas. Masses worth billions of dollars—enough to supply earth's iron needs for decades—might be obtainable (McCord, 1976). Other asteroid materials such as rock, water, and hydrogen may be used for construction of space stations, possibly designed to generate power to be beamed to earth, according to NASA studies in 1977 (Figure 10·13).

Origin of Asteroids

Asteroids are probably planetesimals that never finished forming a planet. Between Mars and Jupiter, as the solar system was taking shape, Ceres-size planetesimals formed from rocky materials that condensed in the inner,

warmer parts of the solar system. Instead of collecting more material and forming a planet, they collided and broke into thousands of fragments. Most of these became belt asteroids; some became Apollos and meteorites. A similar process near Jupiter's orbit may have produced the Trojans.

Meteorites

Meteorites are stony and metallic objects that fall from the sky. Modern evidence is that they are fragments

Figure 10·13 Beyond the moon, the closest planetary bodies are Apollo asteroids. Small Apollos, a few hundred meters across, may approach close enough to the earth–moon system (upper left) to be visited by shuttle-class spaceships. They could become the next planetary bodies to be visited by astronauts, shown here in transit from a "parked" spaceship to the asteroid surface. Some researchers and NASA planners believe their materials could be economically exploited for use on or near the earth. (Painting by author)

of Apollo-type asteroids that cross earth's orbit and eventually collide with earth.

Meteorites as Venerated Objects

Stones from the sky have long been a source of awe. Meteorites have been buried with American Indian artifacts, (Figure 10·14) wrapped in "mummy cloth" (Casas Grandes, Mexico), and kept as a sacred possession of an Alaskan tribe. A stone venerated in the Temple of Diana at Ephesus (one of the seven wonders of the ancient world) reportedly fell from the sky and was probably a meteorite, as is the sacred black stone in Mecca, enshrined around A.D. 600 or earlier (Sagan, 1975). When a meteorite fell at Ensisheim, France, in 1492 (Figure 10·15), the Emperor Maximilian, in residence nearby, decided he should go on a Crusade. The main mass of the mete-

Figure 10·14 **Burial of the Winona stone meteorite in a crypt constructed by prehistoric Indians in northern Arizona.** (Museum of Northern Arizona, Anthropological Collections; photo by author)

orite, suspended in the Ensisheim church, was regarded with reverent awe by the peasants.

Scientific Discovery of Meteorites

In 1794 a German physicist, E.F.F. Chladni, reported that the Ensisheim and other supposed celestial stones seemed similar to each other, and different from normal terrestrial stones. He concluded that these "meteorites did indeed fall from the sky." This started a controversy among naturalists. Upon hearing that a meteorite had fallen in Connecticut, Thomas Jefferson, himself an accomplished naturalist, is supposed to have said, "It is easier to believe that Yankee professors would lie than that stones would fall from heaven."

The French Academy, the scientific establishment of the day, dismissed as superstition the notion of stones from the sky. As luck would have it, a meteorite exploded over a French town in 1803, pelting the area with stones. Cautiously, the Academy sent the noted physicist J.B. Biot to investigate. His report, one of the historic documents of science, methodically constructed an irrefutable chain of evidence, using eyewitness accounts, measurements of the 2 × 6 km area of impacts, and specimens of the meteorites themselves. This report established that stones can fall from the sky.

Meteorite Impacts on Earth

Interplanetary debris collide with the earth at very high speeds, usually 10 to 60 km/sec (22,000 to 135,000 mph). At such speeds, material is

Figure 10·15 **An old woodcut showing the fall of a meteorite near the German town of Ensisheim (left), France, in 1492.**

heated by friction with the air. Dust grains and pea-size pieces burn up before striking the ground. Larger pieces are usually slowed by drag and strike the ground as if they had been dropped from an airplane. They pass through the atmosphere too fast for their interiors to be strongly heated, and stories of meteorites remaining red-hot for hours after falling are untrue (Figure 10·17).

The larger the meteorite, the rarer. Only a few brick-size meteorites are recovered each year. In 1972 an object weighing perhaps 1,000 tons just missed the earth, skipping off the outer atmosphere; it was filmed from the ground and detected by Air Force reconnaissance satellites. Ten-thousand-ton objects, large enough to cause nuclear-scale blasts, like the Siberian object of 1908, may fall centuries apart. Larger blasts may form km-scale craters many thousands of years apart.

Since meteorites are "free samples" of planetary matter, recovering and reporting a meteorite is a rare honor. Meteorite discoveries should be presented, or at least lent, to researchers; for example, at the Smithsonian Institution in Cambridge, Massachusetts and Washington, D.C., or the Center for Meteorite Studies, Arizona State University, Tempe, Arizona.

Physical Properties of Meteorites

Meteorites are probably the most complex materials studied by astronomers and geologists. There are **stony meteorites,** metallic types called **iron meteorites,** and types with both rock and metal, called **stony-iron meteorites.**

Some types have been melted—they could be called igneous rocks—but other types have never been melted and seem to have formed from mineral grains brought together in free space. Some types have mineral grains originally formed in very different environments, but jammed into contact with each other. Other types show shocks and fractures indicating that they have suffered collisions at high speeds.

Origin of Meteorites

All these characteristics, and others, indicate that meteorites come from the interiors and surfaces of larger

◄ *Figure 10·16* **NASA scientists examine the Lost City meteorite, a 10-kg stone that fell in Oklahoma on January 9, 1970. The meteorite was photographed by an automatic camera system, and was found about a kilometer from the impact point calculated from the photographs.** (NASA)

bodies, called parent bodies. Chemical studies show that most meteorites come from as few as 4 to 30 parent bodies. Properties of iron meteorites indicate that their parent bodies were at least several hundred kilometers across.

Spectral studies show that the surface materials of asteroids are very similar or identical to meteorites. All these facts indicate that meteorites originated inside asteroids or asteroid-like bodies which broke apart during collisions and scattered their fragments through interplanetary space, as shown in Figure 10·18.

Meteorites and the Early Solar System

In their chemistry and structure, meteorites reveal conditions that must have existed in the early solar system and inside planetesimals while the planets were forming. Some meteorites have been little altered since that time. Table 10·3 shows a classification of meteorite types, starting with the least-altered kinds, which are especially useful in revealing early conditions.

The **carbonaceous chondrites** are a stony type that has never been subjected to very much heat. The term *chondrite* comes from tiny, BB-size silicate globes, called **chondrules,** shown in Figure 10·19. Chondrules seem to have formed as molten droplets ejected during collisions at the

◄ *Figure 10·17* **The effects of heat on meteorites passing through the atmosphere are illustrated in this portion of an "artificial meteorite," the heat shield of the Gemini 4 space capsule. In this section, roughly $\frac{1}{2}$ m across, burn and flow damage can be seen in the honeycomb structure of the heat shield, radiating from the central, leading point as the capsule reentered.**

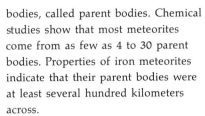

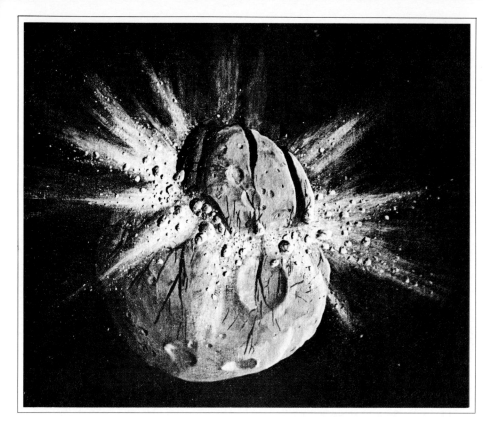

◄ Figure 10·18 **An imaginary view of the collision of two asteroidlike parent bodies of meteorites. The diameter of the larger body is about 50 km; the debris includes pulverized dust and rocks of various sizes that may ultimately fall on the earth and planets. Fractures through the bodies may cause them to fall apart during this or later collisions.** (Painting by author)

very beginning of the solar system.

Carbonaceous chondrites contain large amounts of carbon, water, and other volatiles that would have been driven off with the slightest heating beyond about 500 K. Thus, carbonaceous chondrites are our best samples of **primitive planetary material.** Though they make up only about 6 percent of all meteorite falls, spectroscopic studies of asteroids reveal carbonaceous chondrite compositions for about 80–90 percent of all asteroids in the outer part of the asteroid belt. That such a small percentage reaches earth probably indicates how difficult it is for asteroids to get all the way from the outer belt into earth-crossing orbits.

The type of meteorite that hits earth in the greatest number is the **chondrite,** a stony type with fewer volatiles than the carbonaceous chondrites. These have been heated in their parent bodies to temperatures high enough to drive off volatiles, but not high enough to melt the rock and destroy the chondrules. Temperatures once around 1,000 K have been suggested for some of these. Thus, all chondrites formed long ago into rocky interplanetary bodies that never melted.

Achondrites are stony meteorites that were heated enough to melt them and destroy their chondrules. Some have textures very similar to basaltic lavas from the earth and moon. They give evidence of volcanic processes on

Table 10·3 Types of Meteorites

Meteorite Type	Percentage of All Falls[a]	Remarks
Stony		
Carbonaceous chondrites	5.7	Most primitive, least-altered material available from the early solar system.
Chondrites	80.0	Commonest type. Defined by millimeter-scale spherical silicate inclusions, sometimes glassy, called *chondrules.*
Achondrites	7.1	Most nearly like terrestrial rocks. Defined by lack of chondrules, which have been destroyed by a heating process. Some resemblance to terrestrial and lunar igneous rocks of basaltic type.
Stony-irons	1.5	Contain stony and metallic sections in contact with each other.
Irons	5.7	Nickel-iron material. Museum specimens are often cut, polished, and etched to show interlocking crystal structure.
	100%	

[a]This table is based on meteorites called *falls:* those actually seen to fall. Meteorites found by chance in the soil, called *finds,* are more numerous but less valuable statistically because they are biased toward iron meteorites, which attract attention whenever found in the ground, while stone meteorites eventually weather to resemble ordinary stones.

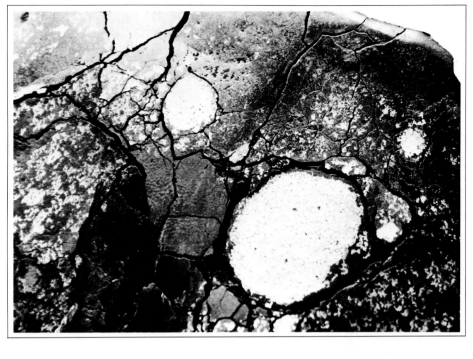

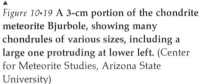

Figure 10·19 A 3-cm portion of the chondrite meteorite Bjurbole, showing many chondrules of various sizes, including a large one protruding at lower left. (Center for Meteorite Studies, Arizona State University)

Figure 10·20 Cut and polished section of a stony-iron meteorite, from the Bondoc Peninsula, Philippines. The dark fractured matrix is silicate rock, possibly fractured in collisions. The bright nodules are nickel-iron inclusions, a few cm across.

Figure 10·21 A cut and polished section of the Campo del Cielo iron meteorite, Argentina. Intersecting metal crystals can be studied to reveal environmental conditions inside the parent body. The dark inclusions are stony silicate bodies, the largest being about 4.5 cm across. This pattern of metal crystals is called the Widmanstätten pattern, after its discoverer.

the parent bodies of some meteorites.

Stony-irons (Figure 10·20) may come from farther within the parent bodies, where melted stony and metallic material coexisted. *Irons* (Figure 10·21) may be pieces of nickel-iron cores formed in the parent bodies, or they may be solidified pieces of molten nickel–iron pools that never drained completely to the parent bodies' centers.

How were parent bodies heated? According to one model, the heat came from radioactivity, and the parent bodies melted from the inside out. Outer regions that never melted are the source of chondrites. Inner regions melted, with heavy molten metal draining to the center, forming nickel–iron cores and achondritic mantles. Later fragmentation of the parent

bodies released the various kinds of material that appear as meteorites.

Dating Interplanetary Material

Meteorites directly dated by radio-isotopic techniques (described in Chapter 3) reveal that meteorite parent bodies formed during a few-million-year interval 4.6 billion years ago. This date is believed to mark the formation of all interplanetary material, including asteroids and comets, as well as the planets, satellites, and the sun itself.

However, the dating techniques indicate that some meteorites took their present form more recently. A small group of meteorites called *nak-*

lites originated in a lava flow on an unknown parent body only 1.2 billion years ago. Other meteorites were broken off their parent bodies in collisions that occurred as recently as a few million years ago, as shown in Figure 10·18.

Zodiacal Light

The smallest particles of the meteoritic complex are microscopic dust grains and individual molecules and atoms, spread out along the plane of the solar system and concentrated toward the sun. If you look west in a very clear rural sky as the last glow of evening twilight disappears (or east before sunrise), you can detect sunlight shining off the cloud of these particles. It appears as the **zodiacal light**—a faint glowing band of light extending up from the horizon and along the ecliptic plane, shown in Figure 10·22. It is brightest closest to the horizon. Measurements made by astronauts indicate that it merges with the bright glow of the sun's atmosphere.

Siberia Revisited: A Comet Impact?

We now find the mysterious Siberian explosion of 1908, which we described at the beginning of this chapter, easier to understand. Interplanetary space contains debris of many sizes. Rocky and icy chunks up to a few kilometers in size cross earth's orbit, and are likely to collide with earth sometime in the future. One of these objects apparently struck the atmosphere over Siberia on June 30, 1908.

Other dramatic events in Russia at this time kept Russian scientists from

Figure 10·22 **Zodiacal light shows as a triangular glow extending up from the western horizon. This 38-minute exposure was made from an altitude of 4 km on Mauna Kea volcano in Hawaii. Venus is the bright object setting in the midst of the zodiacal light. The field of view is about 80° wide. This photo is printed at high contrast but with no other contrast enhancement; a dark, clear horizon is needed in order to see the zodiacal light.**

visiting the site until 1927. During the 1927 expedition and later expeditions, they found that, surprisingly, the object did not reach the ground and form a crater. It apparently exploded in the atmosphere. Trees at "ground zero" were still standing but had their branches stripped by the blast forces in a downward direction. Farther away, trees had been knocked over by the blast out to 30 km from ground zero as shown in Figure 10·23. A forest fire had started and trees had been scorched by the blast out to about 14 km. No stone or iron meteorites were found.

Most investigators believe the object was composed mostly of ices, and was part of a comet, probably some tens

Figure 10·23 **Devastation caused by the great Tunguska meteorite fall of 1908. This view, more than a decade later, shows trees blown over by the force of the explosion.** (Photo from E. L. Krinov, 1966)

of meters in diameter. It injected much dust high into the stratosphere. On June 30 and July 1, sunlight shining over the north pole illuminated this dust during the night, and newspapers could be read at midnight in Western Siberia and Europe.

Such was the effect of a relatively small bit of the meteoritic complex striking the earth. To put this event in perspective, if the same object had exploded over New York City, the scorched area would have reached nearly to Newark, New Jersey. Trees would have been felled beyond Newark and over a third of Long Island. The man knocked off his porch could have been in suburban Philadelphia. "Deafening bangs" might have been heard in Pittsburgh, Washington, D.C., and Montreal.

Figure 10·24 **Schematic histories of cometary and asteroidal material, showing condensation into multikilometer bodies (left) and subsequent perturbation and fragmentation to explain phenomena now observable.**
▼

The fact that such a blast of astronomical origin actually occurred in this century, together with the tenuous international balance of terror and suspicion, leads to serious questions about the consequences of a major meteorite explosion near a city. With each nation's nuclear response set at a hair-trigger level, care would be needed to avoid a response against an imaginary enemy! New reconnaissance satellites may provide early warning against such objects, which may hit earth every century or so.

On the more positive side, distortions of geologic strata by ancient impacts on earth have been economically valuable. Rich iron ore deposits at Sudbury, Ontario, occur in a prob-

able impact crater more than 100 km across and about 1.8 billion years old; the impact exposed native iron-bearing strata. Several oil deposits in North America and the USSR occur in sediments collected in ancient craters.

10

Summary

All of the solar system's small bodies—comets, meteors, asteroids, and meteorites—can be viewed as examples of the planetesimals or their fragments. These were the preplanetary bodies that formed in the solar system during

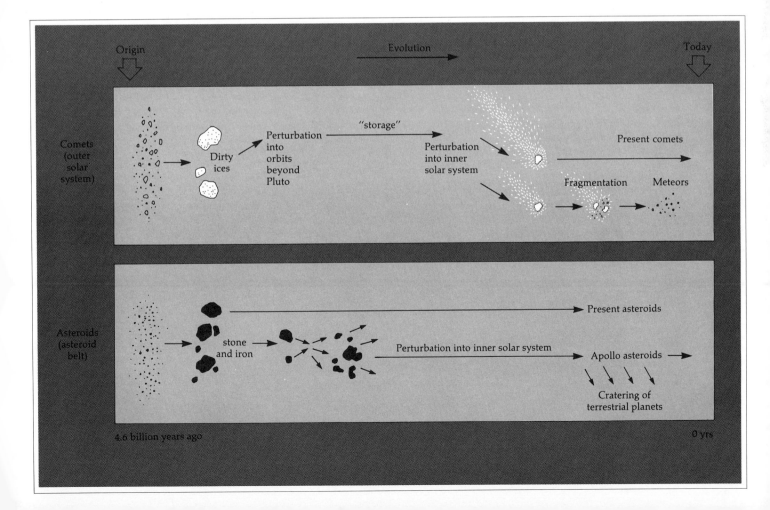

a short interval of a few million to 100 million years, 4.6 billion years ago.

Comets are probably icy bodies that formed in the colder, outer parts of the solar system. Rocky planetesimals probably formed of silicate and metallic minerals in the inner solar system, and the asteroids are their survivors. Most asteroids are now in a belt between Mars and Jupiter; others interacted with planets and were either ejected from the solar system after near-misses, or crashed onto planet and satellite surfaces, making craters.

The process of transforming icy, stony, and metallic planetesimals into the comets, belt asteroids, Apollo asteroids, meteors, meteorites, and craters that we see today is shown in Figure 10·24.

The most detailed information about these bodies comes from meteorites, whose chemistry reveals that they came from asteroid-like parent bodies a few hundred kilometers across. Many of these were melted, causing minerals to separate into iron and silicate phases. Others survive with only minimal heating, providing samples nearly unaltered since the solar system's earliest days.

Thus the small bodies of the solar system provide some of the best clues about the origin of planets, both in our solar system and perhaps near other stars. Recent studies indicate that, in addition, they may be exploited as sources of raw materials for use on earth and for space exploration.

Concepts

meteoritic complex

comets

asteroids

meteoroids

meteors

meteorites

parent body

planetesimals

comet head

comet nucleus

comet coma

comet tail

Oort cloud

spectroscope

spectrum

emission bands

composition of comet nuclei

dirty iceberg model

meteor shower

radiant

fireball

asteroid belt

Apollo asteroids

Mars-crossing asteroids

Trojan asteroids

Lagrangian point

stony meteorites

iron meteorites

stony-iron meteorites

carbonaceous chondrites

chondrules

primitive planetary material

chondrites

achondrites

zodiacal light

Problems

1. If a comet should happen to pass through Saturn's satellite system, why would it probably not be detected from earth?

2. Kepler's third law states that $a^3 = P^2$, where a is the semi-major axis of a body orbiting the sun (expressed in AU) and P is the period (expressed in years). Should this result apply to comets? If a typical comet in Oort's cloud has a semi-major axis of about 100,000 AU (10^5 AU), how often would it return to the inner solar system?

3. Antarctica has been the most fruitful hunting ground for finding fallen meteorites. Why? Does this mean more meteorites strike Antarctica than other continents? Why or why not?

4. In terms of measuring and reporting useful scientific information, what actions would be appropriate if you observed an extraordinarily bright meteor or fireball? In two columns, list examples of useful and nonuseful descriptions of the fireball's speed, brightness, and apparent size. What actions would be appropriate if you saw a meteorite strike the ground?

5. Summarize relations between the bodies and particles responsible for meteors, comets, the zodiacal light, asteroids, and meteorites.

6. Suppose future astronauts could match orbits with a comet and reach its nucleus. Describe the possible surface appearance of a comet. Consider gravity, surface materials, sky appearance, and so on. Which would be easier to match orbits with, a long-period or short-period comet?

7. Typical interplanetary material may move at about 15 km/sec relative to the earth–moon system.

a. *If a kilometer-scale asteroid were discovered on a collision course with earth at a distance of 15 million km, how much warning time would we have?*
b. *What would be the potential dangers?*
c. *If a much smaller asteroid were similarly discovered at the distance of the moon, how much warning time would we have?*
d. *How long would the objects take to pass through the 100-km thickness of the atmosphere?*

8. Based on everyday experience, what is the danger of an event such as in Problem 7, compared with the danger of other natural disasters such as earthquakes? Is a large-scale meteorite disaster a plausible source of myths during the 10,000-year history of humanity? Defend your answer.

Advanced Problems

9. Use the small angle equation to calculate how close a 2-km-diameter asteroid would have to come to earth to allow its shape to be resolved by a telescope that can reveal angular details one second of arc across. How does this distance compare with the moon's distance?

10. If typical long-period comets were found to have periods of the order 10 million years (10^7 years), and about ten such comets pass through the inner solar system each year, how many comets would you expect to exist in the Oort cloud? Could there be many less? Many more? Explain your reasoning. (Similar reasoning led to the hypothesis of the cloud by Oort and his associates.)

11. Suppose a small asteroid of pure nickel-iron, with radius $r = 100$ meters, could be located and exploited. If the value of the alloy were 90¢ per kilogram, what would be the potential

economic value of the asteroid? Compare this with the $20 billion dollar cost of the Apollo program. (Hint: mass in grams $= \frac{4}{3}\pi r^3 p$, where p is the density of the material, about 8 g/cm³. All values in this equation must be in cgs units.)

12. How many asteroids would be required to provide enough material to make one planet the size of the earth? Assume a typical asteroid is 120 km across and the earth is about 12,000 km across.

Projects

1. Use a large piece of cardboard to make a model of the inner solar system out to the orbit of Jupiter. Assume that the planets travel approximately in the plane of the cardboard. Use orbital properties listed in Table 10·1 to cut out scale models of orbits of various interplanetary bodies. (A slit through the first cardboard could be used to show how comet or asteroid orbits penetrate through the ecliptic plane. Note that the sun must always occupy one focus of each orbit.)
a. *Show how the geometry of passage of a comet (or other body) through the ecliptic plane, especially for highly eccentric orbits, depends on the angle between the perihelion point and the ecliptic plane, measured in the orbit. (This angle is fixed for each body, but is omitted from Table 10·1 for simplicity.)*
b. *Show how the prominence of a given comet may depend strongly on where the earth is in its orbit as the comet passes through the inner solar system.*

2. Visit Meteor Crater, Arizona. Why is this feature misnamed? Observe the blocks of ejecta and deformation of rock strata, as explained in museum signs and tapes. What would prehistoric observers, if any (estimated

impact date, 20,000 years ago), have witnessed at various distances from the blast that formed this crater, nearly a mile across? (Other impact sites are known in various states, but most have not been developed for visitors; many are in primitive areas. Visit such sites, if possible.)

3. Examine meteorite specimens in a local museum. Compare the appearance of stones and irons. Are chondrules visible in any stones? Heat damage? In iron samples that have been cut, etched, and polished, look for the crystal patterns that give information about cooling rate and environment when the meteorite formed inside its parent body.

Origin
of the
Solar System

The planets, satellites, meteoritic material, and sun did not exist before about 4.6 billion years ago. Samples from the earth, moon, and meteorites suggest that the solar system formed from pre-existing material within an interval of about 100 million years. Prior to that time, the atoms in the earth, in this book, and in your own body were floating in clouds of thin gas in interstellar space. How did the sun and planetary system form from that material?[1]

---

[1]Note that this question is not addressing the origin of the whole universe, which probably occurred around 13–20 billion years ago, or the origin of our galaxy, which probably occurred around 12–13 billion years ago. These topics will be taken up in later chapters.

Facts to Be Explained by a Theory of Origin

Nobel laureate Hannes Alfvén, who has spent years researching the solar system's origin, once said, "To trace the origin of the solar system is archaeology, not physics." He meant that our ignorance of the initial conditions forces us to work backward through time, reasoning from whatever clues we can find. The most important clues are *facts about the solar system that have no obvious explanation from present-day conditions or known physical laws.*

Table 11·1 lists some of these clues. Facts such as Kepler's laws are not included since they can be explained from Newton's laws of gravity. There are no obvious conditions or laws, however, requiring that the planets' orbits be nearly circular (Keplerian motion also allows ellipses) or aligned in nearly the same plane (Keplerian motion allows high inclinations); these facts are thus included in a list of facts to be explained by a successful theory of the origin of the solar system.

Catastrophic versus Evolutionary Theories

In many fields of science, theories of origin have historically been scattered through a range of possibilities, from so-called catastrophic theories to so-called evolutionary theories. **Catastrophic theories** tend to explain things through sudden or catastrophic events, often rare accidents or uncommon circumstances. **Evolutionary theories** tend to explain things as relatively gradual, normal outgrowths of previous conditions. For example, catastrophic theories have been proposed to explain the origin of mountains, life, biological species, or the

◄ *Figure 11·1* **This photo taken by Apollo astronauts symbolizes the basic problem of solar sytem origin: Why does a typical star like the sun have planets nearby? The photograph shows a solar eclipse caused as the earth (the large crescent) passed between the sun and the spacecraft. Bright spots in the solar glare are reflections in the camera lens. (NASA)**

universe as results of extraordinary conditions; evolutionary theories have explained these phenomena as results of earlier conditions.

In some famous scientific controversies in the past, many theologians have tended to defend catastrophic theories, while many scientists have tended to defend evolutionary theories. In most of these areas, with the exception of the problem of the origin of the universe, modern scientists regard the evolutionary theories as more successful. Laboratory work has demonstrated the likelihood that mountains, life, and biological species, for example, will evolve from a specific set of pre-existing conditions according to certain known laws.

This same division of theories has been important in understanding the origin of the solar system. The first evolutionary theory was proposed by the French philosopher and mathematician, René Descartes, in 1644. He assumed that, regardless of how matter had been created "in the beginning," it was free to evolve according to its incorporated laws of nature. He thus proposed that space was initially filled with randomly swirling gas in which local eddies, or dense regions, evolved into individual stars. He thought smaller eddies might have created planets around stars.

Table 11·1 Characteristics of the Solar System to Be Explained by a Theory of Origin

1. All the planets' orbits lie roughly in a single plane.

2. The sun's rotational equator lies nearly in this plane.

3. Planetary orbits are nearly circular.

4. Planets and sun all revolve in the same west-to-east direction called *prograde*, or *direct* revolution.

5. The sun and all the planets except Venus and Uranus rotate on their axes in the same direction (*prograde* rotation) as well. Obliquity (tilt between equatorial and orbital planes) is generally small.

6. Distances between planets usually obey the simple Bode's rule.

7. Planets differ in composition.

8. Composition of planets varies roughly with distance from the sun: dense, metal-rich planets lie in the inner system, whereas giant, hydrogen-rich planets all lie in the outer system.

9. Meteorites differ in chemical and geological properties from all known planetary and lunar rocks.

10. Planets and most asteroids rotate with rather similar periods, about 5 to 10 hours, unless obvious tidal forces slow them (as in earth's case).

11. As a group, comets' orbits define a large, almost spherical cloud around the solar system.

12. Planet–satellite systems resemble the solar system.

13. The planets have much more angular momentum (a measure of orbital speed, size, and mass) than the sun. (Failure to explain this was the great flaw of the early evolutionary theories.)

The first catastrophic theory was also French. The naturalist Georges Buffon suggested about 1745 that the sun had been hit by a passing object about as massive as itself, and that the debris from the collision condensed into planets.

The reason for stressing the difference between catastrophic and evolutionary theories can be seen from comparing these two examples. The evolutionary theory predicts that planetary systems would be common around other stars. The catastrophic theory assumes an accidental circumstance (collisions between stars would be very rare) and hence predicts that planetary systems would be uncommon. It is thus exciting to investigate which general type of theory is right, since this may help us predict whether other planets like the earth exist elsewhere in the universe.

Several lines of evidence suggest that an evolutionary theory is more nearly correct. As we will see more clearly in later chapters, stars apparently form—even today—from vast clouds of gas in interstellar space. We will next outline how this process occurs and how it leads to the formation of planets, noting the explanation of various items in Table 11·1 as we go along.

The Protosun

Before the sun existed, its material must have been distributed in interstellar space in a large cloud like the interstellar clouds we see today. The sun began to form a little more than 4.6 billion years ago when this cloud (or part of it) became so dense that its own inward-pulling gravitational forces became stronger than the outward force of the pressure generated by motions of its atoms and molecules. The cloud thus began to shrink.

Early Contraction and Flattening

Even if gas in the cloud was randomly circulating initially, the contracting cloud would have developed a net rotation in one direction or another. An experiment to show this can be done with a cup of coffee or a pan of water. If you stir the liquid vigorously but as randomly as you can and then wait a moment and put a drop of cream in it, the cream will usually reveal a smooth rotation in one direction or the other, because the sum of the random motions will usually be a small net angular motion. Because the cloud was relatively isolated from the rest of the universe (as such clouds are today), this rotating motion, or angular momentum, could not easily be transferred anywhere, and thus was *conserved* as the cloud shrank.

This principle of **conservation of angular momentum** predicts that the cloud would rotate faster as its mass contracted toward its center, just as a figure skater spins faster when she pulls in her arms. Mathematical studies show that centrifugal effects associated with the faster spin caused the outer parts of the cloud to flatten into a disk, while material in the center contracted fastest, forming the sun as shown in Figure 11·2. This accounts for item 1 in Table 11·1, the *single plane of all planets' orbits.* Because the sun itself was an integral part of this disk, item 2, *the sun's rotation in the same plane,* is also explained.

At first, when the cloud was large, its atoms were far apart and could have fallen freely toward the center of gravity in the middle of the cloud. This state is called **free-fall contraction.** If free-fall persisted, the cloud could have completely contracted into the sun in only a few thousand years.

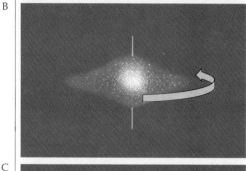

Figure 11·2 **Three stages of evolution of the protosun. In A, a slowly rotating interstellar gas cloud begins to contract because of its own gravity. In B, a central condensation forms and the cloud rotates faster and flattens. In C, the sun forms in the cloud center, surrounded by a rotating disk of gas.**

But eventually, because of their growing concentration and random motions, the atoms in the cloud would begin to collide with each other. Atoms interacting in a gas cause *pressure*. A build-up of pressure in the cloud would, of course, slow the collapse. Thus, the inward force of gravity and the outward force of gas pressure competed. Because the properties of gases are well understood, astrophysicists can analyze the later contraction and evolution of the cloud under these competing forces.

Helmholtz Contraction

Contraction in which the shrinkage is slowed by outward pressure is called **Helmholtz contraction** (after Hermann von Helmholtz, the German astrophysicist who first studied it). In 1871, Helmholtz showed how contraction would cause heat to accumulate in the contracting protosolar cloud (quoted in Shapley and Howarth, 1929):

> If a weight falls from a height and strikes the ground, its mass loses . . . the visible motion which it had as a whole—in fact, however, this is not lost; it is transferred to the smallest elementary particles of the mass, and this invisible vibration of the molecules is [what we call] heat.

Figure 11·3 **In Helmholtz contraction, a slow contraction due to gravity causes the internal temperature to increase, which in turn causes radiation to be emitted.**
▼

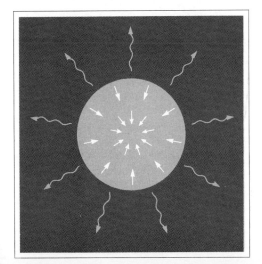

In the contracting protosun, atoms or swarms of atoms would fall toward the center until they collided with other parts of the gas cloud. Temperature would increase inside the cloud. Wien's law guarantees that the protosun would radiate energy (Figure 11·3), warming the surrounding cloud.

Calculations based on the Helmholtz theory indicate the conditions: Eventually the cloud's central temperature would have risen to 10 million degrees Kelvin or more, starting the nuclear reaction that made it a star and not just a ball of inert gas. Meanwhile, and more important for the formation of planets, the outer parts of the cloud formed a disk of gas as big as the solar system, at temperatures of a few thousand degrees Kelvin. *These theorized steps have been confirmed by actual observations of newly formed stars, surrounded by clouds of gas and dust.*

The Solar Nebula

A cloud in space is called a **nebula** (plural, *nebulae*), from a Greek term for mist. The nebula that surrounded the contracting sun is called the **solar nebula.** It is believed to have been a disk-shaped rotating cloud. Molecules of gas or grains of dust must have moved in circular orbits, because noncircular orbits would cross the paths of other particles, leading to collisions that would damp out the noncircular motions. Thus, neglecting small-scale eddies in the gas, broad-scale motions in the cloud were in parallel circular orbits, accounting for items 3 and 4 in Table 11·1. Finally, once the nebula stabilized, its gas began to cool.

Condensation of Dust in the Solar Nebula

The solar nebula was initially gas that Helmholtz contraction heated to at least 2,000 K. At such a temperature, virtually all elements were in gaseous form. Like other cosmic gas, most of its atoms were hydrogen, but a few percent were heavier atoms such as silicon, iron, and other planet-forming material.

How did solid particles form in this gas? The answer can be seen on earth. When air masses cool, their condensable constituents form particles: snowflakes, raindrops, hailstones, or the ice crystals in cirrus clouds. Similarly, in the cooling solar nebula, particles condensed out of the gas in a sequence known as the **condensation sequence.**

Chemical studies show that as the temperature in any part of the nebula dropped toward 1,600 K, certain metallic elements such as aluminum and titanium condensed to form metallic oxides in the form of **grains** or microscopic solid particles, as shown in Table 11·2. At about 1,400 K a more important constituent—iron—would condense. Microscopic bits of nickel–iron alloy would form as grains, or perhaps coat existing grains. Still more important, at about 1,300 K abundant silicates would begin to appear in solid form. For instance, the magnesium silicate mineral enstatite ($MgSiO_3$) would form at about 1,200 K (Lewis, 1974).

Complicated mixtures of magnesium-, calcium-, and iron-rich silicates should have condensed, depending on the temperature, pressure, and composition of the gas at various points in the solar nebula. Since local conditions depended on distance from the newly-formed sun, different compositions of mineral particles may have dominated at different locations, explaining items 7 and 8 in Table 11·1. In particular, the outer parts of the solar nebula reached lower temperatures because they were farther from the sun and because they were shaded by gas and dust grains in the inner nebula (Cameron, 1973, 1975).

At about 300 K, water became an important constituent in the minerals. In the outer solar nebula, abundant ice particles formed. Actual snowflakes of water ice may have appeared. At temperatures of 100–200 K, ammonia and methane also appeared as ices. In the outer solar system these ices were so far from the sun that they could survive even direct sunlight. They may survive today in comets and icy satellites of outer planets. These ice particles were all rich in hydrogen (H_2O, NH_3, CH_4, respectively) and probably contributed to the hydrogen-richness of the outer planets.

The compositions of meteorites strongly support this theory of condensation. The type called carbonaceous chondrite contains microscopic grains of material believed to be among the earliest solid particles in the solar system. These grains are rich in elements that would have condensed first—meaning at highest temperatures—such as osmium and tungsten. Compared to earth-rocks, meteorites contain excesses of minerals formed at temperatures at about 1,450–1,840 K, as is consistent with item 9 in Table 11·1. Also, carbonaceous chondrites contain clumps of microscopic grains of magnetite (Fe_3O_4) and other minerals, shown in Figure 11·4. These may have collided and stuck soon after forming in space. Enstatite is a common mineral in many meteorites, as is predicted by the theory. Water is preserved as a common constituent in minerals in parts of carbonaceous chondrites.

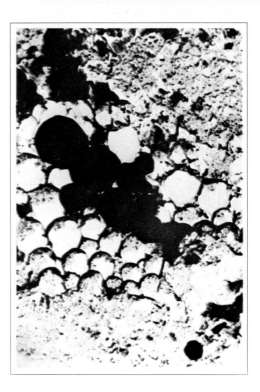

▲

Figure 11·4 **This electron microscope photo shows a cluster of micron-size spheroidal crystals of the mineral magnetite inside a carbonaceous chondrite meteorite. Some researchers believe these may be the original condensed grains, formed in the solar nebula and preserved after incorporation into the meteorite.** (John Kerridge)

From Planetesimals to Planets

Although the preplanetary particles may have formed as microscopic grains, they clearly grew bigger. (Otherwise, there would be no planets!) The hypothetical intermediate bodies, from millimeters to many kilometers in size, are usually called **planetesimals**, as noted in Chapter 10. Evidence that they existed includes:

1. Craters on planets and satellites indicate impacts of planetesimals with diameters of at least 100 km.

2. Meteorites are their surviving fragments.

Approximate Temperature (K)	Element Condensing	Form of Condensate (with examples)	Comments
2,000	None		Gaseous nebula
1,600	Al, Ti, Ca	Oxides (Al_2O_3, CaO)	
1,400	Fe, Ni	Nickel–iron grains	Parent material of planetary cores, iron meteorites?
1,300	Si	Silicate and ferro-silicate minerals (enstatite, $MgSiO_3$; pyroxene, $CaMgSi_2O_6$; olivine, $(Mg, Fe)_2SiO_4$) in form of microscopic grains	First stony material, combined to form meteorites; some still preserved in primitive meteorites
300 ↓ 100	H, N, C	Ice particles (water, H_2O; ammonia, NH_3; methane, CH_4)	Large amounts of ices, still preserved in outer planets and comets

Table 11·2 Condensation Sequence in the Solar Nebula

Source: Data from Lewis (1974); Grossman (1975); and others.

3. Asteroids and comet nuclei reaching more than 100 km in diameter are probably surviving representatives.

But how did microscopic grains evolve to produce 100-km-size planetesimals? If they had circled the sun in paths comparable to present asteroidal and cometary orbits, they would have collided with one another at speeds much faster than rifle bullets. They would have shattered, and the solar system would still be a nebula of dust and grit.

But did dust particles have such high speeds in the early solar nebula? According to dynamical analyses, they collected in a denser swarm of particles with similar orbits in the central plane of the disk, where relative velocities were lower. In low-velocity collisions, some dust grains simply stuck together, held by weak adhesive forces such as gravity or electrostatic attraction. Mutual gravity probably

caused groups of planetesimals to clump together, growing to multikilometer size in only a few thousand years—a mere moment in cosmic time. Actual examples of such clumping are shown in Figures 11·5 to 11·8.

Dynamical studies indicate that coalescing particles would tend to form bodies rotating in a prograde motion (item 5). No one is sure why

the planetary spacings are regular (item 6), but studies suggest that gravitational forces tended to divide the solar nebula into ring-shaped zones, each favoring formation of a planet.

The larger planetesimals collided with smaller ones, knocking off debris. Some small particles fell back on their surfaces, forming a powdery soil layer that, like the lunar regolith, was effective in trapping other small fragments. Thus the largest planetesimals grew fastest, sweeping up the others. These bodies may have grown to 100-km size in a few million years. Some planetesimals, the *parent bodies* of meteorites, were heated, melted, and differentiated into metal and rock portions. Some were shattered by

Figure 11·5 **This chain of microscopic fragments from a spherule of lunar glass extends about 40 microns. The chain may result from electric attractive forces and may be an analog of accumulations of particles in the solar nebula.** (Electron microscope photo courtesy of G. Arrhenius) ▼

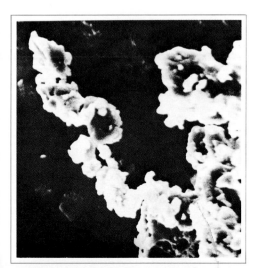

Figure 11·6 **Droplets formed from molten or vaporized material during impacts have welded onto the surface of this 40-micron spherule of glassy material from the moon. Similar welding processes may have occurred in the solar nebula.** (Electron microscope photo courtesy of G. Arrhenius) ▼

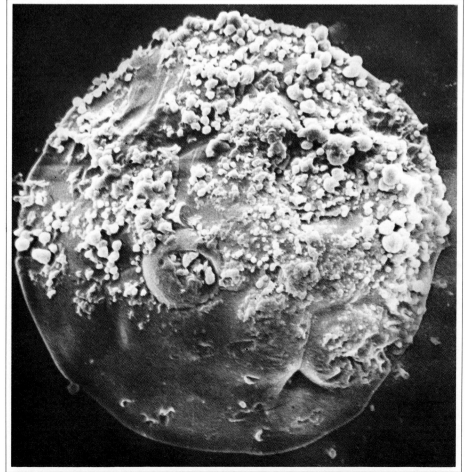

collisions with other large or fast neighbors, freeing iron, stony-iron, and achondrite meteorites from their interiors. But as some shattered, others grew to replace them.

The scene at this ancient point must have been one of floating dust and rocks (Figure 11·9), with scattered larger bodies making occasional collisions or near misses (Figure 11·10).

Once the largest planetesimals reached a few hundred kilometers in breadth, they had strong gravitational fields of their own, and began to pull in the remaining planetesimals. Thus, the largest planetesimals continued to grow fastest. (We call these large bodies with nearly planetary mass **protoplanets**.) Sometimes an approaching planetesimal would score a direct hit on a protoplanet. But more often a near-miss would occur, and the small planetesimal would be deflected

into a different orbit (just as Pioneer 10 and 11 were deflected into the outer solar system by near-misses of massive Jupiter). Icy planetesimals among the outer planets passed near Jupiter and were flung into wide-ranging orbits that account for item 11 in Table 11·1.

What Became of the Planetesimals and the Nebula?

Most of the planetesimals were eventually accumulated by the planets, probably within 90 million years of the solar system's beginning. For another 500 million years, as lunar rocks show, some of the remaining planetesimals rained down onto planetary surfaces, creating still more craters. Others were thrown into the outer solar system, forming comets. Some may have been ejected entirely from the solar system by the gravity of massive Jupiter. Between Mars and Jupiter, the planet-forming process was aborted, and a swarm of asteroids was left as an example of the early solar system.

A few planetesimals may have been captured into orbits around the larger planets, perhaps explaining the outer-

most satellites of the giant planets. The few largest planetesimals to strike each planet may have had a significant effect in determining properties such as tilt of rotation axis, speed and direction of rotation, character of the satellite systems, and geologic differences from one hemisphere to the other. Meanwhile, the smallest particles, such as dust grains, molecules, and atoms of gas, were carried away from the solar system by the outward forces of solar radiation and outrushing solar gas.

The Chemical Compositions of Planets

Why do the planets vary in their chemical composition? Why do meteorites differ from lunar and terrestrial materials? The answers to these questions now become quite clear. The

Figure 11·7 **This photomicrograph of a sliver of the chondrite meteorite Brownfield shows it to be composed of millimeter-scale spherical chondrules (two examples are in the center) surrounded by a matrix of tiny mineral grains. Chondrules may be produced by solidification of molten droplets created during collision.** (Center for Meteorite Studies, Arizona State University) ▼

Figure 11·8 **This glassy particle was formed by the welding of several different-size spherules of molten material ejected from an experimental explosion of 500 tons of TNT. In size and appearance the particle resembles the chondrules in meteorites, which may have formed in meteorite collisions. Millimeter scale is at bottom.** ▼

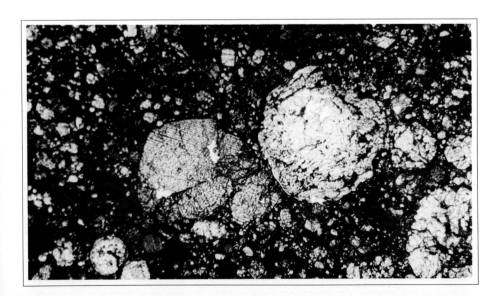

▲
Figure 11·9 An imaginary view inside the solar nebula during the formation of meteoritelike planetesimals. Dust clouds partially obscure the sun. (Painting by author)

Figure 11·10 At an intermediate stage in planet growth, kilometer-size planetesimals formed. Their collisions caused craters or total fragmentation; near collisions caused gravitational accelerations, which changed orbits. In this imaginary view, an observer on the rim of a crater on a one-km-scale body sees another passing about 10 km away, causing a momentary eclipse of the sun. (Painting by author) ▶

Figure 11·11 **Collisions of the planetesimals with planets and satellites left intensely cratered surfaces such as this lunar region. (Lunar and Planetary Laboratory, University of Arizona)**

Figure 11·12 **Terrestrial craters, like the 20,000-year-old Meteor Crater in Arizona, represent only the final stages of sweep-up of solar-system debris. A. Aerial view of the crater. (D.P. Cruikshank) B. Surface view from the rim, showing fragmentary ejected boulders.**

A

condensation sequence shows that different groups of minerals existed at different *temperatures* in the nebula. In the cold outer parts of the nebula, hydrogen-rich ices formed, but in the hot inner parts, only metals and silicates. Massive bodies such as the sun and Jupiter retained their original gases; studies of these bodies show that the nebula was originally mostly hydrogen and helium, with only a few percent of heavier atoms such

as silicon, oxygen, and iron, as shown in Table 11·3.

Table 11·3 shows what happened after the gas condensed into solid planetary materials. Differentiation processes continued. Virtually none of the hydrogen or helium was trapped in the solid materials of the inner solar system. Primitive meteorites of the carbonaceous chondrite and chondrite types show that primitive silicates were composed mostly of oxygen,

iron, silicon, and magnesium. However, as soon as these materials melted, the iron apparently drained into metallic cores resembling iron meteorites, and this differentiation process meant that the surface materials of planets became more iron-poor and silicon-rich than the primitive materials. This sequence can be seen by following Table 11·3 from left to right.

Most differentiated was the earth, where most of the iron disappeared

B

Element		Designation[a]	Sun ("Cosmic" Composition)[b]	Carbonaceous Chondrite (Primitive Meteorite)	Chondrite Meteorite	Earth: Ultra-basic Rock	Mars: Basalt(?)	Moon: Mare Basalt	Earth: Basalt	Moon: Upland	Earth: Granite (Continental Crust)
H	Hydrogen	V	78%								
He	Helium	V	20								
O	Oxygen	L	0.8	34%	35%	43%		43%	44%	45%	48%
Fe	Iron	S	0.04	24	26	9.4		11	8.6	4.6	2.1
Si	Silicon	L		15	18.5	20	15–30%	21	23	23	33
Mg	Magnesium	L		13	15	20	14	5.5	4.6	4.4	0.6
S	Sulfur			8.6	2.3	0.03	2–5	0.2	0.3		0.3
Ca	Calcium	L		1.5	1.3	2.5	3–8	8	7.6	7	1.5
Ni	Nickel	S		1.4	1.4	0.2		0.0001	0.01	0.0001–0.01	0.001
Al	Aluminum	L		1.2	1.1	2	2–7	8	7.8	13	8
Na	Sodium	L		0.46	0.70	0.4		0.4	1.8	0.4	2.7
Ti	Titanium	L		0.08	0.08	0.003	0–1	2.1	1.4	0.3	0.2
K	Potassium	L		0.05	0.09	0.004		0.1	0.83	0.01–0.1	3

Table 11·3 Elemental Abundances in Cosmic Matter (Selected Elements, Percent by Weight)

[a]V = Volatile (driven off by heating).

 L = Lithophile (concentrated in siliceous rocks; sulfur usually classified separately).

 S = Siderophile (iron-affinity elements) concentrated in basic rocks and metallic minerals, as in earth's core and mantle.

[b]This composition is called "cosmic" because it is believed to be the same gas that formed not only the sun but all nearby stars.

Source: Data from Fairbridge (1972); Taylor (1973); Mason and Melson (1970); Page (1973).

into the core, and elements such as oxygen and silicon were strongly concentrated into the upper continental crust. Those processes explain why the rocks beneath our feet are so different from most common materials in the universe. The measurements from Mars and the moon indicate less complete melting and differentiation, so that Martian and lunar surface materials are somewhat closer to the primitive meteoritic matter that condensed in orbit around the sun.

As we discussed in Chapter 9, the outer, giant planets are composed mainly of ices. Did the giant planets reach their present state primarily by gravitational contraction of gas and dust, as the sun did, or by accretion of asteroid-size bodies, as the smaller planets probably did?

The systems of Jupiter, Saturn, and Uranus suggest miniature solar systems with large central bodies like the sun, small inner satellite(s) like the terrestrial planets, larger middle-distance satellites, and small, irregular outer satellites like Oort's comet swarm. Many astronomers have therefore suggested that the giant planets were formed by processes analogous to those that formed the solar system (item 12 in Table 11·1). The origin of giant planets, small planets, and meteorite parent bodies remains an area of active research.

One problem of the composition of the solar system continues to puzzle astronomers—the uneven distribution of isotopes of the same element throughout the solar system. If isotopes have similar chemical properties (although differing in numbers of neutrons), why do parts of the solar nebula have different proportions of the same isotope? For example, the earth's material seems to have one ratio of oxygen isotopes O^{16} and O^{18}, whereas meteorites have a different ratio. Chemists are puzzled, because normal chemical processes in the solar nebula should mix various isotopes of oxygen in constant proportions. Were other isotopes added to parts of the nebula from the outside?

Some astrophysicists believe that batches of isotopes were formed by nuclear processes inside another star near the sun. This star may have exploded and added some of its own material to the solar nebula. This is not too unlikely an occurrence, because stars tend to form in groups, and the early sun may have had many relatively close neighbors. Some such stars would be expected to be unstable, leading to explosions. So there is no guarantee that current theories depict all the complexities of planetary growth.

Magnetic Effects and the Sun's Spin

As noted before, the early sun would have been spinning fast enough after its contraction to deform into a disk. But today it is rotating slowly and is spherical. It has evidently slowed down. What happened to the missing angular momentum? For many years the lack of an answer to this question was the fatal flaw in evolutionary theories of planet formation. However, in the 1950s and 1960s Swedish-American astronomer H. Alfvén and others realized that the early sun probably had a strong magnetic field. The hot surface of the early sun shot out ionized gas, probably at a greater rate than it does today. Because ions must move with any local magnetic field, the rotating magnetic field of the sun tried to drag the ions of the inner solar nebula with it. The gas would thus have exerted a braking force on the rotating sun, just as water does on a spinning tennis ball dropped in a pool.

This **magnetic braking** probably slowed the sun's rotation, transferring angular momentum to the planetary

material in the solar nebula. This theory solves the angular momentum problem of item 13 in Table 11·1. Early theorists, who did not understand magnetic effects, could not have foreseen this solution. Certain types of stars, believed to be very young, display strong irregular flare activity possibly related to strong magnetic fields. Thus, the magnetic transfer of angular momentum may actually be occurring within view of our telescopes, giving further support to Alfvén's hypothesis.

Stellar Evidence for Other Planetary Systems

Among the consequences of this evolutionary theory of the solar system's origin, one is most intriguing: It makes plausible the idea that many stars should have planets. All stars should form from gravitationally unstable, collapsing nebulae. All stars should have some angular momentum and thus produce rotating, flattened clouds around themselves. This cloud should cool as its heat radiates away into space, and dust grains of varied compositions should therefore appear. Many observations of stars indicate that these processes do occur elsewhere.[2]

But would the subordinate material always accumulate into planet-size bodies? Probably not, according to simple observations of stars. Many if not most stars are double or triple

---

[2]Later chapters will add strong evidence supporting this evolutionary picture of star and planet formation. Chapters 15 and 17 describe interstellar clouds and their relation to newly formed stars. Chapter 15 shows that some new stars are still surrounded by clouds of dust resembling the solar nebula, while others appear to be just shedding these clouds.

star systems. In such a system the second biggest object is not a planet, like Jupiter, but a full-fledged star with nuclear reactions and an incandescent gaseous surface. Nonetheless, the evolutionary sequence of events described above indicates that, among the remaining stars, dust particles may very well have accumulated into planet-like bodies. Thus, many astronomers suspect that a sizable number of other stars may have planets orbiting around them. If so, favorable environments for living creatures might not be restricted to our solar system. Even if only 0.1 percent of all stars had planets on which liquid water could exist and life as we know it could form, roughly 100,000,000 inhabited planets could exist in our galaxy! This thought, with its exciting implications, will be deferred to Chapter 25, after we have more critically examined the nature of stars, their formation, history, and multiplicity.

11

Summary

Information about the origin of the solar system has been culled from meteorites, lunar samples, the oldest terrestrial rocks, and chemical and dynamical analysis of planets and satellites. This information indicates that the sun formed from a contracting cloud of gas about 4.6 billion years ago. As outer parts of this cloud cooled, solid grains of various minerals and ices condensed and accumulated into planetesimals during a relatively brief interval ranging from a few million to 90 million years.

During this interval, neighboring planetesimals collided, often gently enough to allow them to hold together by gravity. In this way, small planetes-imals accumulated mass and grew into larger bodies. Sometimes these larger bodies collided at high enough speeds to shatter each other, producing meteorite-like fragments. The larger bodies survived and grew into planets. Continuing collisions caused the large bodies to sweep up most of the smaller bodies; the latter formed craters as they fell onto the surfaces of the planets.

The sun and the outer planets, with their high masses and strong gravity, retained the light hydrogen-rich gases of the original cloud. The lower-gravity terrestrial planets lost these gases. No data about the early solar system restrict these processes to the solar system; thus, many scientists suspect that planetary debris or full-fledged planets may have formed near some other stars besides our sun.

Concepts

catastrophic theories

evolutionary theories

conservation of angular momentum

free-fall contraction

Helmholtz contraction

nebula

solar nebula

condensation sequence

grains

planetesimals

protoplanets

magnetic braking

Problems

1. In view of the theory of thermal escape of gases from planetary atmo-spheres (see Chapter 7), explain the absence of abundant hydrogen in the terrestrial planets' atmospheres. Why does this not need to be listed as a fact to be explained by theories of solar system *origin*, in Table 11·1?

2. What properties of Venus and Uranus are particularly troublesome to theorists attempting to explain the formation of the solar system? How might collisions between these bodies and other large planetesimals during the formative period help explain the present properties of Venus and Uranus?

3. Why do meteorites have different chemical and geological properties than rocks you might find in your own yard, since all planetary material con-densed from the same nebula?

4. Because of heating by the sun and by the contraction process, gases in the inner solar system were prob-ably at higher temperatures when planetary solid matter formed than gases in the outer solar system. In terms of the condensation sequence, relate this to the estimated or observed composition of the planets.

5. Judging from planetary compo-sition, where was the inner boundary of the part of the solar nebula where ices condensed?

6. If planets orbited the sun in ran-domly inclined orbits, both prograde and retrograde, how might theories of the solar system's origin differ?

7. List some observations that help confirm the theory of solar system origin described in this chapter.

8. How are theories of solar system origin different in principle from theories of the origin of the universe?

9. What reasons do we have for believing that the entire solar system formed during a single, relatively short interval?

10. If a five-year-old member of your family asked where the world came from, how would you answer?

Advanced Problems

11. If a planetary system were form-ing around a nearby star, and the star were obscured by a "solar nebula" of dust grains at a temperature of 1,000 K, explain how such a dust nebula might be detected. (Hint: Apply Wien's law and assume that a large telescope is available with infrared detectors.)

12. If dust grains in circular orbits in the early solar nebula collided at a speed equalling 0.1 percent of their orbital velocity, how fast would they have collided near the present orbit of the earth? How fast near the present orbit of Pluto?

1

Stars and Their Evolution

This is Outer Space
Where the atom changes mass
Where the dynamos are fast
Where the Cosmic Hoard is stashed
Where the Carbon Torch is passed
To the Oxygen Orbiters
Humid in gas
From the Hydrogen Past. . . .

Burgert Roberts,
from *Spacewalks*

The Sun and Its Radiation

If we could leave the earth and travel at the speed of light, it would take less than two seconds to reach the moon, at least four minutes to reach Mars, five hours to clear Pluto's orbit, nine months to get well into the Oort cloud of comets, and 4.3 years to reach the nearest star outside the solar system. Most other visible stars could only be reached after decades, centuries, or millennia of travel at the speed of light.

How can we study such remote objects as stars? We can study the closest one, which is nearer to earth than five of the nine planets. To reach it at the speed of light would take only eight minutes. Our eyes are dazzled by it. The earth is bathed in its flow of radiation, washed by the winds of its outer atmosphere, blasted by seething swarms of atoms blown out of it, bombarded by bursts of X-rays and radio waves emitted by it. It is our sun.

Humans pondered the stars for many centuries before they realized that the overwhelming orb of the daytime sky was just another of the pinpoint lights of the nighttime sky. The sun is the only star on which we can see surface details and processes. It is so close that we can watch storms develop on its surface, and track them as they are carried all the way around the sun by **solar rotation,** as shown in Figure 12·1. This rotation takes about 25.4 days relative to the stars (and 27.3 days relative to earth—a longer figure since earth's orbital motion must be added in). As in Jupiter's atmosphere, the equatorial region rotates faster than the polar regions—proof that the sun has a gaseous, not solid, surface.

Spectroscopic Discoveries

The strongest astronomical tool for studying the gas of the sun and other stars is **spectroscopy,** or the study of the colors of light emitted by these objects.

The Nature of Light

Several earlier chapters have described certain properties of light, such as the peak of radiation occurring at certain colors (wavelengths), depending on the temperature of the radiating body (Wien's law, p. 124).

Figure 12·1 **Six photos of the sun in sequence, August 21–26, 1971, showing the rotation of the sun by the movement of sunspot groups.** (W. A. Feibelman)

But as we begin our study of stars, we need to review the nature of light more systematically, because stars, unlike planets, can be studied only from vast distances by the light they radiate. **Light** is energy in the form of pulsations in the electromagnetic fields that permeate most of space. These pulsations move at the **speed of light,** which is constant as perceived by all observers, about 300,000 km/sec. It is often designated c.

Light has some properties that can be visualized by analogy. For instance, light has some characteristics of wave motion, like the waves that travel at constant speed across the surface of a pool. A cork floating on the surface of the pool would start bobbing up and down as the wave came by. Similarly, an electron or other atomic particle is disturbed by passage of a light wave. A poolside observer could describe the water wave either by how fast the cork bobs or by the length of the wave. In the same way, light can be described either by its **frequency** of pulsation or by its **wavelength,** the characteristic length of the disturbance in the electromagnetic field.

As mentioned in Chapter 7, each wavelength, or frequency, corresponds to a different color. This is shown in Color Photo 30, where the colors are arranged in order of wavelength. Our eyes perceive only radiation in a narrow, intermediate range of wavelengths. Gamma rays, X-rays, and ultraviolet light have wavelengths too short to see. Violet is the shortest wavelength we can see; yellow is intermediate; red is the longest we can see. Infrared has too long a wavelength to be visible, and radio waves are still longer (see Figure 12·2).

The analogy between light and waves is not perfect. Energy in waves is spread out along the series of waves,

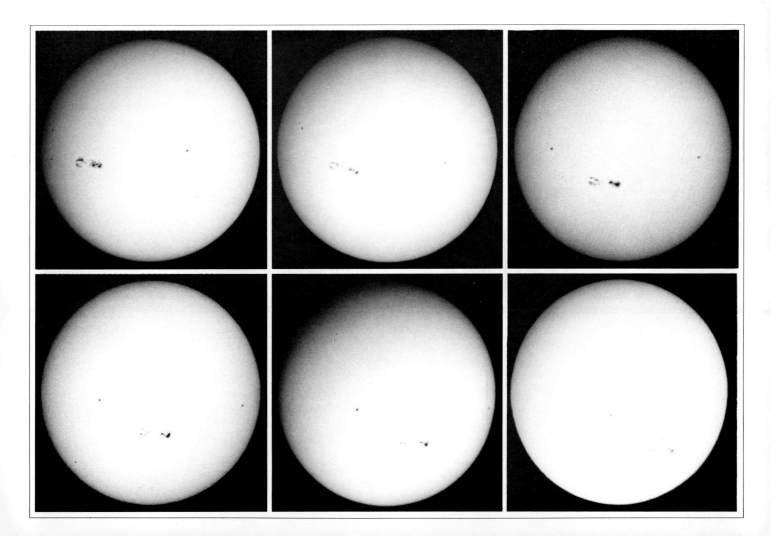

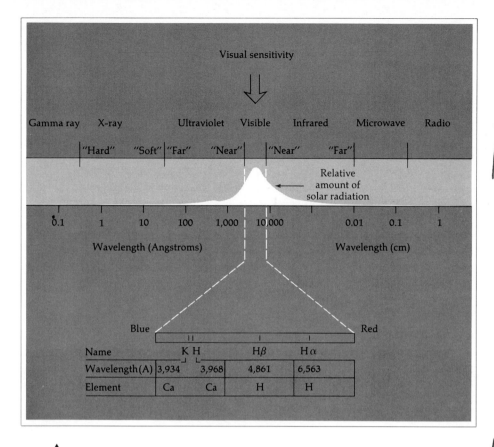

Visual sensitivity

Gamma ray	X-ray	Ultraviolet	Visible	Infrared	Microwave	Radio
	"Hard" "Soft"	"Far" "Near"		"Near" "Far"		

Relative amount of solar radiation

0.1   1   10   100   1,000   10,000   0.01   0.1   1

Wavelength (Angstroms)   Wavelength (cm)

Blue   Red

Name		K	H	Hβ	Hα
Wavelength(A)	3,934		3,968	4,861	6,563
Element	Ca		Ca	H	H

▲
Figure 12·2 **The solar spectrum, showing different regions named according to wavelength. Upper diagram shows how the amount of solar radiation strongly** peaks in the visual region, the only portion detectable by the eye. Lower diagram is an expanded image of this region, showing several prominent absorption lines.

but energy in a light beam is concentrated in discrete units, called **photons.** Photons are individual units of energy carried by the beam of radiation, and they are sometimes visualized as subatomic-size, massless particles. One might think of them as microscopic BBs; when a photon strikes an atom, it may knock out some of the atom's electrons. The amount of energy carried by each photon is inversely proportional to the wavelength of the light being considered. Thus, photons of blue light have more energy than photons of red light—a fact with several consequences we will consider later.

The Solar Spectrum

As Newton showed, sunlight or any **white light** is a mixture of all colors. Sunlight passing through Newton's prism revealed a **spectrum,** or array of colors in order of wavelength, as shown in Color Photo 30. In 1817, German physicist Joseph Fraunhofer found that certain wavelengths were missing from the sun's spectrum, so that it appeared to be crossed by narrow, dark lines, called **absorption lines,** as shown in the bottom part of Figure 12.2. Fraunhofer named them A, B, C, and so on, from red to blue.

What were these lines? By the mid-1800s, scientists discovered that when a given element is burned, it produces certain colors and no others. These color patterns, unique to each element and unmistakable as a set of fingerprints, are called **emission lines.** Some of the emission lines exactly matched the positions of Fraunhofer's solar absorption lines. For instance, his D line (actually a close pair) matched an emission line from sodium; his H and K lines, calcium. Did this mean that the D line in the solar spectrum was caused by sodium in the sun, and the H and K lines caused by calcium?

Kirchhoff's Laws

The answer proved to be yes when in the 1850s, German physicist Gustav Kirchhoff discovered the conditions that produce the three different kinds of spectra: absorption lines, emission lines, or a glow consisting of a mixture of all colors, called a **continuous spectrum.** For instance, when Kirchhoff looked through his spectroscope toward a sodium flame against a dark background, he saw the sodium D line in emission, like the emission lines in the bottom of Color Photo 30. But when he changed the background to a brilliant beam of sunlight passing through the same flame, he saw the D line as an absorption line, like the lines in the middle of Color Photo 30. They came from the same gas.

Thus, in simple terms, the absorption lines in the solar spectrum can be thought of as occurring when photons of solar white light pass outward through the cooler gas of the solar atmosphere. Photons of certain wavelength (such as the D line), when striking certain atoms (such as sodium), will be absorbed from the outgoing beam, causing an absence of that color in the beam.

Kirchhoff reduced such observations to three statements called **Kirchhoff's laws:**

1.

A gas at high pressure, a liquid, or a solid, if heated to incandescence, will glow with a continuous spectrum.

2.

A hot gas under low pressure will produce only certain colors, called bright emission lines.

3.

A cool gas at low pressure, if placed between the observer and a hot continuum source, absorbs certain colors, causing dark absorption lines in the observed spectrum.

Kirchhoff himself found that the absorption lines and emission lines of a given gas have identical wavelengths. What is seen depends on the temperature and density of the gas relative to the radiation coming from behind it, as indicated in the third law. There was an important later modification to Kirchhoff's laws: An absorption spectrum need not originate *in front of*, or in a cooler gas than, the continuous spectrum. Indeed, some solar Fraunhofer absorption lines even originate in the same surface layers that produce the continuous spectrum we call sunlight. These layers are called the **photosphere**—the visible surface layer of the sun. Emission lines also occur there; they can be seen during eclipses but are normally swamped by the bright continuous radiation of the photosphere.

In summary, spectroscopy allows us to determine many things about the properties of the sun. Identification of the spectral lines allows identification of the elements in the sun. Application of Kirchhoff's laws proves that the photosphere is a layer of hot gas. Application of Wien's law (page 124) allows us to use the wavelength of the strongest solar radiation to determine the photosphere temperature. Other spectroscopic principles allow astronomers to measure temperatures and pressures at different depths in the gas near the sun's surface.

Images of the Sun

Normal photographs of the sun are made from light of many colors, since normal photographic films are sensitive to many wavelengths. However, color filters and other devices can restrict the wavelength range. By using a narrow enough range, we can see the sun in only the light emitted by a chosen gas (hydrogen for example) in a specific atomic state, thus tracing the distribution of that gas. The **spectroheliograph** is an ingenious instrument that allows this to be done. It spreads the light into different colors, from which one color alone is selected to form an image. When hydrogen gas is in certain states of temperature and pressures it emits an especially useful red color—the well-known **hydrogen alpha line,** or *Hα emission.* In the resulting image the only bright regions are those where hydrogen of a certain temperature is emitting this specific red glow. Dark parts of the image are regions where little hydrogen is in this state. Figure 12·3 compares a normal solar image with a view in hydrogen alpha light.

Many techniques, illustrated in this chapter, have been used to obtain solar images in the light of other specific gases. These images allow analysis of the distribution of gases and temperatures on the sun.

Figure 12·3 **The left photo shows the sun in normal visible light, with sunspots and limb darkening caused by solar atmospheric absorption of light. The right photo shows** ▼ the same face of the sun photographed in red light emitted by hydrogen (Hα spectral line); bright areas involve intense hydrogen emission. (Hale Observatories)

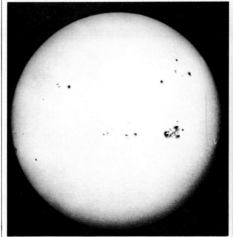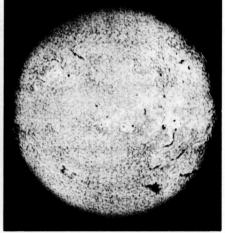

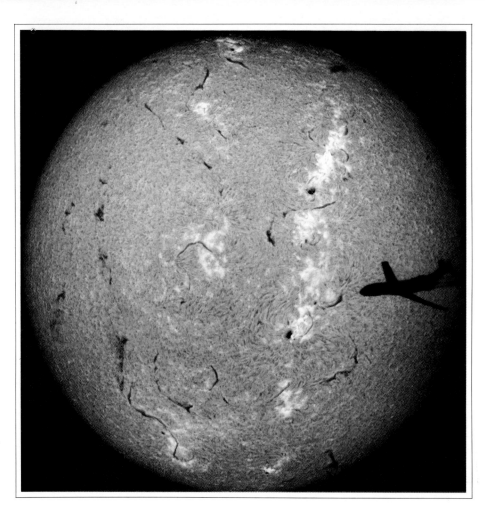

Composition of the Sun

More than a century of spectroscopic study has resulted in accurate knowledge of the sun's composition. The most abundant elements are listed in Table 12·1. An interesting episode occurred in 1868, when the French astronomer Pierre Janssen and the English astronomer Norman Lockyer independently found solar spectral lines corresponding to an unknown element. This element, named *helium*

◄ *Figure 12·4* **A spectroheliograph view of the sun in the light of hydrogen alpha emission happened to catch the silhouette of a commercial jet crossing the solar disk. The solar equator is near-vertical, with parallel bands of dark and bright hydrogen clouds.** (Pennsylvania State University)

Figure 12·5 **A comparison of solar photos from ground-based and high-altitude telescopes. Atmospheric turbulence blurs fine-scale granulations in the ground-based photo (left), leading to a recent emphasis on airborne and orbiting telescopes.** (NASA) ▼

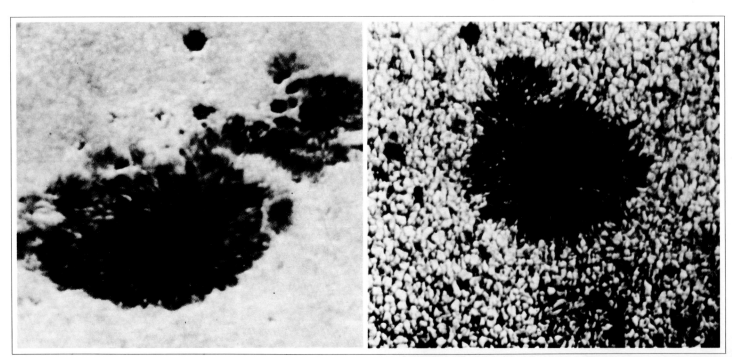

(from the Greek *helios,* sun), was the first to be discovered in space instead of on earth (where it was not observed until 1891). It is the second most abundant element in the sun, after hydrogen.

The sun is about three-quarters hydrogen and nearly one-quarter helium. The heavy elements common in the planets comprise only 2 percent of the sun.

Table 12·1 Composition of the Sun		
Element	Fraction of the Sun (by mass)	Atomic Number
Hydrogen	78.4%	1
Helium	19.8	2
Oxygen	0.8	8
Carbon	0.3	6
Nitrogen	0.2	7
Neon	0.2	10
Nickel	0.2	28
Silicon	0.06	14
Sulfur	0.04	16
Iron	0.04	26
	100.0%	

Source: Data from Gibson (1973).

Solar Energy from Nuclear Reactions

Hermann von Helmholtz calculated in 1871 that the energy output of the sun corresponds to the burning of 1,500 pounds of coal every hour on every *square foot* of the sun's surface. Because no ordinary chemical reactions can produce energy at this rate, the sun's energy could not come from familiar chemical combustion: the sun is not "burning" in the normal sense.

In the 1920s, astrophysicists realized that the energy of the sun and other stars comes from **nuclear reactions,** interactions of nuclei of atoms. Normally, nuclei are protected from interacting by their surrounding clouds of electrons. Familiar **chemical reactions,** such as coal-burning, involve interactions only between *electrons* of different atoms, far outside the central nucleus. However, if temperature and pressure are high enough, atoms collide fast enough to knock away electrons, allowing nuclei to interact.

Nuclear reactions in stars do two important things: They change the star's composition, and they generate energy. The most important reactions in the sun are believed to be a three-part fusion sequence that changes hydrogen to helium. Because the sequence starts with two hydrogen nuclei that consist only of protons, it is called the **proton-proton cycle:**

$$^1H + {}^1H \rightarrow {}^2H + e^+ + neutrino$$

$$^2H + {}^1H \rightarrow {}^3He + photon$$

$$^3He + {}^3He \rightarrow {}^4He + {}^1H + {}^1H$$

In the first reaction, two hydrogen nuclei fuse to form a "heavy hydrogen" (deuterium) nucleus containing one proton and one neutron, liberating a positron (an electron-size positive particle) and a chargeless particle called a neutrino. In the second reaction, the deuterium nucleus unites with another hydrogen nucleus to form a helium-3 nucleus and a **photon,** or unit of radiation. Finally, two of the helium-3 nuclei unite to form a normal helium nucleus and two hydrogen nuclei.

Nuclear reactions also generate energy. For example, at the end of the proton-proton cycle, the total mass left is slightly less than the total mass at the beginning. A small amount of mass, m, has been converted to energy,

E, according to Einstein's famous equation $E = mc^2$. (The constant c is the velocity of light, 3×10^{10} cm/sec. All units must be in the cgs system, with m in grams and E in ergs.) In the proton-proton cycle, about 0.007 grams of matter are converted into energy for each gram of hydrogen processed. This liberates 4×10^{33} ergs/sec inside the sun, and the sun radiates this much every second to maintain its equilibrium. In other units, this corresponds to 400 trillion trillion watts—which equals a lot of light bulbs!

Every second, the sun converts into energy and radiates into space 4 million tons of solar matter. Does this loss of mass mean that the sun will ultimately disappear? At this rate it would take 14 trillion years to use up all the mass. Long before this, the solar core will have converted so much hydrogen to helium that there will not be enough hydrogen left in the core to fuel further reactions. The reactions will probably stop; *stars cannot radiate indefinitely,* but must run out of fuel. According to a recent calculation, the sun won't run out of hydrogen for about 3.5 billion years.

Interior Structure: How Energy Gets to the Surface

Theory and observation dovetail in describing the interior of the sun. Theory allows calculation of the 15-million-K temperature at the center of the sun (or the center of any gas ball gravitationally contracted to the sun's size). This calculation agrees with the laboratory observations on the temperature required to sustain the proton-proton reactions in the sun's central region, or **solar core,** where nuclear energy is generated.

Other conditions between the core and surface can be calculated by using

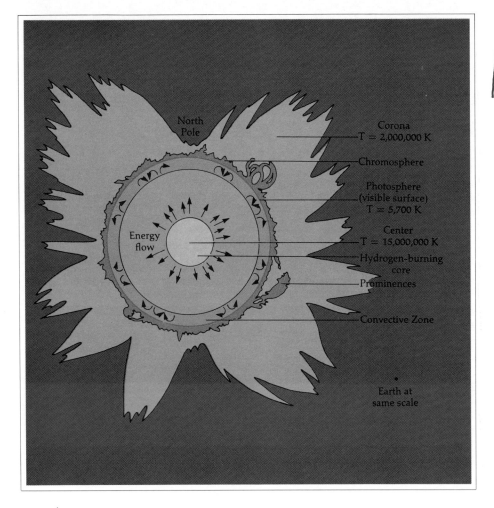

North
Pole

Corona
T = 2,000,000 K

Chromosphere

Photosphere
(visible surface)
T = 5,700 K

Center
T = 15,000,000 K

Hydrogen-burning
core

Prominences

Convective Zone

Energy
flow

Earth at
same scale

Figure 12·6 **A cross-section of the sun and its atmosphere, showing (to approximate scale) the energy-producing core, the outer convective zone, and the thin, outer corona. (T = temperature.)**

equations that describe the pressure at any depth, the properties of gas under different pressures, and energy generation rates at points inside the sun; results appear in Figure 12·6.

These calculations indicate that the gas pressure at the sun's 15-million-K core is about 250 billion times the air pressure at the earth's surface. This high pressure compresses the gas in the core to a density of about 158 g/cm³—158 times as dense as water and about 20 times as dense as

iron. One cubic inch of this gas would weigh nearly six pounds! The core of the sun occupies about the inner quarter of the sun's radius. This 1/64th of the sun's volume contains about half the solar mass and generates 99 percent of the solar energy.

Because heat energy always flows from hotter to cooler regions, solar energy travels outward from the hot core, through a cooler zone of mixed hydrogen and helium, toward the surface. Throughout most of the sun's volume, this energy moves primarily by **radiation.** That is, the energy radiates through the gas, just as light travels through our atmosphere. Very little moves by **conduction,** the mechanism by which a pan on a stove becomes hot.

In the outer part of the sun we find a third mechanism of energy transport—**convection** (page 61, Chapter 3). Convection occurs when the temperature difference per unit length between the hot and cold regions is so great that neither radiation nor conduction can carry off the outward-bound energy fast enough. So-called "cells" of gas, having been heated enough to expand, become less dense than their surroundings and rise toward the surface.

The Photosphere: The Solar Surface

Energy ascending from inside the sun heats the photosphere, or bright surface layer of gas that radiates the visible light of the sun. The photosphere's temperature can be found quickly from Wien's law, which relates the wavelength of maximum radiation to the temperature of the radiating surface. On the solar spectrum the maximum solar radiation is of yellow color, wavelength about 5.1×10^{-5} cm. Application of Wien's law shows that the temperature of the photosphere is about 5,700 K.

If the sun is a giant ball of gas, why does it appear to have a sharply-defined surface? The answer involves the **opacity** of the gas, its ability to obscure light passing through it. Air, for example, has low opacity, whereas smoke has higher opacity. Gas at the bottom of the photosphere has many **negative hydrogen ions** (hydrogen atoms with an extra electron, designated H⁻) that obstruct light and cause high opacity. They produce an opaque layer, beyond which we cannot see. Only a few hundred kilometers above this layer, at the top of the photosphere, there are few H⁻ ions and the gas is clear. Most of the sun's light comes from a layer about 400 km thick, giving the appearance of a sharply defined surface. There is also hydrogen in the solar atmosphere above the photosphere, but there is no abrupt change in the gas density; a solid probe, if it could survive the 5,700 K temperature, could drop directly through the photospheric "surface" and plunge into the sun, like an airplane passing through the "surface" of a cloud.

The convective motions that bring mass and energy from the interior disturb the photospheric surface. A photograph of the sun shows pronounced **granules** in the photosphere, as shown in Figure 12·7. Each bright grain is a convection cell 1,000 to 2,000 km across, rising from the subphotospheric layers. Each grain rises at a speed of 2 to 3 km/sec and lasts for a few minutes. Dark regions between grains mark areas where cooled gas descends again into the sun.

Supergranulation is a larger-scale patterning observed in the granulation. Supergranules average some 30,000 km in diameter with gas flowing horizontally toward their outer edges. Surging wave motions have also been observed in the gas, with wavelengths about 5,000 km, and periods of about 5 minutes. Keep in mind that the moving masses being churned on the surface of the sun are as large as the entire earth!

Chromosphere and Corona: The Solar Atmosphere

Shooting up into the solar atmosphere from among the supergranulations are **spicules:** columns of incandescent gas 10,000 km long. Essentially giant flames, they fade in 2 to 5 minutes and are a transition from the photosphere to the overlying layer called the chromosphere. They are shown in Figures 12·9 and 12·10.

The **chromosphere** (which means color layer) is a pink-glowing region of gas just above the photosphere. Its light is mainly the red hydrogen alpha emission line, mentioned above. E. G. Gibson (1973) has called the chromosphere "froth on top of the turbulent and relatively dense photosphere." The chromosphere can be seen by the naked eye during a solar eclipse. When the moon covers the rest of the solar disk, this thin outer layer is visible as a ring of small, intense red flames. As sketched in Figure 12·6, the chromosphere is a thin layer, about 2,500 km thick. In its upper regions the temperature exceeds 10,000 K.

Above the chromosphere is the rarefied, hot gas of the **corona.** Whereas the gas density is about 10^{-6} g/cm^3 within a few hundred kilometers of

Figure 12·7 A detailed photo of the solar surface in the region of a sunspot. Outside the sunspot, the normal solar surface is mottled by granulations, believed to be convective cells in the solar gas. The main sunspots are comparable to the earth in size, and the large granulations are comparable to continents. (Balloon-borne telescope photo, Princeton University, Project Stratoscope, supported by NSF, ONR, and NASA) ▼

Figure 12·8 Oblique view of convection cells in cumulus clouds from about 10 km over Indiana, showing a convective pattern similar to that found on a larger scale in the solar gas. ▼

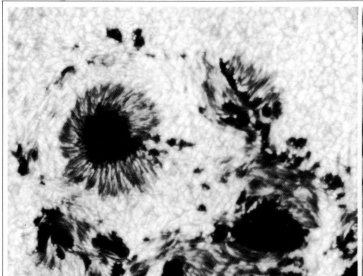

Figure 12•9 Spicules, or clustered jets of solar gas, photographed in the red Hα light of hydrogen. Though the spicules are luminous gas, they appear darker than the bright solar background when silhouetted against the photosphere. (Hale Observatories)

Figure 12•10 A schematic, oblique view of the solar surface, showing the types of features mentioned in the text. ▶

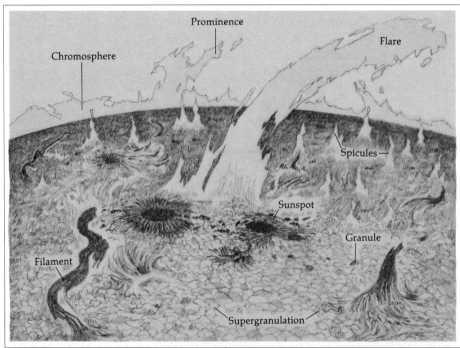

Chromosphere

Prominence

Flare

Spicules

Sunspot

Granule

Filament

Supergranulation

Figure 12•11 **Structure of the sun's corona revealed during the total eclipse of March 7, 1970. Exposure was made with a radial varying filter, designed to offset the brightening of the corona toward the sun.** (Photographed in Mexico by High Altitude Observatory, National Center for Atmospheric Research, supported by NSF)

the photosphere, it drops to 10^{-10} g/cm³ in the middle chromosphere and to less than 10^{-14} g/cm³ in the lower corona. Clearly the corona is only the outermost tenuous atmosphere of the sun (Figure 12•11).

During eclipses, the corona is visible to the naked eye as a pearly glowing gas around the sun. Although Plutarch recorded this, some early astronomers thought it was an optical illusion. Early photographs finally proved its existence, and spectra were obtained in 1869, eventually revealing that the coronal gas has extremely high temperatures of 1 to 2 million K! In 1930, French astronomer Bernard Lyot built an instrument called the **coronograph,** which artificially eclipses the sun and allows the intriguing gases of the corona and chromosphere to be studied at will (Figure 12•12).

Why are both the chromosphere and corona hotter than the photosphere, since they are farther from the sun's internal energy source? Apparently, magnetic effects and shock waves from the violent sub-surface convection transfer large amounts of energy to this gas. The coronal gas expands and merges with the interplanetary gas and dust. While the inner glow of the corona comes from solar gas, the outer glow is really the inner zodiacal light—sunlight reflected from interplanetary dust.

Sunspots and Sunspot Activity

Although sunspots sometimes can be seen with the naked eye when the sun is dimmed by fog or a dark glass, their nature was not realized until 1613, when Galileo reported his telescopic studies:

Having made repeated observations, I am at last convinced that the spots are

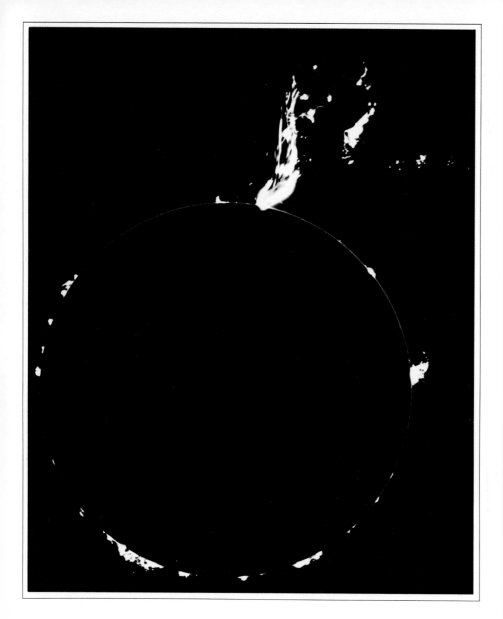

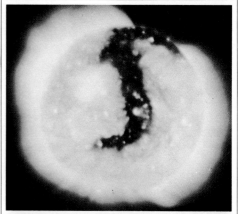

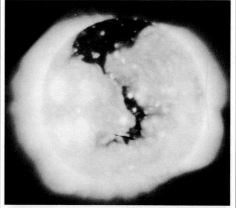

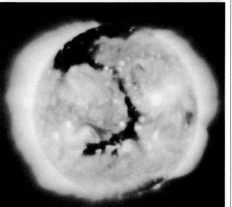

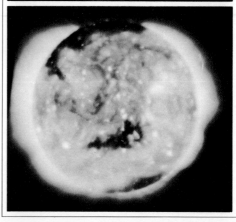

objects close to the surface of the solar globe, where they are continually being produced and then dissolved, some quickly and some slowly; also that they are carried round the sun by its rotation. . . .

A **sunspot** is a storm cloud whose motions are controlled not by atmospheric forces, as with terrestrial storms, but by magnetic fields of the sun. *Ions* (charged atoms or molecules), which are common in the sun, cannot move freely in a magnetic field, but

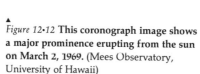

Figure 12·12 **This coronograph image shows a major prominence erupting from the sun on March 2, 1969.** (Mees Observatory, University of Hawaii)

Figure 12·13 **The coronal structure of the sun's atmosphere, photographed by Skylab astronauts using an X-ray-sensitive telescope. This technique revealed rifts in the coronal structure, through which the cooler (darker) surface gases can be seen. The evolution of such a rift is shown in the sequence.** (NASA, American Science and Engineering, Inc.)

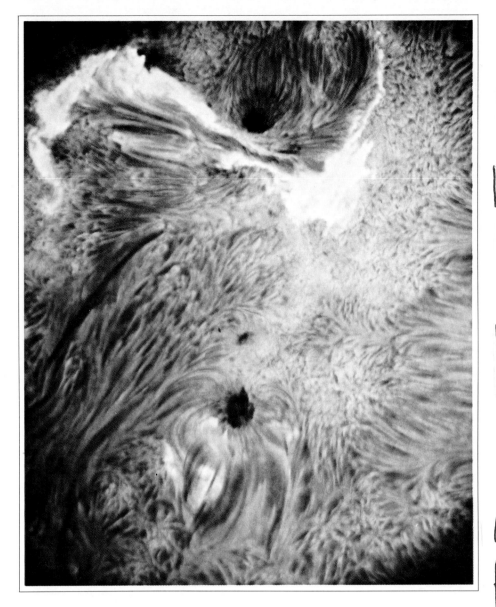

▲
Figure 12·14 **Vertical view of gaseous streamers and solar flare activity (bright region, top) around a pair of sunspots (the dark regions, top and bottom).** (Air Force Cambridge Research Laboratories)

be seen when silhouetted above the solar limb, or edge. *Eruptive prominences* shoot out at speeds averaging about 1,000 km/sec. *Quiescent prominences* are masses of flowing gas held in relatively fixed positions above the sun for hours or days by magnetic fields. The largest blasts of material and their very active sunspot sites are called **flares.** Figures 12·12 through 12·16 show examples of flares, prominences, and related eruptions.

The 22-Year Solar Cycle

Around 1830 an obscure German amateur astronomer, H. Schwabe, began observing sunspots as a hobby. After years of tabulating his counts, he announced in 1851 a **solar cycle**: the *number* and *positions* of sunspots vary in a cycle, as shown in Figure 12·18. This discovery, followed a year later by discovery that terrestrial magnetic compass deviations exactly follow the same cycle, was a key step in understanding the sun and its effects on earth.

The cycle averages 22 years long, and consists of two 11-year subcycles, shown in Figure 12·17. At a time of "sunspot minimum" (when there is a minimum number of sunspots), the few visible spots are grouped within about 10° of the solar equator. When a new cycle begins in a year or so, groups of new spots are at high latitudes, about 30° from the solar equator. The spots often appear in pairs, and the eastern spot in each northern-hemisphere pair is of specific polarity, for example, a north magnetic pole. In the southern hemisphere this polarity is reversed. After a few years, the sunspot number reaches a maximum and the spots are at intermediate latitudes, about 20° from the solar equator. After about 11 years, the

must stream in the direction of the field, for example from the north magnetic pole to the south.

A gas with many ions is called a **plasma,** and unlike a neutral gas, its motions are strongly influenced by magnetic fields. For this reason, plasmas in the sunspots and elsewhere in the solar atmosphere move in peculiar patterns that indicate the twisted patterns of the solar magnetic field. Sunspots look dark only because their gases, at 4,000–4,500 K, radiate less than the surrounding gas at about 5,700 K.

Huge clouds of gas, larger than the whole earth, shoot out of the disturbed regions of sunspots. These are called **prominences** (Figure 12·12) and can

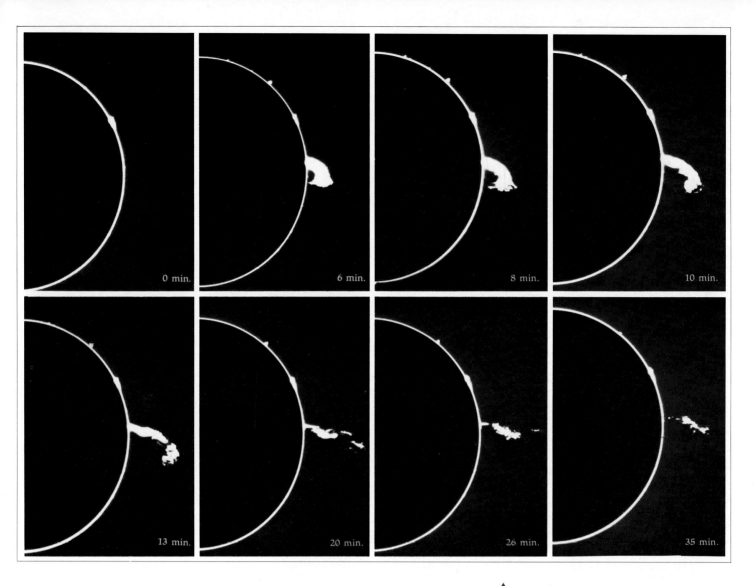

| 0 min. | 6 min. | 8 min. | 10 min. |
| 13 min. | 20 min. | 26 min. | 35 min. |

spots appear mostly about 10° from the equator, and a sunspot minimum occurs again.

Now the cycle begins to repeat, except for a noticeable difference. The new spots forming at ± 30° have reversed polarity. The eastern spots in the northern hemisphere are now south magnetic poles! Thus, it takes another 11 years to complete the full cycle, when all features resume their initial pattern. A sunspot maximum should occur around 1979.

Still more remarkable is the fact that the magnetic field of *the entire sun* reverses about every 11 years, thus participating in the 22-year cycle. Imaginary observers on the sun would find their compasses pointing "north" in one direction for 11 years (subject to disturbances by frequent magnetic storms), and in exactly the opposite direction for the next 11 years. This behavior is not entirely unknown: the earth's field reverses every few hundred thousand to few million years. Both patterns of reversal may involve cyclic flow patterns in the deep liquid or gaseous cores of the two bodies.

According to American astronomers H. Babcock (1961) and R. Leighton (1969), the sunspot cycle occurs be-

Figure 12·15 **This sequence of photos shows a jet of gas blasting off the sun over a period of 35 minutes. The photos were made with a coronograph, which obscures the bright solar disk and allows solar atmospheric activity to be monitored.** (National Center for Atmospheric Research)

cause the sun rotates faster at its equator than near its poles. This causes a shearing and twisting of the magnetic field that controls motions of the ionized solar gas, as shown in Figure 12·19. The field is further distorted by convection, until the twisted pattern

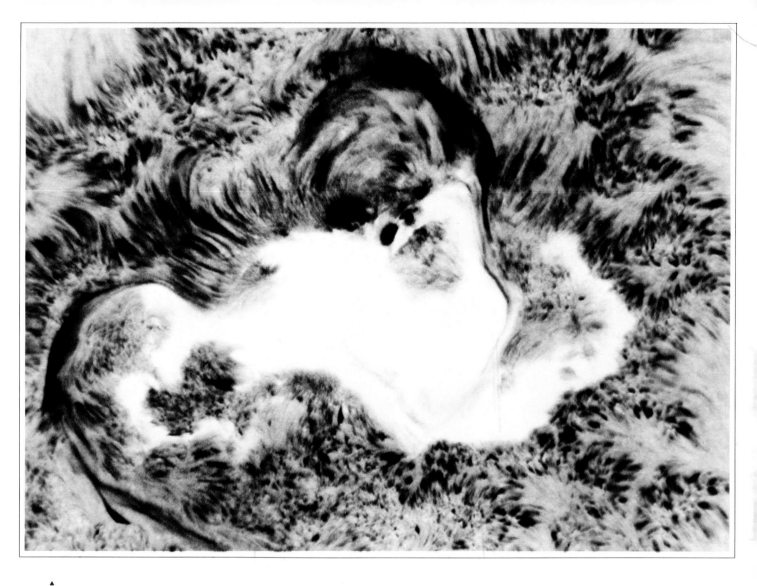

▲
Figure 12·16 **The solar flare of August 7, 1972, photographed at peak intensity in Hα light.** (Hale Observatories)

Figure 12·17 **Sunspot counts since 1700 show the cycle averaging 11 years for sunspot numbers (half the 22-year magnetic cycle),**
▼

with evidence for a longer, 80-year cycle (dashed line). (After data of M. Waldmeier in Gibson, 1973)

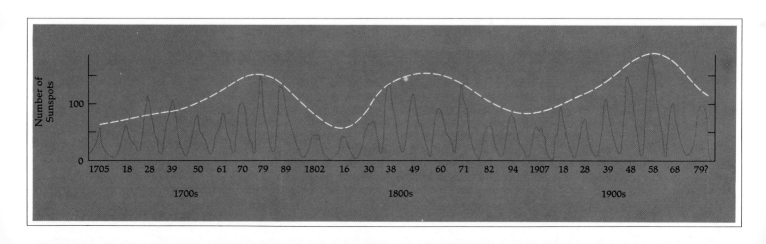

Years	0	2	4	6	8	10	12	14	16	18	20	22
Sunspot number	Minimum		Maximum			Minimum			Maximum			Minimum
Sunspot latitude	High		Medium		Low		High		Medium		Low	High
Flares, storms, prominences	Minimum		Maximum			Minimum			Maximum			Minimum

Figure 12·18 **A schematic sequence of solar changes during the 22-year cycle. Typical coronal appearances are shown for the minimum and maximum of the cycle in the two drawings at left.**

breaks and a new pattern forms, starting a new cycle.

Solar Wind

The importance of sunspots to us on earth becomes clearer when we realize that particles shot out of the sun, especially sunspots, rush outward through interplanetary space. The solar coronal plasma, having been heated to nearly 2 million K by the violence of photosphere convection, expands rapidly into space (limited only by magnetic forces acting on charged particles and gravitational forces acting on all particles). Together these effects cause the **solar wind:** an outrush of gas past the earth and beyond the outer planets. Near the earth, the solar wind travels at velocities near 600 km/sec, and sometimes reaches 1,000 km/sec. The gas has cooled only to 200,000 K, but is so thin that it transmits no appreciable heat to the earth. The solar wind extends at least as far as Jupiter's orbit, according to data from Pioneer 10 and 11.

In addition to the solar wind, solar radiation itself exerts an outward force on small dust particles. This effect, which is greater on smaller particles, is called **radiation pressure.** Together, these two forces blow micrometeorites out of the solar system soon after they are dislodged from comets and asteroids.

Aurorae and Solar–Terrestrial Relations

Solar radiation and particles blasted out of the sun strike the earth. Because 99.98 percent of all energy passing through the earth's atmosphere comes from the sun,[1] it is not surprising that

[1]Based on a tabulation by Hubbert, 1971; the rest is in the form of tidal energy, starlight, cosmic rays (atomic particles from space), and subterranean geothermal energy.

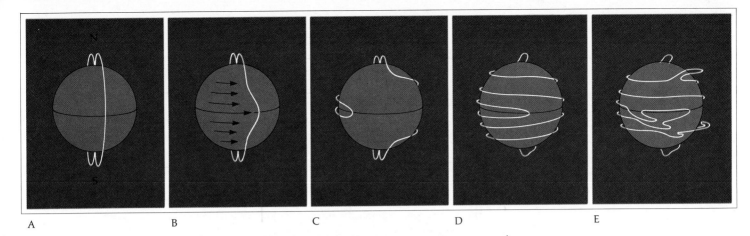

A   B   C   D   E

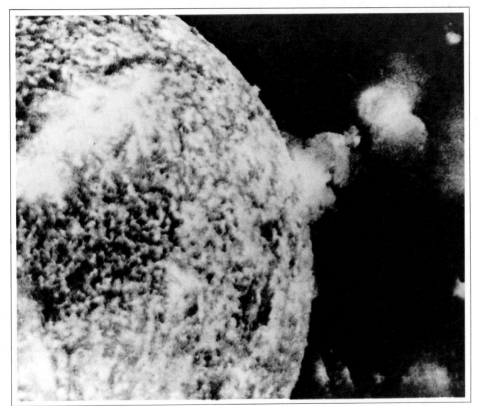

Figure 12•19 **Some features of the solar cycle are explained by this simple model following the history of a "magnetic line," which defines the direction of the solar magnetic field (A). Because of differential rotation of the sun (B), and because the field lines must move with the ionized gas, field lines tend to distort (C), becoming wrapped around the sun (D). Convection and eruptions further distort the field (E). After about 11 years the field becomes unrecognizable and neutralized, and a new field forms.**

◄ *Figure 12•20* **Detached prominences being blown off the sun during storm activity, shown in a spectroheliograph photo. (NASA)**

small "flickers" in solar output cause major effects on earth.

Effects come from changes both in solar radiation and in solar particles. During a solar flare, the total visible radiation from the sun changes by much less than 1 percent, but the X-ray radiation may increase by a hundredfold. The X-rays are energetic, and when they strike the earth's upper atmosphere they change its distribution of ions and affect radio transmission on the ground.

The solar particles that strike the upper atmosphere also change its chemistry and create vivid glows, called **aurorae** (singular: aurora). As shown in Figure 12•23, these are fantastic, colored, swirling patterns of light visible in the night sky occasionally (more often during sunspot maxima) from locations near the north and south poles.

Discovering the cause of the aurora took 250 years. As early as the 1700s, the English astronomer Edmund Halley discovered that auroral forms lie along lines of the earth's magnetic field. Around 1920, Norwegian observer F. C. Störmer showed that aurorae occurred at an altitude usually exceeding 100 km (65 mi).

In 1950 American astronomer Aden Meinel showed that the solar atomic particles enter the atmosphere at several thousand km/sec, disturbing atmospheric atoms at heights of about 100–300 km. Artificial satellites revealed in 1958 that many of these particles are trapped temporarily in the earth's magnetic field, before they hit the earth. Here they form the **Van Allen belts** of high-energy ions, shown in Figure 12·24. Aurorae are produced when swarms of these particles are funnelled by the earth's magnetic field and dumped from the belts into the regions where the belts touch the atmosphere—around the north and south magnetic poles. The ions collide with air molecules, causing spectral emission bands whose colors are determined by the elements and collisional energies involved. Even under ordinary conditions, when no aurora is visible, atomic interactions in the upper atmosphere cause a faint glow called *airglow* (see Figure 12·26).

Major solar flares often cause major aurorae and magnetic disturbances

Figure 12·21 **X-ray image of flares and solar corona made with a specially constructed telescope in the Skylab space station, 1973. X-ray emissions are characteristic of million-degree-K gas found in the corona.** (NASA, American Science and Engineering, Inc.)
▼

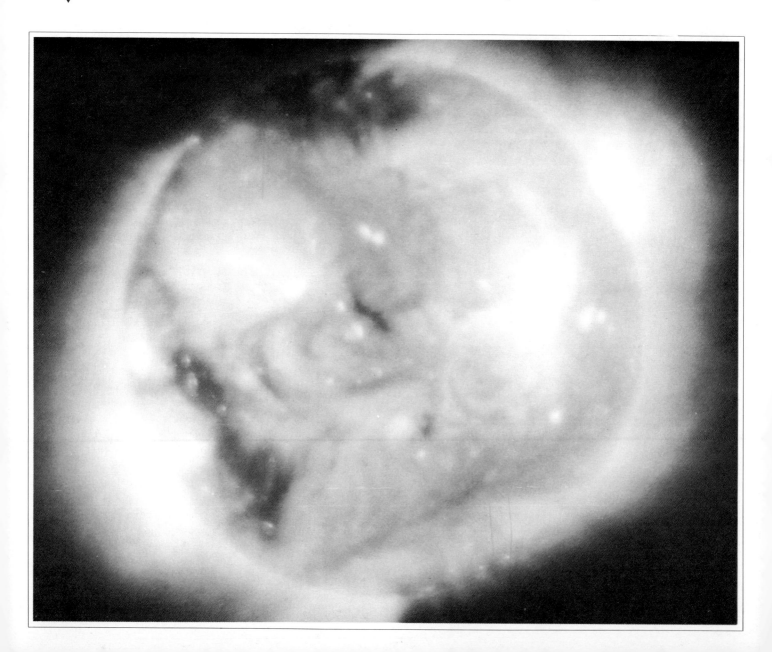

Figure 12·22 **Surface and atmospheric activity of the sun. During the June 30, 1973 solar eclipse in Kenya, the outer corona of the sun was photographed in white light (outer image). About an hour earlier, astronauts in the orbiting Skylab photographed the solar** surface (circular inset) using X-ray radiation, showing centers of flare activity. Major streamers in the outer corona are shown to be aligned with X-ray flares on the surface. (National Center for Atmospheric Research; American Science and Engineering, Inc.; and NASA)

Figure 12·23 **Typical forms of the aurora. A. Vertical rays.** (W. A. Feibelman, Pittsburgh) **B. Overhead "corona" photographed from State College, Pa. C. Curtain form.** (From NASA aircraft over Canada)
▼

A   B   C

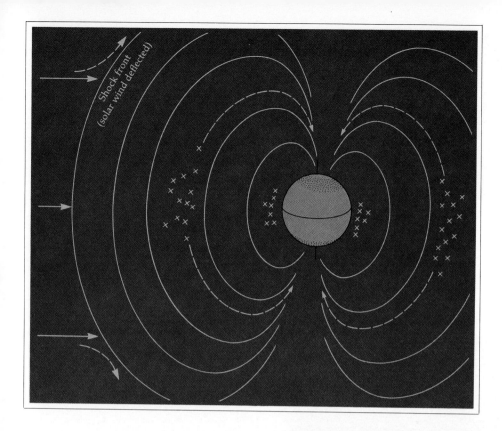

on earth. During these disturbances, compass needles may deviate by a degree from normal, and aurorae may be seen from such low latitudes as Mexico and India. Electromagnetic effects may overload electric power lines, burn out transformers, and disrupt telephone, telegraph, radio, and power transmission.

Is the Sun Constant?

Solar radiation reaches the earth at a rate called the **solar constant**. In cgs units it is 1.4×10^6 ergs/cm²/sec. That is, 1.4 million ergs of energy reach every sunward-facing square centimeter of earth every second. In units used by energy engineers, this is 1.4 kilowatts/meter².

This rate is called the solar constant because astronomers once assumed that it was unvarying. But evidence now indicates that it varies by small

A

B

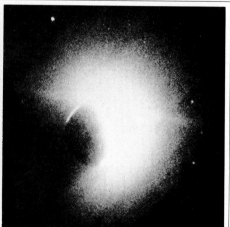

amounts, and that the larger the time period we consider, the larger the variation that may occur. During the few days of a single flare, X-ray radiation may change dramatically, though the total solar output changes by less than a percent. During the 22-year solar cycle, the number of these flares

changes. Moreover, the 22-year cycle itself is erratic. During the period 1645–1715, for example, sunspot numbers were very low, solar flares were virtually absent, and coronal light was pale during eclipses (Eddy, 1976). This period is called the **Maunder minimum,** after an astronomer who recorded sunspots during this time.

The importance of these results is that changes in solar activity can change climatic conditions on the earth (and other planets). Though the amount of climate change is debatable, increasing evidence from tree rings, radio-carbon dating, and other evidence shows that climatic conditions were measurably altered during the Maunder minimum, and that climatic trends correlate with solar changes (Schneider and Mass, 1975). Thus, subtle solar changes could drastically affect food production, because crop yields re-

spond to average temperature changes of a few degrees or less. For example, Figure 12·27 shows a correlation between the sunspot cycle and murderous African droughts, which seem to concentrate during periods of sunspot minimum. Understanding solar variations, therefore, is important to world social stability.

To make matters worse, a monkey wrench was thrown into solar theory in the 1970s when experiments failed to detect the flood of solar neutrinos (subatomic particles) predicted from the proton-proton reaction. If the proton-proton reaction is proceeding inside the sun, there should be many neutrinos striking the earth, but few have been detected at all! Therefore, some astronomers have questioned the entire modern theory of the sun's interior, energy generation, and stability. Some have suggested the possibility that the reactions in the solar core are proceeding at a lower rate than in the past. If so, the sun may have decreased its luminosity in the last 10 million years or so.

Solar variations on time scales of 1 to 200 million years could help explain several solar system mysteries, such as climate variations on Mars and known ancient climate variations on earth. They might even have had biological effects, such as causing the decline of the reptiles about 70 million years ago. Astrophysicist A. G. W. Cameron has even suggested that solar cooling during the last 6 million years may have created a challenging change in environment that put a premium on the biological trait of cleverness, favoring rapid emergence of the primate *Homo sapiens*.

The whole question of solar influence on biology and climate is attracting new research. A Soviet–American agreement has increased international exchange of information about this subject (Wilcox, 1976), and solar studies may have increasing importance for astronomy, meteorology, biology, geology, agriculture, and world economics.

Figure 12·26 **Airglow, the emission of light from the earth's high atmosphere by atoms excited by sunlight, lightens the sky even on dark nights. Auroralike bands are seen in this infrared photo of emissions from OH molecules at wavelengths around 8,000 A. The horizon is at the bottom. The W-shaped constellation Casseopeia dominates this exposure, made with a 35 mm camera with infrared film and a red filter in New Mexico.** (Alan Peterson and Lois Kieffaber)

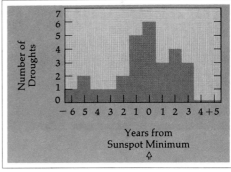

Figure 12·27 **A graph of suspected correlation between terrestrial weather and the solar sunspot cycle. Major droughts in Ethiopia, recorded from 1540 to 1974, appear to correlate with sunspot minima. Such studies may benefit agriculture by helping to predict weather cycles.** (Data from Wood and Lovett, 1974)

Solar Energy
and Other Cosmic Fuels

Solar energy originally stored in organisms has accumulated in the ground as coal and petroleum. Industrial civilization has been built with this fossil solar energy, and a region's standard of living is correlated with its rate of consumption of such energy. Unfortunately, while it took about 600 million years to accumulate the fossil fuels, we have burned through a significant fraction of them in only 100 years. If today's trends continue, fossil fuels will be gone in a few decades.

Thus we face a hurdle. As a planetary culture, we must be sure that the dwindling reserves of fossil fuel are used to ensure production of alternate fuels, because if we run out of fossil fuels, it won't be easy to develop the technology for alternate sources of energy.

With the sun providing every daylit square meter in the earth's vicinity with 1.4 kilowatts, solar energy seems a likely alternative to fossil fuel. The *total* energy needs of the United States (at the use rate of the 1970s) could be met by collectors operating at only 10 percent efficiency, spread in giant solar farms in desert areas covering only one-tenth the area of Arizona. (Engineering plans for such a project were designed by Aden Meinel, the astronomer who first measured velocities of auroral particles.) Still grander schemes have been proposed using the greater collection efficiency possible in the permanent sunlight of space. Engineering studies have indicated the economic feasibility of building orbiting space colonies from lunar or asteroidal materials, with solar collectors that convert solar energy to radio waves and beam it back to earth (Figure 12·28).

Figure 12·28 **A preliminary design study of an earth-orbiting solar power station is shown in this NASA artist's conception. Giant solar-collector panels could be fabricated from materials brought from** earth (or possibly from lunar or Apollo asteroid material). Energy derived could be beamed to earth by microwave and converted to pollution-free electricity.

More down-to-earth is the fact that the roof area of a typical home intercepts roughly 1,000 to 2,000 kilowatt-hours of energy per day, equalling the average daily energy consumption of a U.S. household (Snell et al., 1976). Electricity-producing roof shingles are being studied as one means of using this energy. Many people are just realizing that American homes during the last few decades were designed when energy costs were being kept artificially cheap, due to government policies. Insulation was skimpy because it was more expensive than the electricity to heat or cool homes. Today, as energy runs out and energy costs rise correspondingly, home energy costs in many American homes exceed $1,000 per year and may rise further, so that self-sufficient solar-energized homes are increasingly attractive.

In contrast to fossil fuels, energy from basic planetary or astronomical sources might be called **cosmic fuels.** In addition to solar energy, cosmic fuels include nuclear fission energy (used in present-day reactors), nuclear fusion energy (use of solar-type fusion reactions, possibly using water as a source of hydrogen; still in the experimental stage), geothermal energy (heat from the interior of the earth), and energy from tides and winds. Although nuclear fission energy is the current favorite to get the earth over the short-term hurdle, it has the overwhelming disadvantage of involving, or even creating, deadly radioactive wastes that could wreak health havoc if they escaped into the environment, either by accident or terrorism.

The other cosmic fuels are cleaner than nuclear or fossil fuels. They have an additional subtle advantage: since they can be collected more or less cheaply anywhere in the world, they could make countries independent of each other in terms of energy, allowing a more stable planetary culture. As long as the planet depends on fossil fuels, it can be dominated by the economic power of whatever nation controls the cheapest reserves (the United States once did, but has now burned through them). Thus, it is up to our generation to see that we respond to the coming exhaustion of fossil fuels not by inaction or a panic-stricken rush to control the last reserves, but by a systematic conversion from fossil fuels to safe cosmic fuels.

12

Summary

The sun can be studied both observationally and theoretically. Observational studies emphasize information about solar gas revealed by spectral absorption and emission lines created in the solar spectrum as sunlight from the sun's interior passes outward through the layers of solar gas. These studies reveal, among other things, that the sun is about three-fourths hydrogen; most of the rest is helium, and a few percent is composed of heavier elements.

Theoretical studies reveal that as the sun formed by contraction of an interstellar gas cloud, it got so hot that atoms at the center collided at high speeds. These collisions cause nuclear reactions in which hydrogen atoms are fused into helium atoms, releasing energy. This nuclear fusion is the source of the sun's light and heat. The most important reactions in the present-day sun are a series called the proton-proton cycle.

Transport of this energy from the sun's center to the outer layers causes violent disturbance of the surface, producing phenomena such as granules, prominences, flares, and sunspots. Particles are shot off the sun in outward-moving gas called the solar wind, which interacts with the earth, causing aurorae and other phenomena.

Consideration of the sun in comparison with other energy sources used on earth shows that we are very rapidly consuming our planetary budget of fossil fuels and will soon have to convert to cosmic energy sources, such as solar energy, to maintain our present rates of energy consumption.

Concepts

solar rotation

spectroscopy

light

speed of light

frequency

wavelength

photon

white light

spectrum

absorption lines

emission lines

continuous spectrum

Kirchhoff's laws

photosphere

spectroheliograph

hydrogen alpha line

nuclear reactions

chemical reactions

proton-proton cycle

photon

solar core

radiation

conduction

convection

opacity

negative hydrogen ions

granules

supergranulation

spicules

chromosphere

corona

coronograph

sunspot

plasma

prominences

flares

solar cycle

solar wind

radiation pressure

aurora

Van Allen belts

solar constant

Maunder minimum

cosmic fuels

Problems

1. By what logic do astronomers infer that the sun's energy comes from nuclear fusion reactions of the proton-proton cycle? How do we know it does not come from chemical burning?

2. If you see a cluster of sunspots in the center of the sun's disk, how long would the spots take to reach the limb, carried by the sun's rotation? How long would it take for the cluster to appear again at the center of the disk?

3. Why is the solar cycle said to be 22 years in length even though the number of sunspots rises and falls every 11 years?

4. Suppose you could make detailed comparisons of the appearance of the sun at different moments of time. What variations in appearance would you see if the intervals were:
   a. *10 minutes?*
   b. *1 week?*
   c. *5 years?*
   d. *10 years?*
   e. *100 million years?*
   f. *9 billion years?*

5. Why is a radio disturbance on earth likely to occur within minutes of a solar flare near the center of the sun's disk, whereas an aurora occurs a day or two later, if at all?

6. How much more massive is the sun than the total of all planetary mass? (See Table 6·1 for data.)

7. What is the most abundant element in the solar system? The second most abundant element?

8. If the earth formed in the same gas cloud as the sun, why is the earth made from different material than the sun?

9. Why will the sun change drastically in several billion years?

10. Why is the sun's energy gen-erated mostly at its center, and not near its surface?

Advanced Problems

11. Using the small angle equation, calculate:
   a. *The angular size of a sunspot that has the same linear diameter as the earth.*
   b. *The angular size of the earth as seen by an imaginary observer on the sun.*

12. At what velocity must particles move to escape from the surface of the sun?

13. Use Wien's law to confirm the temperature estimate for the sun's surface, based on a maximum energy emission at wavelength 5.1×10^{-5} cm, as quoted in the text.

Projects

1. According to the principle of the pin-hole camera, light passing through a small hole will cast an image if projected onto a screen many hole-diameters away. Confirm this by cutting a one-inch hole of any shape in a large cardboard sheet and allowing sunlight to pass through the hole onto a white sheet in a dark room or enclosure several feet away (10 feet or more if possible). Confirm that the projected image is round, an actual image of the sun's disk. Are any sunspots visible?

2. Cut a one-inch hole in a sheet of cardboard and use it as a mask over the end of a small telescope. *After* masking the telescope, point it toward the sun and project an image of the sun through an eyepiece onto a white card. UNDER NO CIRCUMSTANCES SHOULD ANYONE EVER LOOK THROUGH THE EYEPIECE, SINCE ALL THE LIGHT ENTERING THE TELESCOPE IS CONCENTRATED AT THAT POINT, AND CAN BURN THE RETINA! Professional supervision of the telescope is suggested. Note also that an unmasked large telescope may concentrate enough light and heat to crack eyepiece lenses.

Are sunspots visible? If so, trace the image on a piece of paper and re-observe on the next day. Confirm the rotation of the sun by following sunspot positions for several days. Class records kept from year to year can be used to record the cyclic variations of numbers of sunspots.

Measuring the Basic Properties of Stars

Because of optical limitations and turbulence in the earth's atmosphere, the world's largest telescope cannot distinguish details smaller than a few hundredths of a second of arc (about ten millionths of a degree). But the stars with the largest apparent angular size are no larger than about 0.04 seconds. So the disks and surface details of nearly all distant stars are hidden from us, in contrast to the great detail we can study on the sun.

In spite of this, we know that there are giant stars bigger than the whole orbit of Mars, stars the size of earth, and stars the size of asteroids. There are red stars and blue stars. There are stars of gas so thin you can see through parts of them, and stars with rock-like crusts that may contain diamonds. There are stars that are isolated spheres, stars with disk-like rings around them, and stars that are exploding.

All of these bodies fit the definition of stars: **stars** are objects with so much central heat and pressure that energy is (or has been) generated in their interiors by nuclear reactions. The most familiar stars visible in the night sky are balls of gas with solar composition and sizes usually a few times smaller or a few times larger than the sun.

How can we know all these details about stars if we cannot even see the disks of stars in telescopes? In the next chapters we will describe the details of familiar and unfamiliar types of stars, but first we will describe *how* we learn about them.

Names of Stars

Stars are named and cataloged by several systems. Because Ptolemy's *Almagest* was prepared in the Alexandrian library and passed on by Arab astronomers, many of the brightest stars have Arab names. Since *al-* is the common Arabic article, many star names start with *al:* Algol, Aldebaran, Altair, Alcor; other scientific "*al* words" also have Arab origins: algebra, alchemy, alkali, almanac. Stars in constellations are cataloged in approximate order of brightness using Greek letters. Thus the brightest star in the constellation of the Centaur is called Alpha Centauri. Fainter stars or stars with unusual properties are often known by English letters or catalog numbers, such as T Tauri or B.D. 4° + 4048.

Images of Stars

Figures 13·1 and 13·2 show typical telescopic photographs of bright stars. The strange patterns of light, such as rays or circles, have nothing to do with the actual shapes of the stars, which would be tiny pinpoints buried in these over-exposed images, but are caused by optical effects in the telescope and film system.

Visual examination of a bright star in a modest telescope shows what appears to be a tiny disk, sometimes surrounded by faint rings, but this is caused by an optical effect called **diffraction.** The disk is sometimes called the **Airy disk,** after the English astronomer who explained it. It again has nothing to do with the real star, and in fact prevents the much smaller real star image from being seen.

In 1975, astronomers at Kitt Peak National Observatory picked one of the stars with largest apparent angular size, the bright star Betelgeuse in the constellation Orion, and used computer techniques to remove the blurring due to atmospheric turbulence and diffraction. Their analysis produced an image of a fuzzy disk, shown in Figure 13·3, which is believed to be the actual disk of Betelgeuse. Analysis showed that this star's actual linear diameter is 500 to 750 times larger than the sun's. This alone proves that not all stars are alike!

Defining a Stellar Distance Scale: Light-Years and Parsecs

The vast distances that separate stars and make them so hard to observe are awkward to express in ordinary units. Astronomers use units appropriate to these distances. The easiest to understand is the **light-year**, the distance light travels in one year, which is about 6 million million miles, or 10^{18} cm.

Remember that the light-year is a unit of distance, not of time. The common mistake of using light-year as if it were a unit of time is like saying the Empire State Building is 6 hours tall.

The nearest star beyond the sun, Proxima Centauri (which is in orbit around Alpha Centauri), is about 4.3 light-years away. The sun could be said to be 8 light-minutes away. The north star, Polaris, is about 650 light-years away. Polaris' light takes about 650 years to reach us, so we are seeing it now as it was in the mid-1300s. If Polaris had suddenly exploded in 1950, we would not know it until about A.D. 2600.

Figure 13·1 **This overexposed image of the star Altair and other stars only a few minutes of arc away illustrates the great differences in brightness among stars. The true disk of Altair would be a pinpoint at the center of the overexposed image. Raylike structures are optical effects produced in the 30-inch telescope used to make the 30-minute exposure. Field of view about 6 × 9 minutes of arc.** (Walter Feibelman, Allegheny Observatory)
▼

▲
Figure 13·2 **A typical overexposed and enlarged stellar image from a reflecting telescope shows a prominent "cross-hair" pattern caused by optical diffraction effects of the secondary mirror's supporting structure. Also visible is a halation ring caused by reflection off the back of the glass photographic plate. Both effects occur inside the telescope. Neither effect relates to the true star, which would be a microscopic dot at this scale. Fainter stars provide too little light to show either effect.** (Mauna Kea Observatory, University of Hawaii)

Figure 13·3 **Reconstructed image of the actual disk of the star Betelgeuse. The star's diameter approximates the diameter of Jupiter's orbit; it is classed as a supergiant. The image suggests limb (edge) darkening, also observed on the sun.** (Kitt Peak National Observatory)
▼

Astronomers more commonly use a still larger unit of distance called the **parsec:**

1 parsec = 3.26 light-years

$$= 3 \times 10^{18}\ cm.$$

In this book we will use the parsec and its multiples (kiloparsecs, megaparsecs) to express cosmic distances as we move to more remote parts of the universe. You can convert parsecs to light-years approximately by multiplying by 3. Another convenient fact to remember is that near the sun, stars are roughly a parsec apart. For instance, Alpha and Proxima Centauri are about 1.3 parsecs away.

▲
Figure 13·4 **The constellation of Orion the hunter (stick figure) showing Greek letter designations and approximate visual magnitudes of selected stars. This exposure (on Tri-X film with a 35 mm camera) records stars to about tenth magnitude.**

Defining a Brightness Scale: Apparent Magnitude

A nearby candle may *appear* to be brighter than a distant streetlight, whereas in *absolute* terms the candle is much dimmer. This statement contains the essence of the problem of stellar brightnesses. A casual glance at a star does not reveal whether it is a nearby glowing ember, or a distant great beacon. Hence astronomers distinguish between **apparent brightness** (the brightness perceived by an observer on earth) and **absolute brightness** (the brightness that would be perceived if all stars were magically placed at some identical, standard distance).

Let us look first at apparent brightness. Because scientists generally use the decimal system, we might expect astronomers to use a brightness scale in which each unit of brightness difference corresponds to a factor of ten. No such luck. Astronomers, too, are victims of history. When Hipparchus cataloged 1,000 stars in about 130 B.C., he ranked their **apparent magnitude,** or apparent brightness, on a scale of 1 to 6, with "1st-magnitude" stars the brightest and "6th-magnitude" stars the faintest visible to the naked eye. This system stuck.

For more precision, nineteenth- and twentieth-century astronomers extended that scale to cover objects brighter than 1st magnitude and dimmer than 6th magnitude. Because Hipparchus' 1st-magnitude stars are about 100 times brighter than his 6th-magnitude stars, they used this as the definition of the brightness system. A difference of 5 magnitudes is *defined* as a factor 100 in brightness. Therefore, any given star is about $2\frac{1}{2}$ times brighter than a star of the next fainter magnitude.[1]

Astronomers have also expressed the absolute brightnesses, or intrinsic brightnesses, of stars by using the magnitude system. A measure of absolute brightness in the magnitude system is called a star's **absolute magnitude,** and is equal to the apparent magnitude a star *would* have if it were located 10 parsecs away from us. The system of absolute magnitudes is defined so that a star with the same energy output as the sun has an absolute magnitude of $+5$. A star with 100 times that energy output is said to have an absolute magnitude of 0. A star 100 times fainter has an absolute magnitude of $+10$.

---

[1]The actual factor is $\sqrt[5]{100}$, or 2.512. This unusual factor as a unit of brightness difference is used in no other science. Physicists are heard to mutter obscenities when they first try to analyze astronomical data expressed in magnitudes. But the 2,100-year-old magnitude system is so ingrained in astronomy that we continue to use it here.

Table 13·1 shows how the apparent magnitude scale has been extended, including fractional magnitudes. Extending the scale beyond the brightest objects included by Hipparchus meant using negative numbers. For example, the planet Venus at its brightest ranks more than minus 4th magnitude (-4.4), whereas the sun reaches be-

Table 13·1 Comparison of Some Apparent Magnitudes

Object	Apparent Magnitude
Sun	-26.5
Full moon	-12.5
Venus (at brightest)	-4.4
Mars (at brightest)	-2.7
Jupiter (at brightest)	-2.6
Sirius (brightest star)	-1.4
Canopus (second brightest star)[a]	-0.7
Vega	0.0
Spica	1.0
Naked eye limit in urban areas	3 or 4
Uranus	5.5
Naked eye limit in rural areas	6 to 6.5
Bright asteroid	6
Neptune	7.8
Limit for typical binoculars	9 to 10
Limit for 6-inch telescope	13
Pluto	15
Limit for visual observation with largest telescopes	19.5
Limit for photographs with largest telescopes	23.5

[a]This lesser-known southern-hemisphere star is used as a prime orientation point for spacecraft. A small light detector on the spacecraft is called the "Canopus sensor."

Source: Data from Allen (1973).

yond minus 26th (-26.5). As the famous astrophysicist Arthur Eddington remarked:

You have to remember that stellar magnitude is like a golfer's handicap—the bigger the number, the worse the performance.

Our accuracy in measuring stellar brightness is only 30 times better than Ptolemy's was 1,800 years ago. One problem is that stars have different colors, and light detectors (the eye, photographic films, photoelectric devices) have different sensitivities to different colors. For this reason, astronomers specify exactly what color any set of measurements refers to. Standards have been derived to express apparent magnitudes measured in blue light, red light, infrared light, and so on. Here, we will usually refer to a system having the same color sensitivity as the human eye, sometimes called **visual apparent magnitude,** and often abbreviated m_v.

Basic Principles of Stellar Spectra

The last chapter showed how the sun's **spectrum,** or distribution of light into different colors **(wavelengths),** gives information about the sun's atmosphere, surface layers, and interior. The same is true of other stars. So critical is **spectroscopy**—the study of spectra—to astronomy that many astronomers devote their entire careers to it.

Spectra can be presented in two ways, photographically or as a chart, depending on the instrument used (Figure 13·6). A **spectrograph** produces a photographic image of the spectrum, usually placed with blue to the left and red to the right. The

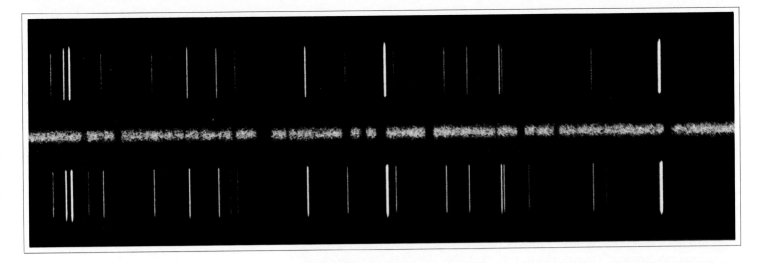

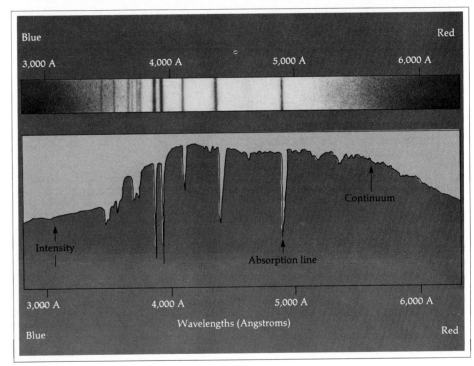

Figure 13·5 **High-resolution spectra photographed with a large telescope usually take this form. The stellar spectrum is the fuzzy horizontal luminous band in the center, crossed by vertical dark absorption lines. It is flanked at top and bottom by emission lines of a known element (usually iron) produced artificially by a spark inside the instrument. These lines, whose positions are known, permit measurement of wavelengths of lines in the stellar spectrum.** (Lick Observatory)

Figure 13·6 **The visual portion of a typical stellar spectrogram of a solar-type star, as shown on a photographic plate by a spectrograph (top), and in a scan by a spectrophotometer (bottom).**

intensities of the photographic print represent intensities of light, or radiant energy. The vertical width of the spectrum is completely arbitrary; sometimes such spectra are artificially enlarged vertically to make an image more readily measurable.

A **spectrometer** produces a spectrum in the form of a chart or graph (Figure 13·6). Wavelength is shown along the bottom, again with red usually to the right, and the vertical direction represents intensity of the light. **Absorption lines** appear as dark vertical lines on a photographic spectrum and as notches or valleys on a scan. **Emission lines** appear as bright

vertical lines or as sharp peaks. The general level of brightness between absorption or emission lines is called the **continuum.**

Spectra of Stars

In 1872, Henry Draper, a pioneer in astronomical photography, first photographed stellar spectra. This

represented a tremendous advance; instead of sketching or verbally describing spectra, astronomers could directly record, compare, and measure them. Spectra of thousands of stars became available for precise analysis.

Such massive amounts of data required a classification scheme. This was begun in the 1880s by Harvard astronomer Edward C. Pickering, and completed by Annie J. Cannon and a group of young women assistants who invented a system of **spectral classes** based on the number and appearance of spectral lines. The classes—A, B, C, and so on, started with spectra with strong hydrogen lines. When Annie Cannon published the Henry Draper Catalog (1918–1924), it contained spectral data on 225,320 stars and became the basis for all modern astronomical spectroscopy.

Further work showed that the classes had to be rearranged to bring them into a true physical sequence based on temperature, as shown in Figure 13·7. The sequence finally adopted begins with the hottest stars—class O—which show ionized helium lines in their spectra. The sequence of classes is O, B, A, F, G, K, M. The M stars are the coolest. About 99 percent of all stars can be classified into these groups.[2] For finer discrimination the classes are sometimes divided from 0 to 9; the sun, for example, is said to be a G2 star.

Spectral Lines: Indicators of Atomic Structures

Under normal conditions an atom consists of a positively charged nucleus surrounded by a number of negatively charged electrons. Each electron can

be pictured as following a specific orbit around the nucleus. Orbits are known as **energy levels.** Only certain orbits are possible. If left to itself long enough, the atom will reach an arrangement called the **ground state** where the electrons fill the orbits closest to the nucleus, known as the lowest-energy orbits. However, if there are disturbances, such as passage of light through the atom, the electrons may absorb energy from the light beam and be **excited,** or knocked to higher energy levels farther from the nucleus. If still more energy is absorbed, the atom may be **ionized,** or have one or more electrons knocked out of the atom altogether. This is the **origin of absorption lines:** energy is removed from a light beam and used to knock electrons to higher energy levels in the atom.

Once the atom is in an excited or ionized state, it will spontaneously revert to the ground state if it is given enough time. Here the process is re-

versed; the electron gives up the energy it had gained, and releases it in the form of new radiation (visible light, radio waves, ultraviolet, or some other form). This is the **origin of emission lines:** energy is released as light when electrons move from higher to lower energy levels.

One of the major discoveries of twentieth-century physics is that the electrons cannot follow just any orbit around the nucleus, as might be expected from Newton's and Kepler's work on orbits. Instead, as shown by Danish physicist Niels Bohr around 1913, the orbits are **quantized,** meaning that only certain orbits are possible, and only certain energy differences exist between them. Each energy difference corresponds to an absorption or emission line of fixed wavelength.

Figure 13·7 **Representations of spectra of stars in the seven spectral classes. Classes A and B show especially well the Balmer hydrogen lines (see Table 13·2).**
▼

---

[2]As Annie Cannon noted, the rearrangement of letters made the sequence harder to remember. American astronomer Henry Norris Russell soon solved this problem with Yankee ingenuity by proposing a sentence as a memory-aid: Oh Be A Fine Girl, Kiss Me! Two later classes, R and N, were easily accommodated by adding the words *Right Now.* However, when class S was added, a controversy broke out between Harvard and Cal Tech over whether it signified the word *Sweetie* or *Smack.* (Classes R, N, S, and some other minor classes contain only a few stars and are seldom used.)

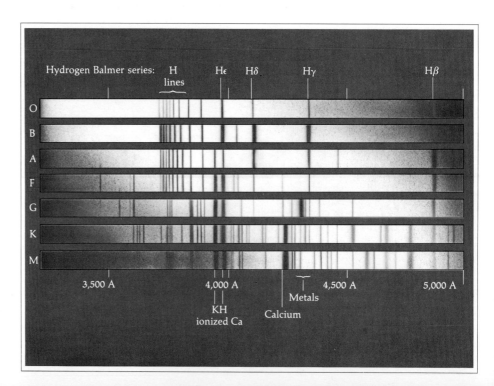

This explains why each element can produce only a certain pattern of lines.

The situation is made clearer by considering the simplest and most abundant type of atom in the universe, hydrogen. As shown in the schematic diagram, Figure 13·8, hydrogen atoms' orbits can be numbered. Since a neutral hydrogen atom has just one electron, an atom in the ground state would have its electron in the orbit n = 1. If the atom had been bumped by other atoms or if it had absorbed radiation, the electron might be in the orbit n = 2, 3, etc. Further absorption of energy might cause it to jump from n = 3 to n = 4, creating an absorption line; or it might spontaneously revert from n = 3 to n = 2, creating an emission line. Each possible transition (1 to 2, 2 to 3, 4 to 2, etc.) creates a different line. As Figure 13·8 shows, hydrogen lines ending at the n = 2 level are the lines prominent in the visible part of the spectrum, called the **hydrogen Balmer series** of lines. The famous red line called **hydrogen alpha**, or **Balmer alpha**, prominent in the solar chromosphere and elsewhere, involves transitions between n = 2 and n = 3 in the hydrogen atom.

Thus we can see that a study of the strengths of different lines found in the spectra gives an indication of how many atoms are in each state. This in turn gives an indication of conditions such as temperature and pressure in the gas of the star.

Spectral Classes as a Temperature Sequence

An important principle is that each spectral class corresponds to a different *temperature* in the gases in the light-emitting layers of the stars. **Temperature** is merely a measure of the average velocities of the atoms or molecules of the gas. The hotter a gas, the faster its atoms and molecules move. If the temperature quadruples, the velocities of the atoms double. The faster the atoms collide the more likely they are to disturb or dislodge each other's electrons. Furthermore, the hotter the gas, the more radiation it emits, which will also tend to disturb electron structures of atoms. Since spectral lines depend on electron structures, and since spectral classes depend on spectral lines, spectral classes form a temperature sequence.

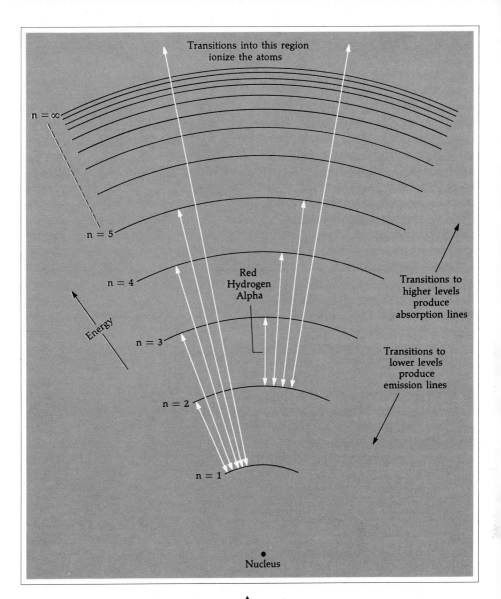

Figure 13·8 **Schematic diagram of a hydrogen atom, showing different possible electron orbits or energy levels (n = 1, 2, 3, and so on). Each change of orbit by the electron produces a spectral line, because energy is removed or added to the light beam. The series of lines starting from or ending on n = 1 occur in the ultraviolet part of the spectrum and are called the Lyman series. Lines starting from or ending on n = 2 occur in the visible part of the spectrum and are called the Balmer series. The well-known red hydrogen alpha involves transitions between n = 2 and n = 3.**

Consider what happens as we decrease the temperature of matter. The hottest stars, the O stars, have temperatures of 40,000 K or more, as shown in Table 13·2. Here, the atoms and radiation are so energetic that even the most tightly bound atoms, such as helium, have had their electrons knocked off, as the table indicates. By class B, the temperature has dropped to around 18,000 K, and helium atoms keep their electrons.

By classes A and F, even relatively weakly-bound hydrogen keeps its electrons. Many metals have at least one electron that can be very easily knocked off the outer part of the atom, and in G-type stars like the sun, at around 5,500 K, lines of ionized metals are prominent along with Balmer lines of hydrogen. By class K, at 4,000 K, even many metals are neutral. By class M, at 3,000 K, energies are so low that different atoms can stick together into molecules; even water molecules have been identified in spectra of such cool stars, as shown in Figure 13·9.

Cooler stars are rarely seen, but the concept may be extended to lower temperatures. In Chapter 11, we discussed how grains of minerals such as silicates can form at low temperatures around 2,000 to 1,500 K; such grains are suspected to exist in the atmospheres of the coolest stars. Planets, of course, have even cooler temperatures, such as 300 K for earth, where liquid water coexists with solid rocks. At cool enough temperatures, near zero K, nearly all matter could assume solid form.

	Table 13·2 Principal Spectral Classes of Stars			
	Spectral Class	Typical Temperature (K)	Sources of Prominent Spectral Lines (Selected Examples)	Representative Stars
Hottest, bluest	O	40,000	Ionized helium atoms	Alnitak (ζ Orionis)
	B	18,000	Neutral helium atoms	Spica (α Virginis)
	A	10,000	Neutral hydrogen atoms	Sirius (α Canis Majoris)
	F	7,000	Neutral hydrogen atoms	Procyon (α Canis Minoris)
Yellowish-white	G	5,500	Neutral hydrogen, ionized calcium	Sun
	K	4,000	Neutral metal atoms	Arcturus (α Bootes)
Coolest, reddest	M	3,000	Molecules and neutral metals	Antares (α Scorpii)

Two Important Laws of Spectroscopy

In addition to the three basic equations introduced earlier (small angle equation, circular velocity, and Wien's law), we now add the final two physical laws that we will use in this book. Both are described in mathematical

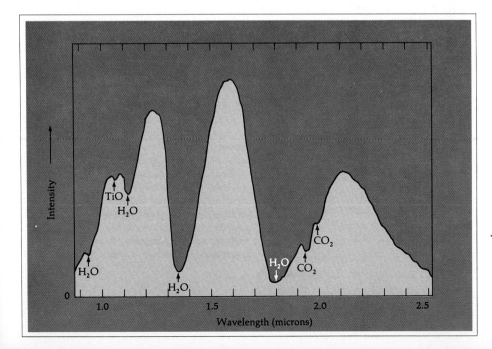

◄ Figure 13·9 A portion of the infrared spectrum of the giant star Omicron Ceti (Mira), showing identification of various molecules, including the first identification of water vapor in a star. (After a diagram by G. P. Kuiper)

detail in the accompanying optional boxes, but they need to be briefly described to all readers, because they are important in interpreting distant stars.

The Doppler Effect *The Doppler effect is a shift in wavelength of a spectral absorption or emission line away from its normal wavelength, caused by motion of the light source toward or away from the observer.* If the source approaches, there is a **blue shift** toward shorter wavelengths, or bluer light. If the source recedes, a **red shift** occurs toward longer wavelength, or redder light. The amount of the shift is just proportional to the approach or recession speed of the source (as long as that speed is well below the speed of light). In the shorthand of mathematics,

$$\frac{\text{shift in wavelength}}{\text{normal wavelength}} = \frac{\text{approach or recession speed}}{\text{velocity of light}}$$

In other words, if the light source were receding at 10 percent the speed of light, the light would be red-shifted by 10 percent its normal wavelength; a line normally found at wavelength 5,000 A would appear at 5,500 A. By measuring such wavelengths, velocities of stars can be studied.[3]

---

[3]The same shift occurs for sound. This explains why the noise of a rapidly traveling automobile, truck, or train changes from higher pitch (shorter wavelength) to lower pitch (longer wavelength) as the vehicle rushes by. The sound a child makes to imitate a race car rushing by is a representation of the Doppler shift.

Optional Math Equation IV · The Doppler Effect: Approach and Recession Velocities

Probably the most important physical phenomenon in astronomical spectroscopy is the **Doppler effect** (named after the Austrian physicist Christian Doppler, who discovered it in 1842). *The Doppler effect is a change in wavelength proportional to any line-of-sight velocity between observer and source.* (That is, the wavelength of a spectral line from a fast-approaching body will differ from that of a fast-receding body *and* that of one slowly approaching.)

Doppler discovered the effect in the light from distant stars orbiting around each other, but it applies to any signal transmitted by waves, including familiar sound waves on earth. The most familiar example of the Doppler effect is the shift in pitch of the sound from an approaching or receding source. As a car or train passes by and recedes in the distance, the pitch of the sound dramatically decreases. As the source approaches, the sound waves rush past the observer apparently more closely spaced than normal, decreasing the observed wavelength. As the source recedes, the waves are perceived to be more spread out, increasing the wavelength. The effect modifies whatever property of the signal is determined by wavelength—the color of light, the pitch of sound.

The equation that describes the Doppler effect is the fourth of the five major equations we will use in this book. It is:

$$\frac{\text{shift in wavelength}}{\text{normal wavelength}} = \frac{\text{relative velocity along line of sight}}{\text{velocity of light}}$$

Thus, if the source of light is not approaching or receding, the Doppler shift is zero. The faster the source approaches or recedes, the greater the Doppler shift in wavelength of the light.

For example, suppose a certain infrared spectral line normally has a wavelength of 1.000×10^{-4}cm. If the light source is receding from the observer at 1/1,000th of the speed of light, the line would appear at the slightly longer wavelength of 1.001×10^{-4}cm, a little toward the red end of the spectrum. If the source approached at 1/1,000th of the speed of light, the line would appear at 0.999×10^{-4}cm.

When source and observer are getting closer together, the wavelength of all light coming from the source is *shifted toward the blue* (shorter wavelengths). But when source and observer are getting farther apart, the spectrum of light from the source is *shifted toward the red* (longer wavelengths).

The two kinds of shifts are loosely referred to as **red shifts** and **blue shifts,** and their sources as red-shifted or blue-shifted. It is helpful to remember that *recession produces red* shifts.

Note that the observer of a Doppler shift cannot say whether it is the source or the observer who is "truly" moving relative to any absolute external frame of reference. According to the principle of relativity, only relative recession or approach of one body with respect to the other can be measured.

The Stefan-Boltzmann Law The *Stefan-Boltzmann law*, discovered by two Austrian physicists about 1880, states that *the higher the temperature of a surface, the more energy radiated by each square centimeter in each second.* The mathematical form of the law gives the total energy E (in ergs) radiated per second by a surface of area A (cm²) and temperature T (degrees K). The law states that $E = \sigma T^4 A$, where σ is merely a constant of proportionality. If the temperature of a source doubles, the amount of energy radiated increases by 2^4, or 16. Thus, while doubling the area of a star would increase its output twice, doubling its temperature would

increase its output 16 times! Therefore, as shown in Figure 13·10, the hotter stars not only radiate bluer light than cooler stars (Wien's law), but also radiate *more* light per unit area. The Stefan-Boltzmann law thus provides us with a way to learn about the sizes or areas of stars once we measure their temperatures and total radiation output.

Measuring Twelve Important Stellar Properties

We now can apply the Stefan-Boltzmann law, the Doppler effect, and other principles to explain how twelve important physical properties

of remote stars can actually be measured. These applications, to be discussed below, are summarized in Table 13·3, p. 248.

How to Measure Stars' Distances

From what you have read so far, can you think of a way to measure the distance of a star? The question is not far-fetched, since stars' distances were first measured more than a century ago. A crude method may be found in Table 13·1. The brightest stars are about 25 magnitudes fainter than the sun. In 1829, the English scientist William Wollaston used this fact to

---

Optional
Math Equation
V
·
The
Stefan-Boltzmann
Law:
Rate of Energy
Radiation

Before describing how to measure specific properties of stars, we need the last of the five basic equations to be used in this book. This is the **Stefan-Boltzmann law,** which describes the total amount of energy radiated in each section from any hot surface.

In studying Wien's law, we saw that any warm body radiates. And the higher the temperature, the bluer is the radiation. About a century ago the Austrian physicists Josef Stefan and Ludwig Boltzmann discovered another characteristic: *The higher the temperature, the more energy is radiated each second,* as shown in Figure 13·10.

Although the subjects of their study were radiating objects in the laboratory, the law applies to stars and all other bodies in the universe.

The law gives the total energy E radiated per second from a body with temperature T and surface area A.

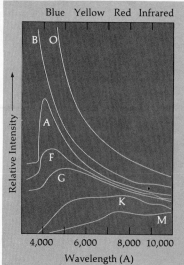

For bodies that radiate efficiently, like stars, the law is:

$$E = \sigma T^4 A$$

Sigma (σ), called the Stefan-Boltzmann constant, equals 5.7×10^{-5} in cgs units. T is given in degrees K, A in cm², and E in the cgs units of ergs/sec.

This equation shows that if the temperature or area of a star increases, the total energy radiated every second will increase. If the *area* is doubled, the radiation rate doubles. The rate is more sensitive to *temperature.* If the temperature is doubled, the radiation rate increases by 2^4, or 16.

◄ *Figure 13·10* **A schematic comparison of the energies emitted by various stars. Curves show the intensity of radiation in the continuous spectrum of each star, neglecting absorption lines. The hottest and brightest are O stars, which emit most of their energy as blue and ultraviolet light. G stars, like the sun, emit most of their radiation in the visible part of the spectrum, especially as yellow light. The faintest stars, class M, emit most light in the red and infrared part of the spectrum.**

estimate typical distances of stars. Since every 5 magnitudes equals a factor 100 (or 10^2) in brightness, and since apparent brightness diminishes according to an inverse square law, Wollaston was able to show that the brightest stars must be about 10^5 (100,000) times farther away than the sun. Since the sun is one Astronomical Unit from us, most stars (if they are sunlike objects) must be *at least* 10^5 Astronomical Units away—or more than 16 million million km.

Because this gives only a typical distance, we need a better technique to measure distances of individual stars. The most important such technique is **parallax** measurement. You will recall that parallax is the apparent shift in the position of an object caused by a shift in the observer's position (page 36). The principle is easy to understand. Hold your finger in front of your face and look past it toward some distant objects. Your finger represents a nearby star; the objects, distant stars. Your right eye represents the view from earth on one side of the sun, your left eye represents the view six months later, after the earth has traveled to the other side of the sun, about 300 million km away (Figure 13·11). Wink first one eye and then the other. Your finger (the nearby star) seems to shift back and forth. Hold your finger only a few inches from your eyes; the shift is large. Hold your finger at arm's length; the shift is smaller. Likewise, the farther away the nearby star, the smaller the parallax. The parallax in this experiment may be measured in degrees, but the parallaxes of actual stars are all less than a second of arc.

Chapter 2 described how Aristotle correctly realized, around 350 B.C., that if the earth moved through space, nearby stars ought to show parallax. Seeing none, he concluded the earth

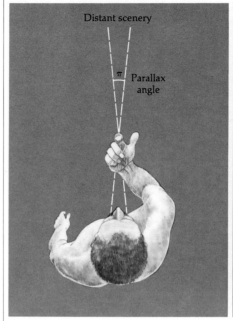

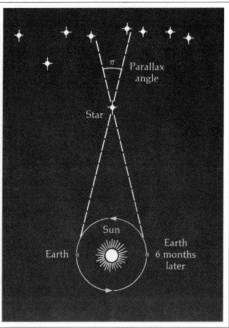

A

B

stood still. Actually, the shift was there, but too small for him to measure. Likewise, around 1600, Tycho looked for stellar parallax with the naked eye and found none. But he concluded (correctly) that movements of the earth were small-scale compared to stellar distances.

In 1837 the German-Russian astronomer Friedrich Struve published an analysis of stars that might be expected to be closest (those most likely to have the largest and most easily detected parallaxes). He chose the brightest stars with the fastest angular motions across the sky, just as on a dark night one might sensibly infer that the closest fireflies are the ones that are brightest and move with the highest angular speeds. (These angular motions, though not apparent to the naked eye, can easily be measured from year to year with the telescope.)

Now the race was on to see who could measure the first true parallax. Credit is usually given to the German astronomer Friedrich Bessel, who in 1838 published the parallax and dis-

Figure 13·11 **The principle of parallax determination, applied to one's finger (A) and to stars (B). Nearby objects observed from two positions appear to shift by a parallactic angle, π, as compared to background objects; measurement of π gives the distance of the object, once the separation of the two positions is known. (By astronomical convention, the angle cataloged as a star's parallax is one-half the angle π shown here.)**

tance of 61 Cygni (star 61 in the constellation of Cygnus, the swan). Bessel found the parallax of 61 Cygni to be $\frac{1}{3}$ second of arc, and the distance to be about 3 parsecs. Most stars are farther than 3 parsecs away and their parallaxes are correspondingly smaller. Parallaxes as small as $\frac{1}{20}$ second of arc have been reliably measured, and the distance of such stars is 20 parsecs.

As the above figures show, the definition of the parsec was chosen so that the distance of the star in parsecs is simply the inverse of its parallax angle in seconds of arc; a parsec is the distance corresponding to a PARallax of one SECond of arc.

The more distant a star, the smaller its parallax. Thus, if a star is too distant, its parallax is too small to be measured. Parallaxes smaller than about 0.05 seconds of arc are difficult to measure accurately. Therefore, *stars farther away than about 20 parsecs are beyond the* **distance limit for reliable parallaxes.**

Knowing accurate distances is the prime requirement of measuring most other properties of stars. Thus, the estimated 1,000 to 2,000 stars that lie within 20 parsecs are our main statistical sample for measuring stellar properties. Other techniques have been devised for estimating distances to more remote stars, but they all depend ultimately on the accuracy of the parallax measures of nearby stars.

*How to Measure
Stars' Luminosities*

The **luminosity** of a star is its *absolute brightness,* the total amount of energy it radiates each second. Unlike apparent brightness (discussed earlier in this chapter), luminosity is intrinsic to a star. The most useful concept of luminosity is **bolometric luminosity:** *the total amount of energy radiated each second in all forms, at all wavelengths.*

A second concept sometimes used is *visual luminosity:* the energy radiated each second in the visible part of the spectrum (the wavelengths to which the human eye is sensitive). Since many stars radiate primarily visible light, the two kinds of luminosity, bolometric and visual, are often roughly the same. In this book *luminosity* will generally mean *bolometric* luminosity. We will abbreviate it L. Similarly, our use of the term *absolute magnitude* generally refers to *absolute bolometric magnitude,* a measure of the bolometric luminosity expressed in the magnitude system.

The bolometric luminosity of the sun ($L_\odot$) is 4×10^{33} ergs/sec. Many stars, of course, are much brighter than the sun. Thus, a star twice as luminous as the sun, radiating 8×10^{33} ergs/sec, would be said to have a luminosity of 2 $L_\odot$.

The most basic method of estimating luminosity derives from the measurement of distance. A faint light in the night may be a candle a few hundred feet away, a streetlight a few miles away, or a brilliant lighthouse beacon 60 miles away. Once we know the distance to an object, we can determine its absolute brightness. Similarly, if we know the distance to a star, we can calculate its luminosity.

This method becomes less accurate for very distant stars. For one thing, as mentioned earlier, parallax distance measures become unreliable beyond about 20 parsecs. More important, there is some interstellar haze. Just as a fog bank might keep you from telling whether a distant light was a streetlight or a more distant lighthouse, variable interstellar haze in uncertain amounts throws doubt on luminosity estimates.

However, other methods are available. For example, very luminous stars have slightly different spectral characteristics than faint stars of the same spectral class. Spectra can therefore be used to measure luminosities (Figure 13·12).

Property	Techniques of Measurement
	Table 13·3 Some Basic Properties of Stars and How They Are Measured
Distance	Trigonometric parallax
Absolute brightness (luminosity)	Distance combined with apparent brightness
Temperature	Color; spectra
Diameter	Luminosity and temperature
Mass	Measures of binary stars—Kepler's laws
Composition	Spectra
Magnetic fields	Spectra, using Zeeman effect
Rotation	Spectra, using Doppler effect
Atmospheric motions	Spectra, using Doppler effect
Atmospheric structure	Spectra, using opacity effects
Circumstellar material	Spectra, using absorption lines and Doppler effect
Motion	Astrometry; spectra, using Doppler effect

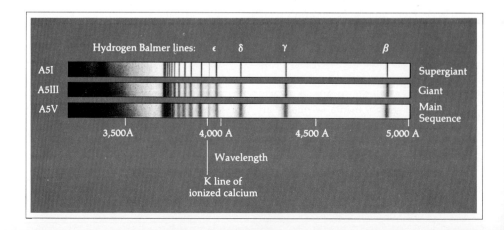

Figure 13·12 **Comparison of spectra of a supergiant, giant, and main-sequence star of constant spectral class. In the larger stars, pressures are less and atoms less disturbed by collisions, resulting in slightly narrower, better-defined absorption lines.**

How to Measure Stars' Temperatures

Here again spectra come to our aid. Stellar temperatures are measured merely by applying Wien's law. By measuring the spectrum, we can see which color is the most strongly radiated (see Figure 13·10). Application of Wien's law gives the temperature. This is often called the **color temperature,** because it is the temperature corresponding to the dominant color radiated by the various hotter and cooler layers of gas from which light reaches us.

Temperatures range all the way from about 2,500 K for the cool M-type stars to 40,000 K or more for hot O-type stars.

How to Measure Stars' Diameters

The diameters of only a handful of very large stars can be measured directly from images (as described earlier in this chapter). Most diameters are measured from temperature and luminosity by applying the Stefan-Boltzmann law. By this law, a star radiating a certain number of ergs per second and having a certain temperature must have a certain area, A. The values can be inserted in the Stefan-Boltzmann equation and the equation solved for A. The star's diameter can be found from its area. Such measures indicate a vast range of stellar diameters, from less than earth-size, through sun-size, to huge stars bigger than the diameter of Mars' orbit!

How to Measure Stars' Masses

According to Kepler's laws, as modified by Newton, when any two cosmic objects are in orbit around each other, the period of revolution (the time it takes to complete an orbit) increases as the distance between the objects increases, and as the sum of the masses of the two objects decreases. It happens that there are many pairs of stars orbiting around each other. For many *binary stars* we can measure both the period of revolution and the distance between the two stars. Thus we can calculate the sum of the masses, sometimes designated $m_A + m_B$, where A and B are designations of the two stars.

But we want to know each individual mass, not the sum of the two. Newton showed that in a system of orbiting bodies, each body orbits around an imaginary point called the **center of mass,** and that by measuring the distance of each star from the center of mass, we can measure the ratio of the masses.

Now we have both the sum of the masses, and the ratio of the masses. From this information, we can get each individual mass. For example, take the double star Sirius, consisting of the bright star Sirius A (which has the greatest apparent brightness of any star in the night sky) and the faint star Sirius B. The above procedure shows that the sum of the masses of A and B is 3 solar masses. The ratio is such that A is twice as massive as B. From these facts, the only possible solution is that Sirius A has a mass of 2 solar masses, and Sirius B, 1 solar mass.

Study of many stars by this technique reveals an important fact: *Stars with nearly identical spectra usually have nearly identical masses.* This fact allows the masses of many stars to be esti-mated from their spectral properties alone.

How to Measure Stars' Compositions

As described in the last chapter, compositions are revealed by spectra. There are two problems: detecting the *presence* of an element, and measuring its *amount.* The presence of an element is detected by identifying at least one—or, preferably, several—of its absorption lines or emission lines in the spectrum of the star.

The amount of the element is indicated by the appearance of the spectral line. Generally, the wider and darker the absorption lines, the more atoms of the element are present. Likewise, the wider and brighter the emission lines, the more atoms. These properties are said to measure **spectral line strength:** The stronger the line, the more of the element is present. Unfortunately, relative strengths of different lines depend also on temperature, pressure, and other conditions in the photospheres of stars. Astrophysicists have developed techniques to measure the true abundances in spite of these problems. The disadvantage is that these problems complicate the measurement of compositions; the advantage is that they allow astronomers to gain information from spectra not only about composition but also about the temperature, pressure, and other properties of stars. Composition measurements show that most nearby stars are approximately sunlike in composition.

How to Measure Stars' Magnetic Fields

Spectral effects can also be used to detect and estimate the strength of magnetic fields near the surfaces of stars. Spectral absorption and emission lines divide into two or more close lines in the presence of a magnetic field, a phenomenon called the *Zeeman effect*. The stronger the field, the greater the splitting. Measurements of this effect reveal that stars like the sun typically have fields about as strong as the sun's, but certain stars have fields thousands of times stronger than the sun's.

How to Measure Stars' Rotation

If a rotating star is seen from any direction except along the rotation axis, one edge will be approaching the observer and one edge will be receding. Light emitted or absorbed at the approaching edge will be blue-shifted. Light from the other edge will be red-shifted. Consider an absorption line being formed in all parts of the star's atmosphere. If the star were not rotating, neither shift would occur and the line would be very narrow. If the star rotates, the line is broadened. The faster the rotation and the closer our line of sight to the equatorial plane, the more **rotational line broadening** occurs (Figure 13·13).

The rotations of stars can thus be inferred to some extent by measurements of the broadening of spectral lines. In general, the fastest rotators are the hot, bluish, massive stars of class O, which have equatorial rotation speeds around 300 km/sec. Slightly cooler stars rotate more slowly, with a sharp break in the trend occurring at stars a little hotter and more mas-

Figure 13·13 **Light from a rotating star would be blue-shifted if from region A, or red-shifted if from region B, or unshifted if from the central region C. The net result is a broadening of spectral lines.**

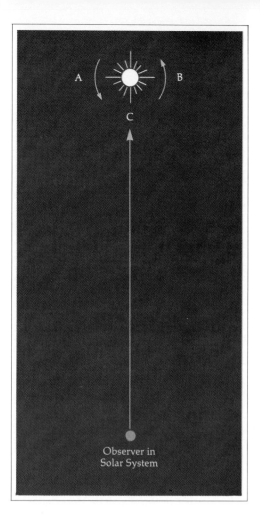

Observer in Solar System

sive than the sun. Cooler stars rotate much more slowly. The sun's equatorial speed is only 2 km/sec.

How to Measure Stars' Atmospheric Motions

If masses of stellar gas rise and fall in convection cells (as they do in the sun), the various cells will display a range of approach and recession velocities. Therefore, blue and red shifts (respectively) would occur, broadening the stars' spectral lines, as shown in Figure 13·14. This is called **turbulent line broadening.** In practice, astronomers have trouble distinguishing this effect from rotational line broadening, though some measures are possible. Especially strong turbulence has been found in the atmospheres of certain cool, large-diameter stars known as red giants. In these stars, masses of gas rise and subside with speeds as high as 40 km/sec.

How to Measure Stars' Atmospheric Structures

Spectroscopy lets astronomers probe the vertical structure of stellar atmospheres. As noted in the last chapter, the **opacity** of an atmosphere determines how far light of a specified wavelength can penetrate it. Generally, a stellar atmosphere has a different opacity at each different wavelength. If the opacity at a certain wavelength is low, we can see deep into the star's atmosphere at that wavelength. If the opacity at another wavelength is high,

we can see only into the uppermost atmosphere.

Thus, phenomena associated with different spectral lines at different wavelengths, such as broadening caused by turbulence, may reflect conditions in different levels of the atmosphere. Important in this regard is still another phenomenon. **Pressure line broadening** is broadening of spectral lines caused by increased collisions between neighboring atoms under high pressure. By sophisticated methods of sorting out other line broadening effects such as rotation and turbulence, astronomers gain information on pressure and other conditions at all levels of stellar atmospheres.

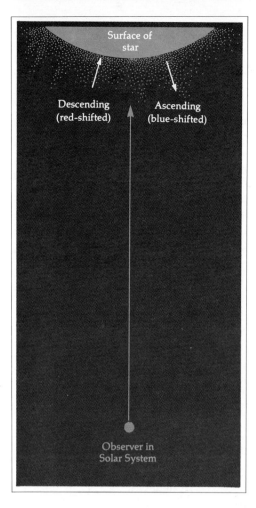

Descending
(red-shifted)

Ascending
(blue-shifted)

Surface of
star

Observer in
Solar System

◀ *Figure 13·14* **Currents of descending and ascending gas in a star's atmosphere cause Doppler shifts that broaden spectral lines.**

Figure 13·15 **Gaseous envelopes expanding ▶ (as shown here) or rotating around stars produce characteristic spectral effects, allowing their detection even if they cannot be seen separately.**

far side of the cloud are red-shifted.[4] These shifted lines reveal circumstellar nebulae (Figure 13·15).

In certain newly-formed stars, circumstellar gas has been observed expanding at 70 to 200 km/sec. Such nebulae are believed to correspond to the late stages of the solar nebula, described in Chapter 11. In gas around certain dying stars, expansion at speeds of 1,000 to 3,000 km/sec has been observed. This gas is believed to have been blown off the star in explosions related to exhaustion of the star's energy supplies.

*How to Detect
Stars' Motions*

The Doppler shift is a very simple way to detect *part* of a star's motion— its **radial velocity,** *or motion along the line of sight.* A consistent blue or red shift of all of a star's spectral lines proves that the star is moving toward or away from us. Among the 50 nearest stars, about half the radial velocities measured are more than 20 km/sec.

The other component of a star's motion is its **tangential velocity:** the motion *perpendicular to the line of sight.* It cannot be measured as simply as radial velocity. In order to measure

---

[4]For a discussion of how such spectral lines are produced, see Kirchhoff's third law, Chapter 12.

*How to Detect
Circumstellar Material*

Unstable stars, such as newly-forming or dying stars, may throw out clouds of gas and dust called **circumstellar nebulae.** Because many of these clouds are nearly transparent (as with flame), we can see light arising from both the near side and far side of the cloud. If such a cloud expands away from the star, absorption lines arising on the near side (where the starlight passes through the gas) are blue-shifted in comparison with lines in the star because the gas is approaching the observer. Emission lines from glowing gas on the near side are also blue-shifted, but emission lines from the

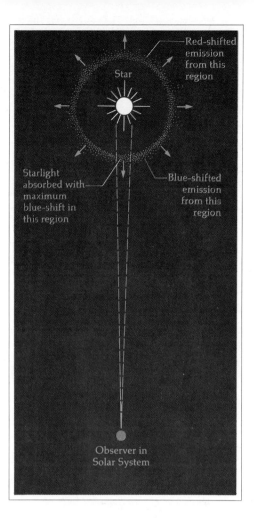

Star

Red-shifted
emission
from this
region

Starlight
absorbed with
maximum
blue-shift in
this region

Blue-shifted
emission
from this
region

Observer in
Solar System

Figure 13·16 **Measurement of radial velocities (V_R) of stars. If the star's true motion V is aligned with the line of sight, it equals the measured V_R, as in case A. More likely, V_R measures only one component of V, as in case B. In case C, V_R may be zero in spite of an appreciable V in the tangential direction.**

▼

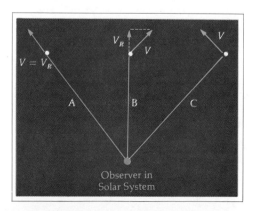

$V = V_R$

V_R

V

V

A

B

C

Observer in
Solar System

lar properties, as summarized in Table 13·3.

Clearly, our journey outward from the earth has taken us far beyond the places that we can investigate in the near future by manned or instrumented flights. Even at the speed of light (which is 10,000 times greater than speeds our instruments have achieved) it would take years to reach the nearest stars. Thus, instead of sending instruments to the stars, we must rely on the messages in the starlight coming to us. In the coming chapters, we will interpret what those messages tell us about the character and histories of the stars that surround us.

▲

Figure 13·17 **A new technique for studying stars and other remote objects is to put large telescopes into space. Two major advantages over earth-based telescopes would be freedom from the blurring caused by atmospheric shimmering, and freedom from absorption of ultraviolet and infrared wavelengths by atmospheric gases. Under planned design concepts, the Space Telescope could be a 2.4-meter telescope delivered to orbit by a space shuttle and operated as an international facility for study of planets, stars, nebulae, and galaxies. (NASA artist's conception)**

tangential velocity, we must measure the distance of the star and its rate of angular motion across the sky, called **proper motion.** (Values of a few seconds of arc per year are common for nearby stars.) These kinds of measurements are often lumped together in a special branch of astronomy called **astrometry.**

If both radial and tangential velocities are known, they can be combined to give the star's **space velocity:** its true speed and direction of motion in three-dimensional space, *relative to the sun.* Space velocities of most stars near the sun are a few tens of kilometers per second, and are nearly random in direction.

13

Summary

This chapter emphasizes concepts and methods of studying stars, rather than the nature of the stars themselves. The chapter has two basic sections. The first section describes concepts: the system of naming stars; the photographic images of stars; definition of light-year and parsec distance units; the system of magnitudes for measuring brightnesses; stellar spectra; the Doppler effect; and the Stefan-Boltzmann law, which describes the amount of stellar radiation.

The second section describes methods for determining twelve basic stel-

Concepts

stars

diffraction

Airy disk

light-year

parsec

apparent brightness

absolute brightness

apparent magnitude

absolute magnitude

visual apparent magnitude

spectrum

wavelength

spectroscopy

spectrograph

spectrometer

absorption lines

emission lines

continuum

Problems

1. If a telescope of 1,000-inch aperture could be put in orbit or on the moon, it could resolve an angle of only about 0.005 seconds of arc. Compare this resolution with the maximum angular size known for stars (0."03). Why would this telescope perform better on the moon than on the earth? Why do astronomers never use the highest magnifications theoretically possible with earthbound telescopes?

2. A star is 20 parsecs away. How many years has its light taken to reach us?

3. What is the chemical composition of most stars?

4. A reddish star and a bluish star have the same radius. Which is hotter? Which has higher luminosity? Describe your reasoning.

5. A reddish star and a bluish star have the same luminosity. Which is bigger? Describe your reasoning.

6. How does the magnitude scale differ from the scales by which we measure lengths (cm or inches), temperatures (°C or K), weights (kg or pounds), or other familiar properties?

7. A certain star has exactly the same spectrum as the sun, but is 30 magnitudes fainter in apparent magnitude. List some conclusions you would draw about it, describing your reasoning.

8. A star has a parallax of 0.05 seconds of arc. How far away is it?

9. In a binary pair of stars, the sum of the masses is found to be 5 solar masses; the stars are equidistant from their center of mass. What are their individual masses?

10. The spectral lines of metals, such as calcium, are prominent in the solar spectrum. Hydrogen lines are less prominent. Why does this not indicate that the sun consists mostly of these elements instead of hydrogen?

11. Each of a certain star's spectral lines is found to be smeared out over a wide range of wavelength. What might cause this?

Advanced Problems

12. How many times brighter is a day-lit landscape than the same landscape lit at night by the full moon? (Hint: see Table 13·1; note that 5 magnitudes = factor 100 in brightness; 1 magnitude = factor $2\frac{1}{2}$.)

13. A certain star has a spectrum similar to the sun's, but all the spectral lines are shifted 1 percent of their wavelength toward the red.

a. *What do you conclude about this star's motion toward or away from the observer?*

b. *Is its speed in this direction unusually fast, average, or slow?*

Project

1. Using a telescope of fairly high magnification (such as 300x or 400x), examine the image of a bright star on several different nights, making sketches. Are there differences in seeing from night to night? Can you identify the Airy disk, diffraction rings, or diffraction spikes? If so, label them on your sketches. Note any shimmering due to atmospheric turbulence or air currents in and around the telescope. Run the eyepiece inside or outside the focus point, making the image a round blob. Shimmering and other turbulent effects are often more evident in this way.

Nearby Stars and Stellar Evolution

As soon as the methods described in the last chapter were available to measure stellar properties, astronomers began pointing their instruments at the brighter stars and compiling statistics about stellar characteristics. This chapter will outline two interesting discoveries: first, stars come in a very wide variety of forms, which must be sorted and classified before they can be understood; second, stars evolve from youth to old age, with different forms at different ages.

In this sense, the population of stars can be likened to that of people. If you travel from country to country and observe people, you tend to see mostly normal, functioning people—youngsters, middle-age people, elderly people. Creation of new people—the birth of babies—is visually secluded in special places such as hospitals, and is not readily visible to the tourist. To push the metaphor a little further, the tourist may see cemeteries where human remains are buried, but the actual deaths of people are usually brief events not commonly seen. This metaphor explains the approach of this chapter and the next two chapters. In this chapter we will take the tourist's approach to the local population of rather ordinary functioning stars and defunct stars, out to a distance of about 100 parsecs or so. The next chapter will seek out the secluded places of star birth, and Chapter 16 will examine some of the extraordinary phenomena of stellar old age.

Making Sense of Stars: The H-R Diagram

Observations based on the techniques described in the last chapter reveal a puzzling variety of star forms. Masses range from about 0.5 to 50 solar masses. Luminosities range from around one-millionth to a million times that of the sun. Temperatures of stars' surfaces range from about a third to nearly ten times the solar temperature.

When astronomers realized they had this variety of star forms, their first step was to devise a sensible way to arrange and study the data about them, in hopes of finding relationships among various forms of stars. This could have been done in various ways, but one method has come to dominate all others. This method was introduced around 1905 to 1915 by Danish astronomer Ejnar Hertzsprung and American astronomer Henry Norris Russell. Their method was to construct diagrams plotting the spectral class of stars versus their luminosity, as shown in Russell's version of the diagram (Figure 14·1) published in 1914. This type of plot is sometimes called the **spectrum-luminosity diagram,** but is more often called the **H-R diagram,** in honor of Hertzsprung and Russell.

Even the earliest H-R diagram revealed an important discovery about stars: for most stars there is a *smooth relation between spectral class and luminosity,* indicated by dashed lines on Figure 14·1. Hertzsprung called stars obeying this relation (falling on the diagonal band of the H-R diagram) **main-sequence stars.** Russell correctly argued that the smooth relation between spectral type and luminosity meant that:

> *The principal differences in stellar spectra, however they may originate, arise in the main from variations in a single condition in the stellar atmosphere.*

As we saw in the last chapter, *temperature* is the principal factor that governs the differences between spectra of O, B, A, F, G, K, and M stars. For this reason, many astrophysicists plot luminosity against temperature. In so doing, they use two objective physical variables, rather than the somewhat subjective variable of spectral class, which depends on the classifier's judgment and is not quantitative. Figure 14·2 is a modern version of the H-R diagram plotting luminosity against temperature. On the other hand, many observational astronomers prefer to plot brightness in magnitudes

instead of luminosity, and color instead of spectral class or temperature, because these properties can be quickly measured by electronic instruments at the telescope. But all these versions of the diagram are called H-R diagrams, and in this book we identify temperature, spectral class, luminosity, and absolute magnitude along the edges of the diagram. Figure 14·2 shows the main sequence clearly as a diagonal band.

Another example of the usefulness of the H-R diagrams was Hertzsprung's

1905 discovery of another important group of stars that do not fall on the main sequence. As shown in Figures 14·1 and 14·2, certain K and M stars have much higher luminosity than the main-sequence K and M stars. That is, these stars radiate more light than others of the same temperature. According to the Stefan-Boltzmann law, they must have more area. Hertzsprung named them **giant stars.** Sometimes they are called red giants, since K and M stars are reddish.

As more data came in, other stars

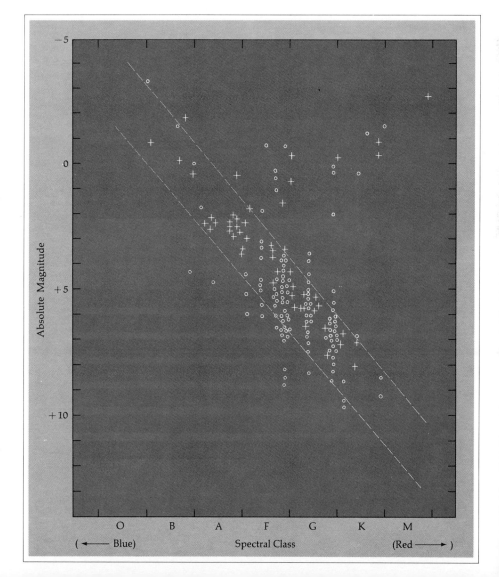

Figure 14·1 **The first H-R diagram, redrawn** ▶ **from Russell's original 1914 diagram. Dots represent stars with directly-measured parallaxes; + represents estimated data from stars in four clusters. Dashed lines mark Russell's identification of the main sequence.**

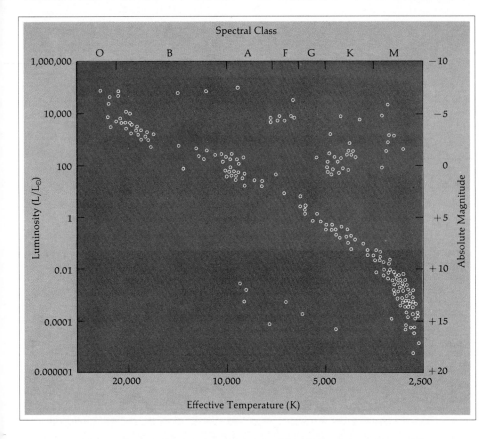

Figure 14·2 **A modern H-R diagram with a sample of well-studied stars (the 100 of greatest apparent brightness and the 100 closest). Spectral class (top) is equivalent to temperature (bottom), while luminosity (left) is equivalent to absolute magnitude (right).** (Data from Allen, 1973)

appeared on the H-R diagram below the main sequence. By the same logic used for giants, these stars were recognized as unusually small. They came to be called **dwarf stars** (or sometimes **white dwarfs,** because their colors were intermediate like the sun's, neither strongly blue nor red). Still other stars were found to be brighter than the giants or the main-sequence O stars; they came to be called **supergiant stars.**

The work of Hertzsprung, Russell, and their followers was important because it showed systematic groupings of stars: one group fitting on the main sequence; a distinctly different group of giants; a group of dwarfs; a group of supergiants. Further work showed that these groups are characteristic of stars in all known regions of space. This meant that universal physical processes could be sought to explain the groups.

Radii on the H-R Diagram

To emphasize the distinctness of these stellar groups Figure 14·4 shows the same H-R diagram as Figure 14·2, but with lines added to show positions of constant radius. The important concept is that *any point on the H-R diagram corresponds to a star of a certain*

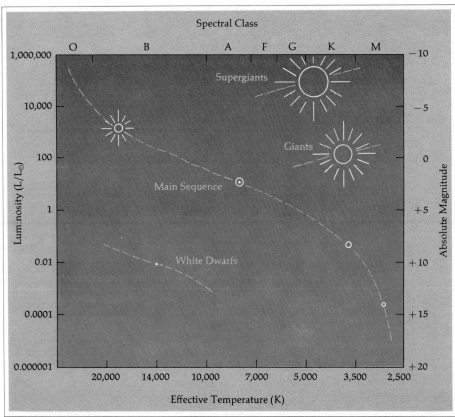

◄ *Figure 14·3* **A schematic version of Figure 14·2, identifying the main sequence, giant branch, and regions of supergiants and white dwarfs. Star symbols represent relative luminosities and sizes (not to scale). This diagram is reproduced as Color Photo 29.**

radius. The reasoning should be clear to you already. Any point on the diagram corresponds to a certain temperature *T*. According to the Stefan-Boltzmann law, any surface at temperature *T* must radiate each second a certain amount of energy per cm². But any point on the diagram also corresponds to a certain luminosity *L*, which is the total amount of energy radiated each second. This luminosity fixes the number of cm² involved; hence the total area; hence the radius.

The diagram confirms that the names of the types are apt. Supergiants can be 500 to 1,000 times the diameter of the sun; giants, 20 times; and dwarfs, only 1 percent of the sun's diameter.

Some well-known stars are marked in Figure 14·4. For example, the sun, Vega (8 parsecs away), and Spica (80 parsecs) are on the main sequence. Arcturus (11 parsecs) and Aldebaran (21 parsecs) are giants. Betelgeuse

Figure 14·4 **An H-R diagram with lines of constant radius, showing the dimensions of stars in different parts of the diagram. Selected well-known stars are marked.**
▼

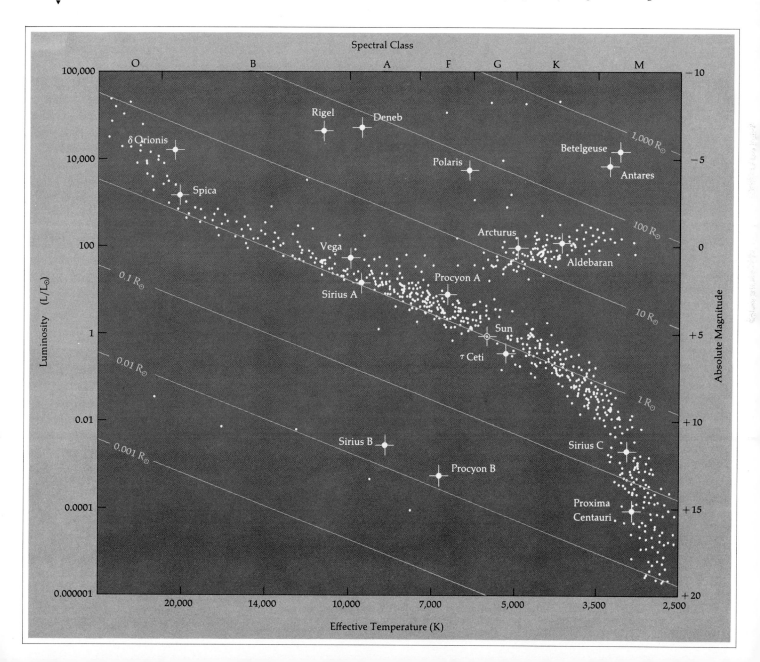

(200 parsecs), the bright reddish star in Orion, is a supergiant. Faint companions of Sirius (3 parsecs) and Procyon (4 parsecs) are dwarfs.

Prominent versus Representative Stars

Before explaining why distinct star-types exist, we must emphasize that the stars we have just discussed do not represent a true frequency of star types in space. Main-sequence stars are much more numerous than giants in any actual volume of space. But giants are so much brighter than most stars that they can be seen a long way off. Giants, Hertzsprung remarked, are like whales among the fishes.

Therefore, if we take the easy route and simply measure the prominent

stars in our sky, we get a bias toward giants. But if we first measure the distances of stars, and then tabulate just the stars in the average volume of space near the sun, we get a truer picture of the nature of stars. This is shown by comparison of Tables 14·1 and 14·2. Table 14·1 shows the **prominent stars**—the 17 stars brighter than 1st magnitude in the *apparent* magnitude system. Of these, 53 percent are main-sequence stars and the other 47 percent are giants and supergiants.

Figure 14·5 **This figure is the same as Figure 14·2, but shows only data drawn from the 100 nearest stars—a representative sample from a volume of space around the sun. This diagram shows that most stars are fainter, cooler, and (as a consequence of the mass/luminosity relation) less massive than the sun.**
▼

Compare this with Table 14·2, which lists **representative stars**—39 stars in 24 systems located within 4 parsecs. Of these, 91 percent are main sequence stars, *none* are giants or supergiants, and 9 percent are dwarfs.

Even this survey may underestimate the number of dwarfs and M stars, because they are so faint. Some astronomers believe that about half of all stars are the faint main-sequence M stars, and that most of the other half may be dwarfs. Figure 14·5 shows the

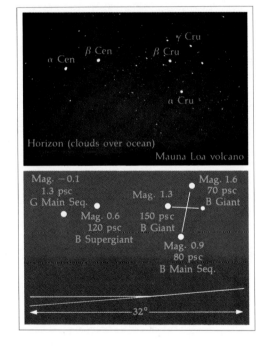

▲
Figure 14·6 **A view of the nearest star system, Alpha Centauri, on the southern horizon of Hawaii. At a distance of 1.3 parsecs, Alpha Centauri (the brightest star in this photo) is the closest star system. Neither of two co-orbiting stars is seen, one (Alpha B) because it is too close to Alpha A, and the other (Proxima) because it is too faint. Alpha and Beta Centauri make a prominent pair and to their right is the Southern Cross. Alpha is a nearby sunlike star; the other bright stars are much more distant and have greater luminosities than the sun.**

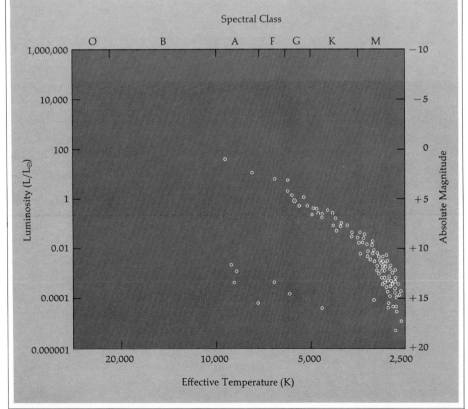

H-R diagram for the 100 nearest known stars, a representative sample of the true population of space. Main-sequence stars brighter than the sun, as well as giants and supergiants, are rare. When you step outside at night, many of the bright stars you see are the superbright beacons of space. But in reality, the whales among the fishes are rare.

Table 14·1 The 17 Stars of Greatest Apparent Brightness
(Stars Brighter than First Apparent Visual Magnitude)

Star Name	Apparent Magnitude (m_v)	Measured Bolometric Luminosity $(L_\odot)$	Type	Radius $(R_\odot)$	Distance (parsecs)
Sun	−26.7	1.0	Main Sequence	1.0	0.0
Sirius (α Canis Majoris)	−1.4	27	Main Sequence	2.0	2.7
Canopus (α Carinae)	−0.7	?	Supergiant	?	60
Rigel Kent (α Centauri)	−0.1	1	Main Sequence	1.2	1.33
Arcturus (α Boötis)	−0.1	130	Red Giant	24	11
Vega (α Lyrae)	0.0	?	Main Sequence	?	8.1
Capella (α Aurigae)	0.1	?	Red Giant	?	14
Rigel (β Orionis)	0.1	?	Supergiant	?	250
Procyon (α Canis Minoris)	0.4	7	Main Sequence	2.2	3.5
Achernar (α Eridani)	0.5	?	Main Sequence	?	39
Hadar (β Centauri)	0.6	?	Supergiant	?	120
Altair (α Aquilae)	0.8	?	Main Sequence	?	5
Betelgeuse (α Orionis)	0.8	85,000	Supergiant	800	200
Aldebaran (α Tauri)	0.8	360	Red Giant	45	21
Acrux (α Crucis)	0.9	?	Main Sequence	?	80
Spica (α Virginis)	1.0	?	Main Sequence	?	80
Antares (α Scorpii)	1.0	28,000	Supergiant	560	130

Source: Data from Allen (1973) and Aller (1971).

Explaining the Types of Stars

Why are stars divided into such distinct groups as main sequence and giants? Why isn't the H-R diagram uniformly dotted with a random variety of star forms? These questions show how the H-R diagram became a pivotal concept for discussing stars.

Some astronomers correctly suspected that theoretical analysis might reveal that stars slowly evolve from one form to another, over periods of billions of years. In the 1920s, the English astrophysicist Arthur Eddington pointed out that we cannot deal with evolution until we understand the **static structures of stars**—that is, the principles that keep each star's internal structure stable during the short term, from year to year. How can we determine the internal structures of stars? Eddington (1926) commented:

> At first sight it would seem that the deep interior of the sun and stars is less accessible to scientific investigation than any other region of the universe. . . . What appliance can pierce through the outer layers of a star and test the conditions within?
>
> The problem does not appear so hopeless when misleading metaphor is discarded. It is not our task to "probe"; we learn what we do learn by awaiting and interpreting the messages dispatched to us by the objects of nature. And the interior of a star is not wholly cut off from such communication. A gravitational field emanates from it. . . . Radiant energy from the hot interior after many deflections and transformations manages to struggle to the surface and begin its journey across space. From these two clues alone a chain of deduction can start which is perhaps the more trustworthy because it [employs] only the most universal rules of nature—

Distance (parsecs)	Star Name	Component	Apparent Magnitude (m_v)	Absolute Magnitude (M_v)	Spectral Type	Mass ($M_\odot$)	Radius ($R_\odot$)	Semi-Major Axis in Multiple Systems (AU)
0.0	Sun	A	−27	5	G2	1.0	1.0	5.2
	(Jupiter)	B	—	—	—	0.001	0.1	
1.3	Alpha	A	0	4	G2	1.1	1.2	23.6 (AB)
	Centauri	B	1	6	K5	0.9	0.9	
		C	11	15	M5	0.1	?	10,400 (AC)
1.8	Barnard's	A	10	13	M5	?		1.3
	Star	B	?	?	?	0.002	?	
2.3	Wolf 359		14	17	M8	?	?	
2.5	+36°2147	A	8	10	M2	0.35	?	0.07
		B	?	?	?	0.02	?	
2.7	Sirius	A	−1	1	A1	2.3	1.8	19.9
		B	9	12	Dwarf	1.0	0.02	
2.7	L 726-8	A	12	15	M5	0.044	?	10.9
		B	13	16	M6	0.035	?	
2.9	Ross 154		11	13	M4	?	?	
3.2	Ross 248		12	15	M6	?	?	
3.3	L 789-6		12	15	M7	?	?	
3.3	ε Eridani		4	6	K2	0.98	?	
3.3	Ross 128		11	14	M5	?	?	

Table 14·2 The Nearest 24 Stellar Systems (Stars out to 4 Parsecs)

Source: Data from Allen (1973).

the conservation of energy and momentum, the laws of chance and averages, the second law of thermodynamics, the fundamental properties of the atom, and so on.

Eddington showed why stars are the way they are. The same two major, opposing influences at work in the formation of the sun compete in any star: Gravity pulls *inward* on stellar gas while gas pressure and radiation pressure push *outward*.[1] In a stable star

these forces are just balanced.

As we have already seen in the discussion of Helmholtz contraction, heat is always produced by gravitational contraction of a star. A stable star is one that contracted only until the inside became hot enough to start nuclear reactions, at which point the reactions in turn produced enough heat so that outward pressure counterbalanced inward gravity. Then contraction stopped.

Although Eddington didn't know it, the main source of energy in the star's interior is nuclear reactions in which hydrogen is consumed. Because stars form with a huge supply of

hydrogen, stars remain stable for a long time at a fixed size. If a stable star were magically expanded, its gas would cool and the reactions would decline, reducing the outward pressure, and the outer layers would fall back to their original state. If it were magically compressed, the inside would get denser and the reactions would increase, raising the outward pressure and expanding the star. The star has to stay in its stable state *as long as its internal chemistry and energy-production rate stay the same.*

---

[1]As mentioned in Chapter 12, radiation exerts a distinct pressure on material through which it passes. In stars, especially massive ones, this pressure is important.

Table 14·2 The Nearest 24 Stellar Systems (Stars out to 4 Parsecs)

Distance (parsecs)	Star Name	Component	Apparent Magnitude (m_v)	Absolute Magnitude (M_v)	Spectral Type	Mass $(M_\odot)$	Radius $(R_\odot)$	Semi-Major Axis in Multiple Systems (AU)
3.4	61 Cygni	A	5	8	K5	0.63	?	85 (AB)
		B	6	8	K7	0.6	?	
		C	?	?	?	0.008	?	
3.4	ε Indi		5	7	K5	?	?	
3.5	Procyon	A	0	3	F5	1.8	1.7	15.7
		B	11	13	Dwarf	0.6	0.01	
3.5	+59°1915	A	9	11	M4	0.4	?	60
		B	10	12	M5	0.4	?	
3.5	+43° 44	A	8	10	M1	?	?	156 (AB)
		B	11	13	M6	?	?	
		C	?	?	K?	?	?	
3.6	−36°15693		7	10	M2	?	?	
3.6	τ Ceti		4	6	G8	?	1.0	
3.7	+5°1668	A	10	12	M5	?	?	
		B	?	?	?	?	?	
3.8	−39°14192		7	9	M0	?	?	
3.9	Kapteyn's Star		9	11	M0	?	?	
4.0	Krüger 60	A	10	12	M3	0.27	0.51	9.5 (AB)
		B	11	13	M4	0.16	?	
		C	?	?	?	0.01	?	
4.0	Ross 614	A	11	13	M7	0.14	?	3.9
		B	14	16	—	0.08	?	

Explaining the Main Sequence

Eddington explained the main sequence by imagining a group of stars with different masses. Since he did not know what the energy-generation process is, he simply assumed that all these stars had the same energy-generation processes, compositions, and energy-transport processes as the sun. His calculations then revealed a **mass-luminosity relation:** each different mass would correspond to a star of a different luminosity in such a way that the group of stars all fall on the main sequence.

As we now know, the energy-generation process is the consumption of hydrogen. In other words, *the main sequence is explained as the group of stars of different masses that have reached stable configurations and are consuming hydrogen in nuclear reactions.* Any hydrogen-"burning" star[2] with one solar mass and solar composition must look like the sun. Lower-mass hydrogen-"burning" stars of solar composition will be

___

[2]"Burn" technically refers to chemical reactions involving only electrons. But astrophysicists use the term informally for nuclear reactions that convert small fractions of the mass of atomic nuclei into energy.

fainter than the sun; higher-mass stars will be brighter.

This explanation of the main sequence helped explain why some stars lie off the main sequence. The same assumptions simply do not apply to them. In other words, giants, supergiants, and dwarfs are stars that do *not* have the same energy-generation process, composition, or energy-transport processes as the sun.

The Russell-Vogt Theorem

As Eddington reached these conclusions, in 1926, the astrophysicists

H. N. Russell and H. Vogt independently derived a related result known as the **Russell-Vogt theorem:**

**The equilibrium structure
of an ordinary star is
determined uniquely by its
mass and chemical composition.**

This says that a certain mass of material with fixed composition—for example, one solar mass of solar composition—can reach only one stable configuration. This is the point on the H-R diagram actually occupied by the sun. But if the composition were altered to one-half hydrogen and one-half helium, the configuration of the star and its location on the H-R diagram would be different. In other words, the Russell-Vogt theorem says, "give me the mass and composition of any star at any given moment, and I can tell you what the star looks like." Using these principles, astronomers have actually calculated what the stars look like—they have made tabulations of the pressures, temperatures, and other characteristics of the interiors, surfaces, and atmospheres of stars.

*Why Main-Sequence Stars
Must Evolve off the Main Sequence*

The Russell-Vogt theorem explains the cause of **stellar evolution** from one form to another form: a star "burning" hydrogen converts it to helium and changes its composition; therefore it must change to a new equilibrium structure. All nuclear reactions cause composition changes and all composition changes cause evolution.

Philosophical Implications of Theoretical Astrophysics

Philosophically, these achievements were profound. Eddington pointed out that humans (or other intelligent creatures), reasoning purely from elementary principles even without telescopic observation, could show that the universe must be divided into objects like stars. Smaller or larger objects would be less likely because smaller objects have too little gravity to assemble themselves from cosmic gas, and larger objects would produce so much radiation that they would blow themselves apart. These suppositions by Eddington have been proved correct by further observation and theory: The most common size of star in the universe has a mass around 0.1 to 1.0 solar masses, as shown in Figure 14·8. Stars smaller than 0.1 $M_\odot$ are less common, and stars bigger than 10 $M_\odot$ are very rare. It has recently been shown that stars larger than some 100 $M_\odot$ will readily blow themselves apart.

Eddington (1927) wrote:

We can imagine physicists working on a cloud-bound planet such as Jupiter who have never seen the stars. They should be able to deduce by [these

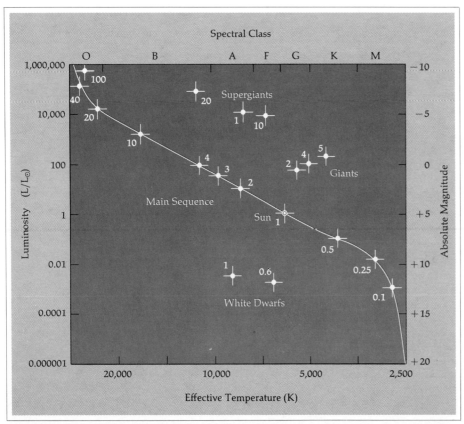

◄ *Figure 14·7* **An H-R diagram with measured masses of stars marked in different parts of the diagram. A smooth progression of masses is found along the main sequence, but masses in other parts of the diagram are mixed because stars of different mass evolve into the same regions—for example, the giant branch.**

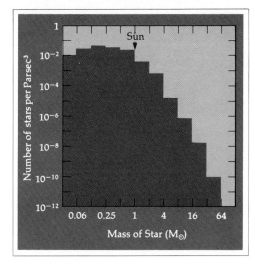

Figure 14•8 **A histogram showing the frequency of occurrence of different stellar masses in the solar neighborhood. The most common stars have about one-quarter solar mass.**

methods] that if there is a universe existing beyond the clouds, it is likely to aggregate primarily into masses of the order of [10^{27}] tons. They could then predict that these aggregations will be globes pouring out light and heat and that their brightness will depend on the mass in the way given by the [mass-luminosity relation].

Eddington became fascinated by the idea that intelligent creatures could mentally derive the appearance of the universe without actually observing more than a few fundamental relations. He asked whether a superscientist, locked forever in a closed physics lab with certain minimal equipment, could discover the nature and arrangement of the universe. Is the arrangement we observe the only possible arrangement? No one knows the answers to these questions, although we can be sure that if some single physical constant—such as Newton's constant of gravitation—suddenly changed, the structure of the universe would dramatically change. For ex-

ample, if gravity suddenly decreased, the earth would expand, causing earthquakes. The earth's rotation and orbit would change, and the moon's orbit would change. The sun would expand, causing a change in luminosity and a change in the temperature of the earth.

It is thus plausible that a single theoretical construct, using the measured values of various physical constants, would allow prediction of everything that can be measured about the universe. Such a theory would allow treatment of matter, gravity fields, electromagnetic fields, and radiation from a single set of equations. Toward the end of his life, Eddington believed he was approaching a theory which could predict the values of the very physical constants themselves, previously thought to be underivable, intrinsic properties of the universe. His last work, however, has not been widely accepted.

Nuclear Reactions and Stellar Evolution

As noted above, the Russell-Vogt theorem guarantees stellar evolution. As hydrogen is converted to helium, the star's composition changes and its structure must change. Some stellar structures are relatively stable, whereas others are transitory, explaining why stars of a single mass may have different forms—main-sequence, giants, dwarfs—as shown in Figures 14•7 and 14•9. Conversion of hydrogen to helium scarcely changes a star's total mass, since only about 0.1 percent of the total hydrogen mass is actually converted to energy. Stars thus evolve *through most of their life-cycle* with virtually *constant mass but changing composition.*

Let us examine the nuclear reactions in detail to see how compositions

change. In main-sequence stars, vast amounts of hydrogen must be consumed in the hot inner core before new stellar forms can evolve.

Main-Sequence
Reaction: Hydrogen Consumption

In Chapter 12 we saw that hydrogen consumption in the sun occurs mainly by the **proton-proton cycle** of reactions:

$$^1H + {}^1H \rightarrow {}^2H + e^+ + neutrino$$

$$^2H + {}^1H \rightarrow {}^3He + photon$$

$$^3He + {}^3He \rightarrow {}^4He + {}^1H + {}^1H$$

The net result is that hydrogen atoms (^1H) combine, forming helium-4 and producing energy in the form of photons—the basic units of radiation.

The proton-proton chain is the primary energy-producing process inside main-sequence stars smaller than F stars of about $1\frac{1}{2} M_\odot$. It dominates if the central temperatures are less than about 15 million K.

In main-sequence stars larger than about $1\frac{1}{2} M_\odot$ where interior temperatures are higher, another reaction series dominates in producing energy. This is the **carbon cycle,** sometimes called the *CNO cycle* to reflect the involvement of carbon, nitrogen, and oxygen. The reactions are:

$$^{12}C + {}^1H \rightarrow {}^{13}N + photon$$

$$^{13}N \rightarrow {}^{13}C + e^+ + neutrino$$

$$^{13}C + {}^1H \rightarrow {}^{14}N + photon$$

$$^{14}N + {}^1H \rightarrow {}^{15}O + photon$$

$$^{15}O \rightarrow {}^{15}N + e^+ + neutrino$$

$$^{15}N + {}^1H \rightarrow {}^{12}C + {}^4He$$

Again, the net result is that hydrogen atoms are used up to produce helium-4 atoms, with associated release of energy. In a sense, carbon acts as

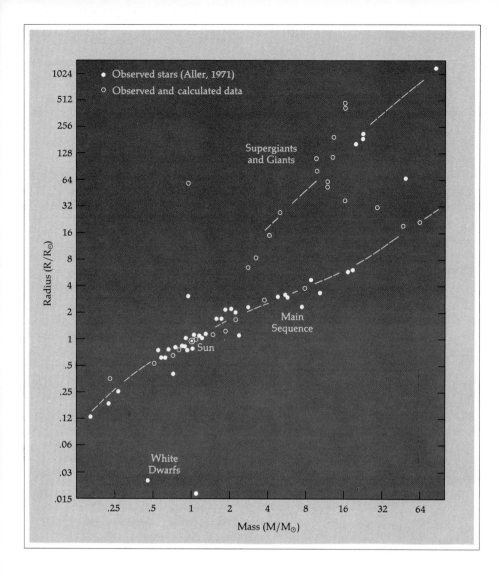

Observed stars (Aller, 1971)

Observed and calculated data

Supergiants and Giants

Main Sequence

Sun

White Dwarfs

Mass (M/M_⊙)

Radius (R/R_⊙)

Figure 14·9 **This figure, showing radii and masses of stars from observations and theoretical models, reveals that distinctly different radii exist among stars of a given mass. This is a result of evolutionary changes; stars in this mass range remain at nearly constant mass during most of their life, but can evolve into states with different radii. Dots are observed stars; circles are observations and theoretical calculations.**

a catalyst (a stimulant of change), because carbon-12 reappears at the end of the cycle, to be used again in the first reaction of subsequent cycles.

Inside main-sequence stars larger than F stars of about $1\frac{1}{2}$ $M_\odot$, the carbon cycle is the main energy-producing process. It dominates if the central temperatures exceed about 15 million K.

Giant Stars'
Reactions: Helium Consumption

What happens when the hydrogen in the star's interior is exhausted? Hydrogen "burning" starts to spread outward to the remaining hydrogen-rich layers. But in the normal main-sequence configuration, the outer regions are too cool to allow much hydrogen "burning." As energy gen-

eration begins to decline in the center, there is too little heat to keep the central regions expanded, and they contract. As always, this contraction raises the interior temperature, as shown in our discussion of Helmholtz contraction (page 203). But at higher temperatures, new reactions occur.

Once temperatures inside any star exceed about 100 million K, other reactions begin to "burn" *helium*, creating even heavier elements. These are the kinds of reactions occurring in old stars, as they evolve off the main sequence. The most important example is the **triple-alpha process** (named after the *alpha-particle*, another name for the helium-4 nucleus):

$$^4He + {}^4He \rightleftharpoons {}^8Be + photon$$

$$^8Be + {}^4He \rightarrow {}^{12}C + photon$$

In this process, three helium-4 nuclei combine to produce a carbon-12 nucleus. (Because ^8Be is unstable, some beryllium atoms may break up before completing the process—but this merely reduces the efficiency of the process.)

The triple-alpha process produces large amounts of energy, which radiate outward through the star. When they reach the outer layers, they cause heating and tremendous *expansion* of the outermost parts of the star. Even though the core has shrunk, the outer surface and atmosphere expand to produce enormous stars—the giants. As the constant supply of gas is spread thinner and thinner over large volumes, the density of the stellar gas decreases, so that the outer layers of giant stars are vast and tenuous.

Arcturus, a giant about 11 parsecs from the sun, radiates 130 times as much energy as the sun and is 24 times as big. Antares, a supergiant about 130 parsecs away, radiates about 28,000 times as much energy as the sun, is about 560 times as big, and would be spectacular from nearby, as shown in Figure 14·10. If it were moved to the center of the solar system, it would engulf Mercury, Venus, earth, and Mars! Its outer surface would lie in the asteroid belt!

Later evolution of such stars may produce unstable stages at which the outer atmosphere expands even more, producing short-lived supergiants. The remote supergiant VV Cephei A, for example, is estimated to be 1,600 times bigger than the sun. If placed in the center of the solar system, it would engulf even Jupiter!

Figure 14·10 **Artist's conception of the supergiant star Antares, as seen from an imaginary planet at the same distance as Jupiter is from our sun. Antares is so big that it would subtend an angle of about 60° as seen from that planet, heating the rocky surface to near-molten conditions. In the upper right is a companion B-type star known to orbit around Antares. The sun would be smaller and fainter than this companion. (Painting by author)**
▼

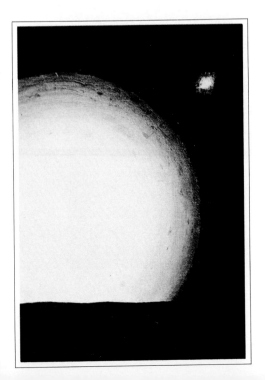

Other Reactions

Meanwhile, in the center of such stars, the small supply of helium does not last nearly as long as the supply of hydrogen lasted on the main sequence. When helium-burning stops, further contraction of the interior core leads to reactions involving even heavier elements. These include the so-called **s-process reactions,** in which neutrons are *slowly* added to nuclei to build up still heavier elements; and the **r-process reactions,** in which *rapid* neutron addition produces very heavy atoms. Thus, the "burning" of various elements of intermediate weights not only consumes the lighter elements and produces energy, but also produces still heavier elements as residues inside the star. The sequence of elements being consumed is shown in simplified form in Table 14·3.

In summary, we have described three important principles that govern stellar evolution:

1. From the time a star forms, its structure is controlled by its mass and composition (Russell-Vogt theorem). At each stage, temperatures in the deep interior determine what nuclear reactions occur.

2. As time goes on, hydrogen is exhausted in the inner regions while heavy elements accumulate. According to the Russell-Vogt theorem, the star's structure must then change as the composition changes.

3. The main sequence is prominent because it corresponds to the long-lasting hydrogen-burning stages of the evolution of stars.

Evolutionary Calculations and Evolutionary Tracks

Because of the work of Eddington, Russell, Vogt, and others, modern

Table 14·3 Elements Important in Nuclear Reactions at Different Stages in Stellar Evolution

	Elements Involved in Nuclear Reactions	Comments
Time	Lithium and other light elements	Relatively unimportant source of energy occurring during formation of star before main-sequence stages.
	Hydrogen	Primary energy source; defines main sequence of stars; proton-proton chain; carbon cycle
	Helium	Important after hydrogen is depleted; involved in reactions during rapid evolution off main sequence and into giant branch of H-R diagram; triple-alpha process
	Heavy elements, including carbon and metals	Energy sources during final period on giant branch and during rapid evolution off giant branch, including supergiant states; s-process, r-process

astrophysicists can calculate models of stars of any given mass and chemical composition. Thus, we can calculate not only what the sun's interior structure is like now, but what its future structure will be when 1 percent of the present hydrogen is gone, 2 percent is gone, and so forth. Knowing the rates of nuclear reactions, we can calculate the time scale associated with this evolution. Using computer modeling, we can estimate what happens during unstable transition states, such as the transition from main-sequence to giant and post-giant stages.

Because a star's temperature and luminosity, as well as other properties, can be calculated for each state of its evolution, the stages of that evolution can be plotted on an H-R diagram. The sequence of such points is called an **evolutionary track,** the set of footprints a star leaves on an H-R diagram as it evolves.

Evolution of the Sun

Figure 14·11 shows the approximate evolutionary track followed by the sun or any star of solar mass. Some parts of the track are more certain than other parts. Consideration of different parts of the track shows clearly why different groups of stars appear on the H-R diagram. As we saw in Chapters 11 and 12, the sun begins its life as a cool, dim interstellar cloud. Therefore, it first appears

on the H-R diagram at the right, with low temperature, low luminosity, and large radius, perhaps exceeding 100 $R_\odot$. It contracts rapidly, becoming quite bright, then reaching a few solar radii in about a million years (only an instant of cosmic time). At this point it is approaching the main sequence. Nuclear reactions begin in the interior, because the central temperature has reached a very high value.

Once the nuclear reactions start, the huge reservoir of hydrogen begins to be converted into helium, and the star reaches a relatively stable, main-sequence configuration. It has taken only about 100 million years to reach

the main sequence, but it is destined to sit on the main sequence for roughly 9 billion years.

Finally, the supply of hydrogen begins to be exhausted. The star's composition having changed, its structure must change, according to the Russell-Vogt theorem. Because there is less hydrogen to fuel the reactions, not enough energy is produced to maintain the outward pressure and keep the star in its main-sequence state. The interior contracts, producing a higher central temperature, and causing new reactions to take place. Helium begins to "burn" in the center, producing more energy. The

Figure 14·11 **H-R diagram showing calculated evolutionary track for the sun (or any star of one solar mass). Dots are placed about 100 million years apart. Track shows evolution from protostar state to main sequence, then to giant, variable, and white dwarf states. (Based on calculations by Westbrook and Tarter, 1975; Iben, 1967; Larson; and others)**

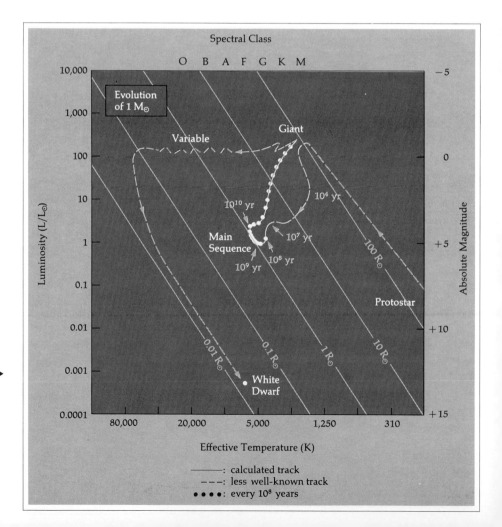

outer layers of the star adjust by expanding. The star rather rapidly moves off the main sequence to higher luminosity and much larger radius. It becomes a giant.

Theoretical calculations can trace the star quite accurately to the giant state, but subsequent states are less well understood, as indicated by the dashed line in Figure 14·11. Evolution now takes the star somewhere to the left. Stars in this region are unstable, often variable, and rapidly evolving. Some stars in this region explode, throwing off masses of gas. Such stars are called **novae** or **supernovae,** and often involve heavier elements in the nuclear reactions at their centers.

Finally, there are no more "combustible" elements in the star's interior. No matter how much it contracts, no matter how hot the insides get, no new reactions produce energy. So the star collapses to a very small size—the white dwarf has been created (or one of several stranger small star-forms, to be discussed in Chapter 16).

The following sequence summarizes the evolution of stars according to the groups found on the H-R diagrams (multiple arrows indicate faster evolution):

Protostar
 ↘ ↘ ↘
  Pre-Main Sequence
   ↘ ↘
  Main Sequence
     ↘
   Giant
    ↘ ↘
    Variable or unstable
       ↘ ↘ ↘
     **White Dwarf
     or other small star**

*Massive
Stars Evolve Fastest*

Stars evolve at different rates. The more massive a star, the higher its interior temperatures, and the faster it uses its nuclear fuel. A sunlike star of 1 $M_\odot$ stays on the main sequence about 9 billion years (9×10^9 years), but a star of 10 $M_\odot$ stays there only about 20 million years (2×10^7 years). The entire history of a star of 1 $M_\odot$ (protostar to white dwarf) takes about 11 billion years, whereas a 10 $M_\odot$ star lasts only about 24 million years.

Thus, the most massive stars stay on the main sequence for only the twinkling of a cosmic eye. Some of them evolve into the supergiant region, and some less massive ones become ordinary giants. All of them quickly evolve to unstable configurations; many may explode; and all disappear from visual prominence.

Determining the Ages of Stars

Consider a group of stars that formed at the same time. After about 10 million years, stars larger than 20 $M_\odot$ will have disappeared from the main sequence. That is, the H-R diagram of the group will contain no main-sequence O stars. After about 100 million years, stars more massive than 4 $M_\odot$ will have evolved off the main sequence, and the H-R diagram will contain scarcely any main-sequence B-class stars. The older the cluster, the more of the main sequence will be gone. The missing stars will have been transformed into giants, white dwarfs, or even fainter terminal objects.

Thus, we reach the important conclusion that *the H-R diagram can serve as a tool for dating groups of stars that*

formed together. This principle will be applied often in later chapters as we probe the **ages of stars** in our galaxy.

Isolated individual stars are harder to date. The sun's age was measured at 4.6 billion years by dating planetary matter, unavailable to us in the case of other stars.

Certain indicators, such as the amount of "unburned" light elements (lithium, for example) in the star's atmosphere, can be used to estimate its age. Thus, by these methods the main-sequence stars closest to the solar system—the Alpha Centauri system (Figure 14·6)—are estimated to be at least 3 billion years old (Boesgaard and Hagen, 1974).

14

Summary

The nearby stars, within about a hundred parsecs of the sun, display a range of masses, luminosities, temperatures, and compositions. These properties are not randomly distributed, but grouped into distinct types. Some stars are on the main sequence, some are giants, some supergiants, and some dwarfs. Forms that a star assumes are controlled by its mass and composition, the most basic properties of a star.

A star that forms with a certain mass and a certain initial composition will evolve through a predictable set of states, in a predictable period of time, and will arrive at a predictable final stage. The reason for the evolution is that nuclear reactions in the star's center constantly change the composition. The Russell-Vogt theorem shows that each combination of mass and composition corresponds to a specific type of star. The evolutionary sequence from one type to the next is usually plotted by astronomers as an evolutionary track on the H-R diagram.

Since stars consist mostly of hydrogen, most stars spend most of their time consuming their hydrogen, in a state called the *main-sequence configuration,* or in a final, stable, dwarf state. A survey of stars within about 100 parsecs of the sun reveals that most stars are either main-sequence stars or defunct dwarf-type stars. Most have masses of about 0.1 to 1.0 solar mass. However, the stars prominent in the night sky as seen from earth are not representative of the true distribution of stars in space; they are biased toward giant and supergiant stars with greater intrinsic luminosity than average stars. Many of these prominent stars are more massive than the sun.

If we survey farther regions of space, hundreds of parsecs from the sun, we encounter mostly the same types of stars found near the sun, but we also encounter unusual star types that shed light on stellar evolution. Some stars are young, perhaps still forming. Some are unstable or dying. Some are closely involved with neighbor stars that affect their evolution in unusual ways. Some are formed from material different from the sun's material. In the following chapters we will describe such stars and their processes.

Concepts

spectrum-luminosity diagram

H-R diagram

main-sequence stars

giant stars

dwarf stars

white dwarfs

supergiant stars

prominent stars

representative stars

static structures of stars

mass-luminosity relation

Russell-Vogt theorem

stellar evolution

proton-proton cycle

carbon cycle

triple-alpha process

s-process reactions

r-process reactions

evolutionary track

novae

supernovae

ages of stars

Problems

1. Suppose you could fly around interstellar space in a "Star-Trek" type mission, encountering stars at random.
   a. *What type of stars would you expect to encounter most often?*
   b. *About what percent of stars would be as massive as, or more massive than, the sun? (Hint: Use statistics from either Table 14·1 or 14·2, as appropriate.)*
   c. *Would the stars encountered be similar to bright stars picked at random in our night sky?*

2. Do giant stars necessarily have more mass than main-sequence stars? Why are they called giants?

3. Four stars occupy the four corners of the H-R diagram. Which ones have
   a. *the highest temperature?*
   b. *the greatest luminosity?*
   c. *the greatest radius?*
   d. *the least probable age since formation?*

4. Two stars lie on the main sequence in different parts of the H-R diagram. Which is
   a. *largest?*
   b. *most luminous?*
   c. *most massive?*
   d. *hottest?*

5. Two stars form at the same time from the same cloud in interstellar space, but one is more massive than the other. Describe differences or similarities at later moments in time.

6. What would an imaginary terrestrial observer see as the sun runs out of hydrogen in the future? If life is confined to earth when this happens, would it perish from heat or cold?

7. Which part of Problem 4 could not be answered just from location in the H-R diagram, if both stars were not on the main sequence?

8. Consider newly forming cosmic objects of stellar composition. What would you expect to happen in terms of stars' life cycles as the masses decrease from one solar mass toward values comparable to Jupiter's mass?

9. Show that astrophysical estimates of the sun's lifetimes are consistent with meteoritic and lunar data on the age of the solar system. Why would the data be inconsistent if astrophysicists calculated that solar-type stars stay on the main sequence only one billion years?

10. Why don't normal stars all collapse at once to the size of asteroids or tennis balls under their own weight?

11. How was the chemical composition of the sun different three billion years ago from what it is now?

12. Red giants have proceeded further in the evolutionary sequence than main-sequence stars, but is it correct to make the blanket statement that red giants are older than main-sequence stars? Explain. Which star is older, the sun or a red giant of 10 $M_\odot$?

Advanced Problems

13. Use Wien's law to show that an M star with temperature about 2,900 K is strongly red in color, and that an O star with temperature about 29,000 K would be strongly blue. Would the strongest radiation emitted by each star be visible to the human eye?

14. For a group of stars of equal radius but different temperatures, T, the Stefan-Boltzmann law shows that the total luminosity will be proportional to T^4.

 a. *Among such stars, how many times brighter would a 20,000 K star be than a 2,000 K star?*

 b. *Using Figure 14·4, confirm your result by reading luminosities on one of the lines of constant radius.*

Project

1. Locate or construct spherical objects that have the same proportional sizes as selected different types of stars, such as an Antares-type supergiant, the sun, and a white dwarf.

The Births of Stars

Somewhere in space at this moment, about 500 parsecs away toward the constellation Orion, a large cloud of interstellar gas is contracting. Its outer parts are attracted toward the center by the gravitational attraction of the atoms in the cloud as a whole. Slowly it will contract and heat up until nuclear reactions begin in the center. A star is being born.

This chapter describes observational proof that this process of star-formation is actually occurring in many regions of space today, and then describes some of the peculiar stellar bodies that are believed to be embryonic stars.

The results are significant from several points of view. Just as new living individuals are being formed from chemical materials against a background of slowly changing genetic pools, new stars are being formed from interstellar gas and dust against a background of slowly changing elemental abundances in the cosmic gas. Locally—and by this we now mean a substantial part of our galaxy—there is a constantly changing environment, with new stars forming and old stars blasting material into interstellar space. From a philosophical point of view, this means that the universe is not static. From a purely astronomical point of view, identifying and understanding newly forming stars is a challenge in our long quest to comprehend the role of stars in our universe.

Three Proofs of Present-Day Star-Formation

It is easy to provide **proofs that stars are forming today:**

Youth of the Solar System The solar system is only about 4.6 billion years old, whereas the entire Milky Way galaxy—our system of 100 billion stars—is at least 10 or 12 billion years old. Thus, our own sun must have formed much more recently than the galaxy, and stars did not all form in one burst at the beginning.

Short-Lived Star Clusters Many young, massive stars are grouped in *open star clusters*. Because of tidal disruptive forces and the tendency of each star in the cluster to follow its own orbit around the galactic center, the clusters dissociate into isolated stars in only a few hundred million years. This is confirmed by analyses of H-R diagrams of clusters, as mentioned in Chapter 14. The famous Pleiades cluster, for example, is only about 60 to 100 million years old.

Short-Lived Massive Stars Chapter 14 showed that massive stars evolve fastest. Calculations show that stars of 20 to 40 $M_\odot$ can last only a few million years in a visible state, yet we see them shining. They must have formed only a few million years ago.

Thus, star-formation has been a continuing process during the whole history of the galaxy, including the last million years. There are stars in the sky younger than the species *Homo sapiens*.

The Star-Forming Environment

The average conditions between stars are a near-vacuum, with very thin, cold gas. Atoms and molecules number only about 1 to 5/cm³. About three-quarters of them are hydrogen (H atoms or H_2 molecules), and most of the rest are helium—as with the atoms in most nearby stars. There are scattered dust particles. The temperature is typically around 100 K ($-279°$F).

Figure 15·1 **The star-forming environment is illustrated by the region of the Horsehead nebula, just south of Orion's belt. In the background (upper part of picture) are veils of glowing gas. The foreground (bottom) is mostly obscured by unilluminated dense clouds of dust and gas. A particularly dense cloud projects to form the horsehead silhouette. A still closer glowing nebula appears at bottom left. The region is part of a star-forming complex roughly 350 parsecs away. The horsehead cloud is roughly 0.5 parsecs across, or about 1,200 times bigger than the solar system. (Kitt Peak National Observatory)**
▼

However, in regions where young stars abound, conditions are different. The main difference is that gas in these regions is much denser, with as many as 10,000 atoms and molecules/cm³. This is still a near-vacuum, of course, compared with ordinary room air, in which there are about 20 billion billion (2×10^{19}) molecules/cm³. There is also more dust in star-forming regions than elsewhere. These findings indicate that stars form in denser-than-average clouds of dust and gas.

Toward an Astrophysical Theory of Star-Formation

How and why do stars form? Chapter 11 already sketched the formation of the sun. A diffuse interstellar cloud contracted and produced a central star surrounded by a dusty nebula. Now we ask for a more general theory that can explain other stars with different masses.

When scientists use the word *theory*, they usually mean a well-tested body of ideas, usually with a mathematical formulation that can be applied to a variety of cases, and usually backed up by observational confirmation.[1] A more general theory of star-formation has been developed. This theory can be approached by asking: What causes contraction of some clouds and not others? Why don't all interstellar clouds contract into stars and be done with it, once and for all? The answer involves the same two opposing forces we have considered before: gravity versus thermal pressure. Gravity

pulls all the atoms in a cloud inward. But even at only 100 K the atoms are dancing, striking each other, creating an outward pressure that opposes the tendency to collapse.

Gravity increases if the density of the cloud increases, since this gets more mass into the same amount of space. Outward pressure increases if the gas heats up, since this makes the atoms and molecules move faster. We can already say, then, that the reason not all clouds contract is that not all clouds are dense enough or cool enough. Furthermore, a non-contracting cloud can be turned into a contracting cloud if it should suddenly experience some turbulence (caused by a nearby stellar explosion, for example) that compresses it.

Astronomers want a more quantitative **theory of star-formation** than this description. For interstellar material of any given density, they have calculated what combination of mass and temperature a cloud would need

to begin its **gravitational contraction,** or shrinkage toward stellar dimensions. The calculations are based on principles sometimes called the *Jeans theory,* after English astrophysicist Sir James Jeans, who first made the calculations. (It is also sometimes called the *virial theorem,* after certain principles in mechanics.)

The results are shown in Figure 15·2. Along the bottom, the figure shows densities ranging all the way from 10^{-35} g/cm³ (perhaps equal to densities between galaxies) to 10^{-3} g/cm³ (about the density of air). This brackets the interstellar density by

Figure 15·2 **This diagram, based on the theory of star-formation, shows the mass likely to be involved in self-gravitating contractions for different gas densities and two gas temperatures. Comparison of different environmental densities (bottom) with observed types of masses (right) shows that many features of the universe can be roughly explained with this simple theory.** ▼

[1]A less complete or less well-tested idea might be called a *hypothesis,* or a *working hypothesis.* An untested idea is often called *speculation.* Unfortunately, newspapers and magazines often publish scientists' speculations, but label them with the more imposing term, *theory.*

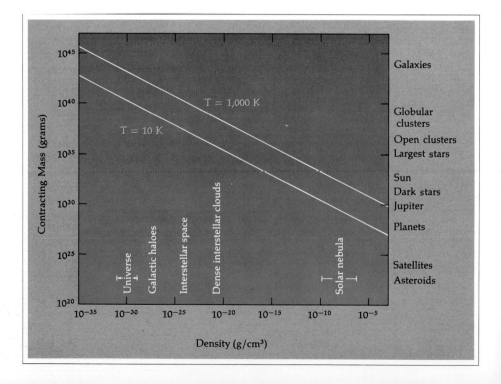

enormous margins, but there is nothing in the theory that says we must apply it only to interstellar gas.

First, let us apply it to the interstellar medium in star-forming regions. What masses will contract there? The figure shows that in the denser interstellar medium, at about 10^{-20} g/cm³, at 100 K, masses contracting would be around 10^{37} grams, or several thousand solar masses, and about 100 times the mass of the largest stars. But our theory was supposed to predict the formation of stars! Is something wrong? In fact, the theory agrees with observations, because, as we noted in the second of three proofs of present-day star-formation, stars apparently form in star clusters that have masses thousands of times greater than the sun. In other words, conditions in the interstellar medium are such that *the gas subdivides into enormous chunks containing enough mass for thousands of stars.* These huge clouds can be thought of as protoclusters.

But how do *individual* stars form? A glance at Figure 15·2 shows that when the protocluster contracts further, smaller, star-size condensations could form inside it. In other words, shrinking protoclusters **subfragment** into individual stars. The Canadian researcher A. Wright (1970), for example, followed the contraction of a 6-million $M_\odot$, 120-parsec-wide cloud and a 500 $M_\odot$, 12-parsec-wide cloud. He found that:

> both large and small density perturbations will grow in size, given the conditions found throughout most of a gas cloud. It would appear that fragmentation is an almost inevitable consequence of the cloud collapse process.

The new fragments become stars, and the whole mass turns into a star cluster. It is not hard to see why the interstellar gas in the disk of the galaxy might begin to contract. The material is not uniform; clots of dust and gas exist and are constantly being stirred by the galaxy's rotation, expansion of hot nebulae, and other influences. Naturally, some clouds accumulate material or become compressed, until their density exceeds the critical density that allows the beginning of contraction.

Comments on the Formation of Just About Everything

Nothing in the theory of star-formation restricts it to masses the size of stars or clusters. That is why Figure 15·2 has been plotted to include such a large range of mass and density. It is provocative to consider the evolution of the whole universe from this point of view.

For example, if all the material in the universe were spread out uniformly, a very low density of roughly 10^{-30} or 10^{-31} g/cm³ would be obtained. The diagram shows that in a more or less uniform, relatively hot gas of this density, the condensations would have the mass of galaxies, about 10^{11} $M_\odot$. In other words, if there was ever a time when no galaxies existed but hot gas filled space, it is reasonable that galaxies should have formed.

Galaxies' outer regions, or haloes, have densities around 10^{-27} g/cm³. At these densities, according to the diagram, we might expect that masses of 10^5–10^6 $M_\odot$ could condense by their own gravity. Exactly such objects actually exist in halolike regions around galaxies (as noted in the Prologue). Called *globular star clusters,* they consist of hundreds of thousands of stars. They are second in size only to galaxies, and our theory makes it quite plausible that they formed where they did through fragmentation of contracting galactic masses.

We have already explained that inside galaxies, where densities are higher, open star clusters form from clouds of about 10^3 $M_\odot$. We also noted that inside those clouds, where it is still denser, individual stars form by further fragmentation involving masses around 1 $M_\odot$.

The same hierarchy of subfragmentation continues to smaller masses. Inside the solar nebula, densities may have reached as high as 10^{-9} to 10^{-6} g/cm³. Under these conditions, according to the same theory, masses about the size of Jupiter might have condensed. Some theorists believe that the miniature solar systems of Jupiter and its satellites, or Saturn and its satellites, formed in this way. Alternatively, as mentioned in Chapter 11, these planets may have grown to their present size by accretion of small planetesimals.

To take another example, small faint stars observed to be companions of larger stars may have formed as gravitationally-contracted masses in nebulae near the larger stars. Finally, Chapter 11 described a current theory that concentrations of dust grains in such nebulae led to even higher densities, where small masses resembling asteroids formed by self-gravitation of dust grains and then accumulated into larger planets.

Thus we see how many loose ends from earlier chapters come together into one simplified theory that accounts for many objects in the universe. The theory as described here is simplified in that it ignores such complications as magnetism, rotation, and the effects of dust opacity. But by so simplifying, astrophysicists can concentrate on the dominant influences to rough out the origin of various objects; astrophysicists today are working out the effects of these complications.

Evolution onto the Main Sequence

We have seen how star-size concentrations of gas arise in space. The questions now are: What steps do these masses go through as they change from contracting clouds into stars? What are the intermediate forms? Where do these forms lie on the H-R diagram? Can they actually be observed among the stars in space? Can we actually see stars being born?

Astrophysicists such as the Japanese theorist C. Hayashi (1961) and his American colleague L. Henyey (1955) pioneered in calculating the evolutionary tracks and appearance of stars contracting toward their main-sequence configurations. Results of such calculations are shown in Figure 15·3, which clearly shows that stars of all masses would approach the main sequence from the right. Newly formed stars would be cooler than main-sequence stars.

The energy during most of this period does *not* come from nuclear reactions, which have not yet started inside the star; instead, heat is generated by the Helmholtz contraction described in Chapter 11. After only a few thousand years of collapse, surface temperatures reach a few thousand K, thus causing visible radiation. Hayashi's work showed that convection would transport large amounts of energy from the interiors of most newly forming stars, making them very bright for short periods, known as the **high-luminosity phase** or the *Hayashi phase*. A star of solar

mass, for example, contracts in less than 1,000 years from a huge cloud to a size about 20 times bigger than the sun with a luminosity about 100 times greater than the present sun's. Figure 15·3 shows it entering the H-R diagram on a rapidly descending track as it fades from its high-luminosity phase.

These calculations have three important consequences. First, they show that stars have complicated, if short-lived, evolutionary histories even before nuclear reactions start. Second, they show that *newly forming stars must lie above and to the right of the main sequence.* (This is important in identifying new stars by direct observation.) Third, the calculations indicate how long stars spend in pre-main-sequence evolution. Generally it is a small fraction of their lifetimes, and motion across the H-R diagram is slower and slower as the star approaches the main sequence, nearly stopping while the star is on the main sequence. A star of solar mass, for example, spends about a thousand years reaching the high-luminosity phase, a few million years fading from

Figure 15·3 **H-R diagram showing pre-main-sequence evolutionary tracks for stars of different mass. Dashed lines show the states reached after the indicated number of years.** (Data from Iben, 1965)

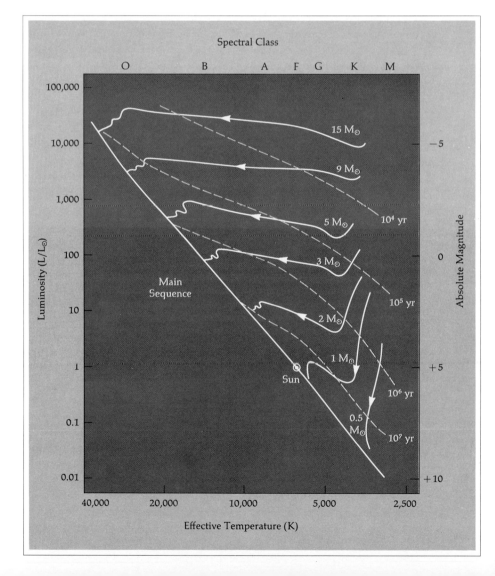

it and perhaps a hundred million years settling onto the main sequence. *More massive stars evolve faster* to the main sequence than less massive stars.

Prediction of Cocoon Nebulae and Infrared Stars

While some theorists were calculating evolving conditions in newly forming stars, others, such as the Mexican astronomer A. Poveda, hypothesized that as young stars form from collapsing clouds, remnants of the clouds might surround and obscure them. As in the early solar nebula, dust would form in the cooling cloud and block the outgoing starlight. The radiation from the new star would heat the nebular dust grains to a few hundred K, and, according to Wien's law, the dust grains would then radiate infrared light. The nebula would eventually dissipate, revealing the star.

The term **cocoon nebula** was coined in 1967 to describe the concept of a star hidden from view during its earliest formative period (perhaps a few million years for a solar-size star), but which later emerges as its surrounding nebula is cast off, just as a butterfly eventually emerges from its obscuring cocoon.

These advances showed the need for new theoretical models, because the earlier calculations had not included effects of surrounding clouds.

Astronomer R. Larson (1969) calculated the evolutionary tracks of stars forming as central condensations inside larger clouds, with results shown in Figure 15·4. The track is more complicated than before, but still approaches the main sequence from the right.

How does the cocoon nebula eventually disappear? This is hard to answer theoretically, being a difficult

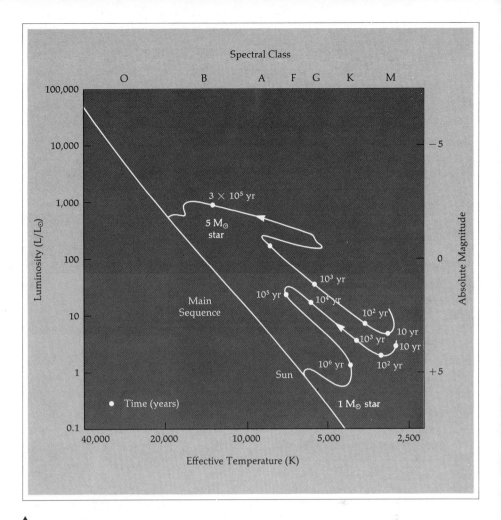

▲
Figure 15·4 **H-R diagram showing pre-main-sequence evolutionary tracks and complex luminosity changes based on a theory modified from that of Figure 15·3 and taking into account a surrounding nebula much larger than the star. Tracks represent only the central contracting star, or "core" of the nebula.** (Data from Larson, 1969)

problem in dynamics of gas, effects of entrapped dust grains, and possible magnetic influences. Larson's calculation of the history of a 1 $M_\odot$ star neglected these effects and indicated that the cocoon nebula simply fell into the star. More realistically, it is probably blown outward by the solar wind from the star and its radiation and heating effects. 1970 calculations

on a 30 $M_\odot$ star as well as solar-size stars suggest that dust particles may eventually be used up as they evaporate in the heat of the star, aggregate in larger bodies, or get blown away, so that the nebula eventually becomes transparent. An observer would then see the star emerge from its cocoon.

But what would the newly forming star and cocoon nebula look like to an observer before the nebula dissipated? Could examples ever be observed? Larson and others tackled this problem around 1972 with calculations of the evolutionary tracks not just of the central star, but of the central star plus the cocoon. Since the star is hidden, the only radiation detectable is the infrared radiation from the dust in the cocoon nebula.

Thus, as shown in Figure 15·5, these objects would have surface temperatures of only about 1,000 K or less, and would lie far to the right of the main sequence. They would be nearly invisible in visual light (wavelength 0.5 micro-meters), but could be detected as **infrared stars** at wavelengths around 1 to 2 micro-meters. As shown in Figure 15·5, they would cross this region of the H-R diagram in less than a million years, and thus would be detectable as infrared stars for a relatively brief period of their lives. They should therefore represent only a small fraction of all stars.

What Happens If the Mass Is Too Small?

If a cloud is dense and cool enough to contract, but has less than a few percent of a solar mass, it will contract but never develop high enough central pressure and temperature to initiate nuclear reactions. It follows an evolutionary track similar to the other protostars we have discussed, but it never makes it as far as the main sequence. After its high-luminosity phase, when its heat from Helmholtz contraction is being radiated, it fades and cools, as shown in Figure 15·6.

Since stars are usually defined as objects generating energy by nuclear reactions, these are not true stars. Objects too small to become stars are sometimes called **black dwarfs**; they are faint, small objects radiating their own heat of contraction in the infrared.

Still smaller objects would look like planets, which radiate little internal energy at all. Figure 15·6 shows that Jupiter (1/1,000 $M_\odot$) can be considered a black dwarf; we have already seen that it radiates its own internal energy at a greater rate than it reflects sunlight (page 166). Astrophysicist P. Bodenheimer has calculated the evo-

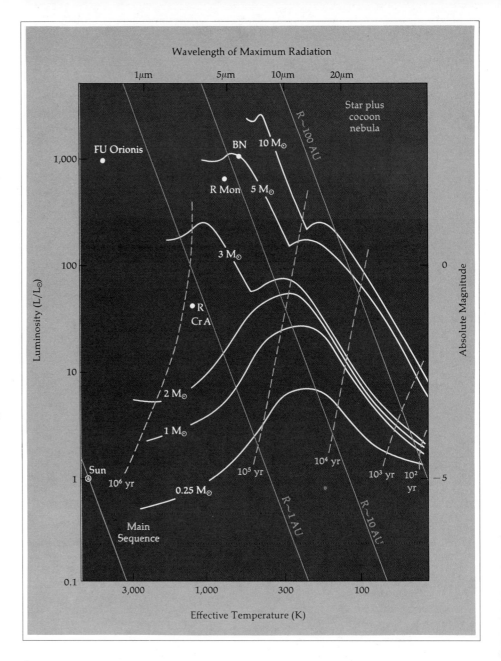

Figure 15·5 **H-R diagram showing apparent tracks followed by pre-main-sequence stars of various masses, surrounded by opaque cocoon nebulae, as viewed from the outside. Unlike the central core stars (see Figure 15·4), the configurations as a whole would be visible only in the infrared for some 10^5 years.** (Data from Larson, 1972)

Figure 15·6 **Extension of the H-R diagram into the low-luminosity, low-temperature region to show evolutionary tracks of newly forming objects of very low mass. A Jupiter-size body and a body ten times as big are shown. These are too small to initiate nuclear reactions, and hence do not become true main-sequence stars. Jupiter may have followed such a track before reaching its present state (lower left).** (Data from Bodenheimer, 1976)

lutionary track shown in Figure 15·6, showing how Jupiter, a "star that never made it," took 4.6 billion years to cool to its present state.

What Happens If the Mass Is Too Big?

If the cloud is dense enough to contract but has more than about 50 to 100 solar masses, the contraction will be violent and produce extremely high central temperatures and pressures. Under these conditions, so much energy will be generated inside the new star that it can blow itself apart almost immediately, without spending much time on the main sequence. This rapid destruction is more appropriate to the topic of the next chapter, the deaths of stars, and will be taken up again there.

Confirming Star-Forming Theory by Observation

We have already mentioned some observational support for the above theory of star formation. First, the youngest stars are found in regions of denser-than-average dust and gas. Second, stars form in clusters, as predicted. There are many other observations of objects that seem to support the theory.

Observations of Infrared Stars and Cocoon Nebulae

In 1966, only a year after some of the early predictions of cocoon nebulae, American astronomers Frank Low and Bruce J. Smith announced infrared spectral measurements of what they interpreted to be a dusty cocoon nebula around R Monocerotis (star R in the constellation of Monoceros, the unicorn), a starlike object about 700 parsecs from the solar system. Low and Smith found that this object showed a large amount of infrared radiation and a faint amount of visible light. They hypothesized that the

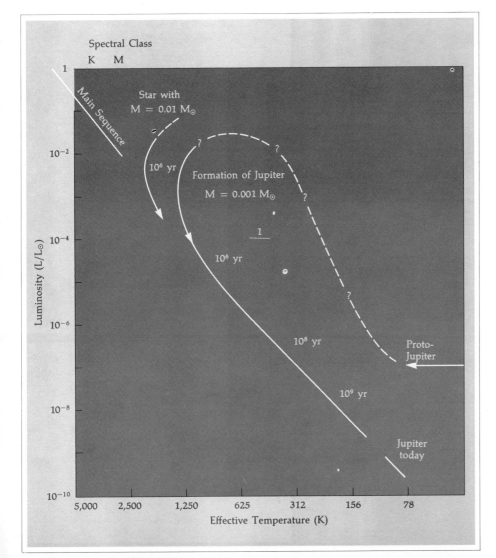

Spectral Class
K M

Main Sequence

Star with
M = 0.01 M_⊙

10^6 yr

Formation of Jupiter
M = 0.001 M_⊙

1

10^6 yr

10^8 yr

10^9 yr

Proto-Jupiter

Jupiter today

Luminosity (L/L_⊙)

Effective Temperature (K)

Figure 15·7 **Nebulosity surrounding the probable young star, R Monocerotis (overexposed and not visible here; see Figure 15·8). The nebula is cataloged as NGC 2261, called Hubble's variable nebula, and probably shines by reflecting light from the newly forming star.** (G. Herbig, Lick Observatory)
▼

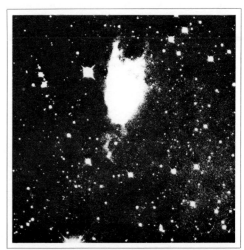

Figure 15·8 **Exposures of Hubble's variable nebula over a 58-year period show changes in form. The exposures are shorter than in Figure 15·7, and cover only its central portion, including the brightest star, left of the nebula. R Mon is the starlike image at the bottom tip of the nebula. Changes in the nebula may reflect changes in the star's illuminating radiation from year to year, suggesting instability in R Mon.** (Lowell Observatory; 1974 photo courtesy Alan Stockton, Mauna Kea Observatory, University of Hawaii)

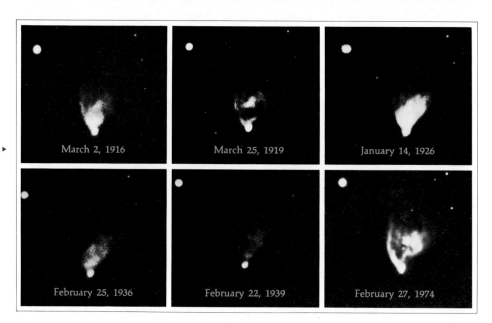

A

visible light came from a new star in the center of a cocoon nebula 200 AU in diameter, which absorbed most of the starlight. The dust grains in the nebula are heated to a temperature measured at about 850 K and radiate infrared light. This object lies at the tip of a triangular nebula about 50,000 AU long, which reflects the light of the young star and its cocoon, as shown in Figure 15·7. As revealed in photos over more than six decades (Figure 15·8), the light, shape, and spectrum of the system change, again indicating that the system may be unstable and rapidly evolving.

Many other infrared cocoon nebulae have been studied, ranging from examples that totally obscure any interior star to examples where most of the young star's light escapes.

Some of the best examples of infrared stars and other probable newly formed stars are found in the **Orion star-forming region,** a large area of the sky around the constellation Orion, shown in Figure 15·9. This region, which includes the R Monocerotis infrared star, is about 400 to 700 parsecs away and seems to be a hotbed of dense clouds and star-forming activity. One very luminous infrared

Figure 15·9 A. A panorama of star-formation in the midwinter evening sky, covering about 80° around the constellation Orion. Many of the features seen are unusually young. (22-minute exposure with a 35 mm camera, wide-angle lens at f2.8, 2475 Recording film) B. Identification of features in A. Distances (parsecs) and ages (millions of years) of selected features are marked. Features in the constellation Monoceros, shown in Figures 15·7 and 15·8, are at top center. Dashed lines show major nebulae; radiating dashes are associations of T Tauri stars, not prominent to the naked eye.
▼

B

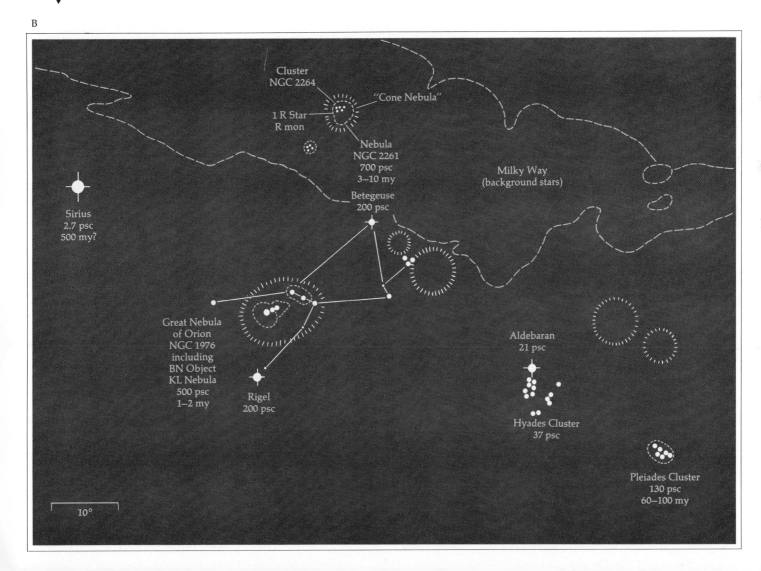

object, the Kleinmann-Low nebula, has been described as a possible proto-cluster of hundreds of stars just forming and still surrounded by a huge, opaque, dusty nebula. Near it is another infrared nebula, the Becklin-Neugebauer object, which is probably a single star surrounded by a dusty nebula.

These and other objects exhibit many features that tie in well not only with the theory of star-formation, but also with evidence about conditions in the nebula that surrounded our own sun as it formed. Among these features are:

1. Spectra indicating the presence of dust, including olivine-rich silicate and ices.

2. Variable brightnesses, indicating instability.

3. Disk-like shapes for the surrounding nebulae, with dimensions from a few AU to 1,000 AU.

4. Detection of molecules of H_2O and OH in some nebulae.

5. Temperatures estimated at about 350 to 1800 K in the dust of the nebulae.

6. Spectral indications of expansion of the nebulae, indicating that they are dissipating.

A close-up view of such a cocoon nebula might resemble the imaginary scene in Figure 15•10. A number of other types of objects have been found in regions associated with infrared cocoon nebulae, and are also believed to be examples of newly forming stars or their debris in various stages of evolution. We will describe some of these types of objects briefly, in their presumed order of evolution.

Masers

Masers are small dense clouds of gas and dust that radiate strong infrared and radio emission lines, but no visible light. Some masers probably are about the same size as the solar

Figure 15•10 **Artist's conception of a newly formed star during transition from an infrared star to a visible star. The central star can be glimpsed through a disk-shaped, obscuring cocoon nebula, but would be obscured by dust if viewed in the plane of the disk. A meter-scale condensed silicate/icy body is seen in foreground, about 30 AU from the new star.** (Painting by author)

system; many are a few thousand AU across. They are strong concentrations of dust grains and molecules, such as H_2O and OH.

The radio emissions are unusually strong spectral emission lines from the molecules. This radiation is far more intense than could reasonably be predicted if the gas were in thermal equilibrium, (that is, if the molecules were energized only by their random collisions with each other). Therefore, some other influence must be *exciting* the molecules to energy levels capable of producing the observed radiations. The probable influence is radio or infrared radiation passing through the cloud. Absorption of this radiation by the gas excites the molecules to higher energy states. This process is called "*m*icrowave *a*mplification by *s*timulated *e*mission of *r*adiation," hence the acronym MASER by which the clouds are known.

Herbig-Haro Objects

Herbig-Haro objects are another group of objects that may be related to newly forming stars. They are small, highly variable nebulae, often found in clusters. They may brighten without warning, remain visible for some years, and die out again as shown by the group in Figure 15•11. In this sense, they may be related to the variable nebulosities around stars like R Monocerotis. Indeed, several

possible Herbig-Haro objects lie near R Monocerotis.

The distinctive feature of Herbig-Haro objects is that no star is visible in any of them, although their spectra may show some emission lines similar to those of the next group of objects, the T Tauri stars. Herbig-Haro objects may evolve into T Tauri stars (Herbig, 1968).

T Tauri Stars

Probably the best-known stars believed to be young are the **T Tauri stars,** named after the variable star T in the constellation Taurus. Although a great many of them are found in the region of Taurus and the neighboring constellation Orion, they can be found in many other parts of the sky. They were first recognized as a group in 1943–44 in work almost unnoticed because the author, K. Himpel, was publishing inside Nazi Germany.

Figure 15.11 **This cluster of Herbig-Haro objects may be a site of star-forming activity. The bottom grouping appears to have changed from two to three objects between 1947 and 1954, while the left-hand grouping changed from one to two objects. These might mark brightenings of unstable young stars, or clearing of obscuring dust concentrations.** (G. Herbig, Lick Observatory)
▼

T Tauri stars may represent a transitional stage between newly forming stars surrounded by opaque nebulae and stable stars that have finished blowing off their cocoons and settled on the main sequence. There are many signs of their youth and instability. They are associated with nebulae and clusters where stars are forming. They vary irregularly in brightness. They lie to the right of the main sequence in the H-R diagram, where young stars are supposed to lie. Many have infrared radiation characteristic of cocoon nebulae as pointed out by the Mexican astronomer, E. Mendoza (1968).

T Tauri stars rotate unusually rapidly for stars of their spectral classes. This fact agrees with their supposed youth, because T Tauri stars are evolving to the left on the H-R diagram and the more leftward stars along the main sequence rotate faster. Thus, a T Tauri star that has reached only 3,000 K and appears to be an M star may actually be a star of two solar masses that will finally evolve into a fast-rotating main-sequence A star of perhaps 9,000 K.

T Tauri stars are surrounded by expanding cocoon nebulae that are apparently in the final stages of being blown away. The gas and dust clouds around T Tauri stars are expanding at velocities of 70 to 200 km/sec. T Tauri stars probably blow off as much as 0.4 $M_\odot$ of gas and dust before they

January 24, 1946

January 20, 1947

December 20, 1954

November 9, 1958

January 5, 1968

reach the main sequence and begin stable radiation of nuclear-generated energy.

T Tauri stars are typically 100,000 to a million years old, judging from all available evidence. Some may be even younger. GW Orionis may be only 20,000 years old, according to one theoretical interpretation. Stars evolved beyond the T Tauri stage would probably be indistinguishable from main-sequence stars.

Young Clusters and Associated T Tauri Stars

T Tauri stars are interesting not only in their individual properties, but also in their association with groups of young stars. In the beginning of this chapter, we described clusters of stars proven by their dynamical properties and H-R diagram position to be very young. Because low-mass stars take the longest to evolve to the main sequence, such stars in these clusters ought to be found in the T Tauri stage. This has been abundantly confirmed by observations.

One of the best-known examples is the young star cluster NGC 2264 (meaning object 2264 in the New General Catalog of astronomical objects), partially shown in Figure 15·12. Most of the low-mass stars (spectral

◄ *Figure 15·12* **Dramatic nebular clouds of dust and gas lie in the star-forming region of the constellation Monoceros (see Figure 15·9). The dark "cone" is a dust-grain and gas concentration estimated to be about 50,000 AU in width, silhouetted against more distant glowing clouds. This region, of angular size about 10 minutes of arc, lies in the outer part of the young star cluster NGC 2264, which is only a few million years old. Molecules of formaldehyde (H_2CO) have been found in the region.** (Lick Observatory)

classes A to K) in this cluster lie distinctly to the right of the main sequence, as shown in Figure 15·13, and many of these are identified as T Tauri stars. From the calculated lines of constant age (dashed lines in Figure 15·13) it can be seen that the

Figure 15·13 **This H-R diagram of stars in the region around Figure 15·12 reveals stars lying right of the main sequence along age lines (dashed), suggesting ages of only a few million years.** (Data from M. Walker, with isochrons from I. Iben)
▼

stars match the positions predicted for an age of 3 to 10 million years. *This demonstrates how the H-R diagram can be used, together with theoretical data, to determine the* **ages of star clusters.** Many further examples appear later.

Another very young group of stars is found in the center of the Orion star-forming region, the great Orion nebula shown in Figure 15·14. Signs of star-formation in this region include T Tauri stars, infrared nebulae, and outrushing gas. The H-R diagram of stars within a one-half degree square of sky in the Orion nebula is shown

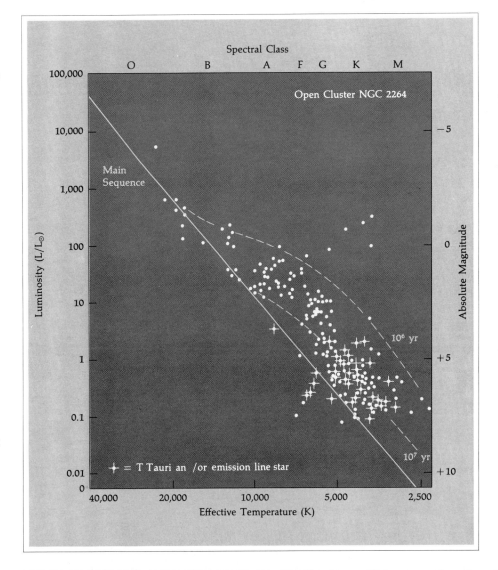

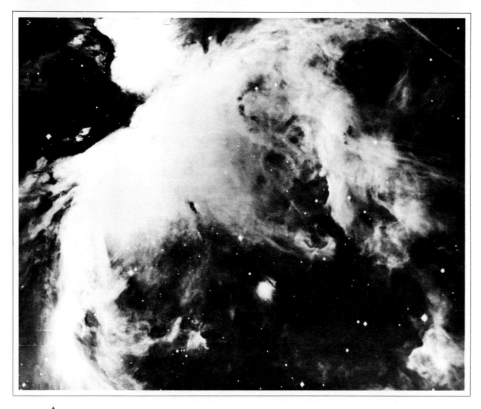

Figure 15·14 **The central part of the Orion nebula, a star-forming region, reveals chaotic clouds of glowing gas and obscuring dark clouds of dust and gas. The region shown is about 4 parsecs across. See also Color Photo 20.** (Official U.S. Navy Photograph)

Figure 15·15 **This H-R diagram for 28 stars within one-half degree of the central Orion nebula (the region of Figure 15·14) shows positions to the right of the main sequence, along time lines (dashed) indicating an age of about one million years.** (Data from M. Penston, 1973)

in Figure 15·15. Here again we see the characteristic departure of the lower-mass stars (spectral classes B to K in this case) to the right of the main sequence. For this group, the theoretical lines of constant age suggest that that the stars average only about a million years in age.

Debris of Star-Formation

In many star-forming regions can be found clouds of gas and dust that may be remnants of the star-forming process, or clouds that did not get quite dense enough to fully collapse. Often, unilluminated foreground clouds are seen in black silhouette against bright star-groups or clouds illuminated by bright stars, as shown in Figure 15·16. Spectacular clouds are often found. Long, tubular dark nebulae called *elephant-trunk nebulae* are sometimes found pointing toward the star-forming region. These may be formed as gas rushing out from the star-forming center sweeps past dense clots of dusty gas.

Another type of dark cloud is the **Bok globule,** named after Dutch-American astronomer Bart Bok, who studied them. Their diameters are mostly about 1,000 to 100,000 AU and their densities reach at least a million times the density in ordinary interstellar space. They are common near young star clusters, but are not always associated with young stars.

A rather large example of such a dark, round cloud is the **Coalsack**

Figure 15·16 **The young star cluster NGC 6611 (upper right) and associated nebulosity (center) show many features of star-forming regions, including hot, massive, blue stars and dark globules rich in dust.** (Steward Observatory)

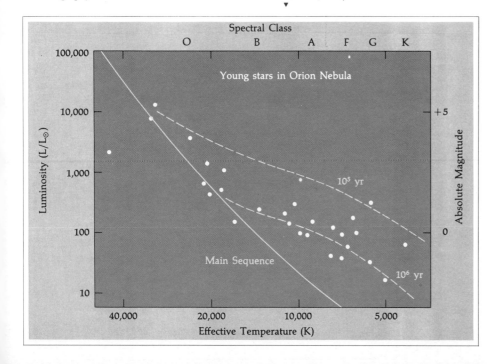

Spectral Class

Young stars in Orion Nebula

10^5 yr

10^6 yr

Main Sequence

Luminosity ($L/L_\odot$)

Absolute Magnitude

Effective Temperature (K)

A

▲

Figure 15•17 **A. The dark round cloud in the center of this view is the Coalsack nebula, silhouetted against the Milky Way in the southern hemisphere. It is estimated to be about 8 parsecs across and 170 parsecs away. This view shows about 30 degrees of the southern Milky Way. (Harvard Observatory)**

B. Identification of some prominent features in A. The blue-sensitive photograph tends to enhance hot blue stars such as Beta Centauri and bright stars in young clusters, but suppresses red stars such as Gamma, in the Southern Cross.

▼

B

Milky Way

Southern Cross

γ

α   δ

Coalsack
Nebula   β

β Centauri

α Centauri

Cluster
IC 2948

Cluster
NGC 3766

Cluster
NGC 3532

Nebula
NGC 3372

Cluster
IC 2602

10°

nebula, shown in Figure 15•17. It is several degrees across and prominent to the naked eye in the southern hemisphere, as it obscures part of the southern Milky Way. Some astronomers think these globules are debris left over after star-formation, but others think some of them may yet contract to form stars.

15

Summary

Stars are forming even today within a few hundred parsecs of the sun and in more distant regions of space. The starry sky is not a static scene, but the site of continual births of new objects. Many stars and star systems are less than a few million years old—much less than one percent as old as our galaxy. Some have become visible since mankind evolved, though prominent newly formed stars have probably not appeared in our sky in historical times.

Theoretical calculations derived in the last two decades have shown how stars form and what features they probably have as they form. Many of these features have been confirmed by observation, especially by infrared equipment, which detects the radiation from relatively low-temperature dust in nebulae around the newly formed stars.

Among the objects revealed in this way are groups of stars evolving toward the main sequence. Many groups of stars and individual stars are surrounded and obscured by cocoon nebulae consisting of dust particles (probably silicates and ices similar to those that formed the first planetary material in our own solar system). More evolved objects, such as the T Tauri stars, appear to be shedding their cocoon nebulae and have almost reached the main sequence.

Concepts

proofs that stars are forming today

theory of star-formation

gravitational contraction

subfragment

high-luminosity phase

cocoon nebula

infrared star

black dwarfs

Orion star-forming region

masers

Herbig-Haro objects

T Tauri stars

ages of star clusters

Bok globules

Coalsack nebula

Problems

1. Did the sun and solar system form in the first half or last half of our galaxy's history?

2. Suppose you magically smoothed out all inhomogeneities in the interstellar gas, so that it was all uniform.
   a. *Would this help or hinder star-formation?*
   b. *What processes would keep the gas from staying uniform indefinitely?*
   c. *Would star-forming conditions tend to return to normal?*

3. Why do stars form in groups instead of alone?

4. How do theories of solar system formation and theories of star formation support each other? Contrast the sources of information on these two subjects.

5. Compare the time scale for significant evolution of massive pre-main-sequence stars with the time during which astronomers have recorded observations of such stars. (See Figures 15·4 and 15·5.) Is it reasonable that some young, pre-main-sequence stars might show evolution-related fluctuations in their properties within the time they have been observed?

6. Suppose a cluster of stars all formed three million years ago. Why would you expect the H-R diagram of the cluster to show no stars on either the very high-mass end of the main sequence or its very low-mass end?

7. Suppose a new star of one solar mass began forming about ten parsecs from the sun. What would earth-based observers see during the next few million years? Consider infrared observers as well as naked-eye observers.

8. Why doesn't gravity immediately cause the collapse of all interstellar clouds?

9. Why does the structure of a star stabilize when it reaches the main sequence?

Advanced Problems

10. Suppose you observe an infrared nebula whose strongest radiation comes at wavelength 10^{-3} cm and which has absorption lines of solid silicates in its spectrum, as well as faint lines indicating a G-type star. Suppose the silicate lines are blue-shifted by about 10^{-3} of their wavelength compared to the star's wavelength. What conclusions can you draw about this system?

Project

1. On a clear night (an early evening in February or a late evening in December is ideal) scan the region of Orion with your naked eyes and compare it to other regions of the sky. Note the concentration of bright blue O, B, and A-type stars (such as Sirius and Rigel) and star clusters (such as Pleiades, Hyades, and the Orion belt region) in this broad area. How do these features indicate that star formation has been going on in this general direction from the sun in the last few percent of cosmic time?

The Deaths of Stars

Stars are born and stars die. By this we mean that the nuclear reactions converting mass to energy in stars have a beginning and an end. The sun, for example, spent some 10^7 years in its formative stages; it will spend about 10^{10} years on the main sequence, roughly 10^9 years in the giant state, and a shorter time in later unstable states consuming heavier elements until energy generation stops (see Figure 14•11). This evolutionary sequence might be likened to the periods of human life: 9 months in the womb, 60 years of normal life, 6 years of rapid aging, perhaps a year of pathological terminal illness. As we will soon see, stars do evolve to pathological terminal states, featuring incredible forms of dense matter and processes unimagined until a few years ago.

Writer Ben Bova (1973) summed up star deaths by quoting one of Ernest Hemingway's characters, who is asked how he went bankrupt. "Two ways," he says. "Gradually and then suddenly." The gradual expenditure of a star's hydrogen is scarcely noticeable to an observer, but then the star suddenly goes into fits of activity, searching for new sources of fuel until, as Bova says, "gravity forecloses all the loans."

Gravity forecloses the loans by causing a continual tendency toward contraction. When a main-sequence star runs out of hydrogen, the energy-producing hydrogen reactions wind down and contraction of the central core begins, raising the temperature until helium nuclei react. After helium, certain heavier elements serve as fuels, but contraction is always required to raise the temperature to the point where these fuels react. Eventually, no more fuels react, and the star has contracted to a very dense state. This chapter will describe this sequence in more detail, and we will see that some of the dense final states of stars are astonishing indeed.

Giants and Supergiants

Calculated evolutionary tracks (Figure 16•1) show the first stages of stellar old age. As described in Chapter 14, a star evolves off the main sequence as hydrogen fuels run out. In what might seem a contradiction of the trend toward contraction, the outer atmosphere of the star expands, taking the star into the giant region of the H-R diagram. But the expansion is only temporary, in response to a sudden flow of energy from the newly-contracted interior. As Figure 16•1

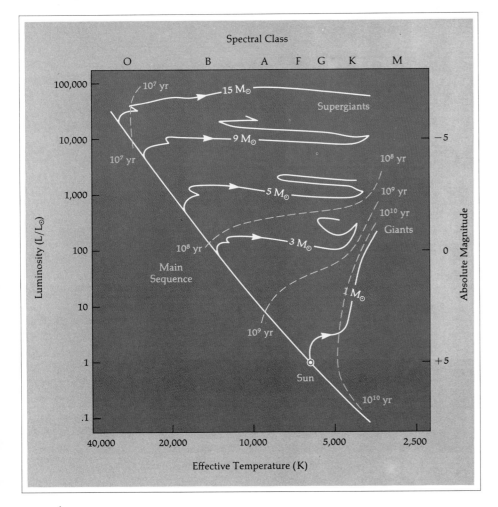

Figure 16·1 **Evolutionary tracks off the main sequence toward the giant region, plotted on the H-R diagram for stars of different mass.** **Dashed lines give the length of time since star formation; massive stars evolve fastest.** (After calculations by Iben, 1967)

shows, stars from all points along the main sequence **funnel** toward the giant region, like patients from all walks of life crowding into a sanitarium, victims of the same disease. By the time stars greater than about 6 $M_\odot$ leave the main sequence, they are already brighter than ordinary **giant stars** and are called **supergiants,** or stars of unusually high luminosity, as they evolve rapidly across the top of the H-R diagram.

Theoretical calculation of the behavior of giants and other stars in old age is difficult because many complex processes occur. The best calculations involve the history of a star of one solar mass. As it evolves toward the extreme right tip of the giant region, it accumulates a large core of helium created from its "burned" hydrogen. After about 300 million years as a giant, the central temperature reaches about 100 million K, high enough to set off the triple-alpha helium-burning process. A so-called **helium flash,** or rapid burning of helium, now raises the central temperature to as much as 300 million K for a few thousand years. This brings the star back into the central giant region, this time burning helium as its major fuel.

During rapid evolution of this kind, some giants and supergiants pulsate or blow gas from their upper atmospheres, causing unstable conditions difficult to predict theoretically.

Example: FG Sagittae

FG Sagittae, a supergiant about 2,500 to 4,000 parsecs away, in the constellation Sagitta, can actually be observed evolving across the H-R diagram; it has the fastest known evolutionary rate. It has brightened by 4 to 6 magnitudes since 1894, peaking in 1971 and fading slightly since. From 1955 to 1972 it cooled steadily from spectral class B4 to F6. It is surrounded by a glowing nebula expanding at about 34 km/sec. This expansion rate indicates that the nebular gas was blown out of the star about 6,000 to 9,600 years ago. From 1969 to 1972, certain heavy elements in the star's atmosphere, such as zirconium, have increased by three to four times in abundance.

Our interpretation is that FG Sagittae was originally a main-sequence star of about 3 to 5 $M_\odot$. It evolved to the tip of the giant region, then underwent a helium flash about 6,000 to 9,600 years ago, blowing off some outer atmospheric mass to produce the nebula we now see and becoming an unstable supergiant. Another helium flash just before 1894 initiated the activity we now see. The shell most recently blown off is opaque, expanding, mixing with gases from the interior, and cooling. These processes account for its rapidly changing spectrum and newly appearing elements.

Variable Stars

A **variable star** is a star that varies in brightness. Most variable stars

Table 16·1 Examples of Stars in Late Evolution

Stage of Evolution (increasing age)	Distance from Earth (parsecs)	Mass (M/M$_\odot$)	Radius (R/R$_\odot$)	Luminosity (L/L$_\odot$.)	Spectral Type	Density[a] (g/cm³)
Main Sequence						
Sun	<1	1	1	1	G2	1.42
α Centauri B	1.3	0.88	0.87	0.36	K4	1.6
Procyon A	3.5	1.2	1.7	7	F5	0.26
Algol	31	4.7	3.1	120	B8	0.23
Giants						
Arcturus	11	~4	23	130	K1	0.0005
Aldebaran	21	~4	45	360	K5	0.00006
β Pegasi	60	~5?	130	1300	M2	0.000003
Supergiants						
Antares	130	~13	~600	28,000	M1	0.0000035
Betelgeuse	180	~15	~700	85,000	M2	0.00000006
VV Cephei A	1200?	80	1620		M2	0.00000003
Abnormal Stars						
Mira A (long per. var.)	40		240	2,500	M6	~0.000001
Polaris (Cepheid var.)	200	~6	~25	830	F8	~0.0005
HD 193576B (Wolf-Rayet)	—	12	~7		"WR"	~0.05
DQ Herculis B (nova)	—	0.2	~0.1	<0.1	M?	284
White Dwarfs						
Sirius B	2.7	0.98	0.022	0.002	A5	130,000
Procyon B	3.5	0.64	0.01	0.0004	F	900,000
Pulsar (Neutron Star)						
Crab nebula pulsar	1100	~2??	<0.00002	High in UV, X-ray	—	~10^{14}?
Possible Black Hole						
Cygnus X-1	≥2500	6 to 8?	<0.000004	0	—	~10^{19}??

[a]For comparison, densities of familiar materials are: atmosphere at sea level, 0.0012 g/cm³; water, 1.00 g/cm³; lead, 11.3 g/cm³.

Source: Data from Peery (1966); Thorne (1968); Aller (1971); Penrose (1972).

apparently represent *post-giant stages* of evolution. Some pulse with constant period; others flare up sporadically, often brightening by many magnitudes and then fading again. Such stars were recorded as long ago as 134 B.C. when Hipparchus recorded the flare-up of a "new star," proving that the heavens were not eternally constant. The star Omicron Ceti was known to the Arabs as Mira (Wonderful) for its brightening from invisibility (magnitude 8–10) to second magnitude every 331 days.

Shakespeare used the simile "constant as the northern star." He might have been interested to know that Polaris is constant neither in brightness nor in marking the north celestial pole. Polaris is a variable star that changes from magnitude 2.5 to magnitude 2.6 in just less than four days, and because of precession (Chapters 1 and 2) is the north star for only a few centuries every 26,000 years.

Some 22,650 variable stars have been cataloged and divided into as many as 28 types. The type most important to astronomers is the **Cepheid variable,** which has regular variations in brightness with periods from 1 to 50 days. In 1784, nineteen-year-old English astronomer John Goodricke discovered that Delta Cephei varies from magnitude 4.4 to 3.7 with a period of 5.4 days; Cepheids were named after that star.

Cepheids are important for two reasons. First, the regularity of their variations allows somewhat better theoretical understanding than for stars whose brightness changes are unpredictable. Second, and more important, the period of variation of each Cepheid directly correlates with its average luminosity. This allows astronomers to recognize the luminosity of any Cepheid at any distance simply by measuring its period. This, in turn, leads to a new way to measure distances of stars, as we will discuss in more detail later.

Other stars that vary irregularly with unpredictable outbursts are called **irregular variables.**

Table 16·2 Types of Variable Stars Shown in Relation to Evolutionary Sequence[a]

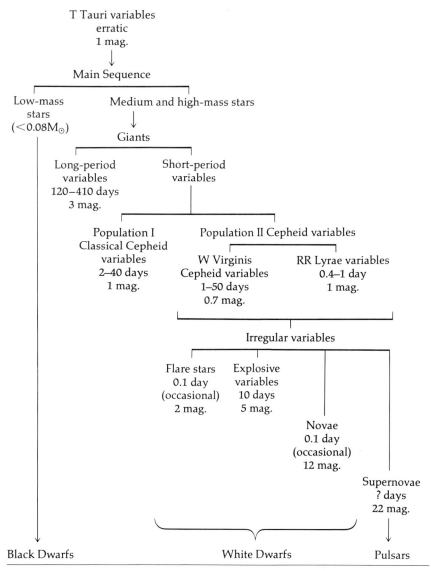

[a]Typical time scales and magnitudes of brightness variations are given. Arrows indicate confirmed evolutionary relations.

Cause of Brightness Variations

Most variables change brightness because of instabilities associated with changes in their interior nuclear reactions or with the flow of radiation out from their centers. Variables vary in physical size and other properties as well as in brightness. Delta Cephei itself varies in diameter (from roughly 38 million to 41 million km) and temperature (from 5,600 K to 6,500 K) during its cycle.

Astrophysicists have long sought the precise cause of the internal physical disturbances that make Cepheids pulsate. Various researchers, especially the Soviet astronomer S. Zhevakin, have established that the H-R diagram contains a near-vertical **instability strip,** shown in Figure 16·2. Under conditions in this strip, ionized helium absorbs outgoing radiation in the star, becoming doubly ionized. In the

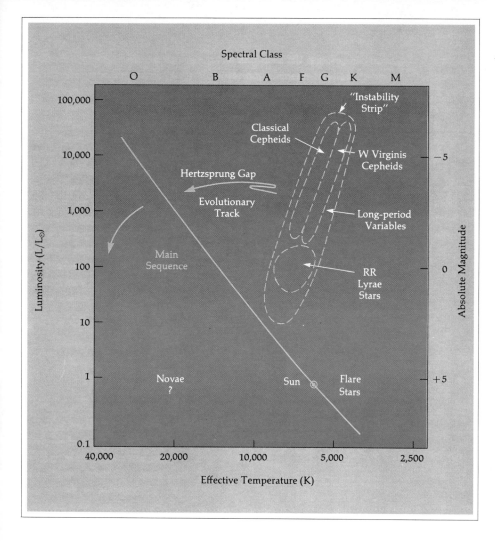

an evolutionary stage so unstable that stars evolve very quickly across it, spending very little time there.

Stars That Blow Off Mass

The instabilities attending post-main-sequence evolution are further exemplified by stars that are blowing off mass. There are many types, including giants and supergiants such as FG Sagittae, mentioned earlier. They provide an excellent example of the application of the Doppler shift. Because the near side of the ejected gas cloud is moving toward the observer, spectral lines originating in this gas are identified by their blue shift.

M giants often have blue-shifted absorption lines, arising in dusty gas blown off the star and expanding toward outside observers.

Wolf-Rayet (rye-AY) **stars** (named after their discoverers) are very hot stars resembling O stars, except for their expanding gaseous shells. About 200 have been found since their discovery in 1867. Their spectra show bright emission lines of ionized carbon, nitrogen, and helium that appear to arise in a gaseous cloud close to the hot star. Blue-shifted from the centers of these lines are dark absorption lines, caused by passage of the starlight through cooler gas in the outer parts of the cloud, as shown in Figure 16·3. Velocities measured from these blue Doppler shifts indicate that the clouds around Wolf-Rayet stars are

instability strip, this absorption occurs not far below the star's surface, where it temporarily "dams up" the radiation.

Just like a bell (or any other object), a star has a certain frequency or time period in which it tends to vibrate in response to a disturbance. If the time required to dam up the radiation is close to the star's natural oscillation period, a repeated oscillation of the star will begin. This is what happens in a Cepheid. As the radiation is dammed, pressure builds up, and the outer layers of the star expand. But once the expansion gets going, momentum and the star's natural tendency to oscillate at this speed carry the expansion too far, allowing radiation to

escape easily. Then the star begins to subside again, and the radiation once again begins to be dammed, restarting the cycle. In other types of stars, where the time scales of radiation damming and oscillation are not synchronized, irregular patterns of variation may develop.

The Hertzsprung Gap

Just to the left of the instability strip in the H-R diagram is the **Hertzsprung gap,** a vertical band nearly empty of stars. Why do no stars have the combination of luminosity and temperature that would put them in this gap? Apparently the gap represents

shooting out, often at nearly 1,000 km/sec! Many Wolf-Rayet stars are members of binary systems (pairs of co-orbiting stars). Their special characteristics probably involve transfers of material as gas blows off one star and strikes the other.

B-type shell stars are stars of spectral type B that occasionally blow off masses of gas. A famous example is Pleione, a star in the Pleiades. Bright emission lines seen in 1888 indicated

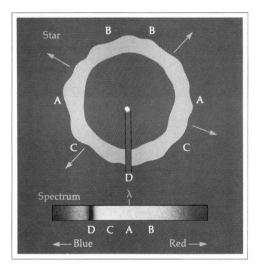

▲

Figure 16·3 Expanding gas shells, such as those around Wolf-Rayet stars or novae, can be detected spectroscopically, even when unresolved photographically. Glowing gas produces spectral emission lines of the shape shown at the bottom. In region A, light with no Doppler shift is added to the continuum spectrum of the star. In region B, red-shifted light is added, and in region C, blue-shifted light. In region D, the maximum blue shift occurs (especially in WR stars), and starlight is absorbed by the nebula, producing a dark absorption band at the blue end of the emission line.

Figure 16·4 Nova Herculis, showing the decline in brightness from March 10, 1935 to May 6, 1935. Later studies showed that this and other novae are binary stars. (Lick Observatory)

that a hot gas cloud had been thrown out near the star. This gas expanded and thinned, and the spectral lines finally disappeared by 1905. In 1938 another outburst appeared, dying away by 1951, followed by a new outburst in 1972. The spectral lines indicate that the star is rotating extremely fast (equatorial velocities near 350 km/sec) and that centrifugal acceleration may aid in blowing out the gaseous shell, which is not expanding very fast.

P Cygni stars are a more spectacular type of eruptive star. P Cygni itself was discovered in 1600 when it suddenly brightened. It has faded, but continues to vary irregularly. There may be several shells expanding at 100 to 200 km/sec. P Cygni's position in the H-R diagram suggests an enormous star of 100 $R_\odot$ and 100 $M_\odot$, about twice the mass normally considered possible for a stable star.

In all these examples, only a tiny fraction of the total mass of the star is being blown away. For example, in 100 years P Cygni loses no more than 2/1,000ths of its total mass. Apparently, the high mass of these stars generates high temperatures and pressures, and during a certain evolutionary stage, the outer atmosphere expands so fast that it escapes.

Novae

A **nova** ("new star"; plural, **novae**) is a *brilliant exploding star* that blows off about 1/1,000th of its mass in a single eruption a few days long. Novae appear to mark a penultimate stage of evolution for certain stars. Their observable history is usually the following sequence:

Prenova state: *faint blue star below main sequence in H-R diagram.*

↘

Sudden brightening: *often 9 magnitudes (factor 4,000) in first day.*

↘

Peak brightness: *often 12 magnitudes (factor 60,000) brighter than originally. Emission lines indicate ejected gas.*

↘

Decline in brightness: *about 3–7 magnitudes over hundreds of days.*

↘

Continuing slow decline for years: *often irregular variations in brightness.*

↘

Postnova state: *expanding gaseous nebula may be visible in large telescope.*

Of the 100 billion stars in our galaxy, probably 30–50 explode as novae each year. Table 16·3 lists some of these "new stars" that have reached second magnitude or greater since A.D. 1000.

The best-observed nova in history was Nova Cygni, which was originally far too faint to see with the naked eye, but brightened to second magnitude and provided Cygnus the Swan with an extra star for a few days in August 1975.[1] Most novae have not been observed during early stages of

the explosion, but during the Cygnus event, amateur astronomers in California and Maryland were photographing the sky and actually recorded the nova as it brightened. This provided a detailed record of the brightness behavior, shown in Figure 16·5. During the explosion, the star was doubling in luminosity about every two hours! Measures of Doppler shifts showed its ejected gas cloud expanded at about 1,000–2,000 km/sec.

Cause of Novae Explosions

What causes novae? Prenova stars, though poorly observed, are believed to be small, hot stars lying in the

Figure 16·5 **The brightness fluctuations of well-observed Nova Cygni in 1975 include observations made both before and after its discovery on August 29. Small dots are visual estimates of brightness by amateur and professional astronomers; larger symbols are more accurate photoelectric or photographic data. Distance of Nova Cygni is estimated at 1,300 parsecs.** (From International Astronomical Union data)

dwarf region below the main sequence, as shown in Figure 16·2. How can such seemingly insignificant stars produce in a few days outbursts that equal the sun's energy output in 10,000 years? A clue came when the American astronomer Merle Walker showed that a 1934 nova was one star of a very close binary pair which co-orbit in only $4\frac{1}{2}$ hours. In 1955, Russian-American astronomer Otto Struve theorized that all novae might be members of binaries. In 1964, Robert Kraft showed by spectroscopy that at least 7 of 10 other novae are binaries. Astronomers now think all novae are members of close binary pairs.

What is it about membership in a close binary that promotes nova explosions? The more massive star of the pair evolves faster and goes through the giant stage, eventually burning all its fuel and contracting into a dwarf. Later, the less massive star expands to the giant stage and begins to blow off gas. Some of the gas may fall onto the surface of the dormant dwarf, adding new hydrogen and causing explosions. The expanding giant may even engulf the dwarf in some cases, causing violent instabilities. There is some evidence that such events may occur in cycles, with hydrogen

[1]Note that it would be incorrect to say "exploded in 1975." Since the nova was estimated to be 1,300 parsecs (4,200 light-years) away, the explosion really occurred around 2200 B.C.! The light just didn't get here until 1975.

Table 16·3 Selected "Guest Stars"[a]
(Novae, Supernovae, and Variables That Have Become Brighter Than Apparent Magnitude 2)

Star	Date Observed	Maximum Brightness (Apparent Magnitude)	Type
Lupus supernova	1006	−5?	Supernova
Crab nebula explosion	1054	−2 to −6?	Supernova
Tycho's star	1572	−4?	Supernova
Kepler's star	1604	−2?	Supernova
Eta Carinae	1843	−0.8	Nova?
T Corona Borealis	1866	+1.9	Recurrent nova[b]
GK Persei	1901	+0.2	Nova
V603 Aquilae	1918	−1.4	Nova
RR Pictoris	1925	+1.2	Nova
DQ Herculis	1934	+1.3	Nova
CP Puppis	1942	+0.2	Nova (or supernova?)
Nova Cygni	1975	+1.9	Nova
o Ceti (Mira)	—	+1.0	Long-period variable (period 331 days)

[a]"Guest star" is the ancient Chinese term for temporarily visible stars. Tentative identification of 14 supernovae ranging back 2,000 years has been made from ancient records. Certain expanding nebulae must also mark prehistoric supernovae.

[b]Second recorded flare-up in 1946 reached only +3 magnitude.

Source: Data from Glasby (1968); Lupus data from *Sky and Telescope* (July 1976).

Figure 16·6 **Gaseous debris blasted out of Nova Persei during an explosion observed in 1901 to form this expanding nebula around the star. Photo was made 58 years after the observed outburst.** (Lick Observatory)

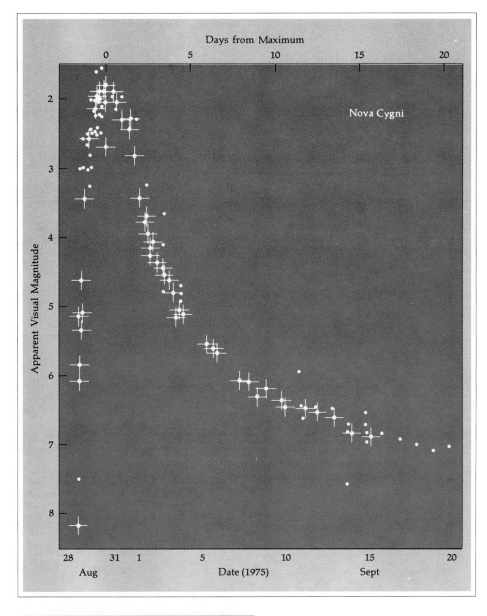

Nova Cygni

glimpsed until 1862, when American telescope maker Alvin Clark detected it. It is almost lost in the glare of Sirius, as shown in Figure 16·8. In 1915, Mount Wilson observer W. S. Adams discovered that it was a strange, hitherto-unknown type. It is hot, bluish-white, and lies below the main sequence on the H-R diagram. It has about the mass of the sun, but is so faint that its total radiating surface cannot be much more than that of the earth.

A star the size of the earth? If a solar mass were packed into an earth-size ball, it would be astonishingly dense. A cubic inch would weigh a ton! A thimbleful would collapse a table!

Dutch-American astronomer W. J. Luyten, who has discovered a number of these stars, described what happened as astronomers first recognized Adams' amazing results (quoted by Shapley, 1960):

. . . the figures were too staggering to be believed without further evidence. Could there be an error somewhere? What had happened? But while we pondered these things, two more stars of the same kind were found: small, extremely faint, and white, and all three were named white dwarfs.

A **white dwarf,** then, is a planet-size, super-dense star. About 200 are known. They apparently represent a terminal stage of evolution after a star has run out of nuclear fuel and gravitationally contracted to very small size.

The explanation of white dwarfs came from theoretical astrophysicists such as Sir Arthur Eddington and the Indian-American theoretician S. Chandrasekhar. Once no more energy is available to generate outward pressure, the star collapses until all its atoms are jammed together to make

building up until nova explosions occur, perhaps 10,000 years apart. Precise details remain unclear.

White Dwarfs

In 1844 the German astronomer Friedrich Bessell studied the motions of the brightest star in the sky, Sirius, and found that it is being perturbed back and forth by a faint, unseen star orbiting around it. This star was not

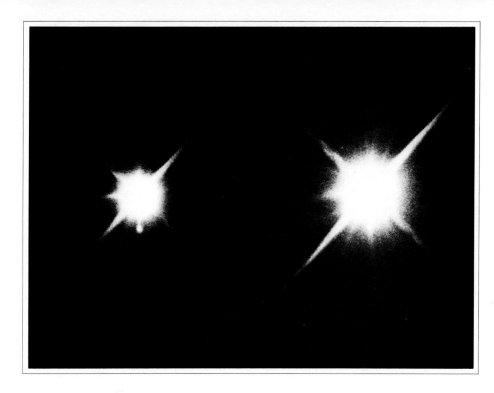

◀ *Figure 16·7* **Nebula NGC 7293 in Aquarius is a glowing spheroidal shell of gas probably blown out of the central star during an explosive event. Delicate filaments radiate from the center.** (Hale Observatories)

▲

Figure 16·8 **Two exposures of Sirius with the Lick 120-inch telescope. Short exposure (left) reveals the faint white dwarf that is a binary companion of Sirius; in the longer exposure (right) the white dwarf is lost in the overexposed image.** (Lick Observatory)

a very dense material. But what does "jammed together" really mean? At these densities, some electrons move at almost the speed of light and matter loses its familiar properties. Stellar matter stops behaving like a perfect gas. In fact, it is no longer either gas, liquid, or solid, but has a new form known as **degenerate matter.** By a physical law called the **Pauli exclusion principle,** only a certain number of electrons can be jammed into a given small volume of space. Thus, the atomic nuclei are held apart by a sea of electrons—a so-called degenerate electron gas—at densities of about 10^5 to 10^8 g/cm³.

Astronomical research on white dwarfs spurred more general physical research in the 1930s on the problem of high-density forms of matter. Luyten has gone so far as to say that analyses of white dwarfs "forged another important link in the chain of events which eventually led to the atomic bomb."

Because pressure at the surface of a white dwarf is not enough to cause degeneracy, white dwarfs may have crusts of ordinary matter about 20 to 75 km thick. It may be gaseous material at the surface and crystalline rocklike solids below the surface. At the base of this crustal layer, densities may be as high as 3,000 g/cm³.

Only stars with mass of less than 1.4 $M_\odot$ can be white dwarfs. Collapsed stars more massive than this limit, called the **Chandrasekhar limit** (for its discoverer), would be still denser, and the white-dwarf model would be invalid. Of course, more massive stars do exist. Here, then, is a tie-in

between white dwarfs and novae. Apparently, many dying stars contract toward white-dwarf states, but if they exceed 1.4 $M_\odot$ they become unstable and blow off mass until they get below the Chandrasekhar limit. Then the explosions stop and they become white dwarfs. Current data indicate that stars that begin with about 0.085 to 0.4 $M_\odot$ evolve relatively smoothly to the white-dwarf state without losing mass.

New light was shed on the problem during the U.S.-Soviet Apollo-Soyuz flight in 1975. A far-ultraviolet onboard telescope detected unexpected, intense radiation from a white dwarf, HZ 43. Application of Wien's law indicated a surface temperature of about 150,000 K, making this the hottest white dwarf known, resembling

the nova-related eruptive variables. HZ 43 is thus intermediate between prenovae and ordinary white dwarfs. It may be a "missing link" in the evolution of ordinary stars toward white-dwarf states.

The theory of evolution to white-dwarf states seems to work best for stars from about 0.08 to 4 $M_\odot$. Stars much smaller become black dwarfs, and stars much more massive become explosively unstable. But what finally happens to the most massive stars?

High-Mass Stars: Supernovae

A **supernova** *is the most energetic of all stellar explosions,* distinctly different from and brighter than a nova. Supernovae are believed to mark the death of high-mass stars. They reach the amazing absolute magnitude of -14 to -20. Although single stars, they blaze for a few days with the light of an entire galaxy! In one year a supernova radiates as much energy (about 10^{50} ergs) as the sun does in a billion years.

Table 16·3 includes some prominent supernovae in our galaxy. The brilliant supernova that looked brighter than the planet Venus in A.D. 1054 was recorded in Chinese documents, and apparently in American Indian rock paintings such as that in Figure 16·9. Today in its position we see an expanding gas cloud known as the Crab nebula, shown in Figure 16·10. Historical records indicate about 14 supernovae in our galaxy during the last 2,000 years, an average of one every 140 years or so. Any of us might live to see one. Supernovae are often seen in other galaxies. In 1885 a supernova in the nearby Andromeda galaxy became barely visible to the naked

Table 16·4 Selected Theoretical Calculations of Star Deaths

Initial Mass of Star ($M_\odot$)	Death Processes	Calculated Mass of Final Remnant Star ($M_\odot$)	Probable Form of Final Remnant Star
<0.085	Gradual cooling	Same	Black dwarf
0.085–0.4	Gradual cooling and collapse	Same	White dwarf
0.4–3	Gradual mass loss	<1.4	White dwarf
3–6	Violent explosions caused by neutrino absorption	0.5–1	White dwarf
12–30	Violent explosions caused by neutrino absorption	1.4–3	Neutron star pulsar
30–100	Very violent explosions; nuclear "burning" of oxygen	3–100	Black hole?
≥100	Catastrophic explosion?	approx. 1 ??	?

Source: Data from various sources, including specific calculations on supernovae from six authors, 1966–1969, summarized by Novikov and Thorne (1973).

Figure 16·9 NASA astronomer John C. Brandt studies an ancient Indian petroglyph in ruins near Zuni, New Mexico. The petroglyph may show the appearance of the Crab nebula supernova, which would have been visible as a brilliant star near the crescent moon during one of the first evenings after it appeared in July, A.D. 1054. The star-moon motif has been found in a number of petroglyphs associated with Southwestern ruins occupied in 1054. (NASA)

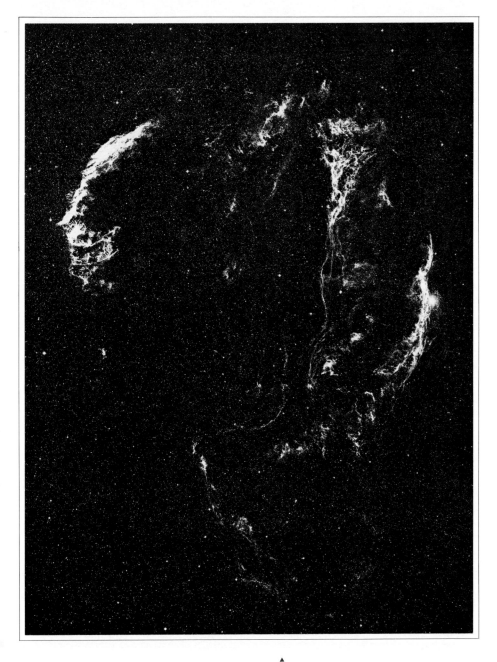

◄ *Figure 16·10* **The Crab nebula, photographed in light emitted by doubly ionized oxygen gas. This image shows the configuration about 920 years after the supernova explosion was observed in A.D. 1054. Compare the filamentary structure with that in Figure 16·7. The nebula is estimated to be 1,100 to 2,200 parsecs away.** (R. Williams, Steward Observatory, University of Arizona)

▲
Figure 16·11 **Wispy glowing clouds in the constellation Cygnus are remnants of a spherical cloud blown into interstellar space by a supernova explosion. Distance is about 500 parsecs and diameter about 22 parsecs.** (Hale Observatories)

eye, briefly doubling that galaxy's brightness. By 1971 some 300 supernovae had been recorded in other galaxies.

Supernovae are the explosions that occur when a very massive star (up to at least 30 $M_\odot$) runs out of its main sources of fuel, but is too massive to collapse smoothly to the white-dwarf state. Table 16·4 shows recent calculations revealing what happens to stars of different mass as they run out of fuel. As the more massive stars try to collapse, the density becomes so high that even subatomic particles called neutrinos—which normally escape easily—are absorbed. The pent-up energy eventually causes an explosion, blowing much of the star into interstellar space.

Neutron Star Pulsars: New Light on Old Stars

What is left after a supernova explosion? As early as 1934, American astronomers W. Baade and F. Zeicky made a correct guess:

> With all reserve we advance the view that a supernova represents the transition of an ordinary star into a neutron star, consisting mainly of neutrons. Such a star may possess a very small radius and an extremely high density.

Nuclear physicists were already working on the problem of very dense forms of matter, and they quickly took up the challenge of dense star-forms. By 1938, J. Robert Oppenheimer[2] and R. Server showed how

[2]Physicist Oppenheimer later became famous for additional nuclear research and his work as leader of the Los Alamos team that developed the atomic bomb in 1945. His physics research was halted when he lost his security clearance in the anticommunist political wrangles of the 1950s.

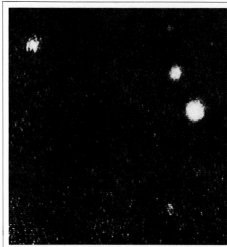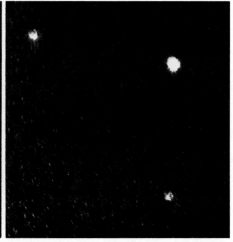

stars, after exhausting their nuclear fuel, could collapse to states much denser than white dwarfs.

The concept of a **neutron star** was merely an extension of the concept of a white dwarf. In an ordinary star, atoms of gas are held apart by their motions, caused by the high temperatures resulting from nuclear reactions. After the nuclear reactions stop, the star forms a white dwarf by collapsing to small size, causing the gravity to be very intense. The temperature is still high, but the gravity is so strong that it jams the atoms together. The electrons forming the outer parts of the atoms interact and are mostly stripped off the atoms, but the nuclei are held far apart from each other by the sea of electrons.

A white dwarf, therefore, might be called an electron star, since electrons control the spacings of particles. If the star's mass is even greater, the gravity is so strong that the repulsion between the electrons is overcome. Now the nuclei themselves are jammed into contact with each other. Since whatever reactions might occur have already run their course, there are no new, energy-generating reactions. Instead, the nuclei may be broken into their constituent neutrons and protons. The protons (+ charge) may coalesce with electrons (− charge) to make more of the neutral-charge neutrons. The important particles are now the neutrons, and the star is a neutron star, smaller than the white dwarf. Its density approaches the incredible densities of atomic nuclei themselves: roughly 10^{14} g/cm^3. A thimbleful would weigh 100 million tons! A skyscraper-full could contain all the mass of the moon! A whole neutron star could contain the mass of the sun but be no larger than a small asteroid—a few kilometers across!

For decades, astrophysicists talked about neutron stars, but, like the weather, nobody did anything about them, because nobody *could* do anything. No known observational technique could detect them and no one could prove they existed.

But in November 1967, a 4.5-acre array of radio telescopes in England detected a strange new type of radio source among many sources recorded during surveys of the sky. Analyzing the surveys (each equalling a 400-foot roll of paper chart), a sharp-eyed graduate student, Jocelyn Bell, was astonished to find that one celestial radio source (about half an inch of data on the chart) emitted "beeps" every 1.33733 seconds!

At first, project scientists speculated that they might have actually discovered an artificial radio beacon placed in space by some alien civilization. By January, another source was found, pulsing at a different frequency, arguing against the beacon hypothesis. Analysis showed that the first source was less than 4,800 km across, much smaller than ordinary stars.

In February, project director Anthony Hewish and his colleagues published an analysis suggesting that the **pulsars**, or pulsating radio sources, might be superdense vibrating stars that could

"throw valuable light on the behavior of compact stars and also on the properties of matter at high density."

The mysterious pulsars turned out to be the long-sought neutron stars. The number of scientific papers on pulsars jumped from zero in 1967 to 140 in 1968, and by 1973, about a hundred pulsars had been discovered. Co-directors of the original discovery project, Anthony Hewish and Martin Ryle, shared the 1974 Nobel Prize in physics.

Why do neutron stars pulse? After collapsing to small size, supernova remnant stars have very strong magnetic fields and very fast spins, rotating once every second or so. Ions trapped in the magnetic fields spin around with velocities near the speed of light. In 1968, Cornell researcher Thomas Gold showed how this can cause strongly beamed radio radiation, so that the pulsar acts like a light-house with a beam turning once every second: about once a second, distant observers see a flash. Among the first 100 pulsars, periods ranged from 0.25 to 1.3 seconds.

Pulsars were definitely linked to supernova explosions when one was identified in the center of the Crab nebula. Careful study of this and other pulsars showed that they pulse

◄ *Figure 16·12* **When the pulsar at the center of the Crab nebula flashes (left), it is brighter than nearby stars; when it is not radiating in our direction (right), it is nearly invisible. This is the star that blew off the expanding gas cloud that forms the Crab nebula (Figure 16·10).** (Television images obtained by Miller and Wampler in 1969, Lick Observatory)

Figure 16·13 **A complete sequence of flashes from the pulsar NP 0532 in the center of the Crab nebula. Portions of the surrounding nebula can be seen. The entire cycle, including two flashes, lasts about $\frac{1}{30}$ second, equalling one rotation of the pulsar.** (Kitt Peak National Observatory)
▼

not only in radio waves, but also in X-rays and visible light, as shown in Figures 16·12 and 16·13. Other pulsars have been found in the centers of other clouds of supernova debris, such as that in the constellation Vela.

Pulsars may have solid surfaces. Abrupt changes in the rotation period by tiny amounts (a ten-millionth of a second in the case of the Vela pulsar) have been intepreted as evidence of "starquakes" inside the pulsar. The stresses set up by pulsars' contraction and rotation apparently cause fractures, rearranging mass and causing slight changes in rotation.

Black Holes

As soon as pulsars were discovered, the search for still denser objects began. From this search came evidence for the strangest of all astrophysical concepts—**black holes.** Black holes are bodies so dense that their gravitational fields can keep most light (or other forms of energy and matter) from escaping.

Black holes can be visualized through an analogy with Newtonian physics made as early as 1798 by the French astronomer-mathematician, Pierre Laplace. He reasoned that some

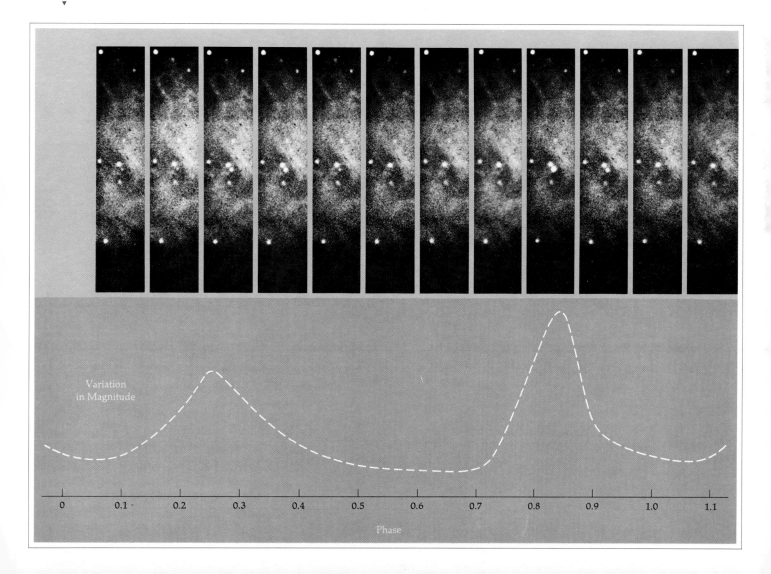

Variation in Magnitude

Phase

bodies might be dense enough to have an escape velocity (at their surfaces) faster than the speed of light. Laplace thus assumed light could never escape from such bodies, and that they would be permanently black and opaque. Although this basic idea is nearly correct, we now know that the situation is more complex, based on Einstein's theory of relativity. It more correctly predicts phenomena involving speeds near that of light. But theorists using Einstein's theory have concluded that black holes probably do exist.

Theorists believe that a black hole would be a very dense mass surrounded by a so-called **event horizon,** or imaginary surface from which no radiation or matter could escape. According to calculations, black holes would form from the collapse of only the most massive post-supernova remnants—probably more than 3 $M_\odot$. Final collapse of these massive objects would bring them to dimensions of only a few kilometers or less.

Two new theoretical results on black holes came in the mid-1970s, especially from the English physicist, Stephen W. Hawking. First, he pointed out that since matter was more densely packed in the earliest days of the universe, objects much smaller than stars could have gravitationally contracted to form black holes as small as subatomic particles, like protons and neutrons. Some of these tiny, primitive black holes might still exist in space.

Second, Hawking (1977) applied the theory of quantum mechanics and showed that *black holes need not be entirely black,* as was once thought. Quantum mechanics shows that subatomic particles often act in ways unpredicted by older theories; thus, proton-size black holes (containing up to 10^{15} grams—microscopic motes with the mass of mountains) could

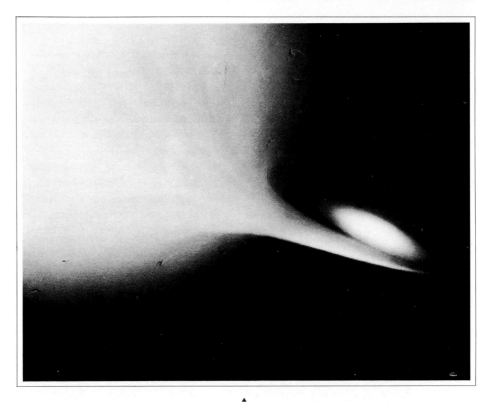

radiate energy, often as an explosion emitting gamma rays. Large, kilometer-scale black holes (containing as much mass as a star, or about 10^{33} grams) would radiate virtually nothing, just as in the original black-hole theory.

If black holes exist, they must have truly mystical and paradoxical properties. A kilometer-scale black hole of stellar mass could be detected by its gravitational influence on a nearby star. But if an instrument probe were dropped toward it, we would stop receiving signals after the probe fell through the event horizon, beyond which virtually no radiation or matter can escape. The probe, then, would disappear from the observable universe. Where would it go?

Pondering this strange question, some scientists speculate that black holes amount to separate universes, or that when mass "pops out of existence" by collapsing to superdensity, new mass or energy might emerge somewhere else in the universe. Could our own universe have begun as a

▲
Figure 16·14 **An imaginary close-up view of a star system, hypothesized to explain the X-rays and gamma rays from systems such as Cygnus X-1. Gas flows from the surface of a supergiant star (left) toward a very dense star, possibly a black hole, where it accumulates into a disk-shape nebula at very high speed, causing extremely high temperatures and emission of X-rays.** (Painting by Adolf Schaller)

black hole in some other "universe"? Do black holes make fact out of the "space warps" invented decades ago by science-fiction writers to allow their spaceships to wink out of sight in one place and reemerge instantaneously at some far-distant point? No one knows. As the British geneticist John B.S. Haldane put it, "My suspicion is that the universe is not only queerer than we suppose, but queerer than we *can* suppose."

Whatever the theoretical possibilities, observational astronomers have found several puzzling celestial **X-ray sources** and gamma rays that may actually involve black holes. The X-rays and gamma rays may arise as follows. In a binary star with two massive stars, the more massive star may have completed its evolution and turned into a black hole. As the second star turns into a giant and blows off material, the material falls toward the black hole, being accelerated to very high energy by the strong gravity. It may collect in a disk-like nebula around the black hole, where further material follows and strikes the nebula as shown in Figure 16·14. These high-energy gas collisions would produce the observed X-rays and gamma rays.

One of the most likely black-hole candidates is Cygnus X-1 (so named because it is the first X-ray source discovered in the constellation Cygnus). Believed to be about 2,500 parsecs away, it has a mass of 6–8 $M_\odot$ and an O or B star of 30 $M_\odot$ as its luminous companion (Bregman et al, 1973). Other proposed black holes include X-ray source HD 77581, believed to

Figure 16·15 **A schematic summary of stellar evolution, showing approximate evolutionary histories for objects with different initial masses, and the associated terms described in this and preceding chapters.**
▼

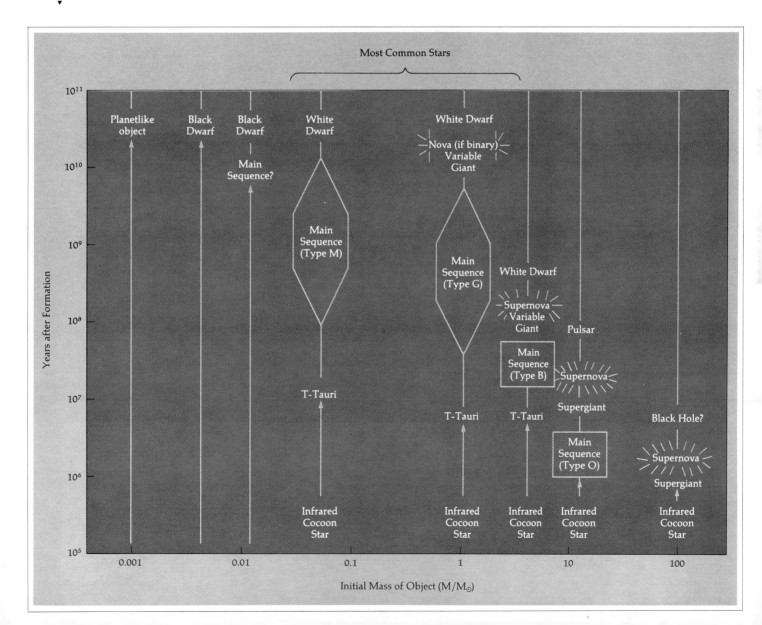

exceed 3 $M_\odot$ with a 50-$M_\odot$ companion; and a companion of the 65-$M_\odot$ star HD 153919.

Black holes and stellar explosions may also account for such unexplained phenomena as strong infrared radiation from the center of certain galaxies; unexpectedly large red shifts of some galaxies; and reported gravitational disturbances in our galaxy. An explosion powerful enough to blow outer layers off massive stars, converting the star to a black hole, could produce enormous amounts of energy. As astrophysicist R. Penrose (1972) notes, "There is no shortage of unexplained phenomena in astronomy today that might conceivably be relevant" to black holes.

16

Summary

The search for the forms of aging stars has yielded some of the most fascinating objects now being studied in physics and astronomy. Stellar old age leads to two basic phenomena: high-energy nuclear reactions, as ever-more-massive elements interact, and inexorable contraction, as energy sources are eventually exhausted.

During the first stages of old age, as stars evolve off the main sequence, the high energy production can cause expansion of stars' outer atmospheres, producing giants and supergiants. As heavier elements go through quick reaction sequences, various kinds of instabilities may produce variable stars or explosive novae. After energy generation declines to a rate too low to resist contraction, low-mass stars contract to a dense state known as a white dwarf, with final mass less than 1.4 $M_\odot$. Rarer massive stars, which start out with as much as 30 $M_\odot$ or more, undergo supernova explosions and blow off much of their initial material, leaving remnants with final masses of about 1.4 to 3 $M_\odot$. These are dense, rapidly rotating neutron stars, known as pulsars.

Still denser and rarer are black holes, which may be supernova remnants with masses greater than 3 $M_\odot$. Although virtually no radiation escapes from their surfaces, they may be detectable by high-energy radiation from material falling into them. The detailed physics of these dense, small star forms is still uncertain and an area of intense current research.

Complications may arise if the evolving star has a nearby co-orbiting companion. Mass may be blown off the one and fall onto the other, changing the second star's mass and causing sudden instabilities. This probably accounts for nova-type explosions and some sources emitting X-rays and gamma rays.

Concepts

funnel effect

giant stars

supergiants

helium flash

variable star

Cepheid variable

irregular variables

instability strip

Hertzsprung gap

M giant

Wolf-Rayet star

B-type shell star

P Cygni star

nova

white dwarf

degenerate matter

Pauli exclusion principle

Chandrasekhar limit

supernova

neutron star

pulsar

black hole

event horizon

X-ray sources

Problems

1. Why do stars just moving off the main sequence expand to become giants instead of starting to contract at once? Why does contraction ultimately win out?

2. Why do high-mass stars tend to produce smaller stellar corpses than low-mass stars? (Hint: Compare gravitational compressive forces for two stars of different mass but same size after they have exhausted their fuels.)

3. Many red giants are visible in the sky, even though the red giant phase of stellar evolution is relatively short-lived. Why are so many red giants visible?

4. Is variability more properly thought of as a property of certain specific stars or as a stage which most stars will pass through?

a. *Could the sun ever become a variable?*

b. *If you could measure the sun's radiative output with any accuracy desired, could you demonstrate that, in some sense, the sun is already a variable star? Explain.*

5. List examples of evidence that certain stars can lose mass.

6. Which stars will eventually become

   a. *white dwarfs?*

   b. *neutron stars?*

   c. *black holes?*

What will be the ultimate fate of the sun?

7. A main-sequence B3 star has about ten times the mass of the sun and therefore has about ten times as much potential nuclear fuel. Why then does it have a main-sequence lifetime only 1/3,000th as long as that of the sun?

8. According to the law of conservation of angular momentum, a figure skater spins faster as she pulls in her arms. How does this principle help explain that neutron stars spin much faster than main-sequence stars?

9. Comment on the roles and relations of theorists and observers in the three decades of work on white dwarfs, pulsars, and black holes. Are black holes fully understood today?

Advanced Problems

10. Suppose a star of one solar mass reached a terminal evolutionary state where it had the same diameter as the earth.

   a. *What would be the velocity of any possible material (such as captured meteoritic debris) in a circular orbit just above the star's surface?*

   b. *What velocity would be needed to blow material off its surface?*

   c. *Compare these values with values for the earth.*

   d. *What type of star would this object be?*

11. Use the Stefan-Boltzmann law (page 246) to prove that the surface of a star such as FG Sagittae, evolving to the right on the H-R diagram (keeping constant total luminosity but decreasing in temperature), must be expanding.

12. White dwarfs have spectral types and surface temperatures similar to A or F main-sequence stars, but are much smaller. Use the Stefan-Boltzmann law to prove that this statement requires white dwarfs to lie below the main sequence on the H-R diagram.

13. Suppose a nova occurred in the nearby star-forming region of Orion, 500 parsecs away. If the cloud expanded at an average velocity of 1,000 km/sec, how long would terrestrial observers have to wait before they could see details of the cloud's shape with telescopes resolving $\frac{1}{2}$ second of arc? (Hints: Use the small angle equation; one year is about $\pi \times 10^7$ seconds).

Projects

1. Locate the star Mira (R.A. = 2^h14^m; Dec. = $-3°.4$) with a small telescope and determine whether it is in its faint or bright stage. If it is bright enough to see with the naked eye, record its brightness from night to night by comparing with other nearby stars of similar brightness. By checking brightnesses of these stars with star maps (1st mag., 2nd, etc.), plot a curve of Mira's brightness over time.

2. With binoculars or a small telescope locate the star Delta Cephei (R.A. = 22^h26^m; Dec. = $+58°.1$) and compare it from night to night with other nearby stars of similar brightness. Can you detect its variations from about 4.4 to 3.7 magnitude in a period of 5.4 days?

Environment and Groupings of Stars

Every cubic inch of
space is a miracle.

Walt Whitman

Interstellar Atoms, Dust, and Nebulae

The preceding chapters discussed stars as if they were isolated individual objects. But they are not isolated. They are engulfed in a thin but chaotic medium of gas, dust, and radiation. They form from this thin material, interact with it, recycle it, and expel it to form new interstellar material.

We sometimes casually say "space is a vacuum," but it is not quite true. While space is a better vacuum than can be achieved in labs (Table 17·1), its material cannot be neglected. What significance can this thin material have for us? For one thing, it dots our sky with **nebulae,** or vast clouds of dust and gas, some dark and some glowing with different colors. For another thing, as Joni Mitchell said in her song *Woodstock,* "we are stardust."

This statement must be true in one sense or another. As Chapter 11 pointed out, our solar system, our earth, and we ourselves are formed from atoms that were once part of the interstellar gas and dust. More provocatively, recent discoveries show that interstellar material contains complex organic molecules, and some scientists have speculated that primitive biological processes, perhaps related to the origin of life, may have occurred in nebulae.

The Effects of Interstellar Material on Starlight

Dispersed particles, whether floating in space or in the atmosphere, interact with radiation. When light from a distant star passes through clouds of interstellar atoms, molecules, and dust grains, the interaction changes the light's properties, such as intensity and color. Several complex physical laws describe these changes in some detail, but the changes can be grouped under two main principles:

1.

When radiation (ultraviolet light, visible light, infrared, radio waves, or any other type) interacts with particles, the type of interaction depends on the types of particles and their sizes, relative to the wavelength of the light.

2.

The appearance of the light and the particles may depend on the direction from which the observer looks.

Two important types of interaction of radiation and matter are interaction with atoms and molecules much smaller in diameter than the wavelength of visible light, and interaction with dust grains about the same diameter as the wavelength of visible light.

Table 17·1 Gas Densities in Different Environments

Locale	Density[a] (g/cm³)	Particles per cm³	Typical Distance between Particles	
Air at sea level	1.2×10^{-3}	10^{21}	10	Angstroms
Circumstellar cocoon nebula	10^{-8}	10^{16}	500	Angstroms
"Hard vacuum" in terrestrial lab	10^{-12}	10^{12}	1	micron
Orion nebula	10^{-21}	10^{3}	0.1	cm
Typical interplanetary space	10^{-23}	10	0.5	cm
Typical interstellar space	10^{-24}	1	1	cm
Interstellar space near edge of galaxy	10^{-28}	10^{-4}	20	cm
Typical intergalactic space	10^{-31}	10^{-7}	2	meters

[a]Average density of matter in whole universe estimated at about 2×10^{-30}g/cm³ according to a 1974 estimate by J. Ostriker and his colleagues.

Interaction of Light with Interstellar Atoms and Molecules

As shown in Figure 17·1, several things can happen if starlight passes through a cloud of interstellar gas. (In reality, the interstellar material is always a mixture of gas and dust, sometimes in dense clouds and sometimes spread smoothly; but it is easier to understand the effects if we imagine clouds of atoms and molecules of pure gas separately from clouds of dust grains.) Suppose a star radiates light of all wavelengths, and the photons of light enter a cloud of gas. Photons corresponding to certain wavelengths will have just enough energy to **excite** this gas, or knock electrons from lower

Figure 17·1 A cloud of gas (atoms and molecules) illuminated by a star and seen by an observer to one side (item 4) and an observer in line with the cloud and the star (item 5).
▼

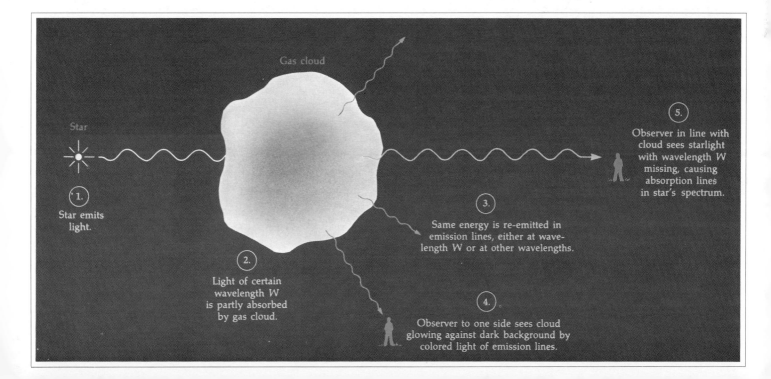

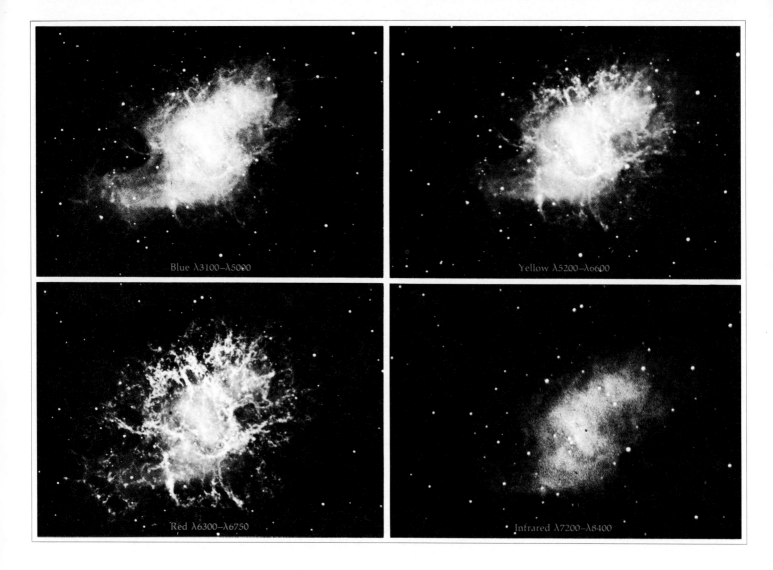

Blue λ3100–λ5000

Yellow λ5200–λ6600

Red λ6300–λ6750

Infrared λ7200–λ8400

to higher energy levels. They may even **ionize** the gas, or knock electrons clear out of the atoms. Each time a photon excites or ionizes an atom, that photon is consumed and disappears from the light beam. Thus, an observer looking at the star through the cloud would see absorption lines created by the interstellar material, as shown in Figure 17•1.

But the energy absorbed by the cloud must be re-radiated to keep the cloud in equilibrium, and this re-radiation occurs as the electrons cascade back down through the energy levels of the atoms, creating emission

lines. The photons in these emission lines leave the cloud in all directions, as shown in Figure 17•1, so that an observer off to one side would see the cloud glowing in the various colors corresponding to the emissions. If a red emission line is especially strong, the cloud might look red. In another cloud, struck by photons of different wavelength or containing atoms at different levels of excitation, a green emission line might be strongest, and the cloud would glow with green light. Different colors reveal different patterns (see Figure 17•2, Color Photos 20 to 25).

▲

Figure 17•2 **Four views of the Crab nebula (M 1), showing different appearance as photographed in light of different colors. Red and yellow views show the outer filamentary structure corresponding to clouds of excited gas giving off spectral emission lines (especially red lines from hydrogen). Infrared view emphasizes inner amorphous clouds, glowing mostly by synchrotron radiation.** (Hale Observatories)

Molecules have similar interactions, except that they produce color absorptions and emissions affecting a range of wavelengths, called bands, instead of the narrow range called lines. Thus a real gas cloud, containing atoms

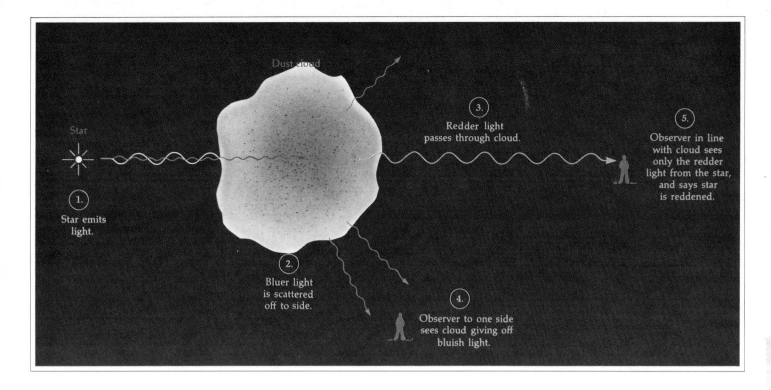

Dust cloud

Star

3. Redder light passes through cloud.

5. Observer in line with cloud sees only the redder light from the star, and says star is reddened.

1. Star emits light.

2. Bluer light is scattered off to side.

4. Observer to one side sees cloud giving off bluish light.

and molecules involving different elements, would produce a variety of absorption lines and bands, and might glow with emission lines and bands of different colors (see Color Photo 30).

Interaction of Light with Interstellar Dust Grains

Whereas atoms are as small as 1/1,000th the wavelength of visible light, and molecules are a few percent of this wavelength, many interstellar dust grains are only a little smaller than light waves, and their interactions are quite different. They affect colors over a much broader range of wavelengths than individual spectral lines or bands. The most important rule describing their interactions is that redder light (longer wavelengths) passes through clouds of dust, whereas bluer light (shorter wavelengths) is scattered out to the side of the beam, as shown in Figure 17·3.

▲
Figure 17·3 **A cloud of dust grains illuminated by a star and seen by an observer to one side (item 4) and an observer in line with the cloud and the star (item 5).**

Thus, the observer who looks through the dust cloud at a distant star sees most of its red light, but not much of its blue light. In this way, interstellar dust makes distant stars look redder than they really are—an effect called **interstellar reddening.** However, an observer who looks at the dust cloud from the side will see the blue light scattered out of the beam, so that a nebula illuminated in this way will have a bluish color.

Why Is the Sunset Red and the Sky Blue?

These same rules apply not just to interstellar material, but also to ma-

terial in our own atmosphere. The lower few kilometers of the atmosphere are full of dust and large molecules, including many that are slightly smaller than the wavelength of light. When we look at the nearest star, our sun, through these particles, the same effects can be seen. If the sun is high in the sky, we look through the minimum amount of dust, as shown in Figure 17·4A; thus the reddening is minimal, and the sun is perceived as white.

At sunset, as shown in Figure 17·4B, the sunlight passes through much more dust, and much of the blue light is lost from the beam, strongly reddening the sun and adjacent parts of the sky. At any time of day, if we look at some other part of the sky, the light we see is the blue light scattered out of the beam of sunlight and then scattered by air molecules and dust back toward our eye. This process, called **multiple scattering,** involves only the blue light, thus explaining why the sky is

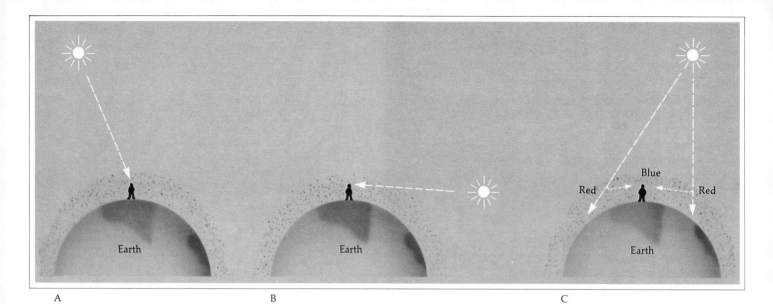

A   B   C

Figure 17•4 **Explaining colors in the sky. A. Light from the sun high in the sky passes through minimal dust and is minimally reddened. B. Light from the sun at sunset passes through maximum amount of dust and is strongly reddened. C. Light from the sky at any time of day is the blue light scattered out of the sunlight beam.**

blue. (The sky on Mars is red because many of the Martian dust particles are red, and bigger than the wavelength of light; hence they do not scatter blue light, but simply reflect their own red color.)

Observed Types of Interstellar Material

Interstellar material includes gas (defined as consisting of atoms and molecules), microscopic dust grains (larger than molecules), and possible larger objects.

Interstellar Atoms

Atoms of interstellar gas were discovered in 1904, when German astronomer Johannes Hartmann detected their absorption lines. While studying spectra of a binary star, he accidentally discovered absorption lines caused by interstellar calcium atoms. Certain other **interstellar atoms,** such as sodium, were soon found to produce additional prominent interstellar absorption lines. These lines are identified as interstellar by the fact that they have different Doppler shifts than the stars in whose spectra they appear.

Further studies have convinced astronomers that, although atoms such as calcium and sodium have prominent absorptions, the most common gas is the ubiquitous hydrogen. Like the sun and stars, interstellar gas is about three-fourths hydrogen and nearly one-fourth helium.

Forbidden Lines from Interstellar Atoms

A scientific mystery soon arose from interstellar gas. All experience prior to the 1920s had indicated that the distant universe was composed of the same elements as found in the solar system. However, as early as 1864, English observer William Huggins had found three greenish and bluish emission lines from the light of certain nebulae. Further work showed that these did not match the spectra of any elements known on earth, and they were assigned to a hypothetical new element named "nebulium." By the 1920s, it was clear that there was no room in the periodic table of elements for a new element like nebulium. What was creating the strange greenish light of nebulae?

In 1927, California astronomer Ira S. Bowen showed that in deep interstellar space, atoms could exist in so-called **metastable states,** which are states with unusual electron distributions that last for several minutes. These states are unknown on earth, because in our dense atmosphere the atoms are struck by air molecules too fast to remain in these states. The spectral lines arising from these unusual states are called **forbidden lines,** because they cannot be found in earth's atmospheric gas. The nebular light turned out to be coming from strong forbidden lines of very rarefied ionized oxygen—a gas familiar in our own air. (The astronomer Henry Norris Russell thereupon quipped that the mysterious nebulium had vanished into thin air.)

Radio Radiation from Interstellar Gas In 1944, the Dutch astronomer H. C. van de Hulst predicted that the most important type of interstellar radiation would be an emission line with wavelength 21 cm, caused by a change in the spin of hydrogen atoms' electrons from one state to another. Such a long wavelength is not visible light, but radio radiation. The predicted emission was confirmed in 1951 when Harvard astronomers detected this **21-cm emission line** of atomic hydrogen, using radio equipment.

This discovery not only confirms the importance of hydrogen as a main constituent of interstellar gas, but gives radio astronomers a tool to detect where clouds of interstellar gas are concentrated. Because of the long wavelength of this radiation, it can penetrate much greater distances through the interstellar gas and dust than ordinary light.

Interstellar Molecules

By 1940, astronomers at Mount Wilson observatory in California built spectrographs that detected absorptions due not only to interstellar atoms, but also to **interstellar molecules,** such as CH and CN. Figure 17·5 shows spectral absorption lines caused by interstellar CH.

Because of molecular structure, a great many of the molecular absorptions lie in the infrared or radio parts of the spectrum. Not until after World War II did astronomers have available the technology of infrared detectors to search for interstellar molecules. After a decade of development in the 1960s, some of the new detectors were put in satellites and large telescopes, sparking an explosion of interstellar discovery in the late 1960s

and 1970s. For a time it seemed that each new issue of astrophysics journals carried news of another molecule identified in interstellar space by spectral studies. Here is a brief summary reflecting the sudden expansion of new astronomical observing equipment and the types of molecules discovered:

1963 *1 molecule discovered (hydroxyl)*

1964–1967 *no molecules discovered*

1968 *2 molecules discovered (ammonia, water)*

1969 *1 molecule discovered (formaldehyde)*

1970 *8 molecules discovered (hydrogen, carbon monoxide)*

1971 *10 molecules discovered (methylacetylene)*

1972 *3 molecules discovered (hydrogen sulfide)*

1973 *1 molecule discovered (sulfur monoxide)*

1974 *5 molecules discovered (ethyl alcohol)*

1975 *6 molecules discovered (sulfur dioxide)*

There is a significant trend toward discovering more and more complex

forms, such as the 9-atom molecule C_2H_5OH (ethyl alcohol).

The atoms recurring again and again in these large molecules comprise the quartet C, H, O, N—the "building blocks of life"! Repeatedly, these "building blocks" have been found in space in complex large molecules, as summarized in the list above. These elements are common, and hydrogen is the most common of all, so the surprise is not that they exist in space, but that some process allowed these elements, in their tenuous environment, to come together to make these complex forms.

Still more provocative is the fact that two of the molecules, CH_3NH_2 (methylamine) and $HCOOH$ (formic acid), can react to form NH_2CH_2COOH (glycine), one of the *amino acids.* These large molecules can join to form the huge protein molecules that occur in living cells.

Thus, two exciting questions have come from research on interstellar molecules. First, does the existence of complex, carbon-rich molecules in space suggest that life has originated and evolved near other stars? The answer may be yes, and we will discuss this possibility in more detail in Chapter 25. Second, how do so many atoms come together to form these molecules? They cannot form in ordinary interstellar gas, because collisions there between atoms are extremely rare. Instead, the molecules are believed to form in the denser

Figure 17·5 **Spectral absorption lines caused by interstellar molecules of CH gas. Background light is part of the blue portion of the spectrum of the star Zeta Ophiuchi. On the original plate, lines of interstellar beryllium atoms are also faintly visible at left.** (Lick Observatory)

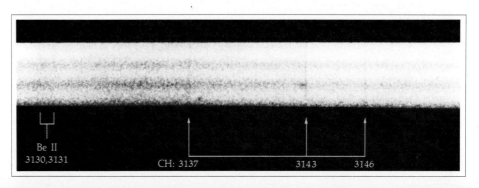

Be II
3130,3131

CH: 3137   3143   3146

clouds, such as the cocoon nebulae around young stars.

Interstellar Grains

There are two observational effects of **interstellar grains:** they cause the reddening already explained, and they also cause a general dimming of starlight at all wavelengths, called **interstellar obscuration.** Some light is lost at all wavelengths due to obscuration. Since grains are widely distributed throughout the interstellar gas, all distant stars are harder to observe because of this effect.

The effect of interstellar obscuration by dust grains played a curious role in humanity's recognition that we do not necessarily live at the center of the universe. After the Copernican revolution, astronomers assumed that the solar system did not occupy a central position. Then, studies from the 1700s to the early 1900s of fainter and fainter stars seemed to indicate that the farther from the solar system we probe, the fewer stars we find. By 1922 astronomers began to wonder: was the solar system in the center of a swarm of stars surrounded by empty space?

In 1930, American astronomer R. J. Trumpler published a very important paper proving that vast quantities of interstellar dust were merely obscuring the more distant stars. Trumpler proved this by studying distant star clusters; their apparent sizes and other properties enabled him to show that the more distant a cluster, the more it is dimmed by dust in the line of sight.

Further research indicates that if we look along the Milky Way plane, dust grains dim stars by an average of about 1.9 magnitudes for every 1,000 parsecs traversed by the beam of light. Of this amount, about 1.6 magnitudes of dimming are caused by grains concentrated in clouds, and about 0.3 magnitudes by grains dispersed among clouds. Over a distance of 10,000 parsecs, stars would be dimmed by 19 magnitudes. No wonder early observers found fewer and fewer stars the farther out they looked!

The Nature and Origin of the Grains Astronomical studies reveal various properties of the interstellar grains:

1. The grains involved in reddening have diameters comparable to the wavelength of visible light, about 0.0001 to 0.001 mm—more like particles in a smoke cloud than like grains of sand.[1]

2. The grains are concentrated in clouds. The dust in different clouds may not even be the same in size and composition.

3. The light scattered through dust clouds is polarized, which shows that the particles are nonspherical. Plate-like or needlelike shapes have been suggested.

---

[1] A few authors prefer the term *interstellar smoke* to the more common usage. Herbig (1974), recognizing the mixture of gas and dust, calls it *interstellar smog*.

Figure 17·6 **A. Unilluminated clouds of interstellar material, named Barnard 68 and 72, obscure distant stars of the Milky Way in the constellation Sagittarius. Similar clouds are often sites of concentrations of molecules and dust grains.** (Steward Observatory, University of Arizona) **B. A long exposure with higher contrast, made with a smaller telescope, enhances cloud silhouettes.** (Hale Observatories) ▼

A

B

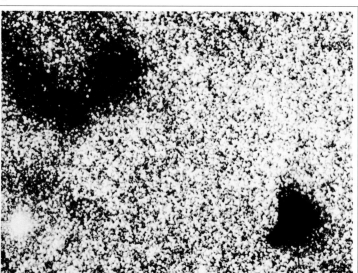

▲

Figure 17·7 **The Horsehead nebula, a famous dark nebula silhouetted against background emission and reflection nebulosity in Orion. The background nebulosity is about 350 parsecs away. The dark horse head, a dense cloud of dust roughly $\frac{1}{2}$ parsec across, is estimated to be about 300 parsecs away.** (Hale Observatories)

4. The elongated grains seem aligned at right angles to the plane of the Milky Way, probably because of the galaxy's magnetic fields.

Grain composition has long been debated. In 1967, Indian astrophysicist N. Wickramasinghe proposed that many grains might be graphite particles condensed in gas blown off carbon-rich giant stars. Silicates might also condense from such gas. Others suggested that grains might be planetesimals blown out of cocoon nebulae of new stars, rather than out of old giants. Silicates, ices, magnetite, and hydrocarbons would be likely in such

nebulae, as proved by meteorites and comets in the solar system. Magnetite grains similar to those in meteorites have been thought to contain 16 percent of all the iron atoms in the galaxy, and to account for the magnetically controlled alignment. Observations are beginning to shed light on the problem. Silicate grains have been identified spectroscopically near some supergiant stars and near some young stars.

Some astronomers suggest that grains in interstellar space may be a mixture of different types from different sources, or may individually contain different materials, such as a silicate core surrounded by "dirty ice," as was postulated for comets. There is still much to learn about materials in interstellar space. As the American astrophysicist John Gaustad (1970) put it:

> . . . If the chemistry of well-studied material like water and carbon is so little understood that new forms are

still being discovered under laboratory conditions, how many more surprises might await us in interstellar space, where timescales and physical conditions are very different from those in the chemical laboratory? We astronomers should keep our eyes open, from the infrared to the ultraviolet. . . .

Interstellar Snowballs?

Astronomers are now looking for **interstellar snowballs**—hypothetical bodies much larger than grains. These could be BB-size, baseball-size, or even kilometer-scale bodies in interstellar space. If light interacted with such particles, neutral obscuration would occur at all wavelengths much less than the particle size. While microscopic grains are revealed by reddening, large particles would be difficult to detect because virtually no color effects would occur. Nonetheless, if single atoms join into molecules, and molecules into dust grains, why not expect still larger particles?

There are some indications that interstellar snowballs exist (Greenberg, 1974; Herbig, 1974). For example, the interstellar gas is strangely lacking in certain elements. Aluminum has not been found at all. Calcium is 400 times rarer than we would predict from its abundances elsewhere in the universe. Titanium, iron, and magnesium are also rarer than expected. Because the interstellar grains apparently do not contain enough of these materials to make up for the atoms missing in the gas, another class of interstellar material—snowballs—may contain the missing atoms. They might look like our present conception of comet nuclei—icy bodies with silicates and other "dirt" mixed in—justifying the term "snowball" coined by Greenberg (1974).

Distribution of Interstellar Material: Causes of Nebulae

Interstellar gas and dust are far from uniform. Different regions have differently shaped clouds with different temperatures and different motions; some regions are relatively free of gas and dust. Most nebular clouds are 5 to 10 parsecs across. Some are bigger and some smaller. Size depends partly on whether thermal expansion or gravitational contraction has occurred. Typical distances separating clouds are about 20 to 50 parsecs, about 20 times the distances separating stars.

Particularly vivid nebulae were cataloged as early as 1781 by the French astronomer Charles Messier, and are thus known by their **Messier numbers** or M numbers. The Orion nebula, for example, is M 42. Others are known by **NGC numbers** or **IC numbers,** based on the more recent New General Catalog and Index Catalog. Tradition also names most bright nebulae according to their appearance in small telescopes; examples include the Crab, Dumbbell, and Ring nebulae.

Why has interstellar material not reached some uniform distribution? Suppose we magically smoothed the interstellar material to uniform density. Supernova explosions, gas ejected from giants, radiation pressure, and differential rotation of the galaxy would quickly create local dense regions. Some of these would be dense enough to contract gravitationally, making still denser clouds. Within a brief 100 million years or so, the material would have re-formed into a new distribution of nebulae.

The Effect of Starlight on Interstellar Gas: HI and HII Regions

One of the most important processes in creating different kinds of regions is the action of light from hot stars on the neighboring gas. The appearance and spectrum of an interstellar gas cloud depend on the states of its atoms. These states depend on the source of illumination and the range of excitations and ionizations it causes in the cloud. Two important physical laws control the effect:

Planck's Law
The bluer the light (that is, the shorter the wavelength) the more energy each photon contains.
Thus, photons of ultraviolet or blue light excite or ionize atoms to a greater degree than photons of red light.

Wien's Law
Hotter sources radiate more blue light than cooler sources.

These two laws predict that the most energetic photons will be encountered near the hottest stars. Thus, excitation and ionization in interstellar gas will generally be greatest near the hot, young stars of type O, and least deep in interstellar space, far from any stars.

Photons of ultraviolet light with wavelength shorter than 912 A are energetic enough to ionize hydrogen. By Wien's law, stars hotter than about 30,000 K produce most of their radiation at these wavelengths, and even stars hotter than about 20,000 K produce many such photons. Therefore, hot blue stars of spectral classes O and

B are surrounded by large regions in which the interstellar hydrogen is ionized, as shown in Figure 17·9. If the gas were uniform, these regions would be spherical. Since the gas is nonuniform, these regions have ragged shapes.

In 1939, the Danish-American astronomer Bengt Stromgren calculated the average radii of the ionized regions around various types of stars, showing that only the hottest stars create large ionized regions. According to Stromgren's calculations, the diameters of the ionized regions in typical interstellar gas would be:

Class	Diameter	
O5	280	parsecs
B0	52	
B5	7.4	
A0	1.0	

Since neutral hydrogen is designated HI and ionized hydrogen is designated HII, and since hydrogen is the most common gas in all parts of interstellar space, regions where hydrogen is ionized around hot stars are often called **HII regions;** all other regions of interstellar space are called **HI regions.** The gas in HII regions is heated by stars to temperatures around 8,000 to 10,000 K. In HI regions, farther from hot stars, temperatures are around 100 K. The most common HII regions are clouds with masses about 10 to 1,000 solar masses.

Figure 17·9 **Creation of an HII region. Ultraviolet radiation from the central hot star is absorbed by hydrogen atoms, ionizing all of them out to a certain distance (dashed line). Recombination of electrons with atoms causes reradiation of various spectral emission lines in all directions, including earthward, causing the HII region to be visible as a glowing nebula.**

In and near HII regions, electrons in various elements lead an up-and-down life. On the one hand, they are likely to be hit by an energetic photon from the nearby hot star, and knocked out of the atom. But on the other hand they tend to associate with atoms and cascade down through the energy levels to reach the ground state. For this reason, gas in the HII regions emits light of varying colors. *Therefore, HII regions are also glowing nebulae called emission nebulae.* Many HII regions are among the most impressive features of the sky.

HII emission nebulae help us understand interstellar material. If a massive new star forms and flares forth, in a few hundred thousand years it heats the surrounding material, creating an HII region that expands into interstellar space. The hot gas may push aside surrounding material, creating density irregularities. HII regions of neighboring stars may expand and meet each other like coalescing

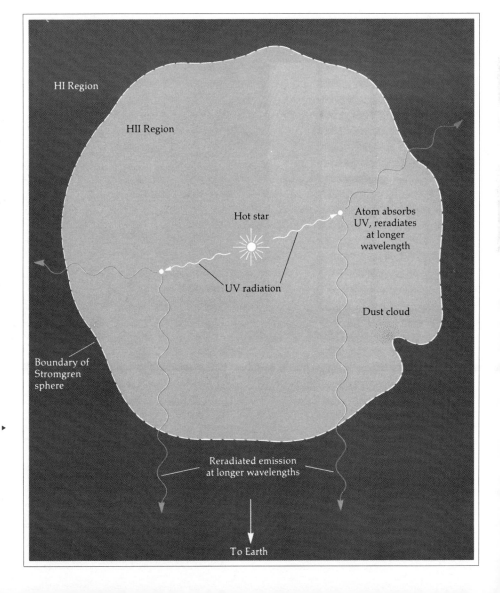

HI Region

HII Region

Hot star

Atom absorbs UV, reradiates at longer wavelength

UV radiation

Dust cloud

Boundary of Stromgren sphere

Reradiated emission at longer wavelengths

To Earth

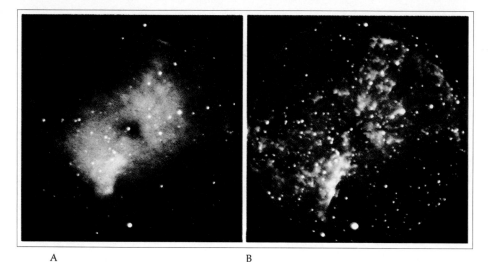

A   B

Figure 17·10 **Two views of the Dumbbell nebula (NGC 6853) show different structure, depending on the wavelength of light used to make the image. A. Light from the Balmer *β* emission of hydrogen reveals** diffuse distribution of excited hydrogen gas. **B. Light from forbidden emission of neutral hydrogen (6,300 A) reveals clumpy distribution of oxygen. See also Color Photo 24.** (R. Williams, Steward Observatory)

Figure 17·11 **Long exposure of the Dumbbell nebula overexposes the central portion shown in the preceding figure and reveals the fainter outer shell, possibly a remnant of an earlier mass ejection from the central star.** (Kitt Peak National Observatory)

bubbles. In a region of star-formation, like Orion, most of space may be taken up by coalescing HII regions, since massive, hot stars abound in such regions. In an older region, like the solar neighborhood, no stars are hot enough to excite the gas, and space is a vast HI region.

Types of Nebulae

By tradition, nebulae are often categorized by their superficial appearance, which in turn depends on whether hot stars are near them, and on whether we happen to see them against a background of dark space or distant bright nebulae.

Figure 17·12 **Trifid nebula (M 20), an emission nebula with dark dust lanes. Estimated distance is about 1,000 parsecs; diameter about 4 parsecs. See also Color Photo 21.** (Steward Observatory, University of Arizona)

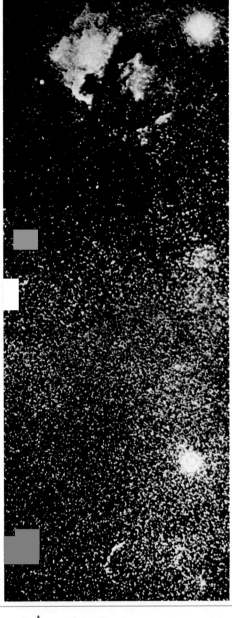

◄ *Figure 17•13* **Several types of nebulae are shown in this portion of Cygnus, covering 16° from north (top) to south. At the top are the North American nebula (left center) and Pelican nebula (right center). These are combined emission-reflection nebulae, estimated to be about 650 parsecs away. At the bottom center is the Veil nebula, a filamentary emission nebula believed to be a supernova remnant about 500 parsecs away. Background stars are part of the Milky Way; near the North American nebula they are obscured by dark nebular clouds 250 and 600 parsecs distant. The bright star in the upper right corner is α Cygnus (Deneb, top of the Northern Cross), a first-magnitude star about 500 parsecs away; the bright star at lower right is ε Cygnus, a second-magnitude star only 23 parsecs away.** (Clarence P. Custer; Aero-Ectar lens)

Figure 17•14 **Three examples of planetary nebulae, showing various shapes. A. Ring nebula, about 0.2 parsec diameter, 700 parsecs distance. Photographed in light emitted by neutral oxygen atoms. See also Color Photo 23.** (Steward Observatory, courtesy R. Williams) **B. NGC 2392, showing two rings, possibly indicating two distinct eruptions. About 0.2 parsec diameter, 1,000 parsecs distance.** (Mauna Kea Observatory, courtesy A. Stockton) **C. Saturn nebula, showing unusual lateral extensions. About 0.1 parsec diameter, 700 parsecs distance.** (Kitt Peak National Observatory)
▼

Dark nebulae: *Clouds rich in dust that happen to be silhouetted against bright background clouds or star-rich regions, thus looking dark by contrast (Figure 17•12).*

Bright nebulae: *Clouds that look brighter than the background by virtue of emitted or scattered light.*

Reflection nebulae: *Subclass of bright nebulae near enough to a star so that starlight illuminates the dust and gas, reflecting off dust grains.*

Emission nebulae: *Subclass of bright nebulae close to a hot star (or stars) whose photons excite or ionize atoms of gas, producing spectral emission lines of different colors (Figures 17•12, 17•13).*

Planetary nebulae: *A class of emission nebulae named for their appearance in early telescopes. The class is ill-named, for its members have nothing to do with planets. They are generally tight, expanding spherical shells of gaseous material blasted off old stars (Figures 17•10, 17•11, 17•14).*

Loop (or veil) nebulae: *Wispy emission nebulae, usually in curved segments and best recognized in wide-angle photographs. They are also believed to be remnant material from stellar explosions, but further expanded than planetary nebulae (Figure 17•13).*

A   B   C

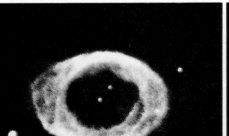

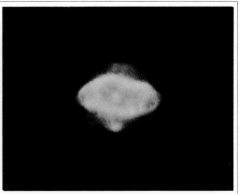

A Nearby Star-Forming Region: The Orion Nebula

A large region of the sky around the constellation Orion, about 400 to 500 parsecs from the earth, is dominated by nebulae and star-forming activity. Orion itself is the constellation of the hunter—a great figure raising a club over his head. As shown in Figure 15·9 (pp. 278–279), three bright stars mark his belt and three below it mark his sword (Figure 17·15). Orion's right shoulder (on the left—he faces us) is the bright red giant star, Betelgeuse, 180 parsecs away. His opposite knee is the blue B-type star, Rigel, 270 parsecs away. The central "star" of the sword is not a star at all, but the brightest glowing core of the **Orion** emission **nebula,** roughly 460 parsecs from the earth.

As shown in Figure 15·9, if you look toward Orion on a February evening, swinging your head back and forth past this part of the sky to compare it with other regions, you will see a great concentration of bright stars. Orion contains 7 of the 100 brightest stars in the sky, and except for Betelgeuse, all of them are massive, hot O- and B-type stars 140 to 500 parsecs away. Since massive stars are short-lived, their mere existence shows that they and other stars must have formed recently in the nebulosity around Orion.

By careful studies of this region, astronomers have actually begun to piece together the arrangement and motions of material in the seething space around Orion. A backyard telescope reveals a quartet of faint stars, called the Trapezium, in the nebula's center (shown in Figure 17·17). The brightest core of the nebula here is about 1 parsec across, but the whole glowing region visible in a small telescope is about 8 parsecs across. Star-

Figure 17·15 **Belt (top three stars) and sword (bottom center) of Orion. The box shows the region of Figure 17·16, containing the Orion nebula (M 42). Vertical length of photo is approximately 16°. The Horsehead nebula (Figure 17·7, invisible here) is just below the left star in the belt. Young stars and nebulosities abound in the region. See Figure 15·9 for broader view of Orion region. Also see Color Photo 20. (8-min. exposure on 35-mm, 2475 Recording film, 135 mm telephoto, f2.8.)**

Figure 17·16 **The Orion nebula (M 42) and adjacent parts of Orion's sword. The nebulosity shown here is about 5 parsecs wide, about 460 parsecs distant. See Figure 17.17. (Lick Observatory)**

formation is probably occurring in this region, where gas densities reach at least 600 atoms/cm³ and dust clouds are scattered. Dust in the nebula has a measured temperature around 70 K.

Figure 17·18 shows a cross section of the inner Orion region. On the far side from the solar system is an especially massive (about 1,000 $M_\odot$) HI cloud of dust and gas, rich in molecules. According to one interpretation, expanding HII gas from the main nebula runs up against the HI cloud, limiting the size of the ionized region (Zuckerman, 1973). Especially dense clots of dust and gas

were discovered inside the HI cloud by infrared observers in the 1960s. One of these, the Kleinmann-Low nebula, may mark a collapsing cloud about to form a cluster of stars. Various clouds move at about 8 to 10 km/sec relative to each other, and backward tracing of their motions suggests some may have formed within the last 100,000 years.

Gas from the Orion nebular region is expanding into a much larger region about 100 parsecs across, as shown in Figure 17·19. This large region contains numerous young stars including T Tauri stars and about 1,000 O and B

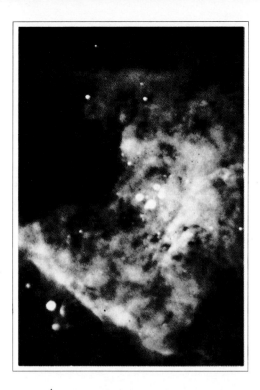

Figure 17·17 The central core of the Orion nebula (M 42). Four stars in the central nebulosity, approximately 0.04 parsecs apart, form the Trapezium (see Figure 17·18). This is part of the overexposed center of Figure 17·16. (Allegheny Observatory, courtesy W. A. Feibelman)

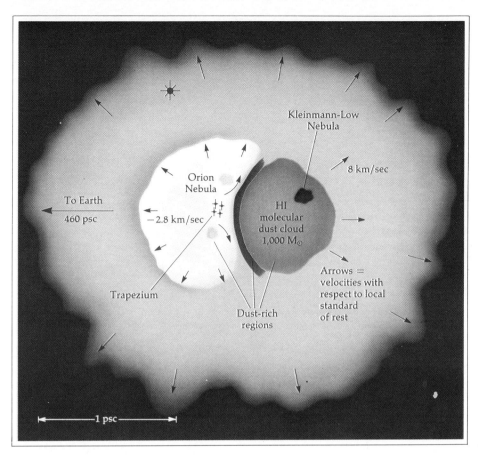

Figure 17·18 A hypothetical side view of the inner Orion nebula, showing Trapezium, dust clouds, and expansion.

stars, centered 400 to 500 parsecs from the solar system. As illustrated in Figure 15·9B, p. 279, the surrounding area contains its own centers of star-formation, such as the nebulosity around R Monocerotis, in the neighboring constellation. The expanding gas is not visible to the naked eye, but can be mapped by its 21-cm radio radiation or UV radiation.

Radio astronomers have estimated that some 110,000 solar masses of HI and HII are involved in the expanding gas shell around Orion. Backward tracing of the motions indicates that the expansion of the shell started about 6 million years ago, when very mas-

sive, hot stars must have formed, heated nearby gas, and created a giant HII region that has been expanding ever since.

A time-lapse movie of Orion beginning about 6 to 10 million years ago, with frames every 10,000 years, would reveal star-formation followed by a cosmic explosion of gas in the region of Orion's sword. The movie would also show certain stars racing out from central Orion like sparks from a blast. These high-speed stars are called **runaway stars.** Three bright O and B stars, for example, are racing out from the nebula at 70 to 130 km/sec, and Betelgeuse itself is moving

away from a region near Orion's belt. Their paths all point like accusing fingers back to the region of dense gas around the nebula. Something unknown happened in that region to create high-velocity stars. They may have been accelerated during near encounters between new stars in clusters, or during disruptions of co-orbiting pairs.

Orion has changed dramatically in the last few million years, since humanlike creatures emerged on the plains of Africa. New stars have blazed out, clouds of hydrogen have been expelled, and star-formation is apparently continuing there today.

Figure 17·19 **Bright stars, nebulosity, and expansion directions in the Orion region.**

Figure 17·20 **Two exposures of the constellation Orion in far ultraviolet light (1,250–1,600 A), a wavelength range that emphasizes extremely hot, recently-formed stars. The 30-second exposure (top) reveals the main stars and nebulosity in the belt and sword (arrows). The 100-second exposure (bottom) reveals an expanding ring of hydrogen, Barnard's nebula (arrows), blown out of the central star-forming region. The old red giant star Betelgeuse, prominent to the naked eye, emits little UV radiation and does not appear (circle, top). The sharp rings and streaks are instrumental effects.** (Photos from high-altitude rocket, 1975, Naval Research Laboratory, courtesy George Carruthers)

A Strange Emission Nebula: The Eta Carinae Nebula

Not many degrees from the Coalsack nebula and the constellation of the Southern Cross[2] lies an object that was recorded in the 1600s as an ordinary 4th magnitude star, which came to be called Eta Carinae. But in the 1800s its behavior was extraordinary. In 1827 it brightened to 1st magnitude and then dimmed. Between 1834 and 1837 the English astronomer

[2]Eta Carinae lies at declination $-59°$ and hence rises above the horizon only for observers south of 31° north latitude.

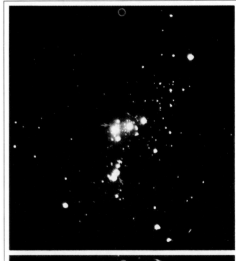

John Herschel, observing from South Africa, recorded it near magnitude +1.2. But on December 16, 1837, he found Eta Carinae outshining Rigel, at +0.1! Within a few weeks, it surpassed Alpha Centauri (−0.3), finally reaching −0.7 in 1843! Observing with a large telescope, Herschel wrote in 1847 (quoted by Lovi, 1972):

It would . . . be impossible by verbal description to give any just idea of the capricious forms and irregular gradations of light affected by the different branches and appendages of this nebula. . . . Nor is it easy . . . to convey a full impression of the beauty and sublimity of the spectacle it offers. . . .

By 1857 the object had faded to +1; by 1900 it had dropped to +8. It has remained variable, rising to +6 in 1967.

It was impossible to interpret this behavior without spectra. The first spectrum, in 1892, showed absorption lines suggesting an ordinary F5 supergiant star. Yet spectra of 1895 and the 1940s showed emission lines indicative of light from nebular gas. Absorption lines and photographs show that the nebula within about 0.05 parsecs of the central star is expanding very rapidly, apparently from the eruptions observed from 1827–1843. Measurements indicate a gas cloud about 0.05 parsecs from the central star and moving away from it at 630 km/sec. Several parsecs from the star, ionized gas is expanding at about 25 km/sec. Neutral gas expands less rapidly.

Modern photographs (Figures 17·21–

Table 17·2 Characteristics of Selected Nebulae

Name	Constellation	Approx. Distance from Earth (parsecs)	Approx. Diameter (parsecs)	Estimated Atoms/cm³	Mass ($M_\odot$)	Spectral Type of Associated Star
Nebulae Probably Associated with Young Objects						
Hubble's (R Mon)	Monoceros	700	10^{-5}	10^{12}	10^{-1}	F
Kleinmann-Low IR	Orion	500	0.1	10^6	100	?
Dark Nebulae						
Coalsack	Crux	170	8	2	15	none
Horsehead	Orion	350	3	25	0.6	B
Emission Nebulae						
Orion (central)	Orion	460	5	600	300	O
Eta Carinae	Carina	2,400	80	200	1,000	peculiar
Lagoon	Sagittarius	1,200	9	80	1,000	O
Trifid	Sagittarius	1,000	4	100	1,000	O
Reflection Nebulae						
Pleiades	Taurus	126	1.5	?	?	B
Cocoon	Cygnus	1,600	2	70	7	B
Planetary Nebulae						
Ring	Lyra	700	0.2	1,000	0.2	white dwarf?
Dumbbell	Vulpecula	220	0.3	200	0.2	white dwarf?
Helix	Aquarius	140	0.5	4,000	0.2	white dwarf?
Supernovae Remnants						
Crab	Taurus	2,200	3	1,000	0.1	pulsar
Veil (Loop)	Cygnus	500	22	?	?	?
Gum	Puppis-Vela	460	360	0.1	100,000	pulsar?

Source: Data from Allen (1973); Maran et al. (1973); and other sources.

Figure 17·21 **The Eta Carinae nebula (NGC 3372) is the site of peculiar brightness changes. Swirling gas, individual bright clouds, and dust lanes are prominent. The box shows the region illustrated in the next figure. The nebulosity extends across about** ▼ **50 parsecs and is about 1,300 parsecs away. Halation rings around the bright star images are produced by the photographic system; the star in the upper left is about third magnitude.** (Cerro Tololo Observatory, Chile; Kitt Peak National Observatory)

Figure 17·22 **The central part of Eta Carinae** ▶ **nebula, showing detail overexposed in Figure 17·21, including large and small dust lanes. The box shows region illustrated in the next figure.** (Cerro Tololo Observatory, Chile; Kitt Peak National Observatory)

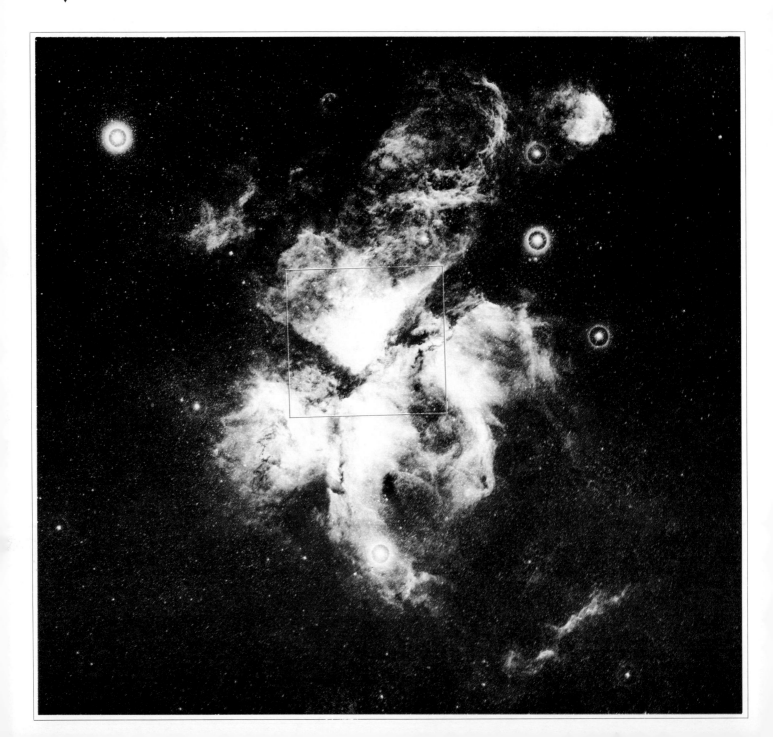

17·23) show that Eta Carinae is surrounded by a large, beautiful, deep-red emission nebula. This is the nebula described by Herschel. It covers nearly 3° and is nearly 120 parsecs across. Its red color comes from the strong light of the *hydrogen-alpha emission line* at 6563 A. The nebula is surrounded by an enormous hydrogen complex, rich in hot blue stars, about 2,100 to 2,700 parsecs from us. Only a few hundred parsecs from the nebula is a cluster of at least 130 O and B stars, known as NGC 3293, and estimated to be only 8 million years old.

If our eyes were sensitive to infrared wavelengths of 10,000 to 20,000 A, instead of 5,000 A, Eta Carinae would be the brightest object in the sky! This infrared radiation is heat from abundant warm (200 K) dust particles in the inner nebula, probably grains that condensed in the material blasted out of the central star.

Like the Orion nebula, the **Eta Carinae nebula** seems to involve star-formation, but much of the most vivid gas may have been thrown off a recently formed massive star that has already reached an unstable, old-age state. But what kind of star? In the center of the nebula is a tiny reddish cloud about 0.02 parsecs across that may be a small nebula containing an unstable star nearing the nova stage. However, the variable star expert D. J. K. O'Connell notes (quoted by Lovi, 1972):

> Eta Carinae is certainly not an ordinary nova. Its behavior is indeed not paralleled by any other known star. Further obser-

◄ *Figure 17·23* **Detail of the Eta Carinae nebula. The frame covers about 4 parsecs, showing near-linear bright and dark wisps and dark clouds less than 0.1 parsec across.** (Harvard College Observatory)

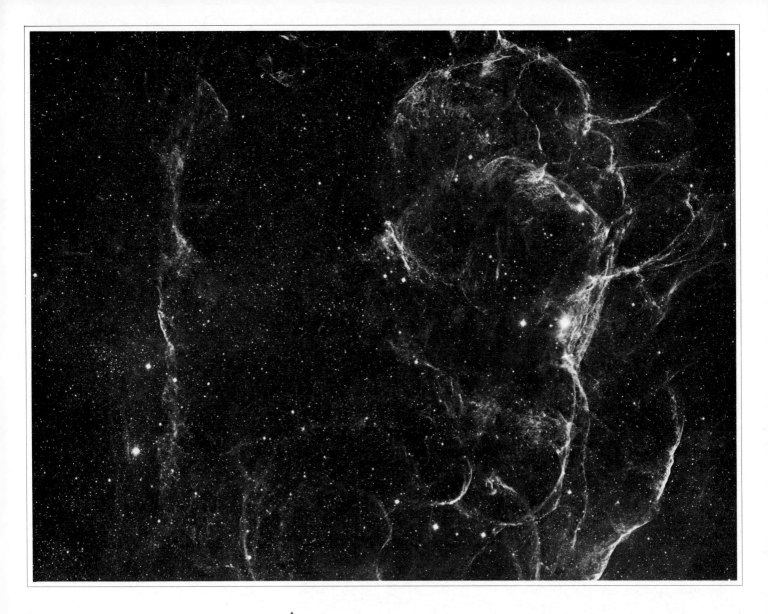

vations are badly needed. . . . It may once again become one of the brightest stars in the sky.

An Ancient Spectacular: The Gum Nebula

In the 1950s, a graduate student named Colin Gum surveyed the southern hemisphere sky with photographs sensitive to hydrogen alpha radiation, revealing HII nebulae. He discovered a ring of nebulosity with

Figure 17·24 **The Gum nebula, here photographed in ultraviolet light, is believed to mark remnants of an expanded supernova shell. A pulsar—the source of the supernova—lies near the center of the roughly circular ring of nebulosity (near the bright filament above the center). Nebular diameter is about 5°.** (Kitt Peak National Observatory, Cerro Tololo Observatory)

the amazingly large angular diameter of 60°, which has come to be called the **Gum nebula,** shown in Figure 17·24. It has been identified as an

expanding bubble-like shell of gas from an ancient supernova explosion.

Figure 17·25 shows its cross section. Its front side lies only about 100 parsecs from the solar system, and its center only about 460 parsecs away, accounting for its large angular size. The supernova remnant, a pulsar, has been found in the center of the expanding bubble.

Apparently, stars formed in this region during the last few million years (as in Orion), but one large star evolved rapidly and exploded.

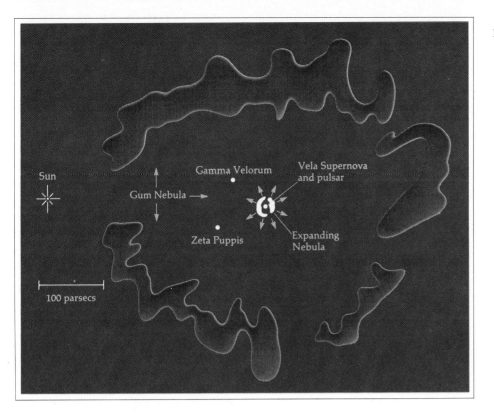

Figure 17·25 **A side view of Gum nebula geometry, showing relative positions of sun, Vela pulsar, and stars Zeta Puppis and Gamma Velorum (compare with the preceding figure).**

Studies of the expansion rate suggest the explosion might have been witnessed around 9000 B.C. Because it was so close (about one-third the distance of the Crab supernova) and had such high energy, it was probably extremely brilliant, reaching an estimated apparent magnitude of −10 (about as bright as the first quarter moon). This raises the intriguing prospect that late Stone-Age humans (south of latitude +47°) may have seen a brilliant star blaze forth for a year or two, rivaling the moon. What impact might this have had on their concepts of the heavens and gods in the sky?

◄ *Figure 17·26* **A planet associated with a newly formed star would be located in a star-forming region and thus would be near or embedded in, major nebulosity and a cluster of young stars. This imaginary view from a crater rim on such a planet shows the sky dominated by bright nebulae and hot, young stars.** (Painting by author)

17

Summary

Space is not empty, but thinly filled with atoms and molecules of gas, grains of dust, and possible bigger debris, which partly obscure distant stars and redden their light. Concentrations of these materials are nebulae.

Nebulae are the media from which stars are born and into which the larger stars blow some of their material when their fuel runs out. Their existence shows that matter in our galaxy has not dispersed uniformly and stably, but rather is continually stirred, formed into clouds, dispersed, and disturbed by influences such as the formation of new stars, the explosions of old stars, and movements of all local material around our galaxy's center.

From nebulae and the general interstellar complex we know that all local material visible from the earth has participated in a vast cosmic recycling. Interstellar matter forms clouds; clouds may contract to form stars; young and old stars blow out their material, creating a new interstellar medium. Nebulae also reveal clear evidence of cosmic events that have markedly changed our celestial environment within the last million years—a period less than 0.01 percent of cosmic time.

Concepts

nebulae

excitation

ionization

interstellar reddening

multiple scattering

interstellar atoms

metastable states

forbidden lines

21-cm emission line

interstellar molecules

interstellar grains

interstellar obscuration

interstellar snowballs

Messier numbers

NGC and IC numbers

Planck's law

Wien's law

HI and HII regions

dark nebulae

bright nebulae

reflection nebulae

emission nebulae

planetary nebulae

loop nebulae

Orion nebula

runaway stars

Eta Carinae nebula

Gum nebula

Problems

1. Since sunlight is white (a mixture of all colors), why does the sun look red at sunset? What happens to the blue light? Why does the part of the sky away from the sun look blue?

2. Why was the sky as photographed on Mars by Viking cameras red?

3. How do massive stars help keep interstellar gas stirred up?

4. How do interstellar molecules help illustrate the fact that complex organic chemistry is likely elsewhere in the universe?

5. How do masses of prominent nebulae compare with masses of single stars? Relate this comparison to formation of stars—are stars more likely to form singly or in groups?

6. Why are O-type supergiant stars likely to be associated with large emission nebulae, whereas solar-type stars are not?

7. Why do planetary nebulae often have simple, near-spherical forms, whereas typical larger emission nebula, such as the Orion nebula, are ragged, irregular masses?

Advanced Problems

8. If the sun's material were re-dispersed into space at a density of 10 atoms/cm³, typical of some clouds, how big a cloud would it make? (Hint: The sun contains about 10^{57} atoms. 1 parsec = 3×10^{18} cm = 2×10^5 AU.)
  a. *Compare with the size of the solar system.*
  b. *Compare with prominent nebulae.*

9. Suppose spectral lines from gas on the near side of the Crab nebula have a Doppler blue shift of 0.5 percent.
  a. *At what velocity is it expanding?*
  b. *If the Crab nebula is 1,500 parsecs away (or 4.5×10^{21} cm), use the small angle equation to derive the expansion velocity of the nebula if it is observed to expand at an angular rate of 0.2 seconds of arc per year. (Hint: One year $\approx \pi \times 10^7$ sec.)*
  c. *Confirm that the above two methods of estimating expansion velocity in the Crab nebula give consistent results.*

Projects

1. Observe a cloud of cigarette or match smoke illuminated by a single light source, preferably a shaft of sunlight in a darkened room, or a strong reading lamp. Compare the color of light transmitted through the smoke (by looking into the beam) with the color of light scattered out of the smoke (by looking at right angles across the light beam). Are there any differences? What can you conclude about the size of particles in the smoke cloud, assuming that the light wavelength is mostly 4,000 to 9,000 A? Compare forms in the drifting smoke cloud with forms of nebulae illustrated in this book.

2. In a dark area away from city lights, by naked eye, observe and sketch the Milky Way in the region of Cygnus. Can you observe the dark "rift" that divides the Milky Way into two bright lanes in this region? The rift is caused by clouds of obscuring interstellar dust close to the galactic plane between us and the more distant parts of the Milky Way galaxy.

3. Observe the Orion nebula with a telescope. Sketch its appearance. Locate the Trapezium (four tight stars near center). The dark wedge radiating from the Trapezium is a dense mass of opaque dust. If a large telescope is available (20–36 inches) look carefully for color characteristics. Generally, the eye is unresponsive to colors of very faint light, but large telescopes gather enough light so that colors can sometimes be perceived, especially with fairly low magnifications, giving a compact, bright image.

4. Observe the Ring nebula in Lyra, or other nebulae such as the Crab (in Taurus), the Trifid (in Sagittarius) or others as available. Comment on differences in form and origin.

Binary and Multiple Star-Systems

We have been conveniently ignoring a fundamental fact: Most stars are not solitary wanderers in space. If we started voyaging out toward the nearer stars from our own system of co-orbiting planets, we would find that the first star we reached, Proxima Centauri, at a distance of about 1.3 parsecs, has two other stars co-orbiting with it, including the bright star Alpha Centauri. Thus the nearest stars to us constitute a multiple system. The next star, Barnard star, is 1.8 parces from earth and is believed to have one small co-orbiting companion, although observations are not conclusive. We next encounter another single star, followed by another co-orbiting pair; followed by another co-orbiting pair including the bright star Sirius, at

2.6 parsecs; followed by yet another pair at 2.7 parsecs.

Surveys of this kind have led to the conclusion that less than half of all stars are truly single; the rest have at least one co-orbiting companion.

Each pair of co-orbiting stars is called a **binary star-system,** or sometimes just a binary star. Each system with more than two stars is called a **multiple star-system.** How do these systems affect our understanding of the universe? For one thing, we would like to know if our own multiple system, the sun and its planets, is related to other multiple star systems, or whether it is a different kind of phenomenon. Secondly, we must be sure that our theories of star-formation and evolution account for binaries.

A review of the preceding chapters will show that in fact the theories are consistent. Chapter 15 showed that stars must form in groups, and Chapter 17 gave examples of groups of young stars (such as the Orion nebula's Trapezium) embedded in star-forming nebulae. While Chapter 14 explained much of stellar evolution by considering only one star at a time, Chapter 16 required companion stars to explain certain facts of late evolution, including transfer of mass in binary systems to explain ordinary novae (not supernovae) and X-radiation caused by dumping of mass from giants onto co-orbiting dense stars or black holes.

Optical Doubles versus Physical Binaries

To understand binaries and multiples in more detail, we face an observational problem. Among the star pairs that appear to be close together in the sky, some are at different distances and merely aligned by chance (as seen from our vantage point in space), not actually co-orbiting. Stars that are far apart in three-dimensional space and merely appear aligned are called **optical double stars.** They can be identified by the absence of any periodic orbital motion around each other (revealed by photos or Doppler shifts) and are of little consequence in astronomy.

Pairs that are close enough to each other to be orbiting around each other are called **physical binary stars.** They are the ones with important consequences for stellar evolution, and the ones we will discuss here.

Early observers thought that all close star pairs were merely optical doubles, but in 1767 John Michell pointed out that so many chance alignments were unlikely, and proposed a physical association. This was confirmed in 1804 when William Herschel discovered that Castor (brightest star in the constellation Gemini) has a companion orbiting around it. In 1827 F. Savary showed that orbital motion

of the physical binary Xi Ursa Majoris is an ellipse fitting Kepler's laws. These results marked the *first discovery of gravitational orbital motion beyond the solar system,* an important confirmation that gravitational relations are universal.[1]

Among physical binaries, the brighter star is usually designated A, and the fainter, B—for instance, Castor A and Castor B. Analysis of orbits reveals which is the more massive star, and it is usually called the *primary.* The less massive one is called the *secondary.* Normally the primary is also the brighter, or star A.

Different Types of Physical Binaries

Astronomers classify physical binaries according to methods of detecting them. To understand what we can learn from binaries, it helps to understand these different methods of detection.

A **visual binary** is a physical binary in which both members can be resolved (seen separately) with the eye, the telescope, or the camera. Some 65,000 have been studied, and a good

◄ *Figure 18·1* **Forty-six years' photography of the visual binary Krüger 60.** Binary pair Krüger 60, on the right, is moving past a background star, on the left. The photos show the orbital motion of one component of the binary around the other with a period of 45 years, and a separation of 1.4 to 3.4 seconds of arc. The two stars have magnitudes about 9 and 11 and masses of 0.3 and 0.2 $M_\odot$. (Leander McCormick Observatory and Sproul Observatory, after J. F. Wanner, *Sky and Telescope,* 1967)

Figure 18·2 **A portion of the spectrum of a single-line spectroscopic binary, Alpha Geminorum.** The bright vertical lines at top and bottom are reference emission lines produced in the spectrograph; the middle bright spectra crossed by dark vertical absorption lines are spectra of the star on two dates. Alternate shifts to the red and blue show that the star is receding and then approaching because it is orbiting around another star. (Lick Observatory)
▼

[1]Herschel realized that this discovery was much more important than his original goal of simply measuring distances, illustrating that important unexpected scientific results often derive from more mundane research on another topic. Herschel reportedly likened himself to Saul, who went out to find his father's mules and discovered a new kingdom.

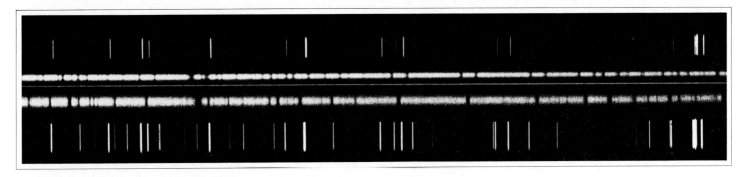

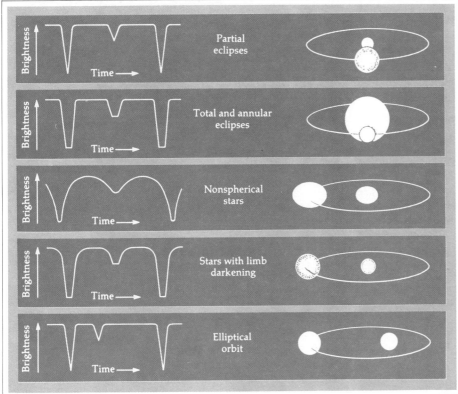

▲
Figure 18·3 **A portion of the spectrum of the double-line spectroscopic binary, Kappa Arietis. Arrangement as in 18·2. Two sets of absorption lines in the top stellar spectrum reveal two stars, one receding (red shifted) and one approaching (blue shifted). The lower stellar spectrum, when orbital motions are perpendicular to the line of sight, shows lines merged with no Doppler shift.** (Lick Observatory)

◄ *Figure 18·4* **Different types of light curves (left) from eclipsing binaries reveal different geometrical properties of the eclipsing stars and their orbits (right), even though the stars themselves cannot be resolved with the telescope.**

produce a marked, short-term decrease in brightness. The star periodically fades by a magnitude or two, then returns to its normal brightness. The most famous example is Algol (β Persei), discovered in 1669. Algol dims to about one-third its normal brightness every 2.9 days.

In an **eclipsing-spectroscopic binary** both Doppler shifts and eclipses can be detected. This is the most informative type of binary, permitting very detailed analysis of motions, masses, and sizes of stars.

An **astrometric binary** is one revealed not by Doppler shifts or eclipses, but by motions measured with respect to background stars. According to Kepler's laws, stars in a co-orbiting system revolve around their center of gravity, as described in Chapter 13. Each star, therefore, describes an

example is shown in Figure 18·1.

In a **spectroscopic binary,** orbital motion is revealed by periodic Doppler shifts in the spectral lines, but the individual stars cannot be resolved. There are two subtypes: those in which only *one* spectrum can be detected (Figure 18·2); and those in which *two* sets of spectral lines are seen (Figure 18·3). The latter, displaying lines of both stars, yield more information. About 1,000 pairs have been measured.

An **eclipsing binary** is a binary pair

(generally unresolved) whose orbit is seen nearly edgewise and is revealed by eclipses. Because our line of sight lies in, or nearly in, the orbital plane, the stars alternately eclipse each other. (Such an event would more properly be called an *occultation,* but the term *eclipse* is universally used.) The eclipses are detected by plotting **light curves,** or plots of brightness versus time, as shown in Figure 18·4. Depending on the relative brightness and size of the stars, the eclipse of the primary may

oscillating path in three-dimensional space, moving back and forth around the system's center of gravity, the so-called **barycenter.** Because *astrometry* is the study of stellar positions and motions, these systems are called astrometric binaries.

Spectrum binaries make up a relatively unimportant class, where the presence of an unresolved binary is revealed by a unique spectrum consisting of a line from stars of two different temperatures, but with no detectable Doppler shifts.

Table 18·1 lists examples of the most important types. The famous star Mizar, in the middle of the Big Dipper's handle, illustrates several types of doubles. Mizar (Figure 18·5) forms an optical double with a fainter star, Alcor, less than a degree away. The Arabs called this pair the "horse and rider" and regarded it as a test of good eyesight. In 1650, the Italian observer Jean Riccioli discovered that Mizar itself is a visual binary with a 4th-magnitude companion, Mizar B, just 14 seconds of arc away. In 1889 Mizar A was found to have double spectral lines with periodic Doppler shifts, revealing that it is a spectroscopic binary. In 1908, the same was

System Name	Distance (parsecs)	Component A Mass (M⊙)	Component B Separation from A (AU)	Component B Mass (M⊙)	Component B Eccentricity	Component C Separation from Primary (AU)	Component C Mass (M⊙)	Component C Eccentricity
Table 18·1 Selected Binary and Multiple Stars								
Visual Binaries								
α Centauri	1.3	1.0	23	0.9	0.52	10,000	0.1	
Sirius	2.6	2.2	20	0.9	0.59			
Procyon	3.5	1.8	16	0.6	0.31			
Eclipsing Binaries								
α Corona Borealis	22	2.5	0.19	0.9				
Algol (β Persei)	27	5.2	0.73	1.0	0.04	2?	?	0.13
β Aurigae	27	2.3	0.08	2.2				
ε Aurigae	1350	35	35	22				
Eclipsing-Spectroscopic Binaries								
α Virginis	71	9.6	0.07	5.8				
β Scorpii	118	13	0.19	8.3	0.27			
η Orionis	175	11	0.6	11	0.02			
Astrometric Binaries								
Krüger 60	4.0	0.26	9.4	0.2	0.41			
Barnard's star	1.8	0.2?	0.05?	0.002?				
Visual and Astrometric Binaries								
L 726-8	2.6	0.12	11	0.11				
61 Cygni	3.5	0.6	83	0.06			0.008	
BD 66°34	10	0.4	41	0.13	0.05	1.2	0.12	0.00
Solar System								
Sun, Jupiter, Saturn	0.0	1.0	5.2	0.001	0.05	9.5	0.0003	0.06

◄ Figure 18·5 **This telescopic photo of Mizar reveals two components of the visual binary, Mizar A and B, about 14 seconds of arc apart. Each component has been found to be a spectroscopic binary in its own right, but the spectroscopic pairs are too close together to be resolved. Other faint stars are background stars that are not part of the system.** (Lowell Observatory)

found true of Mizar B. Thus, Mizar is really a quadruple star system, with one close co-orbiting pair revolving around another close co-orbiting pair.

What Can We Learn from Binary Stars?

The various kinds of binaries yield much information. Take for example an eclipsing-spectroscopic system, where Doppler shifts reveal how fast both stars are traveling. Suppose the smaller star moves in a circular orbit at 100 km/sec, and the time it takes to pass behind the larger star (eclipse duration) is three hours, or about 10,000 seconds. Then the diameter of the larger, obscuring star is 1 million km (10^2 km/sec $\times$ 10^4 sec $= 10^6$ km), or about the size of the sun. This type of observation is the most accurate measure of stellar size, and hence one of the most basic checks on theories of stellar structure.

The same eclipse provides another example of how binaries contribute to knowledge of stellar structure. As the smaller star is eclipsed, the larger star successively obscures different parts of the smaller star's surface. It first covers part of the limb, then sweeps across the central parts, and finally covers the other limb. Thus we can study the brightness *distribution* across the face of the hidden star, thereby learning more about the atmospheric structure of a star.

How Many Stars Are Binary or Multiple?

To answer this question is a challenge to observers. The nearest stars are easiest to observe but give too small a statistical sample to be reliable. At greater distances there are more stars, but faint companions might not be detected. Spectroscopic binary statistics are biased toward pairs with small separation distances, because according to Kepler's laws these have the fastest velocities and greatest Doppler shifts, thus being the most likely to be discovered. Visual binary statistics are biased toward wide separation distances, which make the two stars easier to resolve. All these biases, which tend to make the data non-representative of the whole population, are called **selection effects.**

Table 18·2 lists three estimates of the **incidence of multiplicity** for systems ranging from single to sextuple. (Of 1,200 cataloged binaries and mul-

---

Table 18·2 Incidence of Multiplicity among Stars (Estimated fraction of systems containing *n* members)

n	25 Systems within 4 Parsecs	Average of 6 Estimates by Various Authors[b]	Estimate Favored by Batten (1973)[c]
1	0.48	0.40	0.30
2	0.36	0.40	0.53
3	0.12	0.15	0.13
4	0.04[a]	0.036	0.03
5		0.01	0.008
6			0.002

[a]Solar system.
[b]Data from Batten (1973).
[c]Batten's estimates attempt to average over all stars, using data from various sources. Differences between the estimates are measures of our uncertainty about multiple systems.

---

tiples, only two sextuple systems are known.) By the time we reach 6-member systems, definitions of multiple systems become hazy. It is unclear whether small groups like the Trapezium in the Orion Nebula should be counted as multiple systems, and whether there is a physical relation between large multiple systems and small clusters, or whether they are different phenomena. While Table 18·2 illustrates our uncertainty about binary numbers, it does show that single stars are a minority.

Kitt Peak astronomers Helmut Abt and Saul Levy (1976) surveyed stars to determine incidence not only of multiplicity but also of different masses. They found that about two-thirds of all stars have detectable companions, consistent with Table 18·2. But from statistics of companions' masses, they estimated that the other seemingly single stars probably all have companions too small to detect! Some companions might be black dwarfs with only a few percent of a solar mass; still smaller ones may be planets. According to this estimate virtually all stars have at least one companion.

Thus, although many people mistakenly assume that most stars are single, *multiple systems are more common than single stars.* Obviously, then, we must understand the origin of systems of two, three, four, and more stars if we claim any understanding of stars in general. We will return to the problem of origin after reviewing some related observational data.

Orbits and the Possibility of Planetary Systems

How do multiple star systems relate to our own multiple systems of planets? As shown in Figure 18·6, orbits of binary stars are often highly elliptical, though tidal evolution (sketched in Chapter 4) has circularized some orbits that were originally elliptical. The nine planets of the solar system have nearly circular orbits. Thus, the formation processes of typical multiple star systems and planetary systems probably differ.

Another test of the relation lies in the inclinations of orbits in multiple star and planetary systems. Among the nine planets of the solar system, Pluto has the highest inclination to the common plane, about 17°. None of the other eight is inclined more than 7°. In other words, as Figure 18·7 shows, planets' orbits are highly *co-planar*. Of 10 triple and quadruple star systems studied, no orbit is less than 19° in inclination, and half are more than 40° (Batten, 1973). Another system, BD + 66°34, may have three low-mass stars in co-planar orbits (Hershey, 1973). Nevertheless, most multiple star orbits are not co-planar, again suggesting that typical multiple star systems are unlike planetary systems.

According to Kepler's laws, the motions of co-orbiting bodies are related to the masses of the bodies around which orbital motion occurs. In many multiple systems the stars are of comparable mass, thus disturbing each other's motions. Multiple orbits of similar sizes would result in unstable systems, raising the possibility that perturbations or collisions would change the system. The most commonly observed arrangements in multiple systems are thus not concentric, but hierarchical. For example, in triple systems, such as the Alpha Centauri triple system, two stars commonly co-rotate in a *close* orbit, and the third star revolves around the pair at a *great distance*. In quadruple systems, such as Mizar, two close pairs

are likely to revolve around each other at a great distance.

Thus, in orbital shape, inclination, and arrangement, widely separated binaries appear to be fundamentally different from planetary systems.

Close binaries, however, may be related to planetary systems. Abt and Levy found that while masses of widely separated pairs have the same

Figure 18·6 **Observed positions of the fainter ▸ component of the visual binary Xi Boötes from 1780 to 1950, plotted relative to the brighter component, reveal Keplerian elliptical motion. (The apparent ellipse is the projection of the true elliptical orbit in the plane of the sky.)** (After a diagram courtesy *Sky and Telescope*)

Figure 18·7 **Different types of orbital arrangements known in multiple systems. In the solar system (A) orbits are circular and co-planar, each about twice as far from the sun as its inner neighbor. In triple systems (B) a distant companion often accompanies a close pair in an elliptical, inclined orbit. Quadruple systems (C) often have two remote pairs in elliptical, inclined orbits. Differences between (A) and (B,C) suggest different origins for planetary systems and typical multiple star-systems.** ▾

statistical distribution as masses of ordinary field stars, masses of close companions (within a few tens of AU, or solar system dimensions) have a different mass distribution. Abt and Levy concluded that close binaries, like the solar system, might form by growth of separate bodies in a contracting nebula. Rotational properties of the nebula might determine

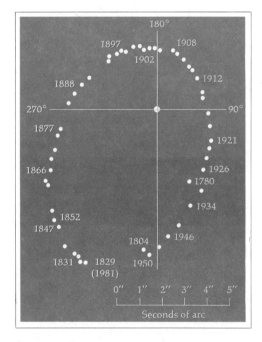

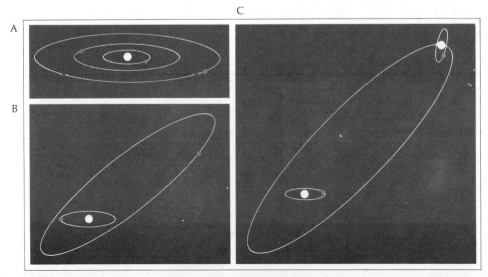

whether two massive stars form, or one star and a group of small bodies.

Evolution of Binary Systems

In 1955 the Czech-English astrophysicist Zdenek Kopal pointed out that binaries can be understood only by considering their evolution as a system, and that this evolution provides a natural way to classify them. The theory of mass transfer between binary members, described in Chapter 16, says that a system of two co-orbiting stars has an imaginary **Lagrangian surface** whose cross section is a figure 8, with one lobe around each star. Slow-moving material inside either lobe is bound to the star in that lobe. Material that moves out of either lobe is not gravi-

tationally bound to either star permanently. It may transfer from one to the other, or, with enough velocity, may eventually leave the system entirely. If one star should turn into a red giant and expand until it filled its lobe, its outer layers would assume the peculiar teardrop shape of the lobe, and the Lagrangian surface would become the real surface of the star. Kopal therefore proposed three **classes of binaries,** shown in Figure 18·8.

1. Systems in which neither star fills its Lagrangian lobe.

2. Systems in which one star fills its Lagrangian lobe.

3. Systems in which both stars fill their Lagrangian lobes; that is, stars in contact with each other.

Bizarre things can happen as binaries evolve. Suppose two stars form in a fairly tight orbit around one another. Tidal effects will tend to circularize the orbits, and the Lagrangian lobes will be well defined. The more massive star, M, evolving faster than the less massive star, m, becomes a giant first. Star M expands to a vast volume, assuming the teardrop shape required by the Lagrange theory in a binary system. Once star M fills its Lagrangian lobe, any further tendency to expand causes matter to be shed, mostly through the point common to the two lobes, as shown in Figure 18·8. Like sand in an hourglass, matter lost through this point enters the lobe

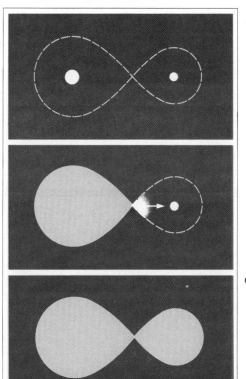

A

B

C

◄ Figure 18·8 Evolution of a binary pair is related to the configuration of Lagrangian lobes. In the first class of systems (A), neither star fills its lobe (dashed line). After the larger star expands to become a giant, it may fill its lobe (B), assuming a teardrop shape and perhaps ejecting some mass through the tip, which interacts with the second star. In the third class of systems (C), both stars fill their lobes.

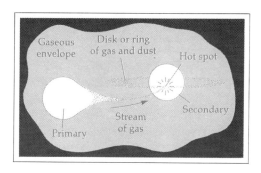

▲

Figure 18·9 **Possible features of a binary system in which the primary fills its Lagrangian lobe and transfers gas to the secondary. A disk of material may form around the secondary and a hot spot may appear where material impacts on the surface. Evidence of such features has been observed.**

of star m. Now star M is losing mass and star m is gaining mass. Obviously, the evolution of both stars is altered from what it would have been if they had been isolated. At some later time, star m is also apt to expand to a giant configuration, either because of mass gain from M, or because of its own normal evolution. If star M is still filling its Lagrangian lobe, both stars may simultaneously fill their lobes, producing two stars in contact. These are called **contact binaries.** They are actually touching as they orbit around each other as a single unit! Thus, a single binary system could evolve through all three classes. Or, if the two stars start very far apart, the giant expansion may not fill its Lagrangian lobe, and the binary may never evolve out of class 1.

Novae and Other Peculiar Stars as Binaries

Now we are better able to explain why in Chapter 16 it was said that **novae** are probably results of binary star evolution. A prenova star is probably a type-2 binary, in which one star has already evolved to the white dwarf state, while the second is expanding into a red giant, filling its Lagrangian lobe. Matter leaks off the red giant and flows onto the small star, as shown in Figure 18·10. The dwarf has exhausted its own nuclear fuel, but now finds itself receiving a new layer of "burnable" hydrogen. This causes catastrophic instabilities in the dwarf, leading to nova explosions. (Supernovae are not related to binaries, exploding instead because of instabilities associated with high mass.)

Observations suggest not only novae but also other unusual stars can be explained by binary evolution. Many cataclysmic irregular variables are binary systems in which matter has streamed from a giant into a disk surrounding a smaller companion. Several stars once thought to be short-period variables have been recognized as probable short-period binaries in which matter is streaming from one star to another, causing irregular variations in brightness. In some systems, matter streaming onto one side of a secondary star makes a hot, bright spot on its surface, causing unusual effects. Certain peculiar stars of spectral type A having unusually strong spectral lines of metals—and therefore known as Am stars—are mostly binaries. Tidal effects in these pairs apparently slow the rotation and cause unusual rates of diffusion among metal atoms in the stars' atmospheres. As we learn more, we may find that many peculiar types of stars are explained by strange processes of evolution in binary or multiple systems.

Contact Binaries—
W Ursa Majoris Stars

Among the most unusual stars are binaries so close that they touch each other. The most famous of the contact binaries are the **W Ursa Majoris stars,** named after the prototype in the constellation Ursa Major (better known as the Big Dipper). These consist of stars of rather similar mass, both filling their Lagrangian lobes so that they would probably look like a glowing figure 8. They have total masses ranging from 0.8 $M_\odot$ to 4 $M_\odot$, and because they are so close together, their common revolution periods are very short, less than 1.5 days. How did such pairs form? Some theorists believe they formed from protostars rotating so rapidly that they split, or *fissioned,* into two components. Other theorists have proposed mechanisms that might cause widely spaced pairs to evolve into contact binaries like W Ursa Majoris stars.

Possibly related are stars with very close orbits and extremely short orbital periods. The shortest known period is that of the peculiar helium-rich star AM Canum Venaticorum. In this irregular variable, hydrogen-rich outer layers have apparently been transferred from one star to a dwarf com-

Figure 18·11 **Possible model of a W Ursa Majoris binary, showing two stars in contact with a common atmospheric envelope. Both stars fill Lagrangian lobes, touching at the midpoint.**
▼

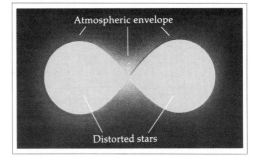

Figure 18·10 **Possible appearance of an imaginary double star-system seen from an associated planet. The red giant star, right, has expanded to fill its Lagrangian lobe, and is transferring mass from its tip to the smaller star, left.** (Painting by author)

panion and "burned up" in nova explosions. As mass was blown out of the system, the semi-major axis and period decreased (by Kepler's laws). Also, as the stars lost mass, the helium-rich cores of the exhausted stars were revealed. AM Canum Venaticorum's revolution period is only 17 minutes 31 seconds. Theoretically, mass loss could produce contact binaries with periods as short as two minutes! Somewhere in our galaxy there might be a planet in whose sky is a giant glowing figure 8 doing two-minute cartwheels like some bizarre advertising gimmick in the sky.

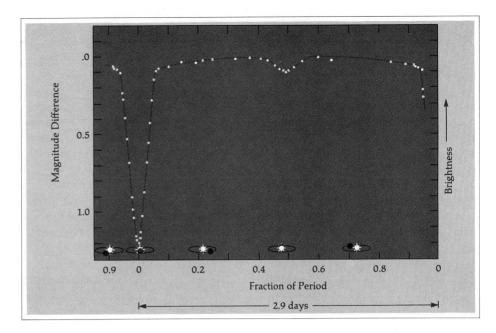

Figure 18•12 **Light curve of the eclipsing spectroscopic binary Algol. The curve plots brightness versus time, and shows that Algol's brightness drops about 1.4 magnitude when the small bright B star passes behind the fainter K giant.**

Examples of Binary and Multiple Systems

Algol and Similar Systems: Altered Masses

Algol, a variable star with a regular period of 2.9 days, is an eclipsing-spectroscopic binary. The primary is a hot blue B8 star of about 5.2 $M_\odot$. The secondary is a K0 giant of about 1.0 $M_\odot$. Here lies a paradox: The more massive B or A star should be the one to expand first; yet the less massive star is the more evolved giant. Why? Is there a fundamental mistake in our ideas of stellar evolution?

Theoretical studies have resolved the paradox: What is now the smaller star was *originally* the more massive star. Reaching the giant stage first, it transferred so much mass to the other star that it became the less massive of the pair. In one theoretical study of such a system, the more massive star starts out with 9 $M_\odot$ and is reduced to a mass of about 2 or 3 $M_\odot$ in only about 50,000 years! During this very rapid mass exchange, the stars reached more nearly similar masses.

Binary X-Ray Sources: Black Holes?

Observations from spacecraft above the atmosphere have revealed binary systems that emit strong X-radiation. Among these are Cygnus X-1 (brightest X-ray source in Cygnus), Hercules X-1, and Centaurus X-3 (third brightest in Centaurus). Each of these is believed to involve matter streaming from an evolving star onto a dense, post-main-sequence companion composed of degenerate gas.[2] As described in Chapter 16 (see Figure 16•14), the streaming gas accelerates and emits X-rays and gamma rays as it crashes into the dense star's atmosphere. The

---

[2]The colorful nomenclature of star types has encouraged shameless wags to concoct such descriptions as "degenerate dwarf in the company of young starlet."

Figure 18•13 **The multiple star-system Beta Lyra. The four brightest images are known members of the system, including the third-magnitude primary (overexposed, center), a 7th-magnitude companion 46″ away (lower left), and 9th- and 10th-magnitude companions 67″ and 86″ away (top). The bright primary is an eclipsing spectroscopic binary. Other faint stars (magnitude 13–15) might be associated with the system, or may be background stars. The primary is a spectroscopic binary, consisting of a 12 $M_\odot$ star transferring mass into a dusty disk-shaped nebula surrounding a star of about 2 or 3 $M_\odot$. The system may be only about 10 million years old. (Allegheny Observatory, courtesy W. A. Feibelman)**

dense star may be a neutron star, or in some cases, a black hole.

Epsilon Aurigae: A Forming System?

Epsilon Aurigae is a strange pair whose brighter component, an F-type supergiant, undergoes an eclipse every 27.1 years by an invisible companion. This eclipse takes 714 days and is "so peculiar as to defy interpretation in terms of any conventional model" (Kopal, 1971). The two stars are estimated to be 35 AU apart, and the primary supergiant is about as big as Mars' orbit. The secondary is believed to be an evolving disk in which a central star is only beginning to form. The disk has a diameter of about 40 AU, and a temperature of about 500 K. Therefore, it virtually duplicates the ancient solar system during planet-formation, except for having an intensely bright sun at the equivalent distance of Neptune. Kopal suggests planets may be forming there. The system is roughly 1,350 parsecs from the earth.

Barnard's Star and Related Systems: Planets?

Barnard's star, the fourth-closest star system, is an unresolved astrometric binary only 1.8 parsecs from the earth. It has caused a flurry of debate because it may contain at least one planet-size companion, thus possibly being another planetary system near our own. In 1963 American astronomer Peter van de Kamp announced that Barnard's star has an unseen companion only 50 percent more massive than Jupiter. Though van de Kamp regarded it as a planet, its orbit appeared highly elliptical, unlike a planet's orbit. In 1969 van de Kamp showed that the data could be interpreted in terms of two planets, with masses about 1.1 and 0.8 times Jupiter's mass, in circular orbits. This interpretation was questioned, because any irregular motions can generally be attributed to a group of small disturbing bodies, such as planets in circular orbits. The situation became more confused as other workers got involved. The October 1973 *Astronomical Journal* carried new observations by Gatewood and Eichhorn indicating no evidence of any planetary system! In the December issue of the journal Jensen and Ulrych reported that van de Kamp's data could just as well be interpreted in terms of four planets, ranging from 1.6 to 0.7 Jupiter masses. In 1975 van de Kamp reassessed the problem and concluded that two co-planar planets of about 1.0 and 0.4 Jupiter masses best explain the observations.

In summary, it appears possible, if not probable, that Barnard's star has at least one Jupiter-size planetary companion.

As for other nearby astrometric binaries, 61 Cygni is a system with two visible components, about 3.5 parsecs away. It has a rather well-measured third, unseen companion about eight times as massive as Jupiter. The star BD + 66°34, ten parsecs from the earth, consists of two main-sequence M-type stars of mass 0.4 $M_\odot$ and 0.12 $M_\odot$, and a third unseen star of mass 0.13 $M_\odot$ (about 130 Jupiter-masses). The orbits may be circular and in the same plane, resembling the solar system's configuration. No star has yet been *proved* to have a companion as small as a planet, but many astronomers believe that further astrometric studies might reveal planets existing elsewhere in the universe.

The Origin of Binary and Multiple Stars

Our hopes of understanding all stars would brighten if we could explain exactly how binary and multiple stars form. Of course, we must explain not only how they form, but how the proportions of each type—binaries, triplets, quartets, and so on—arise. Unfortunately, we cannot.

Astrophysicists are confident that most individual stars arise from contracting gas clouds, as explained in Chapter 15. Co-orbiting stars may be the product of *evolution in the forming clouds,* or of *interactions between stars after they form.* Theories can be divided into three groups reminiscent of the three classes of theories on the origin of the "binary" earth–moon system.

Fission Theories Theories that picture a fast-spinning protostar as splitting in two are called **fission theories.** Many researchers believe that the W Ursa Majoris stars—the group of contact binaries described earlier—are formed in this way. However, these theories do not explain widely separated pairs (such as most visual binaries), triplets, and other multiple systems.

Capture Theories Systems of widely separated stars that share weak gravitational attraction may be chance configurations arising when one star approaches another. Star A could be captured by B only if energy were lost from the pair, but this could happen if a third star or an enveloping gas cloud were present to absorb the energy. The problem with **capture theories** is that the chance of encounters among random field stars is far too low to explain the observed numbers of binaries. Furthermore, randomly paired stars would be of widely different ages, but this is not observed. Where could stars of similar ages interact in a closely packed group? In a newly formed star cluster. Massachusetts astronomers T. Arny and P. Weissman (1973) showed that fully half the protostars in a cluster probably undergo collisions or close encounters. Therefore, some astronomers believe at least some binaries and multiples formed inside newly formed open clusters as protostars approached, and, by various possible gravitational interactions, became bound in orbit around one another.

Common Condensation Theories In these theories, attention is directed to the prestellar collapsing gas cloud from which the stars form. Depending on rotational properties, subfragmentation of a prestellar cloud may produce multiple systems with either two high-mass stars or several low-mass stars in co-planar orbits. Chapter 11's description of our planetary system's origin can be classed as one of these **common condensation theories.** Small groups of several dozen stars, perhaps resembling the Trapezium in Orion, may form near each other and finally break apart into binary and multiple star-systems.

Each of these three theoretical processes may produce binaries of a certain type. A complication in sorting out binaries of these different types is that orbits of binary and multiple stars evolve through gravitational influences. Mass transfer in close pairs can alter orbits. Widely spaced pairs formed inside larger star clusters can evolve into closely spaced pairs as the clusters break apart. In one theoretical study of some 800 imaginary triple star-systems, about 97 percent were gravitationally unstable, eventually kicking out one star and becoming binary systems. Sometimes

the ejection velocities are quite high: 58 km/sec is a typical example. This in turn helps explain runaway stars observed speeding out of some young star-forming areas, as described in Chapter 17. Thus, the observed statistics of binaries and multiples may not reflect their original characteristics.

Nonetheless, astronomers who have attempted to correct for selection effects have found real statistical differences between the orbits of closely spaced systems and widely separated systems; they suggest different formation processes for different types of stars.

18

Summary

At least half of all the seemingly single stars in the sky are binaries or multiple systems. Many of these may have formed by interactions of stars in new clusters, but some may have formed by other means such as fission or common condensation. Disruption in star-forming clusters may cause runaway stars.

All these phenomena produce systems that can be detected in different ways. Binaries were once classified by these different methods of detection, which allow different types of knowledge to be derived from them. Examples include spectroscopic binaries, eclipsing binaries, and astrometric binaries.

A classification based on evolution includes binaries that have filled neither Lagrangian lobe, one Lagrangian lobe, or both Lagrangian lobes. Once one lobe has been filled by expansion of a star to the red giant stage, gas may flow smoothly from one star to the other, causing flare-

ups, X-ray emission, and novae.

Studies of astrometric binaries have revealed unseen companions with masses only a few times greater than Jupiter, and many astronomers believe that companions of some such stars may be true planets. Whether or not companions as small as planets form in a contracting protostar may depend on the protostar's rotational properties. If this idea is correct, there may be other earth-like planets in the universe.

Concepts

binary star-system

multiple star-system

optical double stars

physical binary stars

visual binary

spectroscopic binary

eclipsing binary

light curves

eclipsing-spectroscopic binary

astrometric binary

barycenter

spectrum binary

selection effects

incidence of multiplicity

Lagrangian surface

Kopal's classes of binaries

contact binaries

novae

W Ursa Majoris stars

fission theory

capture theory

common condensation theory

Problems

1. Describe verifications of Kepler's laws other than the planets' motions around the sun. What was the first verification outside the solar system?

2. How do binary and multiple star-systems generally differ from planetary systems?

3. Give evidence that at least a subclass of binary and multiple star-systems might be generically related to planetary systems.

4. How are novae related to binaries? Are supernovae related to binaries?

5. How will the evolution of a 1 $M_\odot$ star in orbit close to a 3 $M_\odot$ star differ from evolution of a 1 $M_\odot$ star by itself?

6. Is it likely that the sun has ever had a stellar (nonplanetary) binary companion? Give arguments for or against.

7. Why are binaries more likely to have formed in star clusters than as isolated field stars?

Advanced Problems

8. If a star of low mass (about 0.05 $M_\odot$, for example) were orbiting in the earth's orbit around the sun, what would be its period of revolution?

9. Use the circular velocity equation to derive the orbital velocity of a Jupiter-size body around a one-solar-mass star if the separation distance is 5.2 AU (7.8 $\times$ 10^{13}cm). (Hint: $M_\odot$ = 2 $\times$ 10^{33} g).
   a. *How does this compare with the actual orbital velocity of Jupiter?*
   b. *What would be the orbital velocity if the central star had two solar masses?*

10. If you observed a 0.5-solar-mass star in circular orbit around a 5-solar mass-star, and could tell from Doppler shifts that the orbital velocity is 47 km/sec, what would you conclude is the separation distance between the stars, in AUs?

Projects

1. Observe Mizar and Alcor with the naked eye. They are the middle "star" (actually a close pair) in the handle of the Big Dipper. Can you see the faint star Alcor? Sketch its position. (Inability to see Alcor may come from insufficiently keen eye-sight, from a hazy sky, or sky illumi-nated by city light, decreasing con-trasts.)

2. Observe the eclipsing binary Algol with a telescope or binoculars each evening for 10–20 days in a row. (This can be done as a class project with rotating observers.) Using neigh-boring stars as brightness reference standards, estimate the brightness of Algol. Can you detect the eclipses, which occur at 2.9-day intervals?

3. The star ϵ Lyrae is famous as the "double double." It consists of a binary pair 208 seconds of arc apart, easily seen in a small telescope. But each of these is a binary only 2 to 3 seconds apart. These pairs are a test of good optics and good atmospheric ob-serving conditions. Does your tele-scope reveal the two close pairs? Sketch them.

Star Clusters and Associations

If you could roam through space to ever-greater distances, you would soon realize that binary stars, triples, or even sextuples are not the major star systems. There are several kinds of larger groups, some involving thousands of stars. They are fundamental; if you backed away far enough, you would lose track of individual stars and see the galactic disk defined primarily by these clusters of stars. Writing about clusters in 1930, Harvard astronomer Harlow Shapley said:

> . . . their problems are intimately interwoven with the most significant questions of stellar organization and galactic evolution. . . .

Yet at the same time Shapley noted that scientific study of them had hardly begun, because nobody knew how to measure their distances or plot their distribution in space until the 1920s. Although some clusters, such as the Pleiades (or Seven Sisters), are easy to recognize with the unaided eye, others are so far away that they require large telescopes to detect. Still others are so close that they cover much of our sky and were not even recognized until recent years.

In other words, our cosmic journey has brought us to clusters so far-flung that they were only in this century recognized as a class. Study of these clusters eventually revealed to us the outlines of our own galaxy.

Types of Star Groupings

The three basic types of clusters are *open clusters, associations,* and *globular clusters.* Examples are listed in Table 19•1, p. 344.

Open Clusters

Open clusters are moderately close-knit, irregularly shaped groupings of stars. They usually contain 100 to 1,000 members and are usually about 4 to 20 parsecs in diameter. Our sun is possibly inside, or on the edge of, a loose open cluster centered only about 22 parsecs away toward the constellation Ursa Major, many of whose stars belong to this cluster, as shown by Figure 19•1. The best-known clusters, the Pleiades and the Hyades, lie 12 degrees apart in our winter-evening sky, about 42 and 127 parsecs away, respectively. About 900 open clusters are concentrated along the Milky Way band, indicating that they lie in the plane of our galaxy.

As described in Chapter 18, most stars form in open clusters, and most open clusters have prominent young stars or associated clouds of star-spawning gas. Then why aren't all stars in clusters? The reason is that open clusters break apart into individual stars within only a few hundred million years because of dynamical forces acting on them. In comparison with most cosmic lifetimes, open clusters are short-lived.

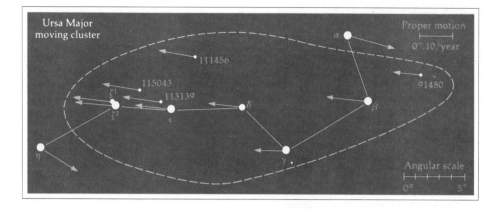

▲
Figure 19·1 **Some members of the closest open cluster, the Ursa Major cluster, form part of the familiar figure of the Big Dipper (solid lines). Arrows show proper motions of the stars, or the rates of angular motion relative to distant background stars. All but two of the stars in the figure are moving together at almost exactly the same rate, thus defining the cluster (dashed line). Stars at each end of the dipper are moving in different directions and are not true cluster members.**

Figure 19·2 **Two open clusters are prominent in this portion of the January evening sky. The Pleiades is the tight group in the lower center, about 127 parsecs away. Closer to us is the Hyades, a V-shaped group 12 degrees to the upper left, about 42 parsecs away. The brightest star, at the upper left tip of the Hyades, is the first-magnitude red giant, Aldebaran. (Ten-minute exposure with 35 mm camera, f1.4, 50 mm lens, 2475 Recording film.)**
▼

Associations

Associations are cousins of open clusters. They often have fewer stars, but are larger in size, with looser structure. Some large associations include an open cluster within them. They may have 10 to a few hundred members and diameters of about 10 to 100 parsecs. They are rich in very young stars, such as O and B stars, which burn their fuel too fast to last long; or T Tauri stars, which evolve toward the main sequence too fast to last long. Associations are classified as **O associations** or **T associations** depending on whether the prominent stars are O and B blue stars or T Tauri variables.

The smallest associations grade into small multiple-starlike groups such

Figure 19·3 **The Pleiades, photographed with a 35 mm camera with telephoto lens. (5-minute exposure, f2.8, 135 mm lens, 2475 Recording film.)**
▼

as the Trapezium in the Orion nebula. These are sometimes called *trapezium systems,* and might be a link between multiple stars and small clusters. Like open clusters, associations are involved with regions of recent star formation and are short-lived. More than 60 associations were tabulated in a 1963 catalog.

Globular Clusters

Globular clusters are quite different from the other two types. They are much more massive, more tightly packed, more symmetrical, and older. They typically contain 20,000 to several million stars, although many of these stars crowd too close to be resolved by earth-based telescopes, especially in the central regions. For example, 44,500 stars were counted in the globular cluster M 3. The individual stars in these clusters must follow complicated orbits among one another!

Typical diameters of the central concentrations range from only 5 to 25 parsecs. To imagine conditions inside a globular cluster, picture 10,000 stars placed around the sun at distances no farther than Alpha Centauri, our nearest star!

◄ *Figure 19·4* **A portion of the Pleiades photographed with a large telescope, revealing clouds of wispy nebulosity concentrated around each star. The black dot shows the location of Figure 19·5. See also Color Photo 19.** (Lowell Observatory)

Figure 19·5 **A high-resolution view of one star in the Pleiades. The overexposed image of the star (with radial spikes caused by diffraction in the telescope) also reveals faint wisps of nebulosity near the star itself, together with additional faint stars that probably belong to the cluster.** (Lowell Observatory)

▼

Table 19·1 Selected Star Clusters and Associations

Name	Distance (parsecs)	Z[a] (parsecs)	Diameter (parsecs)	Estimated Mass ($M_\odot$)	Estimated Age (years)
Open Clusters					
Ursa Major	21	18	7	300	2×10^8
Hyades	42	18	5	300	6×10^8
Pleiades	127	54	4	350	5×10^7
Praesepe	159	84	4	300	4×10^8
M 67	830	450	4	150	4×10^9
M 11	1,710	89	6	250	8×10^7
h Persei	2,250	156	16	1,000	1×10^7
χ Persei	2,400	167	14	900	1×10^7
O Associations					
I Orionis	470	150		3,000	
I Persei	1,900	164		180	
T Associations					
Ori T2	400	132	28	800	
Tau T1	180	52	?	50?	
Globular Clusters					
M 4	2,800	800	9	60,000	$\sim 1 \times 10^{10}$
M 22	3,100	430	9	7,000,000	$\sim 1 \times 10^{10}$
47 Tuc	4,600	3,300	5		$\sim 1 \times 10^{10}$
M 13	8,200	5,400	11	300,000	$\sim 1 \times 10^{10}$
M 5	9,200	6,700	12	60,000	$\sim 1 \times 10^{10}$
M 3	13,000	12,000	13	210,000	$\sim 1 \times 10^{10}$

[a]Z = Perpendicular distance north or south of Milky Way plane.

Even the nearest globular clusters are thousands of parsecs away from us. It is only because they have so many and such very bright stars that we see them at all. Yet a modest backyard telescope can reveal many prominent examples. The total of known globulars around our galaxy is about 120, but if we may judge from the numbers of globulars surrounding other galaxies, the total number near the Milky Way may exceed a thousand. In three-dimensional space they are not confined to the galactic plane, but are distributed in a spherical *halo* surrounding our galaxy. Similar distributions are found around other galaxies. They are among the oldest objects in the galaxy, with ages estimated around 12 billion years.

Discoveries and Catalogs of Clusters

The Pleiades, shown in Figures 19·2 to 19·5, form a cluster easily recognized by the naked eye. They are known as the Seven Sisters, from a myth about the constellation. To be able to count seven stars in the group is a test of good eyesight and good sky conditions. Less obvious as a cluster are the Hyades, also shown in Figure 19·2. The Hyades cluster is less obvious because it is closer, appears more dispersed, and has fewer hot bright stars. Similarly, the Ursa Major cluster is still harder to recognize because it is so close that its stars appear widely spread. Other clusters are so far that telescopes were required to recognize them. Thus, recognition of clusters is an enterprise spread out over many centuries, as shown in Table 19·2.

Between 1864 and 1908 many clusters were cataloged in the New General Catalog (NGC) and the Index Catalog (IC), and became known by their NGC or IC numbers. Not until the 1920s was the difference between various types of clusters and the still more remote galaxies understood, and only then could reasonable research begin. Shapley's 1930 book was the first substantial work on star clusters.

Measuring Distances of Clusters

Because of clusters' enormous range of distances, astronomers have been forced to devise varied and sometimes ingenious methods to determine their distances. Details of all the methods are beyond the scope of this book, but a few examples can be given.

Parallaxes The most basic method of measuring star distances, the measurement of **parallaxes** (described in Chapter 13), is almost useless on clusters because it is not accurate beyond about 20 parsecs. It provides data for a few stars in the Ursa Major cluster and the Hyades, but it also provides a foundation on which other distance-measuring methods are built.

Table 19·2 First Recognition of Selected Clusters

Cluster	Type	First Recognition	
Pleiades	Open	Prehistoric[a]	
Hyades	Open	Prehistoric[a]	
ω Centauri	Globular	Ptolemy ca. A.D. 140[b]	
Praesepe	Open	Galileo	ca. 1611
M 22	Globular	Ihle	1665
M 11	Open	Kirch	1681
M 5	Globular	Kirch	1702
M 13	Globular	Halley	1714
27 Globular Clusters and 29 Open Clusters cataloged		Messier	1781
I Persei	Association	Ambartsumian	1949

[a]First recognition as star clusters similar to telescopic examples was in Messier's 1781 list.

[b]Not resolved into stars, but listed as a bright, fuzzy patch.

Source: Data from Shapley (1930).

Figure 19·6 **Convergent motions of stars in the Hyades cluster, showing that the cluster is receding from us. Arrows represent** **proper motions, based on observations at Leiden Observatory. Brighter stars are indicated by dots.**
▼

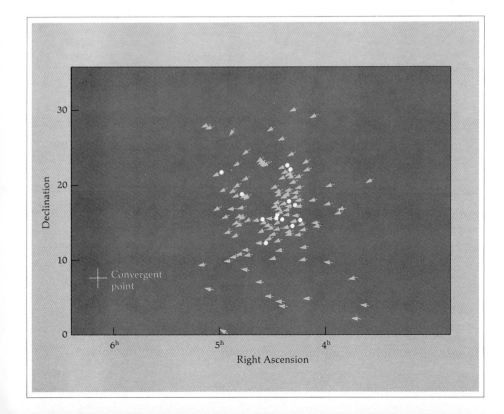

Star Luminosities If observers can determine the luminosity of any star in the cluster, the distance from the earth to the cluster can be estimated—assuming no complications from obscuring interstellar matter. (This method is similar to the example in Chapter 13 of estimating the distance of a certain light if we know that it is a candle and that there is no intervening fog.) One way to determine luminosities is to study the nearby stars whose parallaxes can be determined. Such studies showed, for example, that spectra could be used to identify main-sequence and other stars, and that each type of star has a specific luminosity. Once the spectra of stars in clusters were measured, their luminosities and distances could be estimated.

Cluster Diameters Once the linear diameters (in parsecs) of a few nearby clusters were known and found to be roughly comparable, then distances of remote clusters were estimated from their angular diameters (seconds of arc) by applying the small angle equation and assuming that all clusters of similar appearance have similar linear size.

The Complication of Interstellar Obscuration

Before 1930, these methods were carried out under the assumption that all intervening interstellar space is transparent. However, early estimates of distances, brightnesses, and sizes of clusters gave inconsistent results. In 1930 Robert Trumpler used the cluster results to show that the problem lay in the erroneous assumption that space was clear. He showed that diffuse

▲
Figure 19·7 **The open cluster M 67 (NGC 2682) in the constellation Cancer. The cluster has an angular size of 18′, corresponding to a diameter of about 4 parsecs at its distance of 830 parsecs. It is estimated to be about 4 billion years old.** (The Observatory of New Mexico State University)

interstellar dust dims the stars and clusters more than a few dozen parsecs away. This throws off the estimates of stars' luminosities and distances. Fortunately, the total amount of dimming can be estimated by measuring the amount of interstellar reddening, or color change caused by the dust. Thus, once the obscuration is measured and taken into account, distances can be accurately measured if the luminosity of any star or class of stars in the cluster is known.

Cepheid Variables as Distance Indicators

Luminosity is the key to distance measurement, and this brings us to the *most useful* technique for measuring distances of remote objects. As described in Chapter 16, the periods and luminosities of **Cepheid variable stars** are correlated. This meant that

instead of going through the difficult process of measuring the spectra of faint stars in distant clusters to estimate their luminosities, astronomers could simply search the cluster for Cepheid variables—stars that vary in brightness with periods of 1 to 50 days—and then read their luminosities from Leavitt's tables.

By the 1950s a small complication was discovered and overcome. The Cepheids found in open clusters are a special group called Type I Cepheids and have a special period–luminosity relation; those in globular clusters are called Type II Cepheids and have a somewhat different period–luminosity relation.

Thus, measurement of the distance of a cluster is a process requiring several steps. A Cepheid must be located, its period and type measured, its luminosity read from the appropriate period–luminosity diagram, and this luminosity combined with the apparent magnitude (corrected for interstellar obscuration) to calculate the distance. Few astronomers are equipped to carry out all these observations, such as measuring light variations and periods, determining spectral properties, measuring absolute magnitudes, and measuring interstellar reddening and obscuration. Thus, astronomers have specialized. Some study periods of variable stars; some make photometric measures of absolute brightnesses; some study interstellar reddening. The simple statement that cluster A is *x* parsecs from the earth may represent years of work by many astronomers.

Distribution of Clusters

Once the distances of clusters could be measured, their positions in three-

Figure 19·8 **Map of the sky showing locations** ▶ **of selected open clusters (dots), O associations (O), and T associations (T). The Ursa Major, Hyades, and Pleiades clusters are indicated by U, H, and P, respectively. Open clusters concentrate along the Milky Way, whose equator is the zero-latitude line in this map.**

dimensional space could be mapped simply by plotting their distances and directions. This process began in the late 1920s and 1930s, and striking results began rolling in at once. The distribution of clusters is not random, but defines the shape of our galaxy. The open clusters lie in a disk, and the globulars form a spherical cloud around the disk. Shapley, a pioneer in these measurements, said that the clusters reveal "the bony frame of our galaxy."

Even two-dimensional maps of cluster positions on the sky make the point. Figure 19·8 shows a map of the sky with the positions of open clusters marked. They stretch along the horizontal line that marks the Milky Way—our view along the disk of the galaxy. Figure 19·9 is a similar map of globular cluster positions. It shows that globulars lie in a ball-shape swarm centered on the galaxy's center.

Figure 19·9 **Map of the sky showing locations** ▶ **of the 93 most prominent globular star clusters. The coordinates of this map are the same as those in the previous figure, with the center of the galaxy lying at the center of the figure. Globular clusters form a spherical swarm centering near the galaxy's center, as Shapley found, using the same data in 1930.**

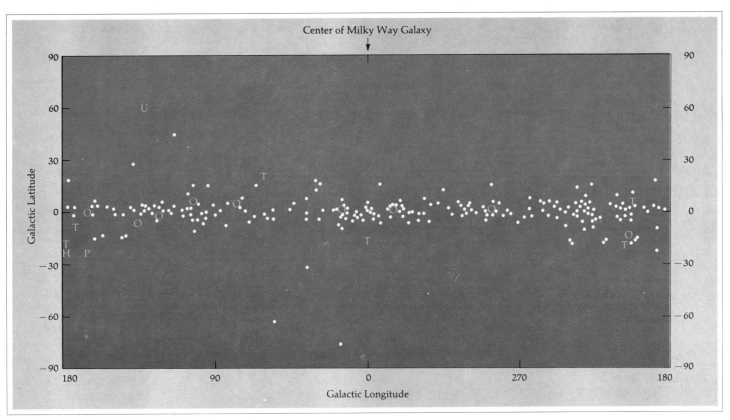

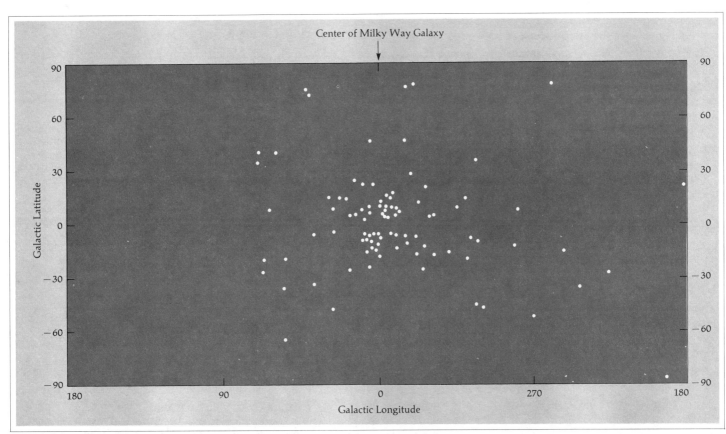

The Nature of Open Clusters and Associations

The preceding discussion mentioned that open clusters and associations are groups of stars formed from the contraction of large gas clouds, which later break apart into individual stars. We will now describe observations and theory shedding more light on the ages, evolution, and ultimate fate of these star-groups.

Open Clusters: Ages and H-R Diagrams

It is relatively straightforward but time-consuming to construct the **H-R diagram of a cluster.** The apparent magnitudes and color temperatures of hundreds of stars must be measured. If the distance to the cluster is known, the apparent magnitudes can be converted to absolute magnitudes, or luminosities. In the 1950s and 1960s, astronomers—and their graduate students—plotted the H-R diagrams of many clusters.

The importance of such work, of course, lies in the evolutionary information it makes available (as discussed in Chapters 14–16). Figure 19·10 brings together data on several open clusters. Along the main sequence is the age (in years) at which various stars evolve off the main sequence. Thus, the point at which stars have left the main sequence in a cluster is a measure of the cluster's age. **Ages of open clusters** range from about 1 or 2 million years for NGC 2362 (among the youngest known clusters), to as much as 10 billion years for NGC 188 (probably the oldest known open cluster). Among published ages for 27 open clusters, about half (55 percent) are less than 100 million years old. The solar system is nearly 50 times as old as this.

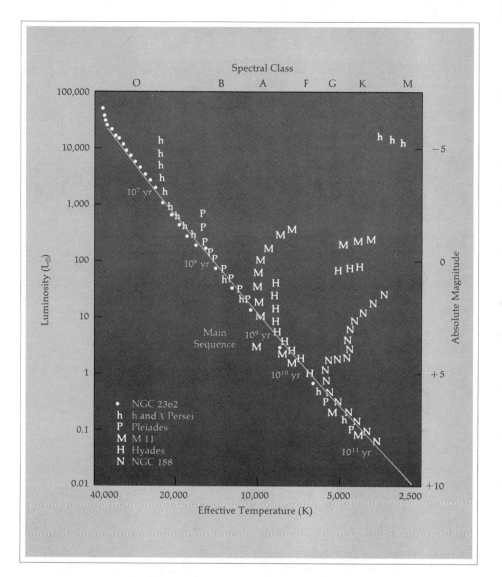

Figure 19·10 **An H-R diagram for several open clusters. Numbers along the main sequence give ages in years for groups of stars turning off the main sequence. Clusters can be dated by measuring these turnoff points.** (Based on the work of A.R. Sandage, O.J. Eggen, and L. Aller)

These results confirm that in cosmic terms, *open clusters are mostly young,* and that (as stated in Chapter 15) star-formation is continuing in open clusters to this day. Often, their brightest stars are the hot blue O and B stars, which formed recently, "burn" their fuel fast, and cannot last long. In Chapter 15 two clusters were mentioned in which the least massive stars have not yet even evolved onto the main sequence. These examples, the Trapezium cluster and NGC 2264 (Figures 17·17, 15·12, and 15·13), are only about 1 and 3 million years old, respectively.

Open Clusters: Disruption and Ultimate Fate

Velocities of stars, measured by Doppler shifts, indicate that some clusters are expanding or losing members, or both. Some high-velocity stars have been found to exceed the escape velocity of their parent clusters. Thus, these clusters are breaking apart as we watch. Clusters tend to disperse after a few hundred million or a billion years. This **disruption of open clusters** occurs by several simultaneous mechanisms. Fast-moving stars, or stars that are accelerated by interaction with others, may escape. This reduces the mass and gravitational self-attraction of the cluster, so that other stars can escape. Tidal forces exerted by the large number of stars near the galactic center, and differences in orbital speed as the stars in the cluster move around the galaxy, tend to shear the cluster apart. The Hyades, for example, are barely stable now, and the outer parts of the Pleiades are already dissipating, though the central, tighter grouping may be stable.

Associations:
Ages and Disruption

Like open clusters, associations are young. Because of their loose structure, they may break up even faster than ordinary open clusters. For example, a T association of 8 T Tauri variables, about 100 parsecs away and 25 parsecs across, has been estimated to be only about 10 million years old; it may soon break apart because of the tidal gravitational forces of the galaxy

acting on the cluster. Another association is a group of possibly thousands of O and B stars surrounding the well-known double open cluster h and χ Persei, shown in Figure 19•11. It has a diameter of about 170 parsecs, whereas the two open clusters inside it are only about 10 parsecs across. It is about 20 million years old. Most associations cannot last more than a few tens of millions of years, because of the disruptive tidal forces and the tendency for their stars to follow individual orbits around the center of the galaxy.

In many associations there are large masses of neutral hydrogen gas, some of which exceed the mass of the stars.

This material adds more gravitational attraction for the member stars, and may help hold the associations together somewhat longer than would otherwise be the case. The hydrogen may be material ejected from the fastest-evolving stars, or debris left over from formation of incorporated stars.

The Nature of Globular Clusters

Careful studies of globulars reveal that they have not only different form, but also different content and ages than open clusters and associations.

Figure 19•11 **The region of the double open cluster in Perseus and its surrounding association of stars. The two clusters, known as h and χ Persei, are nearly 1° apart at the center. Surroundings include members of O-association I Persei. The brightest stars at the top are components of the W-shaped constellation Casseopeia. About 700 stars are involved in the clusters and association, which is about 2,000 parsecs away and estimated to be only about 10 million years old.** (Don Strittmatter)

Globular Clusters:
Ages and H-R Diagrams

The technique of H-R diagram analysis is especially well-suited to determining the ages of globular clusters, because they have well-defined turnoff points along the main sequence. It turns out that globulars are extraordinarily old. Star-formation has ceased in them. In many globulars, the O, B, and A stars have already evolved off the main sequence, and have become red giants. For this reason, *the bright stars in almost all globulars are red, and there are no bright blue stars.* This is well shown in Figure 19·13, depicting the H-R diagrams of three globular clusters.

Assigning numerical ages to globulars must be done with some care, because spectra of their stars show that they contain fewer heavy elements than the more familiar stars of the solar neighborhood. This means that their inner processes of energy generation and transport are different. Hence the evolutionary time scales may be different, and different calculations are used to determine the ages corresponding to various main-sequence turnoff points.

The best techniques set the **ages of globular clusters** at about 12.5 ± 3 billion years old. All of them seem to date from a single, brief period of formation, believed to be the period when our galaxy formed. We will return to this exceptionally important relation between globulars and Milky Way formation in the next chapter.

Globular Clusters:
Dynamics

Many cosmic systems form flat disks because of their rotation—for example, the rings of Saturn, the solar

system, the primeval solar nebula, and even the galactic disk itself. *A disk is the final stable state for a contracting, rotating, self-gravitating system with substantial initial rotation.*

Why, then, are globular clusters not disk-shaped? It is a question of how much rotation (or more exactly, angular momentum) they had when they started to form. The real surprise would be if they had zero rotation, causing a perfectly spherical shape. However, globular clusters are not

Figure 19·12 **Globular cluster 47 Tucanae (NGC 104). Inner regions are overexposed in this photo, which reveals the myriad of outer stars. The cluster is about 10 parsecs in diameter and at a distance of about 5,100 parsecs.** (Cerro Tololo Inter-American Observatory)

spherical but are slightly flattened by various amounts. They are slowly rotating with rotation axes distributed approximately at random with respect to the galaxy.

These results imply that globulars formed from gas clouds with random directions of rotation, but with a distribution of angular momentum different from that in the present interstellar gas that is spawning open clusters. This difference may in turn be due to differences between the present galactic gas and the conditions some 12 billion years ago when the globulars, and the Milky Way galaxy as well, were probably still forming.

Inside a given globular cluster the orbits of individual stars must be very complicated. The cluster's overall gravity field, the spatial distribution of its stars, their relative speeds, and effects of near encounters or even collisions among stars are all important influences. Stars may occasionally collide or transfer mass in the crowded central regions of globulars.

Furthermore, as the cluster orbits the galaxy, it passes through the galactic disk every hundred million years or so, and shock waves from collisions with galactic gas and dust cause important effects, such as sweeping gas out of the globular. (The galactic disk, being much larger, is little affected, though some gas may be dragged out of the galactic plane.)

Because of their dense population of stars and strong gravitational bonds, globular clusters last much longer than open clusters—perhaps 20 billion years or more, consistent with their great observed ages. Nevertheless, a few stars probably do escape from their outer regions, accounting for the few scattered individual stars above and below the galactic plane.

Figure 19·13 **An H-R diagram for three selected globular clusters. Estimated main-sequence turnoff positions are indicated for ages of 10^9, 10^{10}, and 10^{11} years.** (Based on data of A.R. Sandage)

▼

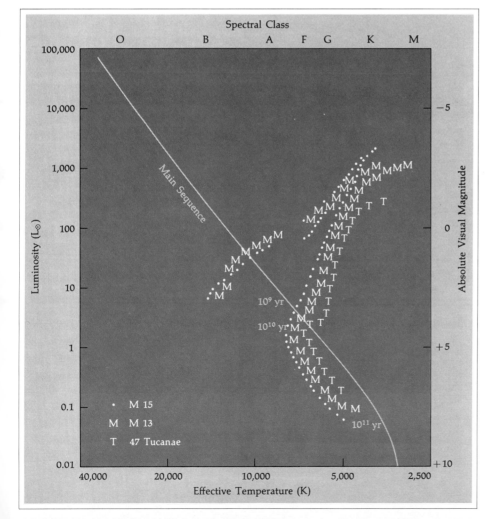

Globular Clusters as X-Ray Sources

Satellites above the earth's obscuring atmosphere have mapped celestial sources of high-energy X-ray radiation. Most sources are in the galactic disk, perhaps associated with binary pairs in which one star dumps material onto another. But by 1975 at least five globular clusters had been found to be sites of strong X-ray radiation. Although the spectra of the X-radiation are similar to those found in other sources, the **globular cluster X-radiation** does not have the smooth periodic variations that are caused by orbital motions in binary pairs. Instead they have irregular variations in intensity, sometimes over weeks or months, but sometimes doubling in intensity within a few minutes. In NGC 6440 rates of energy radiation in the X-rays alone reach as much as 7×10^{37} ergs/sec (17,000 times the sun's total luminosity).

These surprising findings led to several new hypotheses about conditions inside globular clusters. One idea is that globulars may contain many neutron stars or black holes, the remnants of "burnt-out" stars.

▲
Figure 19·14 **Globular cluster M 22 (NGC 6656), in Sagittarius. It is about 9 parsecs in diameter and 3,000 parsecs away, being one of the closest globular clusters. It has an estimated mass of 7 million solar masses, making it one of the most massive clusters.** (The Observatory of New Mexico State University)

▲
Figure 19·15 **Globular cluster NGC 2419, about 32 parsecs in diameter and 6,500 parsecs away. The photo resolves the bright inner regions, while still indicating the abundance of fainter outer stars.** (The Observatory of New Mexico State University)

▲
Figure 19·16 **Globular cluster M 13 (NGC 6205), in Hercules, photographed with an amateur astronomer's 12½ inch telescope. The cluster is about 11 parsecs in diameter and 7,700 parsecs away, with an estimated mass of 300,000 $M_\odot$.** (Exposure 2h 40m on 103 aO emulsion; Clarence P. Custer)

The stars being relatively close together inside globulars, these dense remnants may often capture passing stars into binary orbits. The resulting evolution of the captured star may transfer mass to its captor, and the resulting high-energy impact of gas may produce X-rays.

A second idea is that black holes of 10–100 $M_\odot$ inside clusters may disrupt passing stars by strong tidal forces, and pull mass onto themselves. A third idea is that many neutron stars or black holes may accumulate inside globulars, eventually colliding with each other and with other stars. In an old globular so much "burnt-out" mass might accumulate that a supermassive black hole could coalesce near the center of the cluster. This supermassive black hole would attract a surrounding cloud of accelerated gas which might fall into it, causing X-rays and other radiation. In support of

this model, at least two of the X-ray clusters have been found to have bright concentrated sources of light near their centers. These sources would be glows from the outer, visible parts of clouds surrounding the hypothetical dense objects.

There is little doubt that the large number of stars packed into a few cubic parsecs make the central regions of globular clusters extraordinary stellar environments. New X-ray satellites and other instruments will clarify these conditions and try to prove whether exotic objects such as black holes actually exist in globular clusters.

Origins of Clusters and Associations

Figure 15·2 showed how gravitational contraction could lead to the **origin of clusters** of various mass,

depending on the density and temperature of the original diffuse medium. As was shown there, systems of *globular-cluster* size could form from a gaseous medium as thin as the gas surrounding protogalaxies. And *open clusters* or *associations* could form from interstellar gas in the galactic disk. The first case involves masses of 10^4 to 10^5 $M_\odot$ (typical of globular clusters). The second case involves masses of 10^2 to 10^3 $M_\odot$ (typical of open clusters). Contracting clouds of these masses would ultimately break up into individual stars rather than a single short-lived superstar of, say, 10^4 $M_\odot$.

The American theorists P. Peebles and R. Dicke (1968) and their Russian colleague, T. Ruzmaikina (1972) have suggested that many globular clusters may have formed even before individual galaxies formed, pointing out that different types of galaxies are surrounded by swarms of globulars similar to those around our own galaxy. They suggest that objects of 10^5 M$_\odot$ may be among the first distinct objects to form in a hypothetical ancient universe of hot gas, and, depending on temperature, density, and cooling processes, these could break apart to form clusters, or collapse to form black holes. Thus, the globular clusters could date from the formative era of the universe. Other astronomers argue that the globular clusters formed slightly later, as the galaxies themselves were forming. The spherical halo of globulars around each galaxy could be the product of the first condensation within the huge clouds that eventually flattened and became galaxies. Both these theories are consistent with the great age of all globulars—around 12 billion years.

Figure 19·17 **Imaginary view from a planet about 60 parsecs from a globular cluster. In such a case, the globular cluster would dominate the sky. It may be unlikely for such a planet to circle a star within the cluster, since globular clusters are deficient** ▼ **in the heavy elements that form solid planets; however, such scenes may occur as globular clusters pass through the galactic disk near stars that may (hypothetically) have planets.** (Painting by Chesley Bonestell)

19

Summary

Although three types of star groups have been defined, they can be grouped in two main categories. The open clusters and associations are young groups of about 10 to 1,000 newly-formed stars, located in the galactic disk. In a second category are the globular clusters, which are old groups of 20,000 to 1 million stars; globulars are located within a spherical volume above and below the galactic plane, centered on the galaxy's center.

Star clusters played an extremely important role in mapping our galaxy and clarifying its history. They also led to the important discovery of two different populations of stars in our galaxy. Stars in the disk and in open clusters and associations have solar-type compositions, with a few percent heavy elements, a certain type of Cepheid variable, and other distinctive properties. Stars in globular clusters have very few heavy elements, a different type of Cepheid, and other distinctive properties. The importance of these properties of clusters will be clarified in the next chapter as we turn our attention to our galaxy as a whole.

Concepts

open clusters

associations

O and T association

globular clusters

parallaxes of clusters

Cepheid variable stars

H-R diagrams of clusters

ages of open clusters

disruption of open clusters

ages of globular clusters

globular cluster X-radiation

origin of clusters

Problems

1. Why are O and B stars the brightest in open clusters? Why are red giants the brightest stars in globulars?

2. If you saw the galaxy from a great distance, which would be brightest, open or globular clusters? Which reddest? Which farthest from the galactic disk?

3. Sketch the H-R diagrams of open and globular clusters and associations.

4. Describe a view of the sky near the center of a globular cluster.

Advanced Problems

5. If the Pleiades have 350 stars in a diameter of 4 parsecs, how many stars per cubic parsec are there? Roughly how far apart are the stars? Compare these numbers with those in the neighborhood of the sun. (Hint: Volume of a sphere $= \frac{4}{3}\pi r^3$.)

6. Assuming globular cluster M 3 has 200,000 stars in a diameter of 13 parsecs, make the same comparison with the solar neighborhood as in Problem 5.

7. If a telescope could resolve one second of arc, what would be the smallest details it could reveal in:

a. *an open cluster 1,000 parsecs away?*

b. *a globular cluster 10,000 parsecs away?*

8. If a star were in a circular orbit around an open cluster, what would be its orbital velocity? Assume a cluster mass of 300 $M_\odot$ (6×10^{35} g) and a radius of 3 parsecs (9×10^{18} cm).

a. *Make the same calculation for a star in circular orbit at the edge of a globular cluster with one million solar masses (2×10^{39}) and radius 3 parsecs.*

b. *How might such velocities be confirmed with a spectrometer?*

Projects

1. Observe the Pleiades with your naked eye and make a sketch. How many stars can you count in the group? Can you see all seven sisters? (The number of stars seen depends on keenness of vision, darkness of observing site, and clarity of atmosphere.)

2. Observe the Pleiades and Hyades or open clusters h and χ Persei in a telescope. Move the telescope and compare star fields in and out of the cluster. Estimate how many times more stars are in the cluster compared to the background region.

3. Locate a globular cluster with the telescope. Make a sketch. Can you resolve individual stars? Compare the view in the telescope with photos, where the central region is often overexposed and "burned out."

Galaxies

Geographers . . . crowd into
the edges of their maps parts
of the world which they do
not know about, adding notes
in the margin to the effect that
beyond this lies nothing but
sandy deserts full of wild
beasts, and unapproachable
bogs.

Plutarch, *Lives* (ca. A.D. 100)

The
Milky Way
Galaxy

Our exploration of space has taken us out to distances of a few thousand parsecs. By looking at the distribution of stars and clusters throughout volumes of this size, we begin to perceive the **Milky Way galaxy.** To a remote observer the Milky Way would be a disk with a central bulge. The disk is about 30,000 parsecs across, 400 parsecs thick, and packed with open clusters, individual stars, dust, and gas, mostly arranged in ragged spiral arms. Globular clusters are dotted around the disk in a spherical swarm, concentrated toward the center.

These distances are difficult to comprehend. In a model of the Milky Way galaxy the size of North America, stars like the sun would be microscopic specks less than a thousandth of a

centimeter across and scattered 150 meters apart. The solar system would fit in a saucer.

The view from earth is not from the outside, of course, but from the inside. From a point part way out in the disk, we see a band of unresolved, faint, distant stars when we look out along the plane of the disk. This is the Milky Way.

Discovering and Mapping the Galactic Disk

Even before the invention of the telescope people could plainly see a band of light arching across midnight skies at certain seasons.[1] Democritus (ca. 400 B.C.) correctly attributed this glow to a mass of unresolved stars, which came to be called the *Via Lactea,* or Milky Way. In 1610 Galileo turned his telescope on the Milky Way and confirmed Democritus' idea (quoted by Shapley and Howarth, 1929):

> . . . the galaxy is nothing else but a mass of innumerable stars planted together in clusters. Upon whatever part of it you direct the telescope, straightaway a vast crowd of stars presents itself to view. . . .

In 1750, the English theologian Thomas Wright correctly hypothesized that the galaxy must be a slab-like arrangement of stars. Other theore-

---

[1] The naked-eye prominence of the Milky Way is unknown to most modern urbanites. It can't be overemphasized that faint celestial displays must be viewed away from urban lights. My own astonishment at the Milky Way's clarity was greatest when I saw it from an altitude of about 3 km on a sparsely inhabited volcano on Hawaii. Ragged appendages and great dark rifts caused by dust clouds were easily visible. Such a view makes one believe we really do live in an immense, disk-shape system of dust, luminous nebulae, and distant stars!

▲
Figure 20·1 **Panorama of the Milky Way as seen in summer from the northern hemisphere. This view covers 140° of the Milky Way with a fisheye lens. The bright region lies toward the center of the galaxy** (cross, lower left). The constellation Cygnus (also called the Northern Cross) is right of center (galactic longitude about 70°). Part of the W-shaped constellation Cassiopeia can be seen at upper right (galactic longitude about 120°). (35 mm camera, 15 mm lens at f2.8, 30 min. exposure, 2475 Recording film; Michael Morrow and author)

ticians, such as Immanuel Kant, analyzed this idea more mathematically with Newton's laws in the 1750s and 1760s.[2]

---

[2]In an excellent, nontechnical account of these early theories, C. A. Whitney (1971) points out that Wright's work is now little remembered because it contained much metaphysical speculation. Whitney calls both Newton and Wright "astrotheologians. . . Newton wished to discover God by studying the universe; Wright wished to discover the universe by studying God."

Herschel's Star Counts

Around 1773, a German-born composer and musician in England, William Herschel, bought some astronomy books and began building his own telescopes. Caught up in the fascination of trying to learn what he called "the construction of the heavens," he finally shifted his career from music to astronomy, and within a few years built a telescope with a 48-inch mirror (a size not surpassed until the 1840s). Backed by what we would call federal support—an annual stipend from King George III beginning in 1782—Herschel used the following method to figure out "the construction of the heavens."

He swept the skies, counting stars in each direction. Next, he assumed that the fainter the stars he could see in any direction, the farther away they were. Naturally, his counts showed more stars of a given magnitude in the Milky Way than in other directions, as shown by Figure 20·2, so he

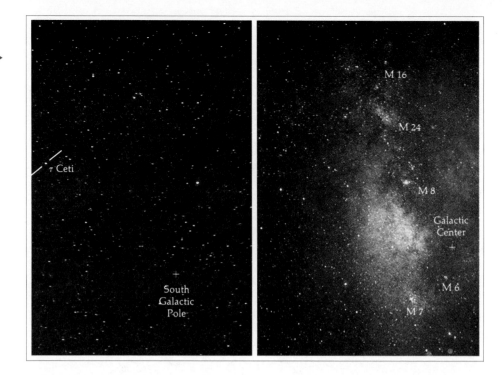

▶ *Figure 20·2* **A comparison of similar views toward the south galactic pole and galactic center, showing concentration of stars and clusters toward center. Left: Region of south galactic pole, also showing nearby solar-type star, Tau Ceti, 3.6 parsecs away. Right: Galactic center about 10,000 parsecs away, obscured by dust and gas. Five open clusters are identified.** (Both exposures: vertical height 30°; 35 mm camera; 135 mm telephoto lens at f2.8; 15 min exposure on 2475 Recording film; Michael Morrow and author)

correctly diagramed a flattened system. In 1783 he wrote:

That the Milky Way is a most extensive stratum of stars of various sizes admits no longer of the least doubt; and that our sun is actually one of the heavenly bodies belonging to it is as evident.

How big was the system? Herschel realized he could not see to the edge in all directions, but he did detect the "upper" and "lower" borders where the stars thin out only a few hundred parsecs from the sun, as shown in Figure 20·3. He also estimated that his 48-inch telescope could probe 2,300 times as far as the nearest stars, making his model at least 5,000 parsecs across. Today we know, as Herschel did not, that interstellar "smog" blocked his view toward the edges of the disk.

Our knowledge of the galaxy had not improved much beyond Herschel's by the early 1900s, as shown in Figure 20·4. Methods of measuring stars' distances had improved, but when Dutch astronomer J. C. Kapteyn tried to apply a method similar to Herschel's in 1918, he still did not know about the obscuring interstellar dust. When he looked around the galactic disk, he estimated that he could see no more stars beyond about 9,000 parsecs in any direction, and so concluded that

the sun was in the center of a disk or donut about 17,000 parsecs across.

Shapley's Studies of the Galactic Halo

The most important advance in the early 1900s came when astronomer Harlow Shapley and others used Henrietta Leavitt's 1912 discovery of a relation between period and luminosity of Cepheids to measure distances of clusters, as described in the last chapter. By 1918 Shapley had published an analysis of positions of globulars, showing that they formed a **galactic halo,** or spherical swarm of globular clusters "above" and "below" the disk, centered on a point in the direction of the constellation Sagittarius. He boldly and correctly hypothesized that *this distant point is the center of the whole galactic system, thus placing the sun well out toward one edge of the disk.* Since the effects of dust and the difference between globulars' Cepheids and open clusters'

Figure 20·3 **A cross section of star densities perpendicular to the plane of the Milky Way. The density of dots is proportional to the density of stars. The concentration of stars within a few hundred parsecs of the galactic disk is pronounced.** (Based on star-count data of Allen, 1973)

▼

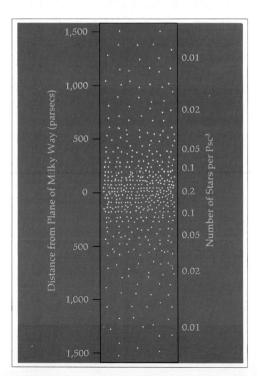

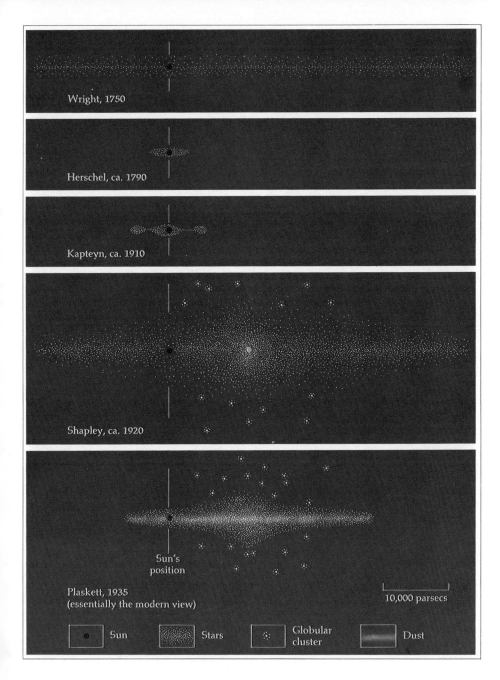

Wright, 1750

Herschel, ca. 1790

Kapteyn, ca. 1910

Shapley, ca. 1920

Sun's
position

Plaskett, 1935
(essentially the modern view)

10,000 parsecs

Sun   Stars   Globular
cluster   Dust

Cepheids were still poorly understood, Shapley overestimated the size of the galaxy by a factor of two or three.

Proof of Other Galaxies

Another important step was recognition of other galaxies like our own. Until 1924, many astronomers thought that certain disk- or spiral-shape glowing patches were nebulae of gas, like the Orion nebula. Many were incorrectly called "spiral nebulae." In the 1920s, however, the new 100-inch telescope on Mount Wilson in California produced photos showing that at least one of these objects, the *spiral-shaped object in the constellation Andromeda, consisted of individual stars, not gas.* Some of the stars were very faint Cepheids, which revealed that the Andromeda object was far beyond any stars in our own galaxy. The Andromeda object and others were soon recognized to be complete galaxies like our own. These discoveries allow us to perceive galactic shapes all at once, instead of probing through the murk from inside the system. As shown in Figure 20·5, photos of other galaxies seen edge-on revealed amazing similarity to our own view of our galaxy, seen edge-on from the inside.

*Correcting
the Galaxy's Size*

As described on p. 314, Trumpler discovered the effects of dust obscuration in the Milky Way about 1930. Astronomers quickly corrected distance estimates for this and other effects, and as early as 1935, many were using the **dimensions of the galaxy** now agreed upon.

According to current estimates, the

Figure 20·4 **Two centuries of conceptions of our galaxy, showing approximate relative sizes of edge-on galactic models, and the position attributed to the sun.** English theorist Thomas Wright speculated that the Milky Way was an indefinitely large layer of stars. English observer William Herschel noted that he could not see stars beyond a certain distance, and considered that he might have detected the nearer edges of the Milky Way's disk. Dutch astronomer J.C. Kapteyn correctly measured stars' distances, but not knowing of the effects of dust, he incorrectly put the sun in the center of the system. Harlow Shapley used star clusters to map the shape of the system, and placed the sun toward one edge, but got the size too big. Modern astronomers have corrected for effects of dust and differences in Cepheids in open and globular clusters, deriving the model shown at bottom.

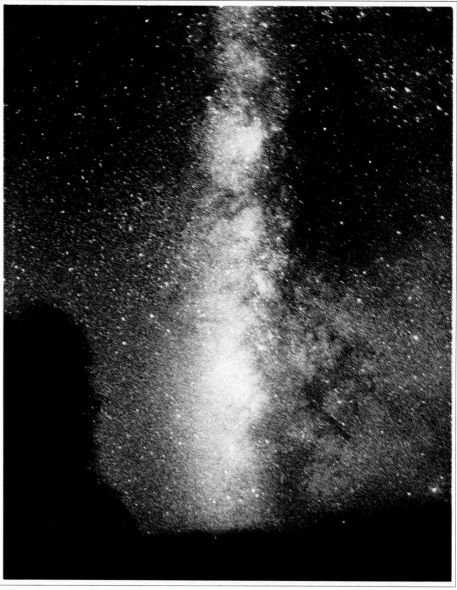

Figure 20·5 **Comparison of a wide-angle view** ▶ **of the Milky Way (A) with a telescopic view of a distant edge-on galaxy (B), showing similarity in form as an observer looks edgewise through a galactic disk. A. About 50° of the Milky Way near the central region in Sagittarius; compare with lower left of Figure 20·1. (35 mm camera; 24 mm lens at f2.8; 16 min. exposure on 2475 Recording film) B. A portion of the galaxy NGC 55. (Cerro Tololo Inter-American Observatory)**

sun is about 8,000 to 10,000 parsecs from the center, and the overall diameter of our galaxy is about 30,000 parsecs.

Galactic Latitude and Longitude

A galactic coordinate system has been defined to help describe the sky and galaxy as seen from earth. The **galactic equator** runs along the center of the Milky Way's band as it crosses the sky. The galactic north pole lies in the hemisphere of the sky containing the north star. **Galactic longitude**, designated l, measures the angular distance around the Milky Way, starting from a zero point defined to lie at the galactic center. The direction $l = 90°$ lies toward the constellations Cygnus and Cepheus, near the tip of the northern cross. Longitude $l = 180°$, away from the center, lies toward Taurus, Auriga, and Perseus, not far from the Pleiades and Hyades clusters; $l = 270°$, lies near Canis Major and Puppis. **Galactic latitude** is designated b. Its zero value lies in the Milky Way on the galactic equator.

The Rotation of the Galaxy

We have noted that cosmic systems of particles tend to become flattened

if they are rotating. The galaxy's flattened shape suggests that it, too, is rotating. All stars, including the sun, are in fact orbiting around the massive central bulge. But which way is the galaxy turning? Are stars in our area moving toward Cygnus or toward Canis Major?

To answer, we need some frame of reference outside the galaxy itself. The distant galaxies provide such a frame, and Doppler shifts reveal a systematic motion of the sun and nearby stars

toward Cygnus ($l = 90°$) and away from Orion ($l = 270°$). The velocity of the sun in this direction is about 220 to 250 km/sec. This velocity is the orbital motion of the sun and nearby stars around the galactic center.

If the sun travels at about 235 km/sec, then the **sun's revolution period,** or time required to travel all the way around our circular orbit, which has a radius of 9,000 parsecs, is about 230 million years.

Nearby stars are at nearly the same

A

B

together, but are ceaselessly changing their positions relative to each other as they move around the center.

Why Is the Galaxy Disk-Shape?

The galaxy's disk shape is tied to the rotation of the system, as was also the case with the disk-shaped solar nebula described in Chapter 11. As shown by the coffee-cup experiment described on p. 202, even a randomly circulating cloud is likely to develop some net rotation as it contracts, and this rotation will result in a flattened shape for the system as a whole. In the case of the coffee cup or pan of water, friction with the sides of the container eventually slows the circulation. But cosmic systems rotate in frictionless space and spin for billions of years, so that galaxies reveal the tell-tale signs of their original spin.

Mapping the Spiral Arms

Does our galaxy have spiral arms, like the Andromeda galaxy and some other nearby galaxies? The objects that need to be mapped in order to seek **evidence of spiral structure** in our galaxy lie in the disk, not in the halo. The easiest ones to map, because they can be seen over the largest distances, are the most luminous stars and the open clusters of luminous stars. Therefore, astronomers began mapping positions of bright, young stars of spectral types O and B, young clusters, and O associations. For example, if a cluster was found to be 1,000 parsecs away toward longitude $l = 100°$ and latitude $b = 0°$, it could be plotted on a map of the galactic plane.

distance from the center as we are, and so they are moving around the center at almost the same speed as our sun. Orbital speeds around the center are different at different distances from the center.

The outer stars of the galaxy show almost exactly Keplerian rotation, since most of the galactic mass is concentrated at the center of their orbits, as is the case in the solar system. However, since the main galactic mass is spread throughout the central bulge, stars inside the bulge do not move under the gravity of a single central mass, and their motions are not Keplerian.

Orbital velocities are actually slower at 5,000 parsecs from the center than at the sun's distance. This difference in speed at different distances is called **differential rotation** of the galaxy. Differential rotation, together with random motions of stars (typically about 20 km/sec) means that stars of the galaxy do not move smoothly

By the 1950s, maps showed that open clusters, O and B stars, nebulae, and other objects lay in bands interpreted as local pieces of our galaxy's **spiral arms,** as shown in Figure 20·6.

Three arms were discerned, lying at an angle to the sun—center line. Such early data, at first controversial, were confirmed by later studies, such as those summarized in Figure 20·7.

The spiral arms sketched in Figures 20·6 and 20·7 are named for the constellations that contain prominent features in each arm. The next arm beyond us (sometimes designated "+1") is called the Perseus arm. Our arm (sometimes designated "0") is the Orion arm, or sometimes the Cygnus arm. The next arm in toward the center ("−1") is the Sagittarius arm, and a suspected arm beyond it ("−2") is the Centaurus, or Norma-Centaurus arm.

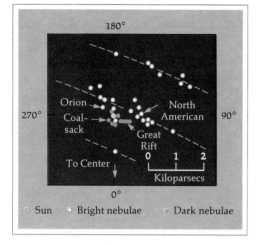

◄ *Figure 20·6* **Early evidence of spiral arms in the Milky Way. The diagram shows positions of bright and dark nebulae within a few kiloparsecs of the sun, plotted on the galactic plane as seen from a position north of the plane. Galactic longitude is shown at the edge.** (From the work of W. Morgan and associates at Yerkes Observatory in the early 1950s)

Figure 20·7 **Modern data on spiral arm features, plotted as in Figure 20·6. Clusters, associations, HII gas clouds, and young stars are concentrated in spiral arms around a distant center toward *l* = 0°.** (Adapted from data of Klare and Neckel, 1970; Moffat and Vogt, 1973; Walborn, 1973)
▼

The View from a Spiral Arm

The maps show us just on the inner edge of a spiral arm. On an evening with a clear sky, rural observers have a commanding view of the galaxy. If you look at the brightest parts of the Milky Way in the southern summer sky,[3] you are looking toward the center of the galaxy, or *l* = 0°, between the constellations Sagittarius and Scorpio. The true center itself is hidden behind 9,000 parsecs of intervening dust and gas.

Higher in the sky, stretching toward Cygnus, the Milky Way is divided by the Great Rift, a band of nearer dark dust clouds lying in the plane (shown in Figure 20·1, and in a more detailed view in Figure 20·8). Overhead on a summer evening, the bright star clouds around Cygnus (sometimes called the Northern Cross) mark our view down our own spiral arm, about 70° from the center. This is somewhat short of the *l* = 90° longitude because the arm is spiral, winding in toward the galactic center in this direction.

Because we live on the inner edge of a spiral arm, our view away from the galactic center toward *l* = 180° directly

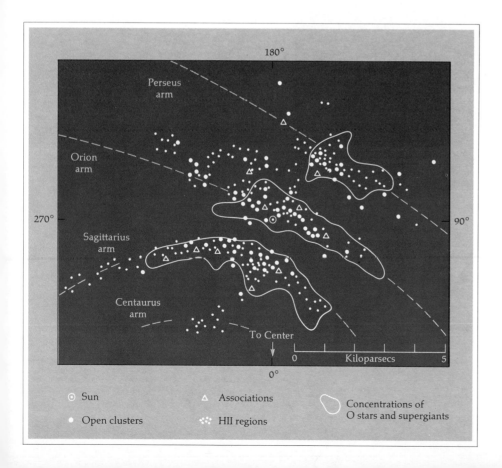

---

[3]The seasonal and directional references are for northern hemisphere observers.

Figure 20·8 **This 60-degree view of the summer Milky Way shows the dark rift in the region of Cygnus (Northern Cross, outlined), caused by obscuring dust clouds between us and the distant parts of the Milky Way. Names identify bright stars comprising the "summer right triangle." Compare with the central part of Figure 20·1. (35 mm camera; 24 mm lens at f2.8; 16 min. exposure on 2475 Recording film)**

crosses our arm. Among features in this direction are the open clusters of the Pleiades and Hyades—regions of recent star-formation within our arm, shown in Figure 20·9. The concentration of bright, bluish stars in the whole region from the Pleiades through Orion to Sirius helps convince us that we are looking into a dense star swarm—our spiral arm—in this direction.

A more spectacular view can be seen on a February evening when Orion is high in the southern sky. When you look toward the brightest star, Sirius, and the nearby Orion grouping of bright stars and nebulae, you are again looking down the axis of our spiral arm (Figure 20·9). Here the arm is winding outward, and the longitude lies in the range $l = 220°$ to $250°$. The central Orion nebula and star associations are located about 500 parsecs away, down the arm.

*Mapping Distant
Hydrogen Clouds by Radio*

All the mapping just described reveals only the closest parts of nearby

Figure 20•9 A panorama of the winter Milky Way, looking away from the galactic center. Sirius and the Orion star-forming region lie in the direction looking down our local spiral arm. The Hyades and Pleiades clusters are in our arm in a direction opposite the center. The W-shaped constellation Cassiopeia lies to the right. (35 mm camera; 15 mm fisheye lens at f2.8; 25 min. exposure on 2475 Recording film)

spiral arms of our galaxy. What about the rest of the galaxy? Is there any way to map the spiral features farther around the disk, or on the other side of the center? Visual light is useless, because we can see only a few thousand parsecs through the interstellar dust. But radio waves pass much farther through the interstellar dust. The **21-cm radio waves** produced by neutral hydrogen, or HI, are especially useful for galactic mapping, because they allow us to detect HI clouds, which are concentrated in the spiral arms.

Suppose we start scanning along the Milky Way with a radio telescope tuned to the 21-cm wavelength. If we find a strong signal near, say $l = 30°$, as in Figure 20•10, a concentration of HI must lie in this direction. To plot it on a map we need its distance. We don't know whether it is nearby (Cloud A) or far away (Cloud B). Once again, the Doppler shift comes to the rescue (Figure 20•10). All clouds lying in the disk are revolving around the center, and objects nearer the center move somewhat faster than objects near the sun. Thus (as shown in Figure 20•10), the velocity of Cloud B would be greater than that of Cloud A. (And the *apparent* difference would be even greater than the actual difference between the orbital velocities, because V_B is directed along the line of sight. V_A, on the other hand, is not along the line of sight, so that terrestrial observers see only one component of the total orbital speed.)

Radio astronomers have used theoretical laws of dynamics, and observations of objects in our galaxy and in other galaxies to estimate the rotation rates in our galaxy as a function of distance from the center. These models give the velocity V_B of a cloud at

position B, or V_A for a cloud at position A, and velocities for objects at all other positions. Thus, once the cloud's velocity is measured, its position can be plotted. Such techniques allow mapping of clouds in the galaxy.

HI clouds have been mapped over a large part of the galaxy. Their positions define, in some detail, what look like spiral arms. Some small details may be falsified by random motions that have been mistaken for orbital velocities, and by imperfections in galactic rotation models. However the main features revealed by hydrogen-mapping, shown in Figure 20·11, probably offer a true view of our galaxy's general shape.

That very little hydrogen is far "above" or "below" the galactic plane again confirms that most of the galaxy's material is concentrated in a thin disk only about 200 to 400 parsecs thick. Only a few thin clouds have been found at distances of about 8,000 parsecs above or below the disk, some falling toward the disk and some moving away. They might be material dragged out of the disk by passing globular clusters, or blown out of the chaotic central regions.

Radio astronomy roughly triples the distances to which we can probe in the galactic disk. We can map structures that lie some 15,000 parsecs away, even farther than the galactic center. Nonetheless, there are regions directly on the far side of the center that we cannot map. It remains difficult to tell exactly how tightly wound the arms are, or where the Milky Way falls in the range of forms of other galaxies.

Why Does the Galaxy Have Spiral Arms?

The origin of the spiral arms is not well understood. Common sense might again suggest analogy with a stirred cup of coffee and a drop of cream. Because the coffee surface rotates faster at the center than at the rim, which slows it, the cream is sheared into long spiral streamers that look like the galactic arms. The trouble is that this idea predicts the arms should be nearly fixed structures almost as old as the galaxy itself, whereas actually they are young regions only a few million years old.

Another common-sense model would be the rotating garden-sprinkler, where sprayed water assumes spiral-arm forms. Matter thrown out of the galaxy's nucleus in directed jets lying

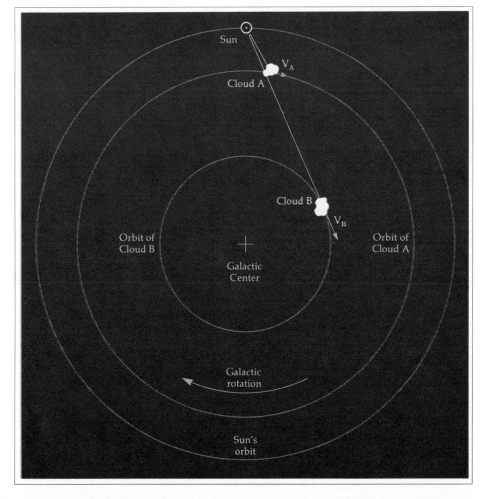

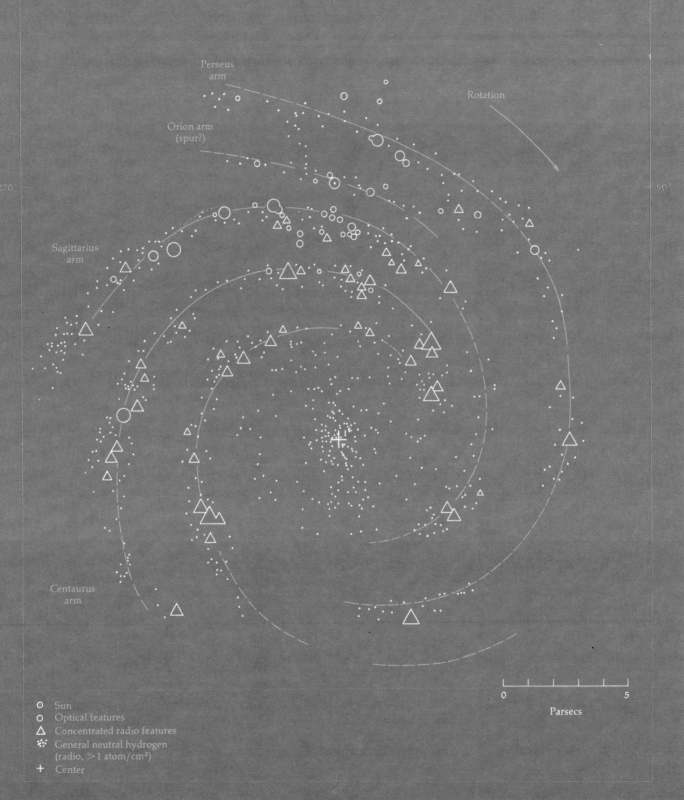

Perseus
arm

Rotation

Orion arm
(spur?)

270°   90°

Sagittarius
arm

Centaurus
arm

180°

○   Sun
○   Optical features
△   Concentrated radio features
✳   General neutral hydrogen
    (radio, > 1 atom/cm³)
+   Center

0   5
Parsecs

◄ *Figure 20·11* **Currently known or estimated features of the Milky Way galaxy, viewed from the north. Features nearest the sun (☉) are the most certain; features on the far side are unmapped.**

in the galactic plane could look like spiral streamers. The trouble is that this model predicts the material must be moving outward, and no such movement has been observed among most spiral-arm material.

Another fatal objection to these two hypotheses is that the **age of the galaxy** is about 12 billion years—time for 52 rotations taking 230 million years apiece; thus if the arms were identified as specific clots of sheared material, they should show some 50 complete windings around the center, instead of one, two, or three windings as observed! Conversely, during the few-million year history of the young material in the arms, the galaxy turns only a few degrees—not enough to cause much of a spiral pattern.

A subtler, more plausible explanation, the **density-wave theory,** explains the galaxy's spiral arms as waves in the galactic material. The crest of an ocean wave consists of certain molecules at one moment and certain other molecules the next, yet an outside observer sees a single wave that seems to have a history of its own. Just so, spiral arms may be persistent concentrations of material, with individual stars entering an arm, passing through, and finally emerging on the other side.

Why should such patterns arise at all? The answer involves subtle effects of different periodicities in the orbital motion of material around the galactic center. Consider stars, dust, and gas near the sun. One obvious period associated with this matter is the 230-

▲
Figure 20·12 **Spiral arms are prominent in some neighboring galaxies, such as M 51 (NGC 5194), as well as our own. These arms are marked especially by bright clusters and associations of recently formed hot stars. This pattern is similar to that in our galaxy, though some astronomers believe the Milky Way's arms are somewhat more tightly wound.** (Hale Observatories)

million-year orbital period. The sun would come back to its starting point every 230 million years, but material slightly inside or outside the sun's orbit, traveling either faster or slower than the sun, would not pass close to the sun every 230 million years, but at a different time. Thus, there is a second periodicity describing how often neighboring particles in certain orbits would pass near each other, causing a pattern of greater concentration of particles. Theorists have found that these two periodicities can combine to produce a third period, which describes the rotation of the *pattern* which comes into being by this process. This pattern is believed to be the spiral arm pattern.

In 1959 the Swedish astronomer Bertil Lindblad discovered that this third periodicity has a constant value

of around 480 million years throughout the region from about 4,000 to 15,000 parsecs from the center—just the region where the spiral arms seem pronounced. An outside observer would thus see the spiral pattern rotating in 480 million years, even though the pattern is created by stars that move around the galaxy at different rates. The idea is thus similar to the ocean wave, which moves with a certain speed, even though it is composed of constantly changing groups of molecules which move at entirely different speeds and do not travel along with the wave itself.

In the spiral-shaped wave, or arm, the gas is about 10 percent denser than in the regions between the arms. Thus, stars form primarily in the arms, which, in turn, defines the spiral pattern even more clearly. The higher density also helps stabilize the pattern because of the increased gravitational attraction toward the arms. Because the spiral pattern rotates only once each 480 million years, whereas our sun and nearby stars circulate in about 230 million years, the sun, stars, gas, and dust overtake and pass through the arms.

Measuring the Galaxy's Mass

Knowing the sun's velocity around the galactic center (about 230 km/sec or 2.3×10^7 cm/sec) and its distance from the center (about 9,000 parsecs, or 2.7×10^{22} cm), we can calculate the amount of mass in the central bulge around which we are orbiting. This is a simple application of the circular velocity equation (Chapter 4, p. 73), giving a result of 2×10^{44} grams, or 10^{11} solar masses.

Because most of the mass is located in the central bulge and little lies outside the sun's orbit, this figure can be considered as approximately the whole **mass of the galaxy.** Since the most common stars have about one-half solar mass, there must be about 2×10^{11}, or 200 billion, stars in the Milky Way galaxy! We live in an enormous star-system!

Comprehending Galactic Distances

We have been casually discussing distances of thousands of parsecs, but it is good to pause and consider what these distances mean. Table 20·1 lists some distances to various objects in our own galaxy. To put them in more meaningful terms, the right column lists the distances in light-years—the time it would take us to reach these objects if we could travel at the speed of light (or conversely, the time it takes their light to reach us). While we could reach the nearby stars in a few years, we would have to build ships to accommodate several generations of travelers to reach well-known open clusters. To reach distant spiral arms, globular clusters, or the galactic center would require more than the entire recorded history of humanity so far! If twentieth-century physics is correct in saying that no matter or energy can ever travel faster than the speed of light, then prospects for astronautical exploration of the rest of our galaxy seem indeed dim.[4]

---

[4]But maybe twentieth-century physics is not the last word. After all, anybody in any other century who claimed to know the ultimate truth about physics and astronomy would later have been proved wrong.

The Kiloparsec

When we consider objects in distant parts of our galaxy, we are dealing with distances of thousands of parsecs. For this reason, astronomers have added a unit of distance still larger than our earlier units of Astronomical Units, light-years, and parsecs. This new unit is a **kiloparsec,** sometimes abbreviated kpc:

1 kiloparsec = 1,000 parsecs

Galactic astronomers say, for example, that we are 9 kiloparsecs from the galactic center, and that our galaxy is roughly 30 kiloparsecs in diameter.

The Two Populations of Stars

One of the most startling discoveries about our galaxy is that it contains star types of different composition, age, distribution, and orbital geometry. These are commonly divided into two major groups called **Populations I and II.** Population I stars are of solar composition, have relatively young ages, and are distributed in near-circular orbits in the galactic disk. Population II stars are nearly pure hydrogen and helium with no heavy elements, are old, and are associated with globular clusters that have orbits taking them

---

Table 20·1 Distances to Selected Destinations in Our Galaxy

Destination	Distance (parsecs)	Travel Time (at speed of light)[b]
Nearest stars	1.3	4.2 yrs
Sirius	2.7	8.8
Vega	8.1	26
Hyades cluster	41	134
Pleiades cluster	125	411
Central part of our spiral arm (Orion arm)	400	1,300
Orion nebula	500	1,600
Leave galactic disk in Z direction (perpendicular to plane)	1,000	3,300
Next-nearest spiral arm (Sagittarius arm)	1,200	3,900
47 Tucanae globular cluster	4,600	15,000
Center of galaxy	9,000	29,000
M 13 globular cluster[a]	11,000	36,000
Far edge of galaxy	24,000?	78,000?
Full diameter of galaxy	30,000?	98,000?

[a]Target of first beamed radio transmission from the earth directed to hypothetical extraterrestrial civilizations, November 1974. Signal will reach the cluster in A.D. 38,000.
[b]These numbers, by definition, also equal the distance as expressed in light-years.

far above and below the galactic plane. You can recall the two types more easily by remembering that Population I stars were the first group we became familiar with, since they are the type located near the sun; Population II stars were discovered *second*.

Evidence for Two Populations

Chemistry In the last chapter we noted evidence that stars in globular clusters have fewer heavy elements than stars in the disk. Figure 20·13 compares the spectra of Population I and II stars. While the hydrogen absorption lines are of equal strength, the absorption lines of metals and other heavy elements are missing in the Population II star's spectrum.

Two Types of Cepheids and Supernovae The Russell-Vogt theorem of Chapter 14 (p. 262) stated that the configuration of a star depends on its mass and composition. If Population II stars have a different composition, they must evolve through different configurations. This explains some facts we have already encountered.

For example, there are two different types of Cepheid variable stars, as mentioned in Chapters 16 and 19. Population I Cepheids are in the disk, and Population II Cepheids, with slightly different period–luminosity relations, are primarily in globular clusters.

A related type of variable star similar to the Cepheids, called an **RR Lyrae star,** has been found associated with Population II and not Population I. Similarly, two types of supernovae, with slightly different rates of brightening and fading, have been found in the two populations. All these differences are believed to be the result of parallel evolution occurring in stars with different compositions. The lack of heavy elements in Population II stars changes the opacity of the layers of gas in the star, as compared to Population I, and thus changes the rate at which radiation flows from the evolving core to the outer layers, producing different behavior in variables and supernovae.

Ages As noted in the last chapter, Population II stars of globular clusters formed mostly about 12 billion years

ago, while stars of Population I have formed more recently, and are still forming.

Motions and Orbits One of the earliest lines of evidence for a second population of stars came from studying motions of stars in our own part of the galaxy. The local stars (mostly Population I) can be thought of as a swarm having near-random velocities of about 20 km/sec, but all moving together around the distant galactic center at a speed of about 230 km/sec. We might represent this by analogy with a swarm of gnats randomly darting among each other at 20 km/sec, while the whole swarm is moving around a distant center at 230 km/sec.

As early as 1785, William Herschel discovered the random motions of stars in our local swarm and measured the direction of the sun's motion in the swarm, relative to the average of all other motions. This drift of the sun with respect to the local swarm is called the motion of the sun toward the **solar apex,** a point in the sky about 10° from the star Vega. The sun is moving in that direction at 19.7 km/sec. In the gnat analogy, if you

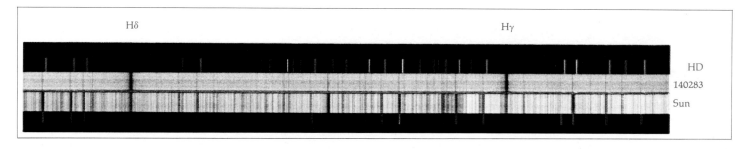

Hδ   Hγ

HD 140283

Sun

▲
Figure 20·13 **Comparison of spectra of a Population II star (upper, HD 140283) and Population I star (lower, the sun) of similar type show strikingly different patterns. (Bright lines at top and bottom are matching** comparison spectra, produced in the laboratory.) Aside from prominent absorption lines of hydrogen (Hδ and Hγ), the sun has many additional absorption lines caused by various heavy elements. In the Population II star these lines are very weak or absent, indicating that the heavy elements are virtually absent from its gases.

(representing the sun) stood still in the middle of the swarm of gnats (stars), you would see them darting about in random directions. But if you started walking toward the east (solar apex), you could detect your motion by a net drift of gnats past you toward the west (away from the solar apex).

Once these measures were made, it was natural to measure other stars' velocities relative to our **local standard of rest,** or **LSR**—an imaginary point in the middle of our local swarm, moving in a circular orbit around the galactic center. The velocity of a star relative to the LSR is called its **peculiar velocity.** Measurement of peculiar velocities revealed Population II stars. Since Population I stars move in circular orbits with the LSR, they have

low peculiar velocities—random speeds of about 20 km/sec. But as shown in Figure 20·14, Population II stars do not share the circular motions in the disk, but have highly inclined elliptical orbits. For this reason, astrometric astronomers in the early 1900s began gathering evidence of a few stars speeding through the solar neighborhood with peculiar velocities sometimes exceeding 120 km/sec. Prominent among these so-called **high-velocity stars** were the RR Lyrae variable stars now known to be associated with Population II. Soon astronomers realized that it is we (the sun and LSR) who are moving around the galactic

Figure 20·14 **Comparison of orbits of disk Population I and halo Population II stars.**
▼

center at 230 km/sec, and the so-called high-velocity stars that are lagging behind on their inclined, elliptical orbits.

Mapping and
Subdividing the Populations

By the 1930s, Harlow Shapley had shown that Population II consists of a diffuse swarm of individual stars and globular clusters orbiting in the regions near the galaxy, but usually outside the disk. In their journey around the galactic center they pass through the disk twice on each orbit.

Figure 20·15 shows a simple demonstration of the populations by plotting the distributions of different types of galactic objects on maps of the sky. Galactic coordinates are used, so that the Milky Way runs horizontally across the diagram with the galactic center in the middle. As expected, we see that the open clusters and dust clouds cling tightly to the galactic equator since they belong to the disk's Population I, while the globulars, in Population II, swarm around the galactic center.

The maps in Figure 20·15 reveal that many objects do not share the extreme properties of "pure" Population I or II. Therefore, as shown

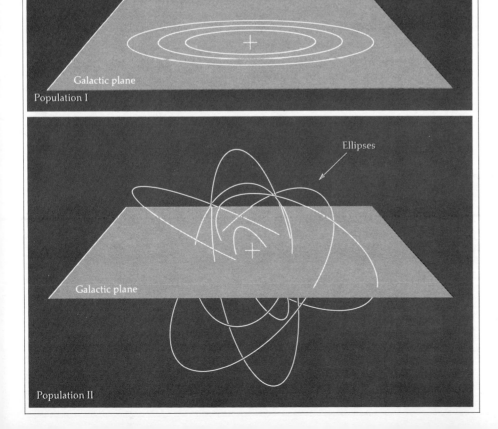

Galactic plane
Population I

Ellipses

Galactic plane

Population II

Figure 20·15 **Maps of the sky in galactic** ▶
coordinates, with the galactic center in the middle of the horizontal Milky Way plane. Maps show different populations of galactic material, starting with Population I and grading into Population II. Population I material, such as young clusters, associations, dust, and the visible stars and nebulae are confined chiefly to the plane of the disk. Intermediate material, such as radio-mapped gas, planetary nebulae, dwarf novae, and X-ray sources are less concentrated. Population II globular clusters are highly scattered.

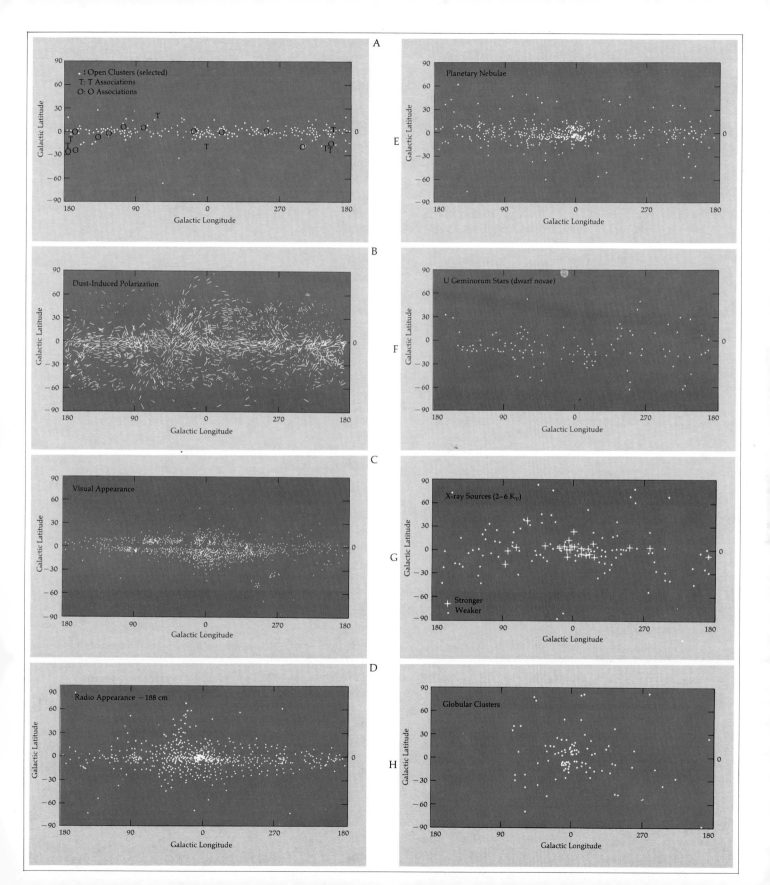

A

: Open Clusters (selected)
T: T Associations
O: O Associations

B

Dust-Induced Polarization

C

Visual Appearance

D

Radio Appearance — 188 cm

E

Planetary Nebulae

F

U Geminorum Stars (dwarf novae)

G

X-ray Sources (2–6 K$_v$)

+ Stronger
• Weaker

H

Globular Clusters

	Extreme Population I	Intermediate Populations			Halo Population II
		Older Pop. I	Disk Pop.	Intermed. Pop. II	
Orbits	Circular	Somewhat elliptical			Very elliptical
Distribution	Patchy, spiral arms	Somewhat patchy			Smooth
Concentration toward galactic center	None	Slight			Strong
Typical Z range[a]	120 parsecs	400 parsecs			2,000 parsecs
Heavy elements[b]	3–4%	0.4–2%			0.1%
Total mass	$2 \times 10^9 \, M_\odot$	$5 \times 10^{10} \, M_\odot$			$2 \times 10^{10} \, M_\odot$
Typical ages	10^8 yr	10^9 yr			10^{10} yr
Typical peculiar velocities	10–20 km/sec	20–100 km/sec			120–200 km/sec
Typical objects	Open clusters Associations Gas and dust HII regions O & B stars	Sun RR Lyrae stars (P<0.4 days) A stars Planetary nebulae Giant stars Novae Long-period variables			Globular clusters RR Lyrae stars (P>0.4 days) Population II Cepheids

Table 20·2 Stellar Populations and Their Properties

[a]Z = distance "above" or "below" galactic plane.

[b]Elements heavier than helium, sometimes loosely referred to as "metals."

Source: Data based on tabulations by D. O'Connell, A. Blaauw, J. Oort, C. Allen, and other sources.

in Table 20·2, astronomers have subdivided the populations into intermediate subclasses, attempting to define a sequence that shows the relation of all types of stars, nebulae, and so on.

Chemistry, Ages, and Origin of Populations

The **composition of Population II stars** *shows virtually no elements heavier than hydrogen and helium.* Yet such elements comprise about 2 percent of the sun and other Population I stars, which include such important planet-forming elements as silicon, oxygen, nickel, and iron. If the surface layers of Population II stars have never been heated to the high temperatures of stellar centers, they have not been much altered from their initial composition. Therefore, astronomers believe that *Population II stars formed from gases with virtually no heavy elements.* Thus, the parent gases of Population II must have been a nearly pure mixture of 75 percent hydrogen and 25 percent helium. This conclusion is especially provocative in view of the fact that Population II objects are older than Population I objects.

What do these facts mean? Astronomers have agreed on the logical implication: When Population II formed, the galaxy (or protogalaxy) consisted of about 75 percent hydrogen, 25 percent helium, and virtually no heavy elements. Some 12 billion years ago the protogalactic cloud must have been an extended spheroidal mass of almost pure hydrogen and helium, matching the volume now occupied by the globular clusters and other Population II objects.

The conclusion that the protogalactic cloud was nearly all hydrogen, and that later stars have more heavy elements, has a striking connection with the subject of element-formation inside stars, described in Chapter 14. Nuclear reactions inside stars fuse light elements into heavier ones through successive stages of stellar evolution. Heavy elements are therefore *created* inside stars and then blasted by disruptive events (such as supernovae explosions) from old stars into interstellar space. Then later generations of stars form from this heavy-element-enriched interstellar gas, and create still more heavy atoms. *Interstellar nebulae and later generations of stars thus continually increase in content of heavy elements.*

The two sets of data fit beautifully together: Observations show that the recently formed population contains many more heavy elements than the older population; and theories of stellar interior processes show why this is so. Observation of star populations thus confirms theories of stellar evolution.

Probing the Galactic Center

Does one part of the galaxy hold a key to the structure and origin of the rest? Observations suggest that the center is the critical region. Much of the light from many distant galaxies, for example, comes from a brilliant central core, or **galactic nucleus.** In the nearby Andromeda galaxy, which resembles the Milky Way, the nucleus, though as small as 5 parsecs across, far outshines the rest of the galaxy.

Figure 20·16 **A 50°-segment of the Milky ▶ Way in the direction of the galactic center. The box shows location of the center and shape of the region mapped in Figure 20·17.** (Hale Observatories)

Figure 20·17 **A map of the galactic center region as observed by radio detectors sensitive at 21-cm wavelengths. Similar detail at similar wavelengths has been found by several researchers. Sagittarius A, probably marking the galactic nucleus, is about 10 parsecs across.** (Adapted from Kapitzky and Dent, 1974)
▼

Does the Milky Way have an energetic nucleus like that in the Andromeda galaxy? The answer appears to be yes. We can't see it, and photo-

graphs of the center (such as Figure 20·16 don't show it, because dust blocks visible light from the central regions, 9 kiloparsecs away. But in-

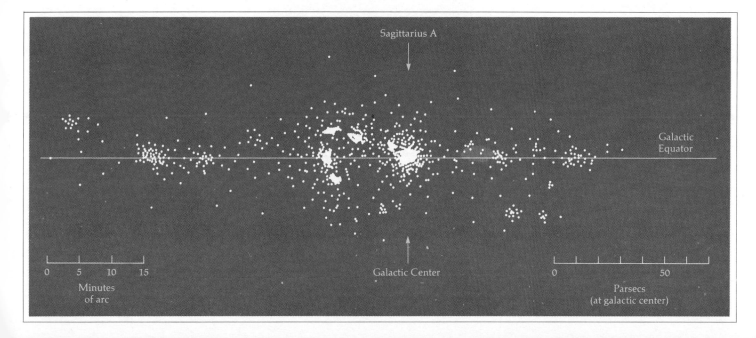

Sagittarius A

Galactic Center

Galactic Equator

0   5   10   15
Minutes of arc

0   50
Parsecs (at galactic center)

frared and radio radiation pass through the dust, bringing information from the hidden galactic center. Infrared and radio astronomers scanning the central region in 1960 found a strong infrared/radio source, known as Sagittarius A, just at the position predicted for the galactic center (Figure 20·17). This object is roughly 10 parsecs across, with a brighter core, and thus resembles the Andromeda nucleus. It is believed to contain some 60 million stars together with heated HII regions and dust. It must be the Milky Way's nucleus (Oort and Rougoor, 1960).

Infrared radiation from hot interstellar dust covers a region of about 35 × 80 parsecs around Sagittarius A. The region around the nucleus is rich in molecules such as OH and NH_3. Nonthermal radiation from a larger region (some 150 by 300 parsecs) is presumed to be synchrotron radiation, resulting from interaction of ionized gas and strong magnetic fields. A magnetic field of about 10^{-5} gauss has been estimated from these observations of the central region—some two to five times stronger than the interstellar field in the solar neighborhood.

Detailed radio and infrared observations in the early 1970s detected a core within the nucleus about one parsec across. It is believed to contain at least five major dust clouds and 1–2 *million* solar masses of Population II stars.

Figure 20·18 summarizes some properties suggested for the galactic

center. The 10-parsec central nucleus may contain 10 times as many stars as a similar-size globular cluster. Thus, such exotic interactions between stars as mass transfer or catastrophic collisions might sometimes occur.

Evidence for High-Energy Events in the Galactic Nucleus

Many observations indicate that the galactic nucleus is releasing prodigious amounts of energy. For example, a mass of hydrogen that resembles a spiral arm is expanding from the nucleus at a distance about 3,000 parsecs from the center. This cloud, called the **3-kiloparsec arm** (or 3-kpc arm), contains about ten million solar masses of neutral hydrogen. It revolves around the center at about 210 km/sec, but also rushes outward toward the sun at about 53 km/sec. Thus, to set this mass of gas in motion required about 10^{53} ergs of energy—over a thousand times

the total lifetime energy production of the sun! Yet, the 3-kpc arm is only about 100 million years old. Apparently, a large energy release occurred recently.

Radiation from the center is further evidence for energy production there. Though we cannot see the nucleus, infrared and radio astronomers observe it as extremely bright. Much infrared radiation, for instance, comes from the hot dust clouds.

Comparison with other galaxies is a third kind of evidence that our own galaxy's nucleus is energetic. Earlier we said that the Andromeda galaxy, structurally similar to ours, has a 5-parsec nucleus that outshines the rest of the galaxy. It probably contains many millions of stars. But other unusual galaxies have even brighter and more exotic nuclei, which fluctuate in brightness. Strong radio radiations from these galaxies give evidence of occasional explosions emitting energies comparable to the 10^{53} ergs needed for our own 3-kpc arm.

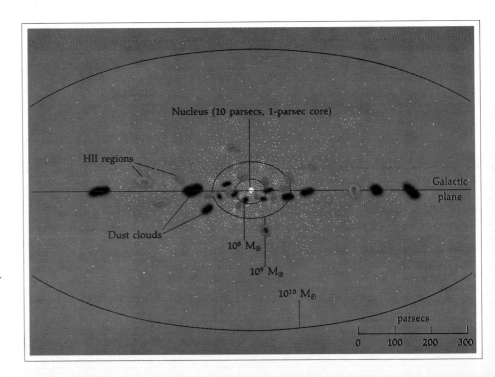

Figure 20·18 **A schematic composite map of features near the center of the Milky Way. Contours contain the stated numbers of stars, indicating the highest densities of stars, dust, and gas clouds near the center.**

*Energy Sources
in the Galactic Center*

What **conditions in the galactic center** could have produced the energy needed for such events? No one really knows, but there are several theories.

Figure 20·19 **Our galaxy as it might be seen from the surface of a planet about 30 kiloparsecs north of the galactic disk, giving a view of the central region and spiral arms. View is based on radio maps, as is Figure**
▼

One promising theory is that enormous, unstable "stars" of, say, 500 or even 1,000 $M_\odot$ may form from the dense dust and gas. As we saw in Chapter 16, such huge bodies would be very unstable and would soon blow themselves apart. Observations of HII

20·11. **(Few planetary systems are likely to have formed in this Population II region, but some may have been expelled into this region by gravitational encounters as globular clusters passed through the disk.)** (Painting by author)

regions and possible giant stars containing heavy elements in these regions also suggest current star-formation. If ongoing gravitational contraction of massive clouds occasionally produces unstable masses, enough energy might be released to create the expanding hydrogen arm. According to one study, the 3-kpc arm may have originated around 67 to 130 million years ago in the last of a series of explosions of enormous short-lived objects of roughly $4 \times 10^8 \, M_\odot$ (Sanders and Prendergast, 1974). These would explode with as much as 10^{59} ergs of energy—equal to the radiation of all stars in the galaxy in ten million years!

A second theory is that similar explosions might occur when Population II stars, in their eccentric orbits, pass through the crowded central regions and collide with stars there.

A third theory is that dust may fall onto high-density, high-gravity objects, causing energetic radiation like that produced when material is transferred to dwarfs or neutron stars in binary systems. According to some models, mass blown off thousands of giants in the central regions may collect in a central layer about 0.1 parsec thick. Italian astrophysicists L. Maraxchi, A. Treves, and M. Tarenghi (1973) estimate that stellar evolution may have produced 20 million neutron stars within 350 parsecs of the galactic center, and infall of dust and gas onto these may have generated about 10^{51} ergs of energy in the lifetime of the 3-kpc arm.

Clearly, astronomers are still groping for explanations of what is happening in the center of the Milky Way.

20

Summary:
Evolution of the Galaxy

Our discussion in the last few chapters can now be fitted into a compelling theory of the origin and evolution of the galaxy. Globular clusters give us the best estimate of the age of the galaxy—about 12 billion years. An independent method based on decay of radioactive elements sampled in the earth gives a similar result (Dicke, 1974).

What was happening 12 billion years ago? Evidently the galactic mass was a single cloud of hydrogen-rich gas, with probably 25 percent helium (by mass) and virtually no heavier elements. The mass of this protogalaxy equaled the present galactic mass, roughly 2×10^{11} solar masses. The protogalaxy must have been rotating and contracting. When it reached a rather spherical shape perhaps 30 to 40 kiloparsecs across, it became dense enough that separate, large, self-gravitating masses formed within it. These became globular clusters with certain characteristics we observe today:

age: *about 12 billion years*

composition: *few heavy elements*

distribution: *spheroidal halo 30–40 kpc across*

mass: $10^5–10^7 \ M_\odot$ *each*

The inner part of the protogalaxy, still containing some $10^{11} \ M_\odot$, continued to contract. Because of rotation and collisions among the atoms and clouds of gas, it flattened into a disk shape. The globular clusters, too far apart to interact, were left behind and did not form a disk. Thus, two populations (with some intermediate objects) arose from the earlier- and later-formed systems.

No one knows what was happening in the center, although violent explosions may have begun taking place. In the disk, nebulae traveling on elliptical orbits quickly collided with other nebulae until the gas, dust, and nebulae all moved together in relatively circular orbits. The spiral arm pattern probably emerged after a number of galactic rotations, perhaps within a billion years. In the spiral arms, the densest clouds contracted and spawned associations and open star clusters. Each group broke apart into scattered stars a few hundred million years after its formation, but new star-groups continued to form, so that the galaxy kept its present general appearance. Perhaps 7 or 8 billion years after the galaxy's formation, in one of the spiral arms, an obscure star formed—and in its surrounding dusty nebula, the earth was born.

Concepts

Milky Way galaxy

galactic halo

dimensions of the galaxy

galactic equator

galactic longitude and latitude

sun's revolution period

differential rotation

evidence of spiral structure

spiral arms

21-cm radio waves

age of the galaxy

density-wave theory

mass of the galaxy

kiloparsec

Populations I and II

RR Lyrae stars

solar apex

local standard of rest (LSR)

peculiar velocity

high-velocity stars

composition of Population II stars

galactic nucleus

3-kiloparsec arm

conditions in the galactic center

Problems

1. During what percent of human recorded history (define as you think appropriate) have people *not* known that we live in an isolated galaxy of stars similar to other remote galaxies?

2. From the appearance of the Milky Way, how do we know that the solar system is in the disk, and not far above or below it? How might the appearance of the central region differ in the latter case? (Hint: See Figure 20·18.)

3. Compare the shapes of the volumes occupied by the swarm of open clusters and by globular clusters. Relate the difference to differences in stellar populations.

4. How do we know the size of the galaxy?

5. How do we know the location and distance of the galactic center?

6. Describe evidence for spiral structure in the Milky Way. Which of the following types of objects reveal spiral structure when their positions are mapped on the Milky Way plane?

a. *O stars*

b. *M stars*

c. *HII clouds*

d. *open clusters*

e. *globular clusters*

f. *star-forming regions*

g. *supernovae*

7. Will the present constellations be recognizable in the earth's sky 100 million years from now? Why or why not?

8. Why is the term "high-velocity stars" a misnomer?

9. Why are no O- or B-type stars found in the galactic halo?

10. Summarize evidence for violent, energetic activity in our galaxy's central regions.

Advanced Problems

11. What is the linear size, in parsecs, of a feature subtending an angle of one second of arc, located at the galactic center?

a. *Could it be seen with a telescope resolving one second of arc?*

b. *Could it be seen with a radio telescope resolving one second of arc?*

12. If the sun moves in a circular orbit at 230 km/sec and is 9,000 parsecs from the orbit's center, calculate the time required to complete one circuit. (Hints: circumference of a circle is $2\pi r$; 1 parsec $= 3 \times 10^{18}$ cm; 1 km $= 10^5$ cm; 1 year is about $\pi \times 10^7$ sec.)

13. Using the relations in Problem 12 confirm the calculation of the galaxy's mass given in the text. Why would it be incorrect (or at least not meaningful) to quote the galaxy's mass to 3 significant figures (such as 2.34×10^{44} g)?

14. What percent of the galactic diameter could be crossed in one lifetime (say, 70 years) if we could travel at the speed of light?

15. If an asteroid could be hollowed out and converted to a spaceship on which many generations of people could live (as in some science fiction stories), how many generations would live (at 20 years each) on the way to the Orion nebula at the speed of light? (Hint: See Table 20·1.)

Projects

1. Compare views of the Milky Way with the naked eye, binoculars, and a small telescope. Scan along the Milky Way with each instrument and record the number of stars/degree² in different constellations (or at different galactic longitudes). Relate these densities to the actual structure of the galaxy. Note that if observations of the summer evening Milky Way can be obtained in the regions of Scorpio and Sagittarius, the direction toward the center can be studied. Compare star counts with each instrument in the Milky Way plane with counts near a point 90° from the plane, where we are looking directly "up" out of the disk. Can you account for the differences?

2. Compare the Milky Way as seen with the naked eye or binoculars:

a. *in the heart of your city.*

b. *on the edge of town.*

c. *as far from any lights as you can possibly get.*

As far as telescopic views of individual bright stars are concerned, city lights have little effect. But when it comes to broad areas of faint nebulosity or unresolved star clouds, even a single street light can illuminate local smog or fog, and cause the iris of the eye to contract, thus destroying faint contrast and the ability to see faint glows.

21

The Local Galaxies

What lies in the vastness of space beyond our own galactic disk and its surrounding swarm of globular star clusters? Our first problem is to see out of the Milky Way at all. So thick and far-reaching are the clouds of gas and dust in the plane of the Milky Way that there is a band around the sky, along the galactic equator, where we see virtually nothing outside our galaxy. This band is called the **zone of avoidance.** At higher galactic latitudes, however, we look "up" or "down" out of the Milky Way disk, so our line of sight passes through few gas or dust clouds. In most directions and at most wavelengths we have a clear view of space beyond our galaxy (see Figure 21.1).

When we scan this part of the sky,

Figure 21·1 **A map of the sky with coordinates of galactic latitude and longitude, schematically showing positions of observed remote galaxies. The "zone of avoidance" along the Milky Way is caused by obscuration by Milky Way dust. Filled circles are selected prominent galaxies; open circles are selected prominent clusters of galaxies. Mapping of the zone of avoidance was an early source of evidence that galaxies are outside the Milky Way system.**

we find, as far as we can see, galaxy upon galaxy—some like our own Milky Way and some not. To continue our pattern of exploring farther and farther from the earth, we will start with nearby galaxies and move to more distant ones, looking for fundamental clues about their nature. The objects we study in this chapter are so remote and faint that their true nature was recognized only a few decades ago. Cal Tech astronomer Allan Sandage has commented:

> *What are galaxies? No one knew before 1900. Very few people knew in 1920. All astronomers knew after 1924.*

Surveying the Nearby Galaxies

In small telescopes (up to about 30 cm aperture) most galaxies look like mere blurs of light. At the time of the Messier, New General, and Index catalogs of celestial objects, galaxies

Figure 21·2 **Patterns of obscuration by dust clouds in edge-on galaxies, illustrating similarities in phenomena in different parts of space. A. A portion of the Milky Way, with its central bright bulge to left. B. Distant edge-on spiral galaxy NGC 4565, showing its central bright bulge and dust-obscured disk. C. Distant irregular or spiral galaxy NGC 4631.** (A and C: Hale Observatories; B: Lick Observatory)

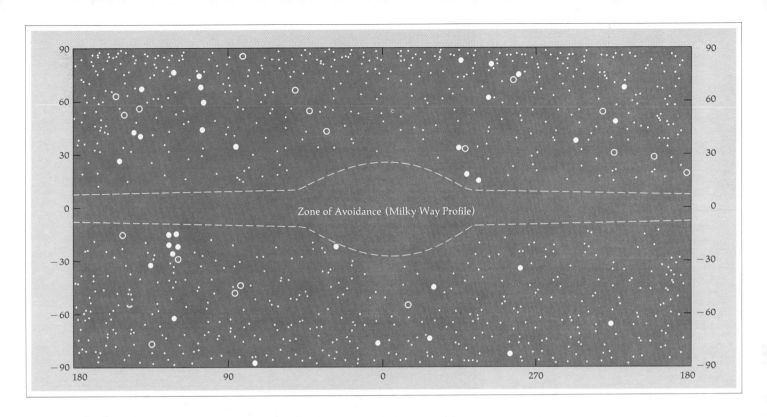

Zone of Avoidance (Milky Way Profile)

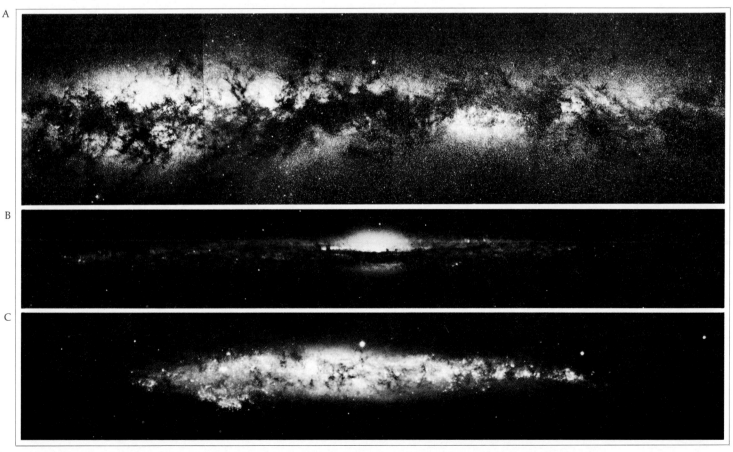

A

B

C

were not recognized as different from nebulae, and so received M, NGC, and IC numbers in sequence with nebulae. The Andromeda galaxy, for example, is M 31.

Later photographs made with large telescopes show some galaxies as pinwheels with beautiful spiral arms—like the Milky Way—but many others as ellipsoidal groups of stars like giant flattened globular clusters, and others as irregular masses of stars and nebulae. These objects were proved to be remote galaxies by identifying individual stars in them and by identifying types of stars of known luminosity such as Cepheid variables or supernovae (Figure 21·3).

To learn about rotations and masses of galaxies, the Doppler shift is, once again, especially useful. By measuring the Doppler shift on each side of a galaxy's center, we can see which side is approaching and which is moving away, thereby determining its rotation. This so-called "rotation speed" for a given star is really the star's speed in its orbit around the center. Combining the rotation speed with its distance from the center, we can apply Kepler's laws and determine the mass of the galaxy, as was done for the Milky Way in the last chapter.

A second method for estimating galaxies' masses can be used if two galaxies are adjacent and orbiting around each other. Again, the velocities can be estimated from Doppler shifts and the masses estimated from Kepler's laws, as for analysis of a binary star.

To learn about the distribution and nature of galaxies, let us start an imaginary trip out from the Milky Way and describe the galaxies we encounter in order of distance. A three-dimensional view of these galaxies is sketched in Figure 21·4, and their properties are summarized in Table 21·1.

Figure 21·3 **Detection of occasional supernovae in other galaxies permits estimates of their distance, because absolute magnitudes of supernovae can be gauged** from their rate of brightness change. The arrow shows a supernova detected in the spiral galaxy NGC 7331 in 1959. (Lick Observatory)

*The Magellanic Clouds:
Two Satellite Galaxies
of the Milky Way*

Medieval Arab astronomers recorded a glowing patch far down in the southern sky, visible from southern Saudi Arabia. Amerigo Vespucci, who sailed farther south, reported that there were really ". . . two clouds of reasonable bigness moving about the place of the (south celestial) pole" (quoted by Shapley, 1957). Since there is no bright south polar star, the clouds helped navigators to mark the pole. Well described by Magellan's around-the-world expedition of 1518–1520, they came to be called the **Magellanic clouds.** They are shown in the wide-angle view of Figure 21·5.

The clouds turned out to be small galaxies, probably moving in orbits around the Milky Way. The Large Magellanic cloud is 52 kpc away and the Small Magellanic cloud is 63 kpc, not much farther than the far edge of our own galaxy.

What kind of galaxies are these

Name	Catalog Number	Distance (kpc)	Diameter (kpc)	Mass ($M_\odot$)	Absolute Magnitude (m_v)	Type[a]
Milky Way		9	30	2×10^{11}	−21?	Sb
Large Magellanic cloud		52	8	10^{10}	−18.7	Ir
Small Magellanic cloud		63	5	2×10^9	−16.7	Ir
Draco system		67	1	10^5	− 8.5	E
Ursa Minor system		67	2	10^5	− 9	E
Palomar 12		75	2			G
Capricorn (Zwicky)		80	10			G
Sculptor system		85	2	3×10^6	−12	E
Palomar 4		120	2.5			G
Palomar 3		130	2			G
Fornax system		170	6	2×10^7	−13	E
Leo I system		230	2	4×10^6	−11	E4
Leo II system		230	1	10^6	− 9.5	E1
	NGC 6822	470	2	3×10^8	−15.6	Ir
Andromeda companion	NGC 205	640	4	8×10^9	−16.3	E5
	NGC 147	660	2	10^9	−14.8	E
	NGC 185	660	3	10^9	−15.2	E
Andromeda companion	M 32	660	2	3×10^9	−16.3	E2
Andromeda galaxy	M 31	670	40[b]	3×10^{11}	−21.1	Sb
Andromeda I		700?	0.7			E
Andromeda II		700?	0.7			E
Andromeda III		700?	0.9		−11	E
Andromeda IV		700?	1.2			E
Triangulum galaxy	M 33	730	18	1×10^{10}	−18.8	Sc
	IC 1613	740	4	3×10^9	−14.8	Ir
Maffei 1		1,000	?	2×10^{11}	−20	S0

Table 21·1 The Local Group of Galaxies (known galaxies out to 1,000 kiloparsecs)

[a]S = Spiral (subtypes 0, a, b, c—see discussion later in chapter)

Ir = Irregular

E = Elliptical (subtypes 0 through 7)

G = Intergalactic globular cluster

[b]Diameter of Andromeda based on angular diameter 3°.1.

Source: Data from Allen (1973); Hodge (1966); van den Bergh (1972).

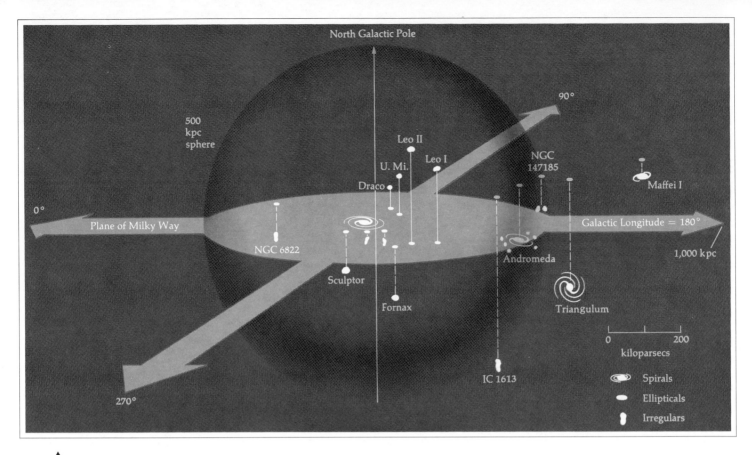

North Galactic Pole

500 kpc sphere

90°

Leo II

U. Mi.

Leo I

NGC 147185

Maffei I

Draco

0°

Plane of Milky Way

Galactic Longitude = 180°

NGC 6822

1,000 kpc

Andromeda

Sculptor

Fornax

Triangulum

0   200

kiloparsecs

IC 1613

270°

Spirals

Ellipticals

Irregulars

▲

Figure 21·4 **A three-dimensional plot of the local system of galaxies. Solid lines extend north of the galactic plane; dashed lines extend south.**

Figure 21·5 **The Large Magellanic cloud (left) ▶ and Small Magellanic cloud (right). The bright star in the upper right is Alpha Eridanus. The angular distance between clouds is about 22°.** (3-hr exposure with 3-in.-aperture lens, made in 1934; Harvard College Observatory, Bloemfontein Station, South Africa)

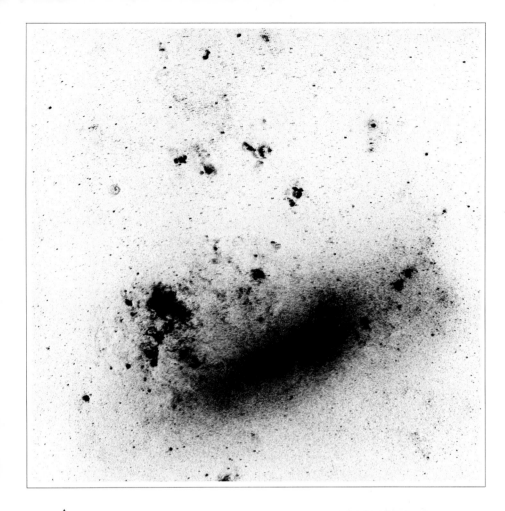

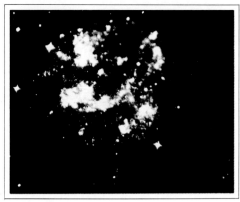

Figure 21·7 **The Large Magellanic cloud photographed by the light of far-ultraviolet radiation from its hot, young stars and associated nebulae. Star-forming, cluster-like regions of Population I stars are prominent. The photo was taken from the moon's surface by astronaut John Young with a special ultraviolet camera prepared by the Naval Research Laboratory. Ultraviolet features are invisible from earth because of atmospheric absorption.** (NASA)

Figure 21·8 **The Small Magellanic cloud, about 63 kiloparsecs away. The photo resolves individual stars and scattered bright nebulae.** (Cerro Tololo Inter-American Observatory)

▼

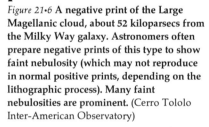

Figure 21·6 **A negative print of the Large Magellanic cloud, about 52 kiloparsecs from the Milky Way galaxy. Astronomers often prepare negative prints of this type to show faint nebulosity (which may not reproduce in normal positive prints, depending on the lithographic process). Many faint nebulosities are prominent.** (Cerro Tololo Inter-American Observatory)

companions of ours? First, they are small. The Large cloud is only 8 kpc across and the Small cloud is only 5 kpc, compared with about 30 kpc for the Milky Way. Star counts and radio measures of hydrogen show that they are only a few percent as massive as the Milky Way.

A second difference is form. The clouds do not show the Milky Way's beautiful spiral structure. They are ragged and are classified as **irregular galaxies.** Each contains a softly glowing barlike structure composed of stars, as shown in Figure 21·6. Figure 21·9 shows hydrogen concentrated in the central regions of the galaxies and also forming a bridge connecting the two.

Can we identify a nucleus in either galaxy? Somewhat off the end of the bar in the Large Magellanic cloud is the **Tarantula nebula,** also known as **30 Doradus** (Figure 21·10). This nebula, although 52 kiloparsecs away, can be seen with the naked eye, so it must be extremely bright. In fact, if it

were moved to the position of the Orion nebula, it would fill the whole constellation of Orion and be bright enough to cast shadows on the earth. In its center is a cluster about 60 parsecs in diameter, containing at least 100 massive supergiant stars. This cluster is several hundred times

brighter than ordinary globular clusters near the Milky Way. The nebula also contains a thermal radio source about 60 parsecs across, surrounded by a source of nonthermal radio radiation, about 200 by 400 parsecs in size. In these respects, the Tarantula nebula resembles the nucleus of our

Figure 21·9 **A representation of neutral hydrogen gas near the Magellanic clouds, showing the connecting "bridge."** (From a 1963 survey of the region by Australian radio astronomers).
▼

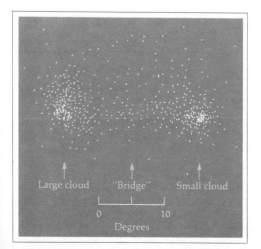

Figure 21·10 **The bright nebula 30 Doradus, believed possibly to be the nucleus of the**
▼

Large Magellanic cloud. (Cerro Tololo Inter-American Observatory)

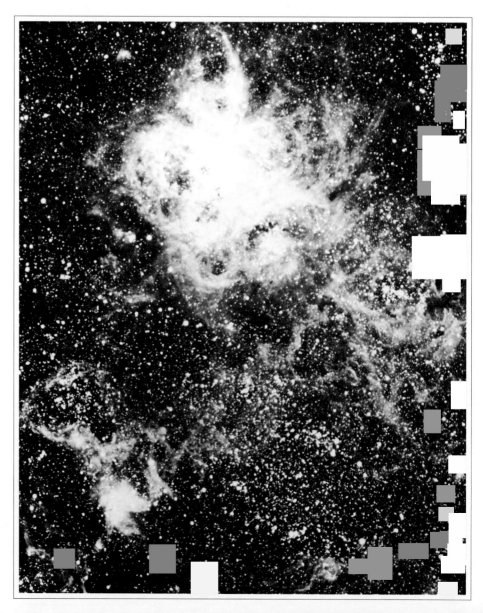

own galaxy. Some astronomers regard it as the partially formed nucleus of the Large cloud, and the barlike feature as similar to the Milky Way's central bulge of stars.

We can best understand other galaxies and their evolution by considering what populations of stars they contain. Are the clouds dominated by the newly formed, hot, blue stars of Population I, or the older, reddish giants of Population II? Both Figures 21·7 and 21·8 show the bright, blue, young stars and nebulae of Population I. Thus some star formation is going on now.

But are there older stars? We can *see* only the brightest stars, as shown in Figure 21·11. But "the spectacular Population I components represent only the frosting on a beautiful cake" (Bok, 1966). Population II might be more important than it first seems. One bit of evidence for this view is that the central bars contain many Population II red giants. Second, Population II globular clusters have been found. Third, there is little dust in the two clouds, except in the prominent young nebulae. In these ways the clouds are like our galaxy's central bulge and unlike our Population I dusty spiral arms.

Judging from age relations in our galaxy, we can guess that the clouds' Population II objects are many billions of years old, though the bright star-forming nebulae of Population I may be only millions of years old. Interpretation of ages is difficult because proportions of heavy elements are uncertain, and thus the accuracy of calculated models of stellar evolution is uncertain.

A strange feature of both clouds are a few **blue globular clusters,** unlike the red-giant-dominated clusters of our galaxy. To contain hot, blue stars (which are massive and have

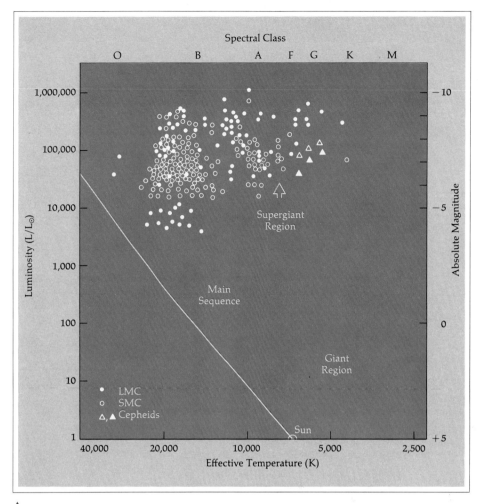

Figure 21·11 **An H-R diagram of measurable stars in the Magellanic clouds. The clouds are so far away that the only stars with** measurable spectra are supergiants. Main-sequence populations are assumed to exist as well.

short lifetimes) these globulars must have formed less than a billion years ago, and others may be still forming. Perhaps they are hitherto-unknown Population I globulars.

The Magellanic clouds are important to modern astronomy for providing a collection of varied objects at essentially constant distance. In the words of South African astronomer A.D. Thackeray:

The opportunity to compare the luminosities of supergiants, giants, main sequence stars of all types, Wolf-Rayet stars, cepheids, eclipsing variables, Mira-variables, RR Lyrae variables, globular clusters, novae, and so on, in an endless sequence all at the same distance simply never occurs in our galaxy.

A bridge of hydrogen called the **Magellanic stream** may connect the two clouds to the Milky Way galaxy. Australian radio astronomers have mapped a long filament of HI extending from the Small cloud in an arc beyond the south galactic pole, and in the other direction possibly reaching into the Milky Way's plane.

It resembles the bridge between the two clouds themselves. Because the two clouds are probably true satellites of our galaxy, their orbits may sometimes take them through the Milky Way disk. The Magellanic stream may thus be a tail of gas drawn out of the Milky Way during such an encounter an estimated half-million years ago.

Elliptical Galaxies and "Intergalactic Tramps"

As indicated in Table 21·1, the next 15 objects beyond the Magellanic clouds include 10 systems known as **elliptical galaxies.** These lie from 67 to 660 kpc distant. All are much smaller than our galaxy, ranging from the mass of the Magellanic clouds down to 1/10,000th of that mass. They resemble giant globular clusters and are more symmetrical than the Magellanic clouds. They lack the disk shape and spiral arms of the Milky Way. Another four, classed as **intergalactic globular clusters,** may be simply smaller examples of elliptical galaxies. Both the elliptical galaxies and intergalactic globulars are dominated by Population II stars and have little gas or dust. Some may actually be globular clusters that escaped from our galaxy. NGC 5694, for example, is moving away from the Milky Way at more than 270 km/sec. These clusters have been called "intergalactic tramps."

Closest of the 15 systems, about 67 kpc away, is the *Draco elliptical galaxy.* It is probably the least massive of the local galaxies, having just less mass than the well-known globular cluster inside our own galaxy, M 3. Because distant galaxies are visible through it, there must be very little dust. Also about 67 kpc away, a very similar galaxy, the *Ursa Minor elliptical galaxy,* contains no globulars, no emission nebulae, and little dust. It resembles a large globular cluster but is flattened, with the short axis 0.45 as long as the long axis.

Four of the next five systems are intergalactic globular clusters from 75 to 130 kpc away. The presence of these little-studied systems in this relatively nearby part of space suggests that other small, faint, intergalactic tramps may be scattered among the more distant galaxies, but may be too small to detect.

More substantial are the *Sculptor* and *Fornax elliptical galaxies,* located 85 and 170 kpc away, respectively. Though more massive than the previous six systems, they still have less than one percent of the mass of the Magellanic clouds. They were discovered by Shapley in 1937–1938 during a Harvard photographic survey of the southern sky from South Africa. Shapley (1961) recounts how the faint image was first thought to be a "dark-room unhappiness—fingerprints during development, perhaps." Some 10,000 stars between magnitudes +17.5 and +19.5 have been counted in the Fornax galaxy, which resembles a loosely bound globular cluster. Sculptor and Fornax are close enough to the massive Milky Way so that tidal forces play a role in defining them: A star on the near edge of Sculptor experiences nearly equal forces of attraction from it and from the Milky Way. Stars more than about 2 kpc from the center of Sculptor are likely to be torn loose to wander in intergalactic space. Observations confirm that the Sculptor galaxy has few stars beyond this limit. The Fornax galaxy has a similar limit about 3 kpc from its center.

The brightest stars in these two galaxies are reddish, metal-poor giants or supergiants, similar to the Population II objects in globular clusters of our own galaxy. The Sculptor system is easier to observe because it is closer. It contains hundreds of variable stars, but no prominent globular clusters, whereas Fornax contains about five globulars that are bigger than those in our galaxy.

The *two elliptical galaxies* in Leo, about 230 kpc away, are symmetrical, flattened concentrations of stars. Neither Leo I nor II has prominent globular clusters, and there is little evidence of gas and dust.

Isolated at about 470 kpc is the nearest analog of the Magellanic clouds, the *irregular galaxy NGC 6822.* It is only a couple of kiloparsecs across. Radio studies indicate that hydrogen gas comprises about 10 percent of its mass.

The Great Andromeda Spiral Galaxy

Finally, at about 660 kpc, we encounter the first **spiral galaxy** truly comparable to the Milky Way. It was first recorded in a star catalog by Arab astronomer al-Sufe in A.D. 964, but must have been known much earlier, being easy to see with the naked eye as a hazy patch on a dark, clear night. The **Andromeda galaxy**

Figure 21·12 **The region of the Andromeda   ▶ galaxy, about 660 kiloparsecs away. In the foreground are stars of the Milky Way galaxy, and the disk of the Milky Way itself (bottom half), with the constellation Cassiopeia outlined (upside-down from its usual W-shape representation). The angular distance from the Andromeda galaxy to Cassiopeia's middle star is about 20°. The box outlines the area shown in the next figure. (35 mm camera; 24 mm lens at f2.8; 22 min. exposure on 2475 Recording film)**

is the one that settled the argument about the nature of so-called "spiral nebulae." Edwin Hubble identified Cepheid variables there and showed that it was far outside the disk of our own galaxy. His announcement of this result at the American Astronomical Society in 1924 caused a sensation and marked the first real understanding of galaxies.

Similarities to the Milky Way
The Andromeda galaxy can hardly be distinguished from the Milky Way in shape or stellar content, though it may be a little bigger. The naked eye sees it as a faint patch about a degree across, but this is really only the brightest, innermost region, about 12 kpc across. Telescopic photographs show that the spiral arms form a disk about 35 kpc across, while outlying associations of B stars suggest a diameter as great as 50 kpc. As with the Milky Way, there are globular clusters and a halo of HI reaching perhaps 100 kpc diameter, as shown in Figure 21·15.

Spiral Arms The spiral arm pattern, shown in Figure 21·16, also resembles that of the Milky Way, though it is much easier to observe since we see it from above the disk. *Inner arms* have conspicuous dust; *outer arms* have less dust but more HII regions and young stars. Most star-formation now occurs in the intermediate and outer arms. Observations seem to support the density-wave theory of spiral arm formation.

Stellar Populations The Andromeda galaxy also played an important role in the discovery of the two main star populations. Hubble's photos in the 1920s with the new Mount Wilson 100-inch telescope revealed bright blue stars in the spiral arms (Figure

▲
Figure 21·13 **The region of the Andromeda galaxy, showing its angular extent exceeding 3°. Vertical height of the photo is about 8°. The bright star, upper right, is Beta Andromedae. The box shows location of Figure 21·14A. (35 mm camera; 135 mm lens at f2.8; 10 min. exposure on 2475 Recording film)**

Figure 21·15 A schematic diagram of radio emission from neutral hydrogen gas near Andromeda galaxy. The hydrogen halo (stippled) is much larger than the visible galactic disk (ellipse). (After Hodge, 1966)

Figure 21·14 Images illustrating the varied appearance of the Andromeda galaxy with different types of photographic and telescopic equipment. A. Short exposure with 6-inch telescope emphasizing the nucleus and barely showing the spiral arms. B. Mosaic of photos with 12½-inch telescope, printed to show maximum detail in both nucleus and spiral arms. C, D. Two prints from the same negative, (C) printed to show bright features of the nucleus and clusters in the arms, and alternately printed (D) to show faint details in the arms. (A: *Sky and Telescope;* B: Clarence P. Custer; C and D: Steward Observatory, University of Arizona)

Figure 21·16 Proof of the Andromeda galaxy's spiral form. The image has been "rectified" by projection on a screen tipped at 13° to the line of sight, giving an impression of the galaxy's spiral pattern as it would be seen from above the plane of the disk.

▲
Figure 21·17 **Resolution of stars in two populations of the Andromeda galaxy. Left photo (blue filter) shows Population I hot blue stars and clusters in the spiral arms. Right photo (yellow filter) shows Population II red giant stars which dominate the elliptical galaxy, NGC 205, a satellite of the Andromeda galaxy.** (200 inch telescope; Hale Observatories)

21·17), but no individual stars in the central bulge. Mount Wilson observer Walter Baade therefore decided to look for possible bright red stars in the bulge by switching to red-sensitive photographic plates. These plates were troublesome because the exposures required as long as nine hours and tended to be fogged by the reflected skyglow of Los Angeles, below Mount Wilson.

A world tragedy came to Baade's aid. During World War II, in 1942–1943, blackouts of Los Angeles produced exceptionally dark skies. In a few months, Baade discovered that the central bulge of Andromeda and its two nearby elliptical galaxies are dominated by, as he put it, "thousands and tens of thousands" of Population II red giant stars (Figure 21·17). This discovery clarified the role of star populations in galaxies and matched the Andromeda pattern to our own galaxy. In spiral galaxies, Population I star-forming regions are concentrated in the spiral arms, which are marked by open clusters, giant nebulae, and massive blue stars. The central bulge is redder in color, as star-formation has ceased. There is less dust, and the dominant stars are evolved red giants.

Dynamics and Mass As in our galaxy, the density of stars is greatest near the center. A region of low rotational velocity and low star density about 2 kpc from the center may be a feature related to our own 3-kpc expanding arm. The total mass is roughly 3×10^{11} $M_\odot$, comparable to the Milky Way's mass.

The Nucleus The Andromeda galaxy gives us a chance to look directly at the light of the nucleus of a galaxy like ours, rather than trying to study it through 9 kpc of dust. The results are intriguing. With a large telescope, long exposures of Andromeda over-

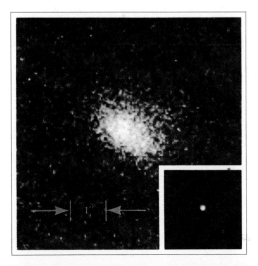

Figure 21·18 **The central region of the Andromeda galaxy, printed to show maximum contrast between small-scale dust clouds and star clusters. The spot in the center shows the approximate size and scale of the central bright nucleus, shown in the next figure.** (Kitt Peak National Observatory)

◀ Figure 21·19 **Highest-resolution photo of the bright nucleus of the Andromeda galaxy, a dense cluster of stars about 3 × 5 parsecs in diameter. (Scale bar: one second of arc.) The photo was made by a 36-inch telescope carried to 84,000 feet by a balloon. The inset at right shows the image of a star made by the same telescope.** (Princeton University, Project Stratoscope, supported by NSF, ONR, and NASA)

expose the nucleus (Figure 21·18), but short exposures reveal that the brightest object in the whole galaxy is a starlike condensation at its center (Figure 21·19). The sharpest photographs show a rather well-defined elliptical nucleus of about 3.2 × 5.0 parsecs. It appears to contain about 10^8 solar masses of stars, crowded at about $3 × 10^6$ stars per cubic parsec. This would put inner stars

typically as close as 0.004 parsecs apart, compared to about one parsec near the sun. If we were located in such a region, our night sky would be ablaze with many stars as bright as the moon and a million stars as bright as Sirius!

The Satellites of the Andromeda Galaxy

Two prominent elliptical galaxies, M 32 and NGC 205, are companions of the Andromeda galaxy; they may be true satellites, orbiting around it in the same way that the Magellanic clouds are believed to orbit our galaxy. Detailed studies show similarity in star content as well: they are typical elliptical galaxies, with Population II stars, poor in heavy elements. The brightest

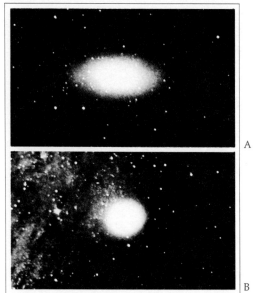

A

B

Figure 21·20 **Two dwarf elliptical galaxies, NGC 205 (A) and M 32 (B). NGC 205 is isolated, whereas M 32 is superimposed on the outer spiral arms of the Andromeda galaxy. Both are about 650 kiloparsecs away.** (Steward Observatory, University of Arizona)

of these are reddish giants. These two galaxies contain little gas, and they seem to illustrate again the similarity of globular clusters, elliptical galaxies, and central regions of spiral galaxies.

Two other similar-size elliptical galaxies, NGC 147 and NGC 185, lie not far from Andromeda (Figures 21·21 and 21·22). Four more small elliptical galaxies, each about 1 kiloparsec across, also lie near Andromeda. They have been named Andromeda I, II, III, and IV. As can be judged from Figure 21·4, these eight galaxies seem to form a cluster centered on the Andromeda galaxy.

The Triangulum Spiral Galaxy

In the inconspicuous constellation Triangulum, only 14° from the An-

Figure 21·21 **The dwarf elliptical galaxy NGC 147, about 660 kiloparsecs from the Milky Way. Individual stars are resolved (but bright stars are foreground objects in our own galaxy). This and NGC 185 (see next figure) are a pair roughly 12 kiloparsecs apart and some 90 kiloparsecs from the Andromeda galaxy.** (Hale Observatories)

Figure 21·22 **The dwarf elliptical galaxy NGC ▶ 185, about 660 kiloparsecs away. Individual stars are resolved. The prominent dark dust cloud is unusual in an elliptical galaxy, most of which are relatively dust-free.** (Lick Observatory)

Figure 21·23 **The Triangulum galaxy, a spiral** ▶ **system about 730 kiloparsecs away.** (Hale Observatories)

dromeda galaxy, is the next galaxy beyond. M 33, or the **Triangulum galaxy,** another large spiral, is about 730 kiloparsecs from us and about 18 kiloparsecs across. Its spiral arms are not as tightly wound as the Milky Way or M 31. Its nucleus is less pronounced, and its general color is bluer because of the greater number of hot, young Population I stars and star clusters in its spiral arms. It has more than a dozen known globular clusters. These differ from the globulars of the Milky Way and Andromeda galaxies: they are probably smaller and their brightest stars are blue instead of red. Causes of these differences are unknown.

IC 1613 is a small, isolated, irregular galaxy resembling the Magellanic Clouds. It has an ill-defined barlike structure, several groups of hot, young, blue stars, and a more diffuse system of older stars. This suggests a patchy Population I distribution, and an older Population II background, which appears somewhat elliptical in outline.

Maffei 1

In 1968, the Italian astronomer P. Maffei used infrared photographs that penetrate the dust in the Milky Way plane to search for galaxies in the zone of avoidance. One galaxy he discovered is a tightly wound spiral perhaps 1,000 kiloparsecs away, which has come to be called Maffei 1.

Classifying Galaxies

Our survey of space out to 1,000 kpc from the Milky Way netted 26 galaxies or intergalactic clusters. These are known as the **local group.** What can we learn from this census of objects in the well-observed region near our galaxy? What types are there? What do they tell us about the nature of galaxies in general?

Clustering of Galaxies

First, galaxies are not randomly distributed, but tend to *cluster in different-size groups,* called **clusters of galaxies.**

Most galaxies in the local group are clustered near either the Milky Way or Andromeda, as can be seen in the diagram of the system in Figure 21·4. These *satellite galaxies* range from ragged irregular galaxies (the Magellanic clouds) to tightly packed elliptical systems resembling glorified globular clusters.

Is the local group itself a distinct cluster of galaxies? Possibly. As we will see in the next chapter, clusters of galaxies are common in other parts of the universe. The clustering of galaxies is an important clue about the history of matter in the universe. It indicates that matter has, under the influence of gravity, aggregated into hierarchies, of which we have already encountered single stars, binary and multiple stars, star clusters, galaxies, and groups of galaxies.

A Classification System: Making Sense of Galaxies

There are several distinct types of galaxies. Most of the known types appear within the local group (Table 21·1), which takes us out to 1,000 kiloparsecs. Even if we expand our horizon to, say, 15,000 kiloparsecs, as in Table 21·2, we discover only a few new types. M 87, for example, is a *giant elliptical* much larger than the ones mentioned so far; it is about 13 kiloparsecs across and more massive than the Milky Way. It contains about 4×10^{12} $M_\odot$, or perhaps 4,000 billion stars! Another class is illustrated by M 83, a spiral galaxy whose central region is a bright, barlike object instead of an ellipsoidal bright area; such galaxies are called **barred spiral galaxies.**

How can we make sense of these different shapes of galaxies and their different types of stellar populations? The standard scientific approach is to try to classify the different types of objects and then arrange them in a system that shows smooth transitions from one type to another. Then this system can be studied for possible causes of the transitions, such as

evolution, mass differences, or rotational differences. For example, elliptical galaxies with different degrees of flatness seem to form an obvious sequence, perhaps related to rotation.

The best-known classification scheme was developed by Edwin P. Hubble in the 1920s and 1930s and extended by his colleagues after his death. The system was based purely on the shape

Table 21·2 Selected Galaxies out to 15,000 Kiloparsecs

Name or Catalog Number	Distance (kpc)	Diameter (kpc)	Mass ($M_\odot$)	Absolute Magnitude (m_V)	Type	Radial Velocity (km/sec)
NGC 55	2,300	12	3×10^{10}	−20	Sc	+ 190
NGC 253	2,400	13	10^{11}	−20	Sc	− 70
M 82 (NGC 3034)	3,000	7	3×10^{10}	−20	Ir	+ 400
M 81 (NGC 3031)	3,200	16	2×10^{11}	−21	Sb	+ 80
M 83	3,200	12	?	−21	SBc	+ 320
M 51 "Whirlpool"	3,800	9	8×10^{10}	−20	Sc	+ 550
NGC 5128 (Centaurus A)	4,400	15	2×10^{11}	−20	E0p[a]	+ 260
M 101 "Pinwheel"	7,200[b]	40	2×10^{11}	−21	Sc	+ 402
M 104 "Sombrero"	12,000	8	5×10^{11}	−22	Sa	+1,050
M 87 (NGC 4486)	13,000	13	4×10^{12}	−22	E1	+1,220

[a]Centaurus A is a strong radio source, appearing as an elliptical galaxy with a peculiar dense dust lane across its face.
[b]M 101's distance was revised by A. Sandage and G. Tammann (1974) from 3,800 to 7,200 ± 1,000 kpc.
Source: Data from Allen (1973).

Figure 21·24 **NGC 3992, an example of a barred spiral galaxy, showing spiral arms originating from a region near the ends of a bar (or elliptical ring?) in the central region.** (Hale Observatories)

of the galaxies, not on measured properties such as spectra or mass. The scheme, shown in Figure 21·25, involves three main types of galaxies, *ellipticals* (designated E), *spirals* (S), and *irregulars* (Ir). Each type is subdivided according to form. For example, as shown in Figure 21·25, ellipticals are subdivided according to apparent flattening, which may be due either to our angle of view or to true flattening of the system. They range from E0 (circular outline) to E7 (the flattest known), with the index number related to the ratio of the major and minor axes.[1] Forming a transition between ellipticals and spirals is a special class, S0, resembling a spiral's disk shape but lacking clear

---

[1]The index number is $10(a-b)/a$, where a is the major axis and b the minor axis.

spiral arms. Spirals are subdivided according to how tightly their spiral arms are wound. Sa spirals have the tightest arms and Sc spirals the loosest. Spirals are subdivided into two parallel sequences—barred and normal spirals—giving Figure 21·25 its common designation as the **"tuning fork" diagram.** Thus, a tight spiral with a central *bar* is designated Sba, instead of Sa. (Some astronomers use a still finer sub-

◄ *Figure 21·25* The simplified classification scheme for galaxies known as "Hubble's tuning fork diagram" after the originator of the system.

Figure 21·26 A nearly edge-on view of the spiral galaxy M 104, about 12,000 kiloparsecs away. This galaxy has an unusually large diffuse central region, resembling an elliptical galaxy, and a dust-rich surrounding disk that appears thinnest in the central regions. Some globular clusters in a surrounding halo are visible, though the brightest nearby objects are foreground stars in our galaxy. (200-inch telescope photo; Hale Observatories)

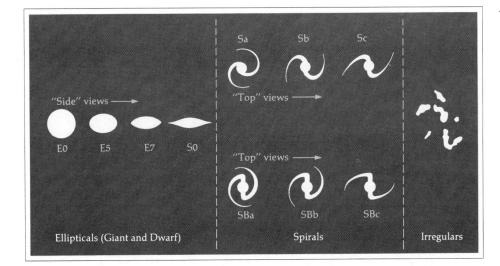

division of spirals to indicate presence or absence of a ringlike configuration of the innermost spiral arms.) Irregular galaxies are subdivided into Ir I (prominent O and B stars, emission nebulae, and Population I) and Ir II (amorphous appearance, no sharp nebulae, Population II).

This classification system correlates well with some physical properties of galaxies, as summarized in Table 21·3. For example, from ellipticals through irregulars there is a progression from older to younger populations of stars and from less to more dust and gas. The implication is that star-formation is still occurring in spiral arms and some irregulars, but not in ellipticals. Another systematic trend is that the amount of mass re-quired to produce the same amount of stellar radiation, called the **mass/luminosity ratio,** is high in ellipticals and low in irregulars. This ratio implies that much of the mass in ellipticals is tied up in low-luminosity stars.

The distribution of the different types of galaxies surprises many people who tend to think of all galaxies as beautiful spirals. Just as giant and supergiant stars are over-represented if we scan the prominent stars in the sky, spirals are also over-represented because they are bright and spectacular. For example, an early survey of prominent galaxies by Hubble netted 80 percent spirals, and a sample of 100 photos in three recent astronomy texts included 67 percent spirals. But in fact, the actual distribution of galax-ies is more like the following (based on the local group and counting inter-galactic globulars as small ellipticals):

Ellipticals	70%
Spirals	15%
Irregulars	15%

Evolution in Galaxies

Lurking in astronomers' minds has always been the idea that if we could just classify the galaxies in the right way, we would automatically have them arranged in order of age. Some early investigators thought that the "tuning fork" classification had arranged old galaxies at one end and newly formed galaxies at the other

Table 21·3 General Characteristics of Galaxies

	Ellipticals	S0	Sa	Sb	Sc	Irregulars
				Spirals		
Dust	None			Some		Some
Percent neutral hydrogen (mass HI/total mass)	0		1	3	9	20
Most prominent populations	II			II (halo, center); I (arms)		I
Typical rotation periods (million years)		59	63	140	200	300?
Dominant color	red			red (halo, center); blue (arms)		blue
Spectrum of central regions	K		K–G	G	F	F–A
Luminosity ($L/L_\odot$)	Giants $\leq 10^{10}$ Dwarfs $\geq 10^{5}$			10^{8}–10^{10}		10^{7}–10^{9}
Diameter (kiloparsecs)	Giants ≤ 200 Dwarfs ≥ 1			5–50		1–10
Mass (solar masses)	Giants 10^{9}–10^{12} Dwarfs 10^{6}–10^{9}		5×10^{11}?	3×10^{11}	10^{11}	10^{9}
Mass/luminosity ratio (solar units)	Giants 5–80 Dwarfs 1–5	50	20	10	5	3

Source: Data from Allen (1973), de Vaucouleurs (1974), Smith and Jacobs (1973).

(reminiscent of the star classification from main sequence through giant stages to dwarf stages). Today this idea seems much too simple; it is ruled out by the discovery that all types of galaxies have at least a few old Population II stars. Thus, no single type of galaxy could have formed just recently. Furthermore, no known concentrations of intergalactic gas are big enough to indicate current or recent galaxy-formation.

Instead, all galaxies are probably billions of years old and the "tuning fork" classification may relate more to the rate of evolution of stars inside the galaxies, as affected in turn by initial properties such as mass distribution and rate of rotation. For example, as shown in Table 21·3, there is a smooth progression of rotation rates along the "tuning fork." Also, in the ellipticals, most of the dust and gas seems to have been consumed and concentrated into stars, stopping the star-forming process; in spirals' arms and in irregulars, much dust and gas are left to form stars.

Theoretical work indicates that the different shapes—elliptical, spiral, and irregular—must arise from different initial conditions of mass, rotation rate, turbulence, and so on. For example, some studies indicate that fast rotation of a protogalactic cloud favors formation of a bar-shaped region, which might explain barred galaxies.

Figure 21·27 **Optical and radio images of spiral galaxy M 81, about 3,200 kiloparsecs away. The optical image (A) shows stars and star clusters. The radio image (B) shows radiation from neutral hydrogen gas. Scale is about the same in both images. Important results are that (1) hydrogen gas coincides with the position of Population I stars in the spiral arms; (2) there is virtually no neutral hydrogen in the center; (3) the hydrogen spiral pattern is much larger than the visible star pattern.** (Leiden Observatory)

A

B
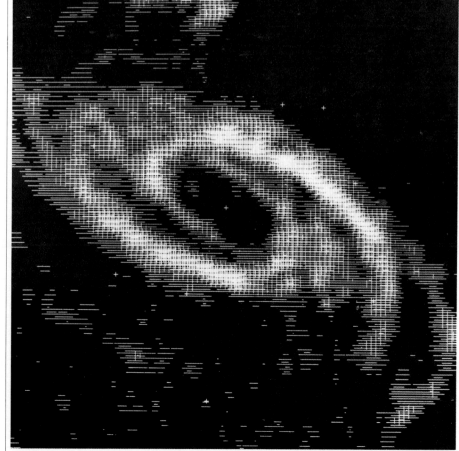

Study of stars' orbits in the bar suggests that some stars, dust, and gas
may be ejected from the ends of the
bar, providing material to form spiral
arms. Slower rotation of protogalaxies
probably favors disk-shaped rather
than bar-shaped masses. Formation
of an elliptical rather than a spiral
may depend on whether the concentration of gas into stars is complete
by the time the protogalaxy reaches its
maximum contraction and minimum
size (Gott, 1974).

Even within single galaxies, evolu-

Figure 21·28 **Optical (A) and radio (B) images
of the spiral galaxy M 101, the "Pinwheel
galaxy," about 7,200 kiloparsecs away.
Comparison shows concentration of gas in
the arms and lack of gas in the center.** (A:
Kitt Peak National Observatory;
▼

A

B: Westerbork Synthesis Radio Telescope, radiograph by R.J. Allen of Kapteyn Astronomical Institute, with assistance of E.B. Jenkins of Princeton University Observatory)

tion of the material proceeds at different rates in different regions. This is made very clear by comparing optical and radio views of galaxies, as in Figures 21·27 and 21·28. The optical views emphasize the brightest *stars*, while 21-cm radio mapping emphasizes the hydrogen *gas*. The figures show that the brilliant Population II central regions of spiral galaxies are virtually devoid of gas. All the evidence agrees that if clouds of free gas are not present, there will be no material from which stars can form.

B

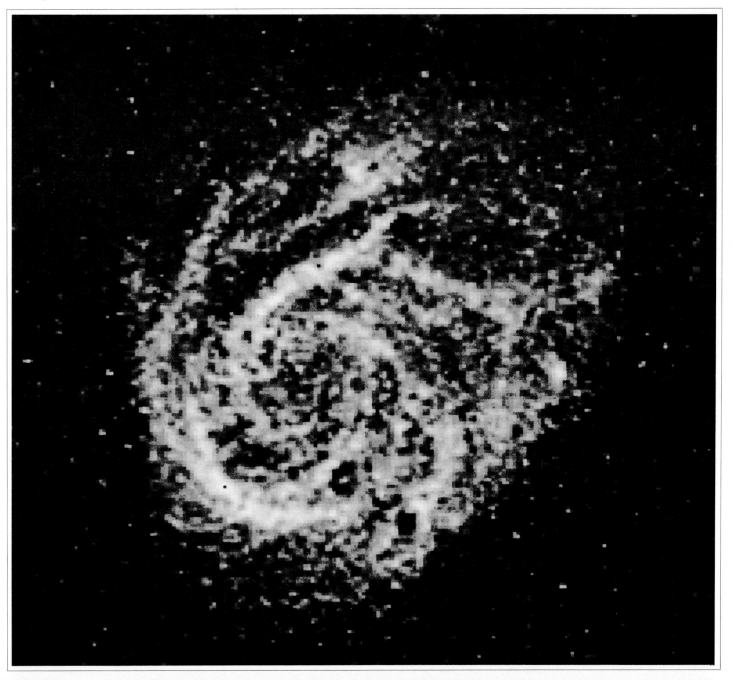

21

Summary

The local group of galaxies, out to about 1,000 kpc, contains a variety of types and sizes of galaxies. Combining observation of these galaxies with data from more distant galaxies, we can say that most galaxies can be classified as elliptical, spiral, or irregular galaxies. Many of these are clustered in groups. A more elaborate classification scheme, the "tuning fork" diagram, subdivides them, including barred as well as unbarred spirals.

These galaxies might be called "normal galaxies." (In the next chapter we will encounter some more unusual types with extreme energy production.) The observations discussed here can be combined with the theories of stellar evolution, discussed in Chapters 14–16, to explain how some galaxies or parts of galaxies have evolved and used up their gas and dust, thus producing different **stellar populations in galaxies.**

Our galaxy (and presumably most others) formed from nearly pure hydrogen, with virtually no heavy elements. As the galaxy contracted from a protogalactic cloud, individual stars formed by gravitational contraction of individual hydrogen clouds. If all the hydrogen had gone into stars of half solar mass, then all those stars would have lasted on the main sequence for tens of billions of years; we would observe such a galaxy today and find that it consisted only of pure-hydrogen Population II main-sequence stars, with no dust and no star-formation going on today. These might resemble elliptical galaxies.

However, stars actually form with a variety of masses. A small amount of the original mass would probably have been incorporated in large stars of 5, 10, 20 $M_\odot$ or even more. These highly unstable stars would have exploded a few million years after their formation. Nuclear reactions inside them would have created heavy elements, and the explosion would have strewn the heavy elements across space as gas and dust. If huge stars of, say, 10^4 $M_\odot$ exploded in central regions of ancient galaxies, they might have blown much of the gas and dust clear out of the central regions, perhaps providing material to form spiral arms. This dust-rich gas would mix with any leftover hydrogen to form new nebulae and a second generation of stars with more heavy elements than the first. This second generation, with its dust and heavy elements, would be Population I.

When the massive stars of the second generation exploded, the spiral arms of the galaxy would become still richer in heavy elements, gas, and dust. Galaxies seen in this state would have central regions where the initial hydrogen stars (Population II) were prominent, and other regions where only dust and newer stars (Population I) were prominent. These could resemble spirals and some irregulars.

If, after a few generations of star-formation, no supermassive stars formed, supernova explosions would cease and gas dispersal would cease. If, at any given moment, all the stars were of one solar mass or half solar mass, they would be stable for billions of years and never explode, so that no new gas and dust would be produced. Eventually, gas and dust would be used up and star-formation would cease. Such galaxies would appear to have a large Population II component and only a small Population I component.

Thus, the theory of stellar evolution is successful in explaining how similar galaxies of great age (perhaps equal age) could have widely different populations of stars. However, no completely accepted theory explains all varieties and properties of galaxies, nor should we try to construct such a theory until we probe farther into the depths of intergalactic space, for there we will find peculiar types of galaxies and new facts bearing on the very origin of the material from which galaxies are made.

Concepts

zone of avoidance

Magellanic clouds

irregular galaxies

Tarantula nebula (30 Doradus)

blue globular clusters

Magellanic stream

elliptical galaxies

intergalactic globular clusters

spiral galaxies

Andromeda galaxy

Triangulum galaxy

local group

clusters of galaxies

barred spiral galaxies

"tuning fork" diagram

mass/luminosity ratio

stellar populations in galaxies

Problems

1. Which type of galaxy tends to be biggest? Brightest? Contains fewest young stars? (See Tables 21·1, 21·2.)

2. Of giant ellipticals, ordinary ellipticals, spirals, barred spirals, and irregulars, which type is most common? (See Tables 21·1, 21·2.)

3. How many years would a radio signal take to reach the Andromeda galaxy?

4. If two very faint and distant galaxies were detected on photos, but were too distant to allow identification of spiral arms and other typological features, and if spectra showed that one was reddish with a spectrum of K type stars, while the other was bluish with a spectrum of B and A stars, what types would you expect the two galaxies to be?

5. If you were to represent the Milky Way and Andromeda galaxies in a model by two cardboard disks, how many disk-diameters apart should they be to represent the true spacing of the two galaxies?

6. In earth's geography, we usually orient ourselves with the "up" direction defined by the N rotation axis. In discussing the solar system, we use the N ecliptic pole, defined by orbital revolution. In discussing the Milky Way, we use the N galactic pole defined by galactic rotation. Has any asymmetry or special direction appeared in this chapter that would define a preferred orientation for discussion of the distant galaxies outside the Milky Way? If so, describe.

7. Why are small ellipticals and "intergalactic tramp" clusters not found in catalogs of the more distant galaxies, such as Table 21·2?

8. Irregular galaxies are dominated by stellar associations, open clusters, and gas and dust clouds, all of which indicate stellar youthfulness. Does this prove that the galaxies themselves have only recently formed?

9. Using principles of star-formation and stellar evolution from Chapters 14 and 15, explain why the prominent light from star-forming regions in galaxies comes from massive, hot, blue stars.
  a. *Why are these stars not seen in regions where star-formation has ended?*
  b. *What population is indicated by hot, blue stars?*

Advanced Problems

10. Use the small angle equation, p. 38, to solve the following problems:
  a. *What is the angular diameter of the main stellar part of the Andromeda galaxy, if it is 35 kpc across and 660 kpc away? How does this compare with the angular size of the moon?*
  b. *What is the angular diameter of the 3-parsec-diameter bright nucleus at the center of the Andromeda galaxy?*

11. If spectral lines of stars observed on the right side of a galaxy were red-shifted 1.65 A relative to those at the galaxy's center, while those on the left side were blue-shifted by the same amount, what would you conclude to be the rotation velocity of these stars? Assume that the spectral lines normally occur at wavelength 5,000 A.

12. If the stars in Problem 11 were measured to be on the outer edge of the galaxy being observed, at a distance 4 kpc from its center, what would be the mass of the galaxy? (Hint: Use the circular velocity equation, p. 73.)

Projects

1. Locate the Andromeda galaxy by naked eye. Compare its visual appearance with its appearance in binoculars and telescopes of different sizes. Across what diameter can you detect the galaxy? ($1° = 12$ kiloparsecs at the Andromeda galaxy's distance.) Why do most photographs show a large central region of constant brightness, while visual inspection reveals a sharp concentration of light in the center? Can dust lanes or spiral arms be observed? (Check especially with low magnification on large telescopes of 20–36 inches aperture, if available.)

2. If photographic equipment is available, take a series of exposures with different times, such as 1 min., 10 min., 100 min. Describe some of the differences in appearance. What physical relations are revealed among different parts of the galaxy?

3. Make similar observations of other nearby galactic neighbors of the Milky Way.

The Expanding Universe of Distant Galaxies

The galaxies we have seen up to now vary in size, shape, and distance, but are basically understandable. They have the same two populations of stars found in our galaxy, and we can at least partially explain their varied forms by imagining different conditions of mass and rotation in local regions of the early universe, as matter broke into protogalactic clouds.

But as we go still farther away from the earth we discover strange objects that strain all our existing theories. This process has happened before in history. For example, medieval people developed a theory in which all worlds move around the earth, but the telescope showed the revolu-

tion of Jupiter's moons, and destroyed the idea that all motions are earth-centered.

Similarly, in the 1600s Newton and other scientists worked out a new way of picturing the behavior of material in the universe, but by 1900 new observations conflicted with the Newtonian picture. For example, it was discovered that all observers perceive all light beams as traveling at the same speed regardless of the observer's own relative motion. This discovery led to the concept that matter cannot travel faster than light, which conflicted with the Newtonian concept that matter can travel at any speed. Einstein and his generation developed the *relativistic physics* of matter and energy. We live in the relativistic era, and our relativistic amendments to Newtonian physics give reasonably satisfying explanations of most phenomena that we have encountered in the universe so far, such as the motions of planets, the generation of nuclear energy inside stars, or even the evolution of massive stars into black holes.

But in this chapter we have to ask whether our relativistic picture of the universe really works. Is it consistent with what we see, or think we see? Although no one has yet proved our current physics wrong, some observers of very remote galaxies suggest interpretations that require new physical relations. Thus, the basic structure of the universe is not entirely clear.

The Largest
Distance Unit of All

In dealing with stars, we used the parsec as a distance unit. In dealing with the Milky Way and nearby galaxies, we used the kiloparsec, equal to

1,000 parsecs. The distances to remote galaxies are so great that we must use the **megaparsec,** which is a million parsecs or 1,000 kiloparsecs. A megaparsec (mpc) is about 3×10^{19} km, or 30,000,000,000,000,000,000 km!

Clusters of Galaxies

Among distant galaxies, as among our own local group, clustering is the rule. As shown in Figures 22·1 and 22·2 galaxies often appear in groups ranging from pairs to collections of several hundred, with ellipticals, spirals, and irregulars mixed together in a volume about 5 megaparsecs across. **Clusters of galaxies** boggle the mind. They are so huge that most of them subtend an angle of more than a degree, even though they are up to 100 megaparsecs away. Thus a distant cluster may cover a patch larger than the full moon, but be too faint to be seen by the naked eye. Figure 22·3 shows one of these large, remote clusters. Some astronomers suggest that even clusters of galaxies may be grouped into larger associations called **superclusters.**

Clustering helps observers to interpret galaxies, because all galaxies in a given cluster are at the same distance. Thus, observers can compare sizes and brightnesses of different types without the complications of trying to correct for differences in distance.

Discovering the Red Shift

Routine cataloging of the Doppler shifts of galaxies led to an unexpected and amazing result. By 1914, Doppler shifts had been published for 13 of the brightest galaxies. Strangely, most

▲
Figure 22·1 **A pair of spiral galaxies, NGC 5432 and 5435. The difference in orientation shows that they are not co-planar. Faint filaments connect them, due possibly to tidal interactions.** (Lick Observatory)

Figure 22·2 **A cluster of at least five galaxies, including NGC 6027, in the constellation Serpens.** (Hale Observatories)
▼

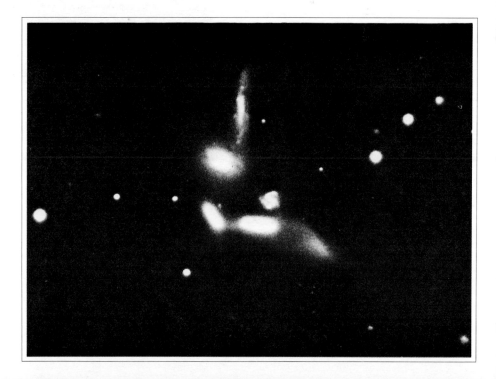

were red shifts, rather than a random mixture of red and blue shifts. This phenomenon, called the **red shift of galaxies,** apparently indicates that most are moving away from us.

Around 1920, Edwin Hubble began further Doppler-shift observations with the 100-inch reflector at Mount Wilson. Hubble soon got new results. By 1925, about 45 galactic Doppler shifts had been published, and Hubble (1936) noted that, again, red shifts completely dominated the list:

> *The numerical values of the new velocities were found to be surprisingly large and of an entirely different order from those of any other known type of astronomical body.*

Red shifts corresponding to recession speeds as high as 1,800 km/sec appeared on this early list. As Table 22·1 shows, dramatically higher speeds have since been measured.

*Red Shift
Correlated with Distance*

In 1928, American physicist H. P. Robertson discovered another curious fact: The more distant the galaxy, the greater the red shift. And the relation is linear; the red shift is directly proportional to the distance. These statements are supported by a variety of observations. For example, Figure 22·4 shows five galaxies, their distances, and their red shifts. Additional studies show that the greater the red shift, the fainter the stars in the galaxy (indicating greater distance),

the smaller the angular diameter of the galaxy (again indicating greater distance), and the smaller the angular size of the cluster of galaxies in which the observed galaxy lies.

**Interpreting
the Red Shift:
An Expanding Universe?**

According to everything we have learned so far, a red Doppler shift

means that the source is moving away from us. Therefore, astronomers concluded from the observations that all distant galaxies are moving away from us, and the farther away they are, the faster they are receding. In 1933, English astronomer Arthur Eddington titled his book on this subject with the famous phrase, **The Expanding Universe,** the name which has been given to this concept ever since, though the phenomenon might better be called **mutual recession of galaxies.**

Hubble's Relation and Recession Velocities

Hubble's relation expresses how the recession velocity increases with distance. Hubble found a constant proportion between velocity and distance, known as **Hubble's constant** H. Recession speeds of various galaxies at different distances yield a current value of $H = 57$ km/sec per megaparsec (with an uncertainty of about 6 km/sec per megaparsec; Sandage and Tamman, 1974). This means that galaxies one megaparsec away are receding from us at about 57 km/sec on the average. Galaxies twice as far away recede twice as fast. Galaxies at 10 mpc recede at about 570 km/sec; galaxies at 100 mpc recede at about 5,700 km/sec; and so on.

Conversely, now that this relation has been measured, we can use it to estimate distances. A galaxy with a red shift corresponding to recession at 5,700 km/sec is estimated to be about 100 mpc away.

Are We at the Center of the Universe?

If all distant galaxies are moving away from us, does this mean that we are located at the *center* of a vast explosion—that is, at the *center* of the universe? That would mean we had come full circle since the Copernican revolution removed us from the center of our solar system, destroying the theory that humanity occupies a special position. The answer is that *we are not necessarily at the center*. Observers in any other galaxy would also see galaxies moving away from them, because *all galaxies are moving away from each other*.

In other words, we may very well be riding out the aftermath of a vast explosion, but the view from one flying spark is like the view from any other. There is no outside reference point to define who is closest to the center, unless we found we were close to the edge of the swarm of sparks. But no astronomer has yet found an "edge" to the swarm of galaxies. They seem to go on and on.

Red-Shift Theory I: Cosmological Red Shifts

Everything we've said so far assumes that the observed red shifts should truly be interpreted as **Doppler shifts**. The Doppler effect has been confirmed in earthly laboratories, with planets, and with stars. But the idea that it applies to *all* galaxies as well is only an assumption, called the **theory of cosmological red shifts**. (This assumption, incidentally, underlies the last column of Table 22·1.) But we don't know for certain that the Doppler effect explains the red shifts observed in the spectra of all distant galaxies. Some recent observers think it does not. There are other possibilities.

Red-Shift Theory II: Noncosmological Red Shifts

Some physical effect besides motion away from us could also cause red shifts. Then the shifts would be ambiguous: Part might be caused by recession and part by something else. And in fact we know at least one other possible cause of red shifts. The theory of relativity shows that light emitted from regions of very high gravity, such as neutron stars or the environs of a black hole, will be red-shifted.

Table 22·1 Selected Clusters of Galaxies

Cluster Name	Estimated Distance (mpc)	Diameter (kpc)	Number of Galaxies	Galaxies per mpc³	Recession Velocity (km/sec)
Local group	0.6	1,000	≥26	50	0
Virgo	19	4,000	2,500	500	+1,180
Pegasus I	65	1,000	100	1,100	+3,700
Cancer	80	4,000	150	500	+4,800
Perseus	97	7,000	500	300	+5,400
Coma	113	8,000	800	40	+6,700
Hercules	175	300	300	20,000	+10,300
Ursa Major I	270	3,000	300	200	+15,400
Leo	310	3,000	300	200	+19,500
Gemini	350	3,000	200	100	+23,300
Boötes	650	3,000	150	100	+39,400
Ursa Major II	680	2,000	200	400	+41,000
Hydra	1,000	?	?	?	+60,600
3C 123 and cluster	2,500	?	?	?	+135,000?

Figure 22•4 **Photographic proof of red shifts of remote galaxies.** The left column shows galaxies (note decreasing angular size, interpreted as caused by increasing distance). The right column shows spectra: white lines at top and bottom of each spectrum are emission lines produced in the instrument for comparisons, being similar in each spectrum. A pair of dark absorption lines (the H and K lines of gaseous calcium) can be detected in each galaxy's spectrum, above the head of the white arrow. The pair of lines is farther right (red) in each succeeding galaxy. The center column shows the inferred distance. (Hale Observatories)

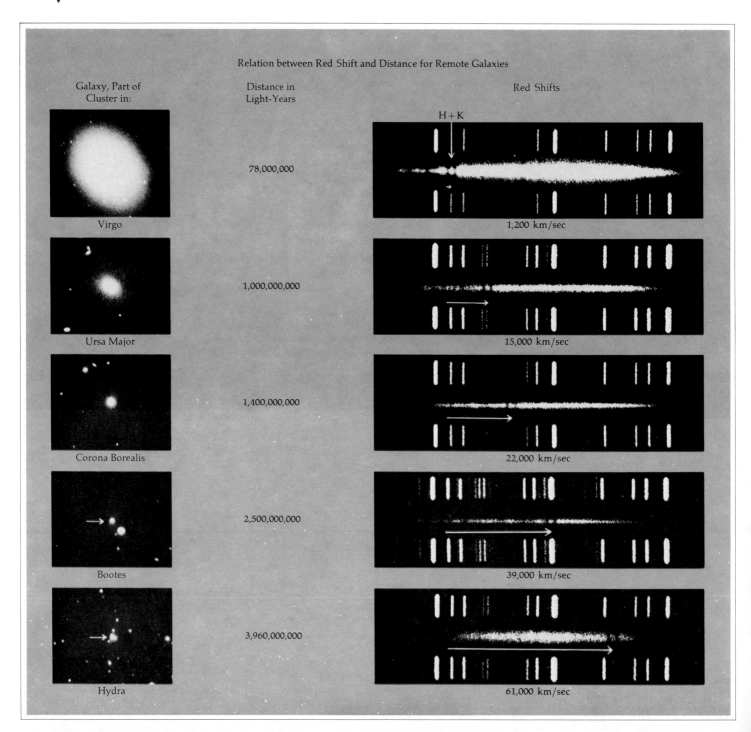

Relation between Red Shift and Distance for Remote Galaxies

Galaxy, Part of Cluster in:	Distance in Light-Years	Red Shifts
Virgo	78,000,000	H + K 1,200 km/sec
Ursa Major	1,000,000,000	15,000 km/sec
Corona Borealis	1,400,000,000	22,000 km/sec
Bootes	2,500,000,000	39,000 km/sec
Hydra	3,960,000,000	61,000 km/sec

Thus, at least some of the shifts we observe may be **gravitational red shifts.** Some astronomers have suggested still other sources of red shifts in the spectra of galaxies. Photons, for instance, might lose energy in their megaparsec journeys. These ideas are grouped as the **theory of noncosmological red shifts.**

If there is at least one nonrecessional cause for red shifts, why not abandon theory I altogether? The answer lies in Hubble's and Robertson's discovery that among nearby galaxies, red shift is correlated with motion and distance, not with mass, gravity, or other properties. Some nearby galaxies even show blue shifts, indicating approach velocities not explained by theory II.

Which theory is right? Is the universe really expanding, or are the more distant galaxies red-shifted for some other reason? The answer is difficult, but some additional data come from a different class of galaxies, which we next consider.

Intensely Radiating Galaxies

As we probe farther into the universe, we discover that a number of galaxies are peculiar in form, wavelengths of radiation emitted, and amount of radiation. Many radiate thousands of times more energy than ordinary galaxies, allowing them to be detected from greater distances.

Many of these galaxies have strong radio emissions, intensely bright nuclei, or evidence of explosions in their nuclei, and are called **galaxies with active nuclei.** From a sample of 126 well-studied nearby galaxies, the French-American astronomer, Marie-Helene Ulrich (1974) found that at least 5 percent of all galaxies have active nuclei. Here we will describe some types that have clarified (or

perhaps muddled!) our understanding of the most distant parts of the universe.

Radio Galaxies

In the late 1940s, when the sky was first mapped with sensitive radio receivers, several of the brightest radio sources were found to be galaxies, called **radio galaxies.** For example, as shown in Figure 22·5, radio source Centaurus A coincides with galaxy NGC 5128, which looks like an elliptical galaxy with a strange dark lane of obscuring dust along its front edge. Another radio source, Virgo A, coincides with NGC 4486 (M 87), an elliptical galaxy with a peculiar radial jet extending from its bright core. One of the two strongest such sources (Cygnus A) is a galaxy with an unusual double nucleus. Thus radio galaxies are unusual both in their radio radiation and in their forms.

In the mid-1950s, astronomers found that the radio radiation was **synchrotron radiation,** produced by a hot gas in which ions in magnetic fields are accelerated to nearly the speed of light. Some very energetic process must be going on in these galaxies to get the gas so hot and so agitated. Compared to the energy generated by a normal galaxy at radio wavelengths, about 10^{38}–10^{40} ergs/sec, strong radio galaxies generate some 10^{40}–10^{45} ergs/sec, hundreds or millions of times more.

When mapped, the radio-emitting regions of radio galaxies are usually larger than the optically visible galaxy, and often the radio emission is concentrated in two lobes, one on either side of the galaxy, as shown in Figure 22·5. No ordinary spiral galaxies are strong radio sources, but many giant ellipticals are. Other emitters include

irregulars, rare galaxies with double nuclei, or other types with unusual shapes.

The radiation from most of these galaxies probably comes from energy released during violent explosions in their centers in the past, as will soon be clearer. However, before pursuing that evidence, we turn to a source of energy that may explain the emissions from a few other galaxies.

Interacting Galaxies

Some radio and other kinds of radiation may be produced by **interacting galaxies,** or galaxies experiencing collisions or very close approaches. Collisions occasionally occur within clusters of galaxies close enough together to interact. Stars in galaxies are so far apart that the colliding galaxies could move through each other without many stellar collisions or gross disruption of the star patterns; but diffuse clouds of gas would collide, with two results.

First, the gas would be heated and the atoms would be excited, causing the observed radio emission. Second, collisions or near-misses could tear clouds of gas out of the galaxies, producing long filaments observed between some pairs of galaxies (see Figures 22·6 and 22·7). If galaxies merely pass near each other, as seems to be the case in Figure 22·1, tidal forces will cause distortions, just as the moon distorts the ocean surface on the nearby earth. By the time the galaxies are moving apart, these forces may have pulled huge streamers of gas from both. Remarkable computer simulations of such events closely match the observed forms of galaxies shown in Figure 22·8.

As we mentioned in Chapter 21, the Magellanic stream that links our

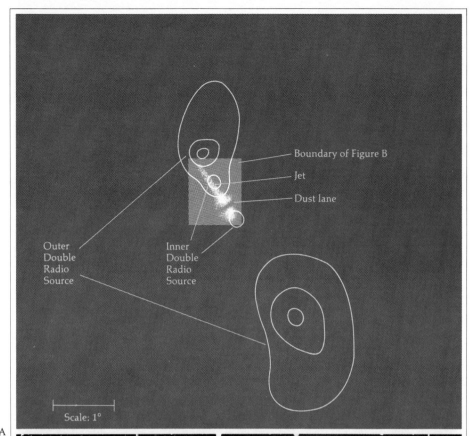

Boundary of Figure B

Jet

Dust lane

Outer
Double
Radio
Source

Inner
Double
Radio
Source

Scale: 1°

A

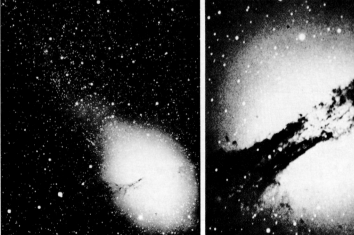

B   C

Figure 22•5 **The unusual galaxy 5128, coincident with intense radio source Centaurus A, about 4.4 megaparsecs away. The galaxy appears to be a giant elliptical, but with an unusual dark dust lane superimposed across its center. A. Optical features (light, stippled) compared to radio features (contours), showing optical jet and two pairs of radio-emitting hydrogen** clouds, possibly caused by violent ejection of gas from the nucleus, nearly perpendicular to the plane of dust. B. Long exposure showing outer regions of the galaxy and faint, luminous jets extending toward upper left. Same orientation as A. See also Color Photo 28. (Cerro Tololo Inter-American Observatory) C. Enlarged shorter exposure of central bulge and dust lane. (Hale Observatories)

Figure 22•6 **Radio galaxy NGC 2623, another example in which long filamentary tails may be material ejected by tidal interactions of two close galaxies.** (Hale Observatories)

Figure 22•7 **Computer processing sometimes displays details too faint to show in normal photographic images. Special processing was used on this image of two distant galaxies to emphasize a diffuse filament of material connecting the two galaxies. The pair is receding from the Milky Way at 5,000 km/sec and is probably about 1,000 megaparsecs away.** (Cerro Tololo Inter-American Observatory)

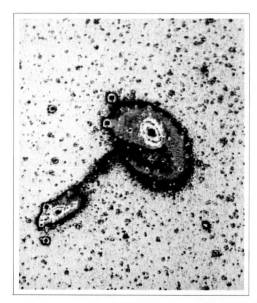

A

Figure 22·8 **A. Galaxies NGC 4038/9 have prominent curved streamers. Known as the Antennae galaxies, they may be an interacting close pair about 15 megaparsecs away.** (Hale Observatories) **B. Computer analysis of a near-collision between two galaxies, designed to simulate the appearance of NGC 4038/9. Circles are material from one galaxy; stars are material from the other. Tidal forces eject a streamer from each galaxy. Successive images are spaced about 200 million years apart.** (Alar and Juri Toomre, 1973)

galaxy to the nearby Magellanic clouds may be an example of a similar streamer.

Thus, galaxies may interact often enough to explain some of the more unusual galactic forms and also to explain some cases of radio emission. However, collisions are probably not common enough to explain all energetic radio emissions from galaxies. Another cause is in evidence, as we will now describe.

Seyfert Galaxies: Galaxies with Bright Centers

A step toward understanding radio galaxies came from reconsidering a peculiar type of optical galaxy discovered some years before radio astronomy flourished. In 1943 astronomer C. K. Seyfert studied about a

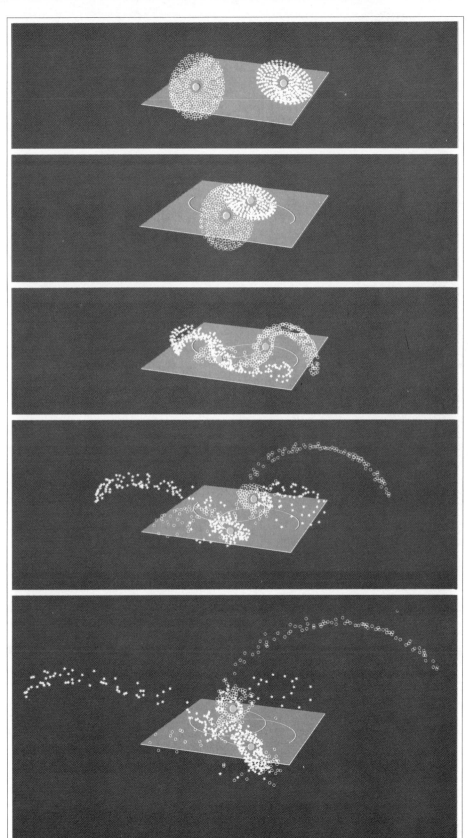

B

dozen unusual galaxies with small, intensely bright nuclei; bright spectral emission lines; and brightness fluctuations. Seyfert concluded that the nuclei of these galaxies had hot gas in violent motion, often with velocities of several thousand km/sec. About two percent of all galaxies have these properties. They came to be called **Seyfert galaxies.** Figure 22·9 is an example.

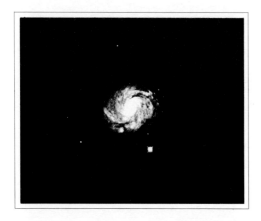

Figure 22·9 **Seyfert galaxy NGC 1068 (M 77) resembles an ordinary spiral except for the small, intensely bright nucleus, overexposed in this image. The ratio of brightness of the central region to the spiral arms is much greater than in normal spirals.** (Lick Observatory)
▼

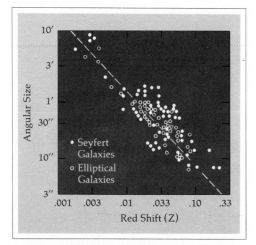

In the 1960s Seyferts were found to have radio and infrared emission similar to that of radio galaxies. Thus, they might be a "missing link" between ordinary and radio galaxies. As shown in Figure 22·10 they share the general trend of increasing red shift with increasing distance.

Seyferts' unusually large and varying amounts of energy imply very energetic explosive bursts in their nuclei. Of course, even ordinary galaxies have strikingly energetic events occurring in their centers—synchrotron radiation and expanding hydrogen clouds occur in our own galaxy as well as others, in a core region about 100 parsecs across. But the curiosity of Seyferts is that they are even more energetic, especially at radio and infrared wavelengths. Could Seyferts represent some transient evolutionary state of ordinary galaxies? Some investigators suggest that the Milky Way itself may have been a Seyfert galaxy a few hundred million years ago during the hypothetical explosions that launched the expanding 3-kpc spiral arm.

The case for a link between Seyferts and ordinary galaxies improved in 1974 when Soviet astronomer E. Y. Khachikian and American D. W. Weedman found that out of 71 studied Seyferts, 42 percent have possible or proven spiral arms, relating them to the "tuning fork" sequence of ordinary galaxies. Also, 58 percent of the Seyferts have a jetlike or filamentary

◄ *Figure 22·10* **Evidence that increasing red shift in Seyfert and elliptical galaxies corresponds to increasing distance. Z is the red shift, expressed as a fraction of the original wavelength. Galaxies with the greatest red shift have the smallest angular size, indicating they are farthest away.** (After data of Khachikian and Weedman, 1974)

structure near their centers, suggesting violent or explosive events in the central regions.

The Mystery of the Galactic Nucleus

Figures 22·11 to 22·13 show examples of galaxies with active nuclei. No one has completely explained the cause of the energetic explosions that have probably occurred in such galaxies. Collisions of large stars, or supernovalike explosions of very massive stars, may create the necessary conditions, including gas temperatures of 5 million K and expansion speeds around 1,000 km/sec. The perplexity of theorists is expressed by astrophysicist W.C. Saslaw:

> What generates the enormous energies that pour from galactic nuclei? How do they evolve? And are new physical laws needed to understand them? Even though I've only asked three questions so far, already the ratio of questions to answers is infinite! At present there is no complete, comprehensive, and compelling theory of galactic nuclei.

The real problem is what goes on in the central 20 or 30 parsecs of galaxies, which is difficult to interpret from photographs alone, as shown in Figure 22·14. It remains a problem because distant galaxies are so far away, and the center of our own galaxy, though only 9 kiloparsecs away, is obscured by dust. Many peculiar galaxies, including some with missing nuclei, suggest mysteries associated with the nucleus (Figure 22·15). Saslaw quotes the poet Robert Frost:

> We dance around in a ring and suppose, But the Secret sits in the middle and knows.

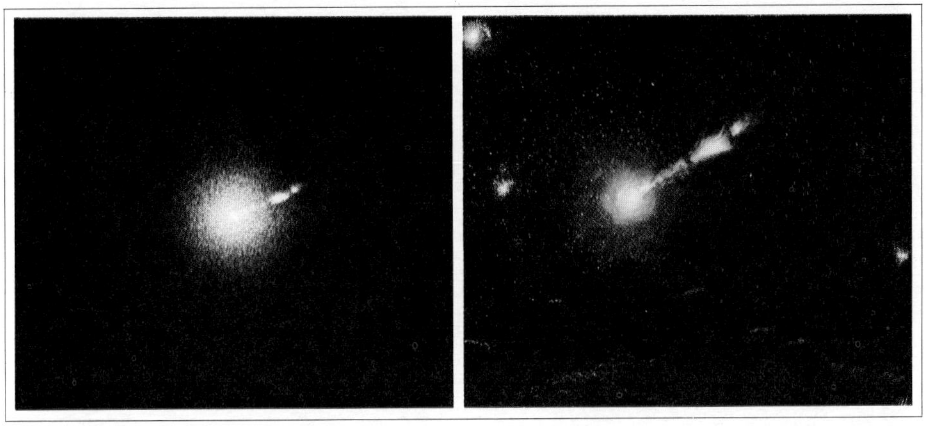

A B

▲
Figure 22•11 **A. The central core of the elliptical galaxy NGC 4486, coincident with the strong radio source, Virgo A. This short exposure shows an intensely bright nucleus from which a jet of glowing material appears to be emerging. The jet glows by synchrotron radiation, caused by hot gas with electrons moving at nearly the speed of light. The data suggest violent, energetic events in the nucleus of the galaxy, about 13 megaparsecs away.** (Hale Observatories) **B. Imaginary view from a planet in the radio galaxy NGC 4486 (M 87), shown in preceding photograph. The jet of material being ejected from the bright nucleus is prominent from the hypothetical planet's desolate surface.** (Painting by James Hervat)

The Discovery of Quasars

Quasars (an acronym for *qua*si-*s*tell*ar* radio source[1]) are yet another strange type of object that may (or may not) represent a still more energetic type of active galaxy. Quasars were discovered when optical astronomers began to photograph visible objects that could be identified with the sources of certain radio signals. By 1960, although many radio sources had been photographed and identified as galaxies, a few seemed to coincide only with faint, starlike objects. Were

[1]Many astronomers prefer QSS for *q*uasi-*s*tellar *s*ource.

Figure 22•12 **Studies of the irregular galaxy M 82 (NGC 3034) showing different internal details. The galaxy is about 3,000 kpc away and shows evidence of explosions in its center about a million years ago. A,B. Short exposures showing extensive clouds of dark, obscuring dust, in light of neutral oxygen. C. White-light exposure showing irregular outer structure. D. Long exposure in red light of hydrogen emission, showing vertical streamers and tangled gaseous filaments extending about 3 kpc from the galactic center, evidence of explosive eruption of hot hydrogen.** (A,B: R. Williams, Steward Observatory, University of Arizona; C,D: Hale Observatories)
▼

D

A

B

C

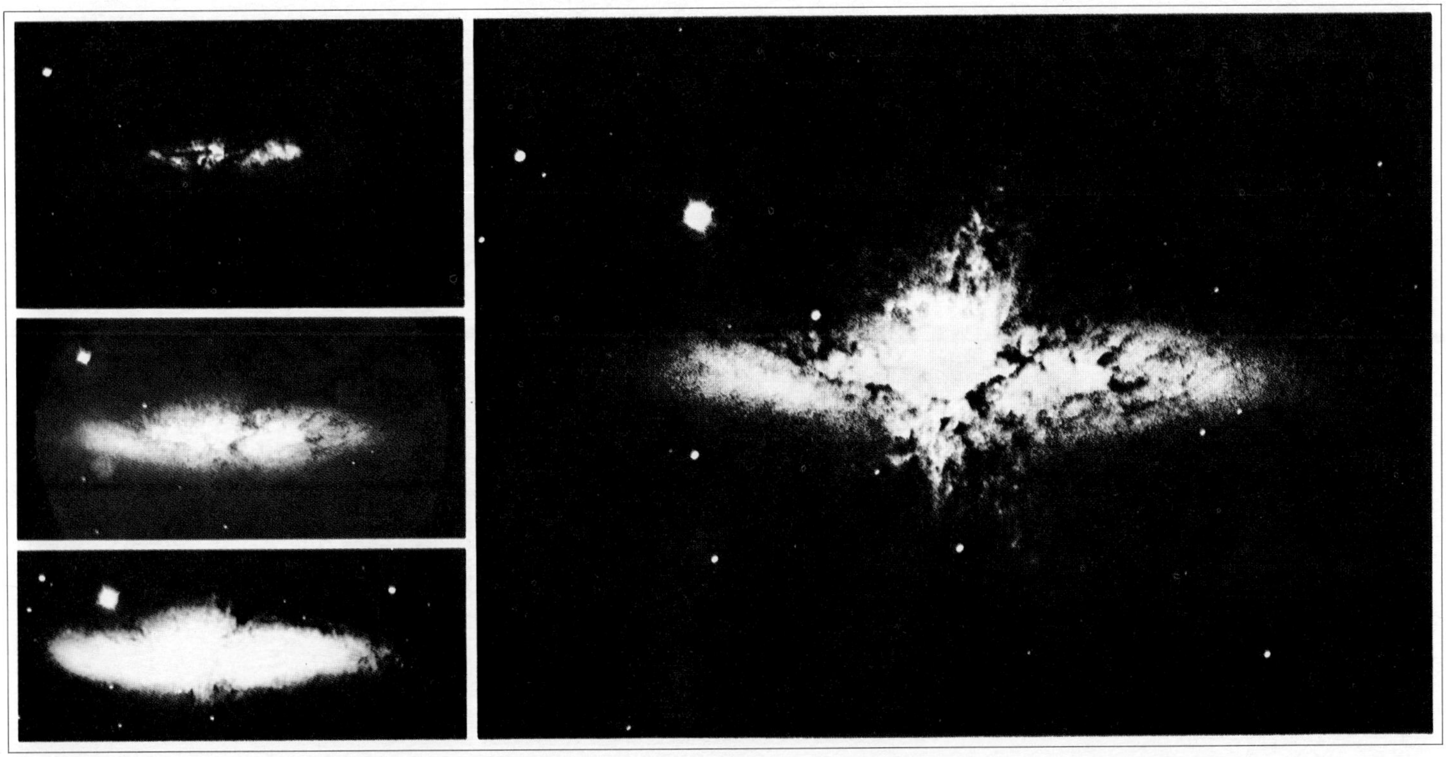

▲

Figure 22·13 **Evidence for explosive activity in the nucleus of Seyfert and radio galaxy NGC 1275. In the light of hydrogen alpha emission, the central region is resolved into semi-radial streamers, resembling much** smaller-scale supernova debris as in the Crab nebula. Several apparent elliptical galaxies (fuzzy ovals) are nearby (sharp circular images are foreground stars in our galaxy). The galaxy lies about 50 megaparsecs away; filaments extend about 14 kiloparsecs from the galaxy. (Kitt Peak National Observatory, courtesy Roger Lynds)

Figure 22·14 **These two prints from the same ▶ negative of NGC 1398 illustrate the difficulties of analyzing galactic forms from images of the nucleus. When only the bright inner structure is visible, the galaxy seems to consist of a bright bar and ring or tight spiral arms; the second print with overexposed nucleus reveals faint, open spiral arms. NGC 1398 is an example of class SBb(r), a barred intermediate spiral with ring structure.** (Original image from Hale Observatories, reprinted by author)

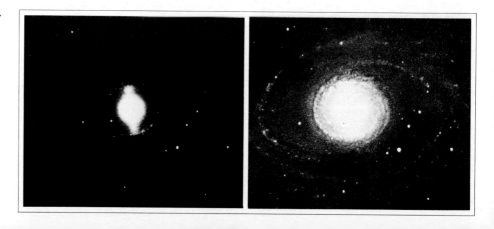

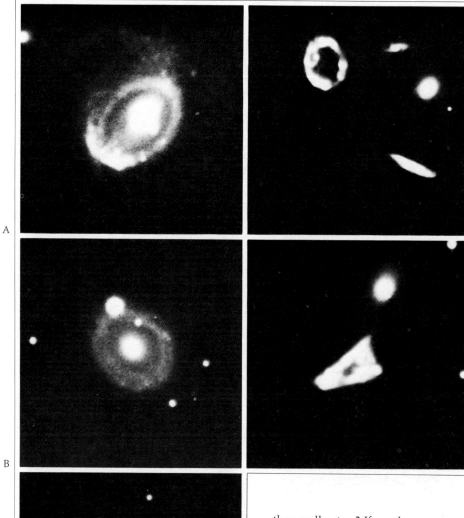

A

B

C

D

E

▲
Figure 22•15 **Peculiar galaxies, including examples with "missing nuclei." Causes of these forms are unknown; evolution from A to E is a possibility.** (Kitt Peak National Observatory)

they really stars? If so, they were unusual, sometimes with unrecognizable emission lines in their spectra, and sometimes accompanied by faint wisps of nebulosity, as shown in Figure 22•16. Hence they were designated *quasi-stellar.*

A check of old photographs of quasar 3C 273 (number 273 in the Third Cambridge Catalog of radio sources) showed that it has varied irregularly in brightness. Palomar observer Martin Schmidt discovered that its spectral lines showed a whopping red shift of 0.16 of the normal line wavelength. This was the first proof of *large red shifts for quasars.*

Most other known quasars were about 100 times fainter than 13th

magnitude 3C 273, but in a few years their spectra were also obtained, and many of their Doppler shifts were found to be even larger. For some, red shifts of 1 or 2 were found, meaning that the shift in wavelength was equal to or twice the original wavelength.

In 1965 other objects were found that appeared to be faint bluish stars and had spectra similar to those of quasars, but were not strong radio sources. These were identified as radio-quiet quasars, sometimes called **QSOs**, or *quasi-stellar objects.*

How were quasars and QSOs to be interpreted? As Figure 22•17 shows, they are very difficult to observe. The interpretation is based in part on spectra. The spectra of some quasars have absorption lines as well as emission lines. Generally the absorption lines are less red-shifted than the emission lines, indicating that a surrounding cloud of absorbing gas is moving outward from a bright emitting region in the quasar center. Also, forbidden lines sometimes appear in quasar spectra, indicating rarefied gas associated with the objects.

Some quasars seem to be related to Seyfert galaxies. Similarities include small angular size for the radio-emitting regions (less than 0″.001 in some cases); their spectra; and variations in brightness. Quasar 3C 273 (Figure 22•16) has a faint radial jet extending from a bright nucleus, reminiscent of the brilliant radio galaxy NGC 4486 (Figure 22•11).

Changes of brightness in quasars help us to interpret them. Quasar 3C 446 varied by a factor of about 20 in a year. Others have changed by about a factor of 2 in a few months; and others have changed by a few percent in as little as 15 minutes. Radio variations are also known, and need not be accompanied by optical variations. The time periods of varia-

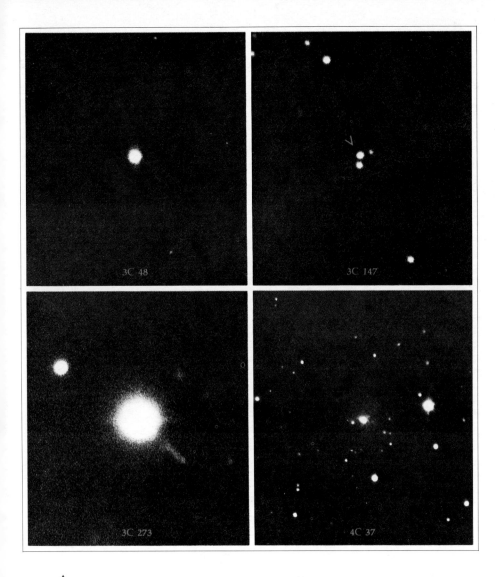

T times the velocity of light. By this rule, 15-minute variations must originate in a region less than 1.8 AU across, or about the size of the inner solar system. The main light-emitting parts of quasars must be less than a few percent of a parsec across. These small regions are often surrounded by larger, expanding envelopes of diffuse gas, and probably have intense magnetic fields and high-energy particles, which account for synchrotron radiation and spectral features.

Of course, we want to go beyond this crude description. How much energy do quasars radiate? What is the source of the energy? Are quasars really galaxies? To answer these questions, we need to know how far away quasars are—which requires us to interpret their immense red shifts, and brings us back to the problem of our two theories of red shifts.

Quasars:
Interpretation I

Interpretation I is the more conventional of the two, using the theory of *cosmological red shifts*. We apply the conventional conversion of red shift into distances and find that quasars, moving away at 80 or 90 percent the speed of light, must be at enormous distances—thousands of megaparsecs, as shown in Figure 22·18. If they are that far away, they must have incredible luminosities to be visible to us at all. As shown in Table 22·2, radio and optical data yield luminosities up to 10,000 times those of normal galaxies, and hundreds of times those of strong radio galaxies. They cannot be single stars, but must be galaxies, or galaxylike groupings.

According to this interpretation, then, quasars are the most energetic of all known galaxies. Galaxies would

▲
Figure 22·16 **Examples of quasars. 3C 48 and 3C 147 are typical examples, showing quasi-stellar images. 3C 273 is the brightest and probably closest quasar, and is accompanied by an apparent jet of gaseous material. 4C 37 also shows some indication of nebulosity immediately adjacent to the** image; the fainter image to its left is a galaxy. Distances may be around 1,000 megaparsecs. (First three photos, Hale Observatories; 4C 37, Mauna Kea Observatory, University of Hawaii, courtesy Alan Stockton)

tions help set limits on the *size* of the region that is varying in brightness. For example, if a brightening takes one year, the region must have a front-to-back dimension no larger than one light year.

The reasoning is as follows: Say the region were bigger. Then, if the whole region brightened simultaneously with the front side, an earth-bound observer would have to wait *more* than a year after the frontside signal arrived for the backside signal to arrive. The general rule is that if a variation takes time *T*, the size of the varying region should be less than

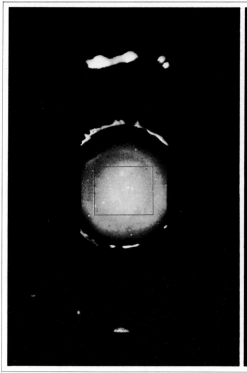 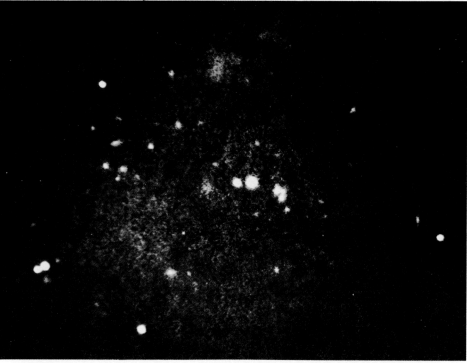

▲
Figure 22•17 Difficulties in studying quasars are indicated by these images of double quasar 4C 11.50. The left image is at the true scale of the original 1-inch wide photographic plate, with its circular field of view in the center: galaxy images are less than ½ mm across. The black box shows the portion greatly enlarged at right. Enlargement reveals that quasars (the bright pair, center) lack detail. The complex image just right of the quasars is a cluster of galaxies; this cluster and the brightest quasar have red shifts of 0.43. Left quasar has a red shift of 1.90. Discrepancy in seemingly associated quasars suggests red shifts may be noncosmological (see text). The faint honeycomb pattern is from fiber optics in the image intensifier used to make the photograph. (Mauna Kea Observatory, University of Hawaii, courtesy Alan Stockton)

Table 22•2 Energy Relations between Quasars and Galaxies		
Type of Galaxy	Total Luminosity (radio + optical)	Total Energy Involved in Gas Motions and High-Velocity Particles
Normal galaxy	10^{43}–10^{44} ergs/sec	10^{54}–10^{55} ergs
Strong radio galaxy or Seyfert galaxy	10^{43}–10^{45}	10^{56}–10^{60}
Quasar (if distant) (Interpretation I)	10^{46}–10^{47}	10^{60}–10^{61}
Quasar (if near) (Interpretation II)	10^{43}?	10^{56}?

thus range from ordinary (nonactive) galaxies through the active Seyfert and radio galaxies to quasars.

How might a galaxy reach such an energetic state? One hypothesis is that huge concentrations of dust and gas could contract to form enormous superstars. Such supermassive stars would be unstable and would create super-supernovae, which might explain the strong radiation. However, some theorists have questioned whether even super-supernovae could generate sufficient energy to be visible at such distances. Other hypotheses proposed include:

1. Supernova chain reactions (one supernova setting off another nearby).

2. Gas and dust falling onto pulsars or black holes.

3. Rapidly rotating massive objects with strong magnetic fields.

4. Collisions of massive stars.

5. Collisions of matter and hypothetical antimatter (matter thought to be made from positive electrons and negative protons).

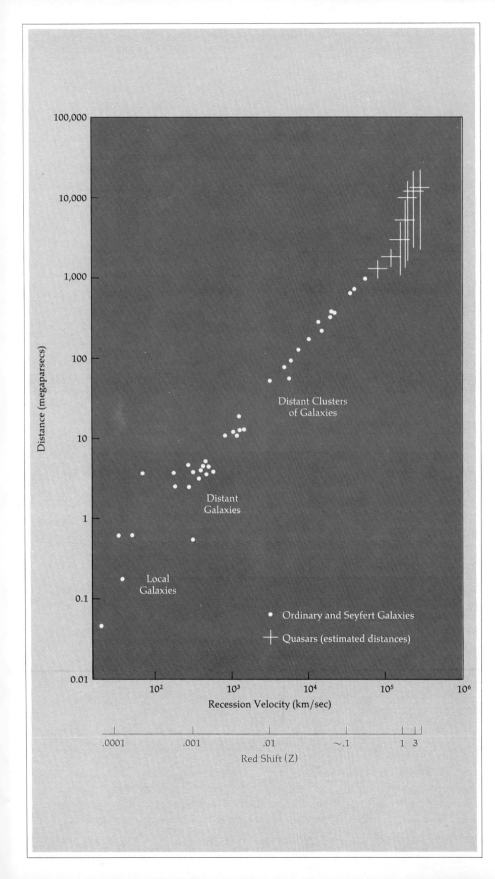

Figure 22·18 **Estimated distances of galaxies and clusters of galaxies with various observed red shifts and recession velocities. Vertical bars for quasars suggest uncertainties in their distances, based on "Interpretation I."**

6. Mathematical paradoxes in relativity theory.

7. Physical laws as yet undiscovered.

According to all of these interpretations, quasars are so far away that their light takes billions of years to reach us. Thus, we would be *seeing them as they were billions of years ago,* perhaps at about the time our own galaxy was forming. Their radiation, then, may represent conditions uncommon in galaxies today, but common during the early stages of galaxy formation. Some theorists think that physical conditions or even basic physical "laws" may at that time have been radically different from what we observe today. In any case, if Interpretation I is correct, quasars bring us light from the distant past. As astronomer Arthur Eddington said:

> . . . *cosmic radiation is a museum—a collection of relics of remote antiquity. These relics are stamped with an inscription indicating the dimensions of the (cosmic) world in its earliest ages. Whoever ultimately identifies the subatomic process originating the rays will be able to read the inscription. . . .*

We have not yet learned to read the inscription on the photons from radio galaxies, Seyferts, and quasars. All we can make out is a collection of disjointed passages—perhaps in different languages—hinting at violent events.

The strongest point in favor of Interpretation I (that quasars are very distant, very violent objects) is its consistency with the theory of the expanding universe and the conventional interpretation of red shifts. The weakest point is that the required energies are difficult to explain. Nevertheless, most astronomers favor Interpretation I.

Quasars: Interpretation II

Supporters of the noncosmological theory of red shifts argue that quasars are not very far away. Although Interpretation II has not been proved correct, it is supported by four arguments.

First, if quasars are not far distant, they would not need extremely high energies and radiation rates to be visible.

Second, a surprisingly large number of quasars lie within a few minutes or seconds of arc from easily recognized galaxies of much lower red shift. Some even appear to be *connected* to these galaxies by bridges of faint gas. Eight such connections have been advanced by astronomer Fred Hoyle to argue that quasars are objects attached to nearby galaxies. Even among relatively ordinary radio galaxies there is evidence that compact clouds have been ejected from galactic centers along the galaxies' polar axes, and some observers think quasars might be special clouds of matter shot out of galaxies. Unfortunately, these arguments are only statistical: The apparent alignments of quasars, galaxies, and filaments of gas might be merely by chance.

A third argument for Interpretation II is that when we compare quasars' apparent magnitudes (corrected for red shift) with their red shifts, we do not find the kind of dimming with increasing red shift that we would expect if the shifts were caused by

recession. A rebuttal to this argument is that the state of evolution of stars in the primitive quasar-galaxies could explain the observed relation.

But the fourth and strongest argument for Interpretation II is the discrepancy in red shift among different galaxies that seem to be members of the same cluster, such as those in Figure 22·19. Table 22·3 lists five

Figure 22·19 **Stephan's quintet, a cluster of galaxies (NGC 7317–20), gives evidence that some red shifts may not be directly related to distance. The galaxies are all believed to lie at the same distance, which can be estimated from various indicators, such as** ▼

angular size. Red shifts of the four small galaxies, however, differ from that of the large galaxy (upper left) and are larger than expected from distance estimates. (Lick Observatory)

Table 22·3 Discrepancies in Red Shifts within Galactic Clusters

Cluster	Number of Galaxies	Lower Red Shift[a] (no.)	Higher Red Shift[a] (no.)	Discrepancy (km/sec)
Stephan's quintet	5	0.0027 (1)	0.02 (4)[b]	+ 5,200
Seyfert's sextet	6	0.015 (5)	0.066 (1)	+ 15,300
VV172 system	5	0.053 (4)	0.12 (1)	+ 20,000
NGC 7603 pair	2	0.026 (1)	0.053 (1)	± 8,000
NGC 4319/Makarian 205	2	0.006 (1)	0.07 (1)	± 19,000
NGC 2903	4	0.002 (1)	0.01–0.03 (3)	± 6,000

[a]Red shift = shift-in-wavelength/wavelength.

[b]Radio and other studies indicate the correct distance is low, corresponding to low red shift. Thus it would appear that in all of the clusters, the higher red shift is anomalous, even when several galaxies have higher red shifts.

Source: Data from Balkowski et al. (1974); Hoyle (1972); Hodge (1972).

instances in which one galaxy differs in velocity from other members of the same cluster by 5,000 to 20,000 km/sec. If these are truly examples of large red shifts in nearby galaxies, then some large red shifts may be caused by something besides recession, and Interpretation I is weakened.

But what else could cause red shifts? One answer is gravity. According to Interpretation II, most of the light from highly red-shifted galaxies and quasars must come from regions of intense gravity—perhaps near extraordinarily massive, unstable objects that will explode, or perhaps have already exploded. Such explosions might even throw the objects out of their galaxies, producing quasars linked to galaxies by faint filamentary tails, as some observers claim.

Evidence that some galaxies eject radio-emitting masses comes from so-called *head–tail galaxies,* which have a bright "head" and a second component at the end of a diffuse "tail." Dutch radio astronomers who studied seven of these concluded that at least one explodes every few million years with an energy of 10^{55} ergs, ejecting pairs of radio sources. Between eruptions, the galaxies' nuclei are active (Miley et al., 1974). Such an interpretation could explain the pairs of radio sources often found on either side of peculiar galaxies, as described in Figure 22·5.

22

Summary

Once we pass beyond the local group and other nearby galaxies, we encounter still more galaxies in large clusters at distances of some tens of megaparsecs. Such clusters, and perhaps clusters of clusters, extend as far as we can see, which is at least some thousands of megaparsecs. As we probe to farther distances, we encounter serious problems of interpretation, which might be summarized by the following list of observations:

Fact: *Galaxies at greater distance than about one megaparsec have red shifts.*

Virtually certain: *The farther the galaxy, the greater the red shift.*

Very probable: *Among the nearer galaxies, which seem to have ordinary forms and luminosities, the red shifts indicate recession. These galaxies are moving away from us and from each other.*

Virtually certain: *Some kinds of galaxies have active nuclei and emit much greater amounts of energy than normal galaxies.*

Likely: *Some faint objects with extremely high red shifts may be intensely luminous galaxies at distances of thousands of megaparsecs, sharing in the recession and moving away from us at appreciable fractions of the speed of light. They are seen now as they appeared billions of years ago, shortly after they formed.*

Nearly all astronomers accept the first three or four steps in this scenario, and agree that the universe is expanding, in the sense that the galaxies are rushing away from each other. Astronomers are still debating whether some part of the largest red shifts might be caused by something other than recession. The most red-shifted objects are the puzzling quasars. Although many astronomers (probably most) think quasars are very distant galaxies with extremely high luminosity, some think that quasars may be nearer, perhaps associated with galaxies with exploding nuclei.

As we approach the frontiers—the farthest objects that astronomers can detect—we find many strange phenomena. However we interpret them, many galaxies apparently generate extraordinary energies. Explosions in their centers, possibly millions of years apart, may cause strong radio and optical radiation, and may expel filaments or condensations of matter. Observing distant galaxies, we see how these objects looked billions of years ago. Interpreting these observations has created some of the most exciting controversies in astronomy today. With research on these distant frontiers, we are probing the very nature of the universe itself.

Concepts

megaparsec

clusters of galaxies

superclusters

red shift of galaxies

The Expanding Universe

mutual recession of galaxies

Hubble's relation

Hubble's constant

Doppler shifts

theory of cosmological red shifts

gravitational red shifts

theory of noncosmological red shifts

galaxies with active nuclei

radio galaxies

synchrotron radiation

interacting galaxies

Seyfert galaxies

quasars

QSOs

Problems

1. How many miles is a megaparsec?

2. Suppose observers located in the Coma cluster of galaxies observe Doppler shifts in the spectra of our local group of galaxies, including the Milky Way.

a. *Would they see a red shift or a blue shift?*

b. *What sizes of shift would they observe? (Hint: See Table 22·1.)*

c. *What would they conclude about our group's velocity if they agreed with the theory of cosmological red shift?*

3. In what ways do Seyfert galaxies bridge the gap between ordinary galaxies and quasars?

4. Explain how the existence of groups of galaxies having members with widely discrepant red shifts, such as Stephan's quintet, may imply that quasars are not at the extreme distances usually assumed.

5. Since our galaxy's nucleus is a radio source, why is the Milky Way not considered to be a typical radio galaxy?

6. Why is the study of the most distant galaxies we can see related to the study of conditions around the time that our galaxy (and perhaps others) was forming?

Advanced Problems

7. For velocities much below the speed of light, the formula for Doppler shifts is:

$$\frac{shift\ in\ wavelength}{wavelength} =$$

$$\frac{velocity}{velocity\ of\ light} = \frac{v}{3 \times 10^{10}\ cm/sec}$$

Explain why Doppler shifts don't cause significant drifting or detuning of car radios as one drives toward or away from radio station transmitters.

8. Suppose a galaxy of stars emitted most of its light at wavelength 5,000 A, where the eye is sensitive. Suppose it were as far away as the quasar OH 471, which has a red shift of 3.4 (*shift* in wavelength = 3.4 × original wavelength).

a. *Assuming that red shifts are cosmological, find the wavelength at which most of its light would appear.*

b. *If it were an ordinary galaxy, would it look brighter or fainter than quasar OH 471?*

c. *Comment on how results of a and b would affect attempts to detect the galaxy.*

9. A certain faint galaxy is found to have a recession velocity of 5,700 km/sec. How far away is it if its red shift is cosmological?

10. Using the small angle equation, estimate the angular size of the galaxy in Problem 9 as seen from earth, if it has a diameter of 30 kpc, like the Milky Way.

Frontiers

Swim o'er the stars now,
Spawning man,
And couple rock, and break
forth flocks of children on the
plains
On nameless planets
Which will now have
names . . .

Ray Bradbury

Cosmology: The Universe's Structure

We are approaching the limits of our ability to probe the skies. The last chapter took us to galaxies so remote that they are not only almost too faint to see but also so red-shifted that much of their radiation is in infrared or radio radiation. What can be said about still more remote regions? Are they like ours? Do galaxies exist there? These questions belong to the field of **cosmology:** the study of the structure or "geography" of the universe as a single, orderly system.

In the prologue we compared humanity to explorers on a strange island in an unknown sea. To extend the analogy, this chapter finds us at a point in our explorations where we know the island is made of grains of sand, and we know how big it is, and we know that the sea is large, and that there are many other islands out there in the distance. And now we ask: Do the islands go on forever? Are there other types of land forms? Is the world flat? If so, does it have an edge, or is it infinite? Or is the world round?

The same kinds of questions are asked in cosmology. Are galaxies dotted across space indefinitely? Does space itself go on indefinitely, or does it come to some kind of end? Although these are exciting questions to debate, they are risky, because they tempt us to leapfrog beyond the limits of available observations and speak of concepts such as "infinity." Cosmology tempts scientists to speculate about the **universe,** defined as all matter and energy in existence anywhere, observable or not. Yet scientists cannot be sure how concepts of "infinity" apply to the real universe. And, as J. D. North (1965) remarked in his history of cosmology:

> It is easy to speak of the infinite. . . but it is difficult to speak of it meaningfully.

Why should normally cautious scientists attempt such speculations? There are several reasons. People have been asking cosmological questions and related religious questions for thousands of years. It is valid to do our best to answer these questions. Innate in the character of curious, restless humanity is the desire to know our surroundings. Unfounded speculation should not be part of science, but scientists can legitimately take the principles we have learned about matter and energy in our part of the universe and then ask, "What would happen if these principles apply in all space and time?" or, "What would happen if certain principles were different in early time or in distant space?" By such questions, cosmologists have been led to some startling conceptions that lie far outside everyday experience.

The Most Distant Galaxies

How do we really know that a certain galaxy is 1,000 megaparsecs away? To say we know by the red shift is not a fundamental answer, because no one has been able to prove conclusively that the Hubble relation of red shift and distance applies to all galaxies—although most astronomers think that it does. Furthermore, the Hubble constant rests on interpretation of Cepheid variable stars and other distance indicators. And these, in turn, rest on other techniques, primarily trigonometric parallaxes of stars close to the solar system. In other words, we have built a whole ladder of distance indicators, each rung resting entirely on the security of the preceding rung, as shown in Figure 23·1. We have been tacking the ladder together as we climb it. We use this ladder because we think it is sound, but we use it only with healthy skepticism.

The distances we have derived for distant galaxies have interesting ramifications about seeing back into ancient time. Table 23·1 shows the estimated time required for light to reach us from different places in the universe. These figures reveal a whole new problem about studying remote parts of the universe. We see the Andromeda galaxy as it was over two million years ago. We see the Virgo cluster as it was when the dinosaurs were dying.

Figure 23·1 **The astronomical distance scale.** ▶ **The accuracy of distance measurement at each level depends on the accuracy of measurements at lower levels.**

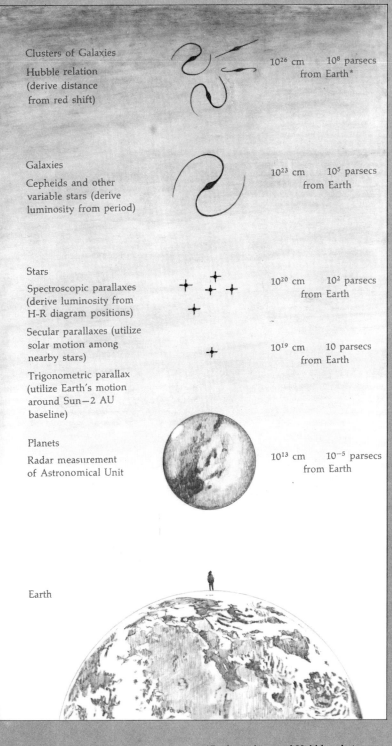

Clusters of Galaxies

Hubble relation
(derive distance
from red shift)

10^{26} cm 10^8 parsecs
from Earth*

Galaxies

Cepheids and other
variable stars (derive
luminosity from period)

10^{23} cm 10^5 parsecs
from Earth

Stars

Spectroscopic parallaxes
(derive luminosity from
H-R diagram positions)

10^{20} cm 10^2 parsecs
from Earth

Secular parallaxes (utilize
solar motion among
nearby stars)

10^{19} cm   10 parsecs
from Earth

Trigonometric parallax
(utilize Earth's motion
around Sun—2 AU
baseline)

Planets

Radar measurement
of Astronomical Unit

10^{13} cm 10^{-5} parsecs
from Earth

Earth

*Farthest objects, if Hubble relation
is valid, are ~ 10^{28} cm distant (about
10^4 megaparsecs)

Table 23·1 Travel Times of Light from Distant Objects

Source of Light	Distance	Travel Time at the Speed of Light
Typical naked-eye stars	100 psc	326 years
Center of Milky Way	9 kpc	29,000
Magellanic clouds	60	196,000
Andromeda galaxy	670	2,200,000
Edge of local group	1,000	3,300,000
M 51 Whirlpool spiral	3,800	12,000,000
Centaurus A radio elliptical	4,400	14,300,000
M 87 Elliptical	13 mpc	42,000,000
Virgo cluster	19	62,000,000
Coma cluster $(Z = 0.02)$[a]	113	368,000,000
Hydra cluster $(Z = 0.2)$[a]	1,000	3,300,000,000
High-Z quasar $(Z = 2)$[a]	5,000?	16,000,000,000?

[a]Z = red shift, expressed as fraction of original wavelength.

We see the Coma cluster as it was when early fishes were appearing in murky seas. We see the Hydra cluster as it was when lavas were filling the Sea of Tranquility on the moon. We see some galaxies as they appeared before the solar system formed.

But suppose we try to look at galaxies so far away that their light takes 12 to 16 billion years to reach us. We would be seeing them only as they existed that long ago. Our galaxy, and probably others, was just forming at that time! Some quasars, for example, may represent objects in the early history of the universe. If no galaxies are older than about 12 to 16 billion years, there may be no visible objects beyond 16 billion light-years (about 5,000 megaparsecs); if they exist, their light may not have reached us yet.

In other words, the universe seems uniformly filled with clusters and superclusters of galaxies as far as we can now see, but somewhere near those limits we perceive a strange and different region, and we do not expect to be able to see much farther. This region at the limit of our vision contains clues about how the universe evolved.

Cosmology and Cosmogony

We can now see how cosmology, the study of the universe's structure, is related to another area: **cosmogony,** the study of the universe's origin. One might think that we could study the present structure entirely separately from the ancient history of the system. In this chapter we will attempt to do this as much as we can, focusing mostly on the "geography" of space. But always in astronomy, the farther we probe in distance, the further back we look in time. Ultimately, mapping the remotest regions means mapping the early history of the system.

Early Cosmologies

To understand how cosmologists have arrived at their present concep-

tions of the universe, it is natural to begin with the earliest models and see how current models developed. In each case, we will first present the model and then discuss some of its limitations, which led to new models.

Cosmology 1: Ancient Models (ca. 3000 B.C.)

Even the earliest known philosophers 5,000 years ago speculated about the nature of the universe. Knowing nothing of the physical laws which govern matter, they tried to make sense of the universe by discussing nonmaterial attributes, or "essences," of things, as we saw in Chapter 1. Early writers often gave these essences names, imaginary personalities, or abilities to control the affairs of the universe.

Early writers also speculated on the arrangement of the universe. A tradition from India, for example, pictures the universe as a giant egg, containing land, waters, animals, gods, and so forth, all created in primordial waters by Prajapati. A Tahitian tradition says that a creator, Taaroa, existed in the immensity of space before any universe existed, and that he later constructed the heavens and rocky foundations of earth. A well-known book compiled around 500 B.C. considers both origin and structure:

In the beginning God created the heavens and the earth. The earth was without form and void, and darkness was on the face of the deep; and the spirit of God was moving over the face of the waters. And God said, "Let there be light!"

Thus, many old traditions trace both the structure and origin of the universe back to some underlying early spirit or principle. Sometimes this spirit was given personality and described as animate gods, or one God. Sometimes it was described as a fundamental principle from which all else followed, as in the Book of John (ca. A.D. 100): "In the beginning was the Word. . . ." "Word," in this passage, was the Greek term *logos*—a principle of rational logic—a concept common in Indian and Mesopotamian philosophy. So this famous cosmological thought might be translated, "In the beginning, underlying everything, was rational order." In many ancient cosmologies based on these ideas, the structure and events of the universe (including those involving living beings) were viewed as controlled by the characteristics of this initial creative essence.

Limitations Those kinds of cosmogonies have retained their appeal for thousands of years, and provide us with an enduring link to our early ancestors. Historically, they helped set the stage for later investigations about initial conditions in the universe, and about immaterial phenomena such as energy. They have the virtue of bringing us to the realization that there are forces in the universe greater than ourselves. The power of their poetry gives a beneficial respect for the sheer vastness and mystery of the universe.

But this kind of cosmogony can produce the stultifying belief that all that *can* be said about the universe *has* been said. Such a belief can rob entire cultures of the motivation to question, explore, and see what they can learn for themselves. Also, such cosmogonies are of little value in predicting or interpreting the phenomena that we actually see in the distant universe. Increasingly sophisticated astronomical instruments and

techniques reveal that there is much more we can say about the universe.

Cosmology 2: The Newtonian–Euclidean Universe (ca. 1700)

Around 1680, when Isaac Newton described how every particle in the universe gravitationally attracts every other particle, he realized that this principle might allow a simple description of the structure of the whole universe. First, he assumed that the principles of Euclid's geometry, such as the relations between angles, lines, and planes, would work just as well over the vast distances of universal space as they do in the farmyard or among the nearest stars. Euclidean geometry and Newton's gravitational law, then, allowed description of the separations and forces between particles. These "particles" can be viewed as galaxies, or even clusters of galaxies. Newton pictured the universe as infinite in extent and filled with these randomly moving "particles," a view called the **Newtonian–Euclidean static cosmology.**

In 1755, Immanuel Kant realized that although the Newtonian-Euclidean universe remained constant on the long term, many individual stars or galaxies might come and go. Kant thought that the *present* system of stars would burn out and pass away, but others might form from debris of former systems, just as we have described stars forming from the debris of earlier stars. The universe was thus endlessly recycling—

> a Phoenix of nature, which burns itself only in order to revive again . . . through all the infinity of times and spaces.

The Newtonian–Euclidean universe was thus static (not expanding or contracting), but evolving.

Limitations Newton was right in predicting many suns far beyond our own, all obeying gravitational relations. And Kant was right in imagining worlds forming and reforming. But these ideas do not adequately describe the universe as we know it today. First, the recession of distant galaxies was neither predicted nor assumed in this theory. Second, why should matter in the universe remain dispersed? Gravitational attraction between particles might cause all the mass in a given region to collapse into a single star or galaxy. Mathematicians in the 1800s studied this problem extensively, and even talked of a hypothetical non-Newtonian repulsive force that might be important only over long distances, thus holding the "particles" apart and keeping the universe static. But the insurmountable objection is that the Newtonian-Euclidean cosmology fails to explain a problem known as Olbers' paradox.

Olbers' Paradox

Sometimes the simplest questions promote the most profound thoughts. The simple question, "Why is the sky dark at night?" leads to an astonishing paradox. For example, the night sky in the Newtonian–Euclidean universe ought to be ablaze with light! Astronomers are not sure who first discovered this logical implication, but Edmund Halley indicated he had heard the idea as early as 1720. The idea was more carefully developed in 1823 by the German astronomer Wilhelm Olbers, whose name finally became attached to it (see North, 1965 for a more complete history).

Olbers' paradox arises as follows. Assume that space extends indefinitely and is filled only with stars resembling the sun. If we look at the sun, we see that each unit of angular area of the sun's surface (a square second of arc, for example) is intensely bright. If we now gaze in some other direction, our line of sight must ultimately intercept the surface of another star, because stars dot space to infinity. Each unit of angular area in that direction, therefore, would have the same surface brightness as the sun. The whole day or night sky should look as bright as the surface of the sun!

What is wrong with this argument? One of the assumptions must be wrong. One early suggestion was that the dust in space simply obscures the distant stars. But interstellar dust does not explain Olbers' paradox, because the dust should absorb the stars' radiation and heat up. According to the Stefan-Boltzmann law of Chapter 13, the dust reradiates, no matter what its temperature. Its temperature keeps increasing until the amount of radiation emitted equals the amount absorbed and a constant equilibrium temperature is attained. Even if the reradiated radiation were not visible light, there would be infrared radiation that we would sense as intense heat. Because the dust is not radiating this intensely, this explanation fails.

A second suggestion was that there are no galaxies beyond a certain distance, so that there are simply no more light sources beyond a certain "horizon." However, current studies indicate galaxies reach as far as we can see.

The modern response to Olbers' paradox is more subtle, invoking the recession of distant galaxies. The red shifts of their light mean that the apparent energies of the photons we receive from them are reduced from high energies (short wavelengths) to low (long wavelengths). Photons received from galaxies receding at almost the speed of light would be strongly

reduced in apparent energy. If Hubble's relation between red shift and distance can be extended indefinitely, there may be a distance somewhat beyond 5,000 megaparsecs where galaxies are receding so fast that energy emitted by them virtually never reaches us. (Hubble's relation, literally interpreted, would give recession speeds greater than the speed of light beyond this distance. Even if that were possible, photons emitted from such galaxies would not reach earth.)

Another peculiarity of this relativistic "horizon" somewhere beyond 5,000 mpc is revealed in Table 23·1. It would take light 16 billion years or more to reach us from objects beyond this "horizon," meaning that in probing this region we would be looking back in time at least 16 billion years. Many cosmologists think the universe was empty, or at least significantly different, in those days. So, probing the universe of that era may mean probing a universe with no radiation.

There are further subtleties to this concept. For instance, there has probably not been enough time in the history of the universe (estimated at around 16 billion years) for radiation from within the "horizon" to permeate the available volume, heat up the dust and other material, and achieve the equilibrium mentioned above. Reaching this equilibrium might require 10^{24} years, after which the sky would radiate with radio waves similar to that from a body at around 20 K, according to a study by Massachusetts physicist, E. R. Harrison (1974).

Olbers' paradox, then, is probably explained by the red-shift "horizon" and the lack of sufficient time for radiation to permeate the universe. Other explanations have been proposed, usually in connection with specific cosmological theories. *Any*

valid cosmological theory must be able to explain Olbers' paradox. This test is *not* met by Cosmology 2, the Newtonian–Euclidean concept of an extremely ancient, nonexpanding, static, infinite universe filled uniformly with stars.

Modern Mathematical Cosmologies

Unlike the sometimes tedious business of telescopic observing, modern cosmological theorizing has some aspects of a mathematical game. The rules are: (1) invent some hypothetical natural laws, and (2) show that they could, in theory, govern a complete universe, but (3) also show that they do not disagree with any observations so violently that astronomers can casually dismiss them. More than most theories, modern cosmologies involve many assumptions, because only poor statistics are available about the universe out beyond about 1,000 mpc. Many modern cosmologies are essentially deductions from initial postulates, rather than experimental or observational tests of nature. As one researcher (Schatzman, 1965) has noted:

. . . *once the basic ideas are understood, the deduction of their consequences is a simple exercise in geometry or algebra, which does not increase our knowledge of the properties of matter.*

Non-Euclidean Geometries

In the late 1800s, mathematicians in Germany, Italy, and Russia became fascinated with geometries quite different from those of everyday experience. These so-called non-Euclidean geometries served as stepping-stones to modern cosmologies.

Non-Euclidean geometries can be described by an analogy. Suppose we represent the three-dimensional volume of space by the two-dimensional surface of a chessboard. Instead of being able to move in any of the three dimensions of ordinary space (N–S, E–W, and up–down), light waves, sound waves, and inhabitants in the chessboard-world could only travel along the two-dimensional (N–S, E–W) surface of the board. Imagine that the board is vast, covering many kilometers. Inhabitants of this world might be visualized as tiny ants, who can move and see only along the surface of the chessboard. The chessboard represents all space.

Now suppose that the chessboard is flat. Then the ants would find that the principles of Euclid's plane geometry are satisfied. For example, parallel lines would never meet. The sum of the angles in a triangle would be 180°.

But now suppose that the chessboard covered the whole surface of the earth. It can still be called a two-dimensional surface, because the ants (and light waves) can travel only in two dimensions (N–S, E–W). As long as the ants operated only in a small region, the surface would seem flat to them, and Euclidean geometry would seem true, within the limits of accuracy of observations. But if the ants probed large enough regions, they could discover that their surface, or "space," is curved. For example, as shown in Figure 23·2, they would discover that the triangle defined by the equator and two latitude lines 90° apart has three 90° angles whose sum is 270°!

If the chessboard covered the whole earth, the ants might discover another interesting thing. They would find that their universe, instead of being infinite (as the ant-Newtons and the ant-Kants might have supposed), had

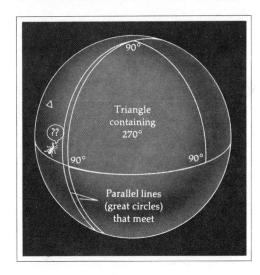

Figure 23·2 **Ants living on a large spherical world would find that the small figures, such as the small triangle, approximately obey Euclid's geometrical laws. But large figures, such as the large triangle, violate Euclidean geometry. In the same way, studies of very large volumes of intergalactic space might reveal departures from Euclidean geometry.**

a finite amount of space,[1] even though it had no edge.

Mathematically, all these ideas can be extended to real three-dimensional space. Following Euclid's laws of geometry, one can write equations describing spatial relationships for either two-dimensional or three-dimensional worlds. For example, in a two-dimensional plane the Pythagorean theorem gives the length of the hypotenuse x in a right triangle with sides a and b by the famous rule:

$$x^2 = a^2 + b^2$$

And, similarly, in ordinary three-dimensional space, the length of a diagonal x in a box with sides a, b, and c is:

$$x^2 = a^2 + b^2 + c^2$$

What the nineteenth-century mathematicians did was to generalize three-dimensional geometry to allow for the mathematical possibility of **curved space.** If the surface, or the space, is curved, neither of the two equations just given is true. Different mathematical relations apply. Geometries of curved surfaces or curved spaces are called **non-Euclidean geometries.** After some controversy, cosmologists accepted the proposition that the real universe *might actually* be curved, or non-Euclidean.

---

[1]The total area would be $4\pi r^2$, where r is the radius of curvature (radius of the earth, in this example).

Of course, everyday experience gives us no clue as to whether space is curved or not. We normally interact with far too small a region to detect even a slight curvature. The evidence for slight curvature of intergalactic space would come only from studying a very large part of space, including the remotest galaxies.

Furthermore, different kinds of curvature are possible, each with its own properties. Examples of different types, shown in Figure 23·3, are based on analogies with surface. Part A shows a surface representing uncurved space, which could extend to infinity in all directions (dotted lines). Euclidean geometric laws would apply everywhere. Part B shows space with a slight curvature, either positive or negative. These spaces could also extend to infinity. Part C shows an example of complex curvature, which could also extend to infinity. These types of space could have infinite volume.

Part D is an especially interesting case in which the curve has closed on itself, therefore having finite rather than infinite volume. If space is really curved in this way, the universe could have finite volume, but still no boundaries, like the universe of the ants who lived on a globe. A traveler who went far enough in one direction

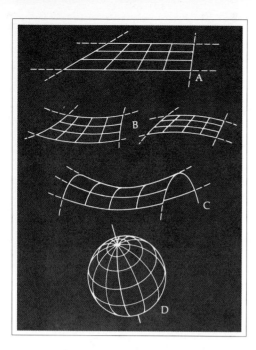

Figure 23·3 **Examples of different curvatures of space, using surfaces as analogs of the volume of three-dimensional space. A is uncurved and infinite in extent; B and C are curved and infinite; D is curved and finite.**

would eventually come back to his starting point.

Universes with infinite volume are called **open universes.** Those with finite volume are called **closed universes.** Thus, in Figure 23·3, A would be called uncurved and open; B and C are curved and open; D is curved and closed.

Cosmology 3:
Static, Curved Universes (1917)

In 1917, Albert Einstein, who had just developed the theory of general relativity, tried to see how it applied to the universe as a whole. He made some basic assumptions that seemed reasonable at the time:

1. The universe is **homogeneous** (meaning that all its particles—stars,

galaxies, clusters of galaxies, or whatever—are uniformly distributed on the large scale) and **isotropic** (meaning that the view is similar in all directions from all galaxies). This assumption is sometimes called the **cosmological principle.**

2. Space is curved, as in non-Euclidean geometry, and the curvature is constant (the surface analog would be a sphere).

3. The universe is static, which means that the mean density of matter is constant and the radius of curvature does not change.

But when he tried to solve the equations representing such a universe, Einstein found that it was impossible! *No static solution fitted the principles of general relativity.*

Still assuming that the universe must be static, Einstein introduced a so-called cosmological constant—a hypothetical repulsive force between material particles, important only over long distances. The repulsion,

Figure 23·4 **Albert Einstein (1879–1955)**

similar to that suggested for Newtonian–Euclidean theories, could be chosen to balance the gravity of whatever mass was in the universe, so that a static state resulted, but only so long as there were no disturbances.

The mathematician W. de Sitter added some interesting variations, also in 1917. The amount of curvature and other properties were related to the mean density of matter in the universe. A universe with no matter in it could easily be static, he found. (A universe with no matter in it may not seem relevant to the real world, but don't laugh—see Cosmology 8!) A few particles (galaxies) put into such a universe would accelerate away from each other. In one de Sitter model, light from such galaxies would no longer reach us after they had receded to about 1,600 megaparsecs—a horizon beyond which we could see nothing.

Limitations The original static model was scrapped in 1929 when Hubble proved that the galaxies were rushing away from each other. The de Sitter model with a few galaxies in it actually resembles the observed recession! However, there is no independent evidence for the repulsive force the model requires. And as we will see next, Einstein, de Sitter, and others soon dropped the idea of repulsive force.

Cosmology 4: Friedmann
Evolving Universes (1922–1924)

In 1922–1924, the Russian mathematician Alexandre Friedmann studied a model that was evolving, not static. He simply assumed that the curvature of space varies with time. For example, space could increase in volume if the radius of curvature increased. In such geometries, the distances between all points continually increase, and galaxies would be seen to recede. The models were not widely discussed until Hubble published his discovery of actual galactic recession in 1929. Einstein, de Sitter, Eddington, and others soon concluded that the Friedmann models might account for an expanding universe without any repulsive force.

An analog of this model would be the surface of an expanding sphere, like a balloon. Galaxies could be imagined as dots on the surface of the balloon. The analogy is not perfect, because the dots would expand with the rubber, whereas real galaxies might maintain their integrity as units of matter by their internal gravitational forces. Antlike observers on any dot would see other dots receding from them as the balloon expanded, just as we would see galaxies recede if the radius of curvature of space increased.

Limitations When Hubble discovered that galaxies are actually receding from each other, theorists were delighted that observations seemed to bear out the expanding model. However, in retrospect, astronomers do not say that the expansion of the universe was *predicted* in the normal scientific sense, because the observed expansion was only one of several possible types of Friedmann evolution dependent on the initial assumptions, and because the Friedmann models were developed almost completely outside the normal astronomical context. They were almost completely mathematical models, with little to say about physical questions such as the distribution of galaxies or density of matter in space. Thus, the next step was to connect these exciting and provocative models with astrophysical reality.

Figure 23·5 **A, B, C, and D show four possible histories of the universe often considered by cosmologists, in addition to the static model discussed as "Cosmology 3." Cases A—D all correspond to an expanding universe as observed today, with higher densities of matter in the past.**

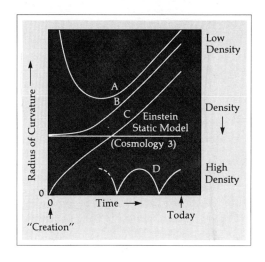

Cosmology 5: The Big Bang (1927—   )

Around 1927, Belgian mathematician-astrophysicist Georges Lemaître began to put more astronomy and physics into cosmology: He pointed out that if galaxies are now flying away from each other, there probably was a time in the past when all matter was closer together. He suggested that all matter exploded from this condition of high density. Lemaître thus became known as the father of what is popularly called the **big-bang theory** of cosmology and cosmogony, which is the most widely accepted theory today. Several models of expansion from that condition are possible. Figure 23·5 clarifies the relations between models discussed in this chapter.

The universe may first have contracted to a point of maximum density, and now may be expanding from that point (model A). It might have started out as in Cosmology 3, Einstein's original static model, but because that model was unstable, a slight internal disturbance might have triggered an expansion (model B). It may have simply started (no one knows how) from a supercompact mass of zero volume and near-infinite density (model C). Or it may be periodically expanding and contracting, with the radius never getting large enough for permanent expansion to set in (model D).

The moment of maximum density is known as the *big bang*. Many people regard this as the actual creation of the universe (as in model C). Modern evidence suggests that this moment occurred around 16 billion years ago. The universe was supposed to have begun from an extraordinary state in which the radius of all space had shrunk to near zero, and all matter was concentrated in a virtual point, with virtually infinite density. Lemaître liked to call this the *primeval atom,* because all matter would have existed only as subatomic particles collected in a single unit—an "atom" of nearly infinite atomic weight! Lemaître pictured this "atom" as flying apart and condensing into galaxies.

Lemaître's picture of a primordial expanding space filled with matter and energy excited physicists and astrophysicists, who recognized that they could analyze the evolution of matter in such a model. The Russian-American physicist George Gamow was well known for this work. In 1948 Gamow and his associates showed that the observed abundances of atoms of the different elements nearly match the abundances calculated to form in an expanding, hot, ancient fireball!

In the next chapter we will return to other successes of the big-bang theory. Here we continue to explore some other modern ideas about the universe's structure.

Limitations The big-bang theory says nothing about how (not to mention why) the "primeval atom" or initial fireball came into existence. It has been quite successful in saying that *if* there ever were such a fireball filling expanding space, then certain *observed* phenomena would result. It does not deal with the philosophical question of conceiving a beginning to the universe—which is perhaps beyond astronomy or science in general.

Cosmology 6:
Changing Constants (1930–   )

In the 1930s physicists began to consider the possibility that certain fundamental properties of nature, normally considered constant, might be slowly changing. Examples could be the speed of light, the charge of the electron, or Newton's gravitational constant, G. For example, the English physicist Paul Dirac postulated that G might decrease with time. This would weaken attractive forces between galaxies, allowing them to recede from each other, as observed. Fred Hoyle revived this idea in 1972, pointing out that evidence for certain changes in the tidal history of the earth–moon system and in solar luminosity might be explained in this way.

Limitations Direct attempts to measure G and other constants have revealed no verified changes, though the proposed changes might be too small to observe. Eddington and others developed elaborate cosmologies based on certain numerical relations between the constants, but these have been viewed as cryptic by most cosmologists. Thus, these cosmologies have little relation to observations, though further work might be productive.

Cosmology 7:
The Steady State (1948–   )

By the 1940s the big-bang theory was a favored model, supported by observations of galactic recession and the abundances of elements. However, partly to stimulate healthy debate and partly to solve certain problems then believed to weaken the big-bang theory,[2] researchers such as Thomas Gold, H. Bondi, and Fred Hoyle proposed, in 1948, a new cosmology— the **steady-state cosmology.** They proposed that the universe—*on the large scale—looks the same not only in each region of space, but also in different eras of time.* This assumption is sometimes called the **perfect cosmological principle.** According to this "steady state" idea, the universe has looked about the same forever. Accordingly, new galaxies must constantly be forming in the spaces vacated as old galaxies move apart. What material could they form from? The steady-state cosmologists made an additional assumption—the **theory of continuous creation:** New atoms of hydrogen spontaneously pop into existence from time to time, mostly in the space between galaxies. At first, some scientists saw this hypothesis as an outrageous violation of the concept of conservation of mass, but defenders of the theory pointed out that it is hardly more outrageous than imagining all the mass of the whole universe appearing at once, as in the big-bang theory!

In the continuous-creation, steady-state theory, the universe is infinitely old. It has been here always, made

---

[2]For instance, in 1948, there was thought to be a discrepancy between the big-bang age of the universe and the age of the elements. The discrepancy, however, is not substantial, according to current data.

up of galaxies and clusters of galaxies always, and will always look about the same, though individual stars and galaxies will form, evolve, move apart from each other, and reach old age.

The steady-state theory, or variations of it, has been the major alternative to the big-bang theory since the 1950s. It was well constructed, with physical considerations and mathematical detail.

Limitations The steady-state theory hit a major snag when astronomers discovered what seems to be radiation left over from the primeval fireball described by the big-bang theory (see also next chapter). This radio radiation is not predicted at all by the original version of the steady-state theory. Therefore, most astronomers have concluded that the steady-state theory must be rejected. Others have tried to modify the steady-state theory to fit the observations, but the effort is not promising.

Cosmology 8:
Hierarchical Universe (1970–   )

Since the early days of Olbers' paradox, some astronomers have questioned the assumption, made in many cosmologies, that all parts of the universe are so much like our part that we can measure the mean density of matter simply by making an inventory of the galaxies we see. As you will recall, the mean density is important in controlling properties such as the curvature of space. In particular, French-American astronomer Gerard de Vaucouleurs has emphasized that the universe may be a series of clusters, clusters of clusters, and so on, so that we might be in a denser-than-usual part of space. This is called the *theory of the hierarchical universe,* because the universe would be arranged in hierarchies of clusters, rather than in a random distribution of galaxies.

We know this picture has at least limited truth: The mean density of material in our galaxy is larger than the density if we include surrounding space. Similarly, the mean density of our local group is probably greater than the mean density out to, say, a megaparsec. In fact, from the atomic particles all the way to clusters of galaxies, the highest densities achieved in the universe decline as we consider larger and larger volumes of space. Some astronomers have found evidence for clusters of clusters of galaxies, on a scale of 10 to 300 megaparsecs, as shown in Figure 23·6.

De Vaucouleurs (1970), particularly, has criticized the fact that so many cosmologies simply *assume* the uniformity expressed in the so-called cosmological principle. His criticism is a good note of caution that must be considered not only in cosmology, but in all science and all life:

> *I am concerned with an apparent loss of contact with empirical evidence.*
>
> *. . . there is a serious danger that the constant repetition of what is in truth merely a set of a priori assumptions (however rational, plausible, or otherwise commendable) will in time become accepted dogma that the unwary may uncritically accept as established fact or as an unescapable logical requirement. There is also the danger inherent in all established dogmas that the surfacing of contrary opinion and evidence will be resisted in every way.*

Counts of galaxies in the largest volumes that we can explore suggest a mean density of about 10^{-30} or 10^{-32} g/cm^3. But the hierarchical theory raises the possibility that if we could explore vaster spaces, the mean density might be found to be as low

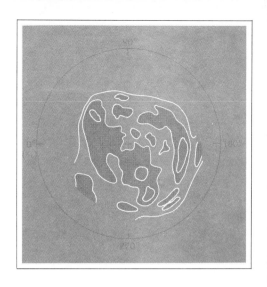

Figure 23·6 A view of the sky north of the ▶ galactic equator (outer circle) after de Vaucouleurs (1970). Contours of the frequency of faint galaxies define patchy clusters. The darkest areas are groupings that may correspond to "super-superclusters" of very remote galaxies, perhaps 300 megaparsecs across.

as 10^{-33} or even 10^{-40} g/cm^3. Conceivably, as we consider larger and larger volumes, the mean density might approach zero, so that the curious "empty universe" of Cosmology 3 might have some real meaning!

Limitations Hierarchical universes are more complicated mathematically than homogeneous universes, and have been little studied. Observations do not strongly support the hierarchical model. For example, some astronomers attribute the 300-mpc superclustering to mere statistical effects. Furthermore, no clustering has been found in the background radio radiation that permeates space. On the positive side, the hierarchical theory has encouraged more careful observational tests of the assumption that the universe is uniform, and these may lead to new discoveries about superclusters.

Direct Cosmological Observations

It is difficult to see enough of the universe to get good cosmological information, aside from the recession of galaxies, revealed by red shifts of nearby and distant galaxies. Studies of superclusters are still controversial. Studies of abundances of elements, radiation probably coming from the ancient big-bang fireball, and ages of old globular clusters give some information about the earliest days, but

they don't tell us much about whether the universe is open or closed, curved or uncurved.

One additional cosmological observation is simply *counting* galaxies at different distances. If galaxies were distributed uniformly with similar luminosities and with no obscuring intergalactic dust, then a certain smooth relation would exist between distance and the number of galaxies within that distance. We can make the counts (Figure 23·7) and see if the relation is satisfied. The relation is satisfied in nearby space, but at the greatest measurable distances, galaxies are either too bright or too numerous to agree with the assumption of homogeneity. Most astronomers believe this is yet another encounter with the distant "horizon" at which we see perhaps 16 billion years back in time and encounter fundamentally different conditions. We may be seeing galaxies as they looked just after their formation. If so, gravitational contraction or explosions of short-lived massive objects might have made newly formed galaxies more luminous than they would be later. In fact, this greater luminosity alone (not to mention possible presence of intergalactic dust) might make galactic counts useless as a cosmological test. We know only that

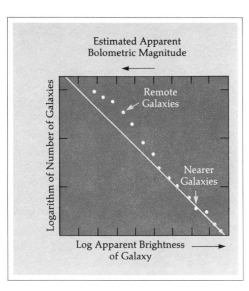

Estimated Apparent
Bolometric Magnitude

Logarithm of Number of Galaxies

Remote
Galaxies

Nearer
Galaxies

Log Apparent Brightness
of Galaxy

▲
Figure 23·7 **If space is Euclidean and galaxies are similar in all regions, the relation between logarithms of numbers of galaxies and their brightnesses should follow a straight line. The predicted relation is found for nearer galaxies, but not for remote ones, whose light is billions of years old. One intepretation is that space is Euclidean, but primordial galaxies billions of years ago were brighter than galaxies today.** (Data after Ryle, 1968)

Figure 23·8 **Dashed lines show predicted differences between the relation of red shift and distance for three different cosmological models. If we could detect remote enough galaxies, we could test which model is right. Present observations, although inadequate, slightly favor the open, Euclidean universe.**
▼

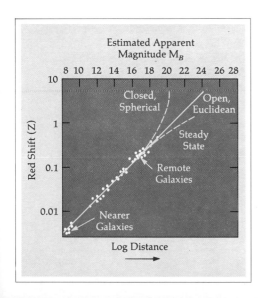

Estimated Apparent
Magnitude M_B

8 10 12 14 16 18 20 22 24 26 28

Red Shift (Z)

10

1

0.1

0.01

Closed,
Spherical

Open,
Euclidean

Steady
State

Remote
Galaxies

Nearer
Galaxies

Log Distance

◄ *Figure 23·9* **Galaxies without end—a portion of the cluster of galaxies in Hercules, about 175 megaparsecs away. The cluster contains at least 300 galaxies and is receding from us at 10,300 km/sec.** (200-inch telescope; Hale Observatories)

if we look out to a great enough distance, and thus far enough back in time, we see something peculiar.

A second, similar test of cosmology is to plot red shifts (assumed to represent recession velocities of galaxies) versus brightness of galaxies. Different cosmological theories (curved space, uncurved space, and so on) predict different relations at very great distances (high red shift and low brightness). As Figure 23·8 shows, we cannot see quite far enough to be sure which model is correct. The objects with highest red shift, the quasars, seem not to follow the simple relation defined by other galaxies—but, of course, we are not certain they are galaxies in the first place. The data do seem inconsistent with the steady-state model. There is some indication that *the real universe is uncurved, open, and Euclidean.*

Studies of all data on the density and curvature of the universe—from galaxy counts to data on the origin of elements—indicate that the density of matter is less than one-tenth the value required for a curved, closed finite universe (Gott et al., 1974)—again suggesting that the *universe is open and infinite in volume.* Since galaxies are certainly rushing apart through this open space, the universe seems to be an evolving, infinite system.

23

Summary

Cosmology is a risky business, and it is unwise to put too much faith in such generalizations about the universe when there is still so much more observing to be done with new and larger instruments. The French-American astronomer G. de Vaucouleurs has said:

Less than 50 years after the birth of what we are pleased to call "modern cosmology," when so few empirical facts are passably well established, when so many different over-simplified models of the universe are still competing for attention, is it, may we ask, really credible to claim, or even reasonable to to hope, that we are presently close to a definitive solution of the cosmological problem?

On a more hopeful note, Gott and his colleagues, who found that data collected up to 1974 indicated an open, infinite universe, have said:

The objections to closed universes are formidable but not fatal; a clear verdict is unfortunately not yet in, but the mood of the jury is perhaps becoming perceptible.

The mood of the jury is that clusters of galaxies are receding from each other through space that is uncurved, or only slightly curved, and is infinite. If it is hard to conceive of a universe with no boundary, perhaps the best way is to imagine traveling forever and coming to no end, no edge. Most of the astronomical jury also agrees that the present universe evolved from an earlier configuration when matter was more densely packed—probably a primeval explosion known as the big bang. The galaxies formed from that matter and they are now rushing through the emptiness. One of them is taking us along for the ride through a universe of which we are a living part, whose beginning we are unsure about, whose end we don't know.

Concepts

cosmology

universe

cosmogony

Newtonian–Euclidean static cosmology

Olbers' paradox

curved space

non-Euclidean geometries

open universes

closed universes

homogeneous

isotropic

cosmological principle

big-bang theory

steady-state cosmology

perfect cosmological principle

theory of continuous creation

Problems

1. Progress in many scientific fields, such as studies of stars, plants, or animals, has come by classification of types, followed by comparisons. How does a cosmologist suffer a disadvantage in this regard?

2. Telescopes much larger than present-day designs, perhaps located in space, would have much more resolving power and light-gathering ability, and could reveal much fainter objects. Give examples of how this would clarify current cosmological problems.

3. How is the real universe different from the Newtonian–Euclidean static model that dominated literary and cultural concepts in the 1700s and 1800s?

4. Why is the sky dark at night?

5. Many cosmologies, such as the big-bang model, assume the cosmological principle that the universe is homogeneous at any given time, given large enough scale. Yet quasars seem to be more common per unit volume at very great distances than near our galaxy. Does this refute the big-bang theory? Why or why not?

6. Give examples of observations that seem to refute:
a. *the Newtonian static model of the universe.*
b. *the steady-state model.*

Advanced Problems

7. If an explosion in a very active galactic nucleus or quasar produced abundant radiating gas at a temperature of one million K, at what wavelength would the maximum radiation occur?
a. *Considering the types of light that are absorbed in the earth's atmosphere, how might a large telescope in space have an advantage in studying such phenomena over a large telescope on the ground?*
b. *Would a quasar-type object (or any other observable object) be likely to have enough red shift to bring this radiation to visible wavelengths?*

8. When objects have speeds approaching the velocity of light, the Newtonian equation for their Doppler shifts cited in Chapter 13 is no longer valid, and must be replaced by a relativistic equation, according to which:

$$Z = \frac{\text{change in wavelength}}{\text{wavelength}}$$

$$= \frac{1 + \dfrac{v}{c}}{\sqrt{1 - \dfrac{v^2}{c^2}}} - 1$$

a. *According to the Newtonian equation, $Z = \frac{1}{2}$ when the velocity v is half the velocity of light c. What would be the correct value of the red shift Z in this case?*

b. *As the velocity of recession of a distant galaxy approaches the velocity of light, what does the observed red shift approach? Comment on how this would affect attempts to observe the galaxy.*

Cosmogony: A Twentieth-Century Version of Creation

How did the universe begin? When did it begin? Or do these questions have any real meaning? **Cosmogony** is the attempt to decipher the origin of the universe and its major parts, such as galaxies. *Cosmology,* though focusing on the universe's structure, often itself involves cosmogony, as we saw in the last chapter.

The big-bang theory is both a cosmogonical and a cosmological theory, because it accounts for both the origin and present structure of the universe. It has come to be strongly favored among all such theories because it is based on an interlocking pattern of supporting observations, some of which will become clearer in this

chapter. Therefore, instead of comparing different theories as we did in the last chapter, we now assume that the big-bang theory is essentially true, and use it to describe how the universe may have formed and evolved.

All cultures have their creation myths, and the big-bang idea is ours. The word "myth" does not mean a falsehood, but a scenario widely repeated and widely believed. The big-bang theory almost certainly gives us an indication of some events that actually occurred long ago, but the word "myth" might be kept in mind to remind us that this is only our best response to a grand mystery—what one satirist called "the Vienna Philharmonic of scientific questions."

The Date of Creation

At least three kinds of observation indicate that the universe as we know it began roughly 16 billion years ago.

Age of Globular Clusters In Chapter 19 we saw that H-R diagrams allow estimates of the **ages of globular clusters.** Results indicate that the clusters formed about 12 ± 3 billion years ago. This period is believed to mark the formation of our galaxy out of hydrogen gas that formed in the big bang. The big bang must have happened at least that long ago.

Hubble's Constant Because **Hubble's constant** measures how fast the galaxies are now rushing away from each other, we can calculate how long it has taken them to get this far apart—if they have been receding at constant speed. This value (which is equal to $1/H$) is called the **expansion age.** The value of Hubble's constant and its uncertainty convert to an age of 16 to 19 billion years since the expansion began. Because of their mutual gravitational attraction, the galaxies probably have slowed their recession since the big bang, favoring an expansion age toward the lower end of this range.

Age of the Elements In Chapters 3, 5, and 11 we discussed how radioactive elements can be analyzed to reveal how long ago they were incorporated in the rocks of the solar system. This date is estimated to be about 4.6 billion years ago. The universe is obviously older than that. But the elements can also be used to estimate the time since they first formed. The big-bang theory allows physicists to calculate relative abundances of different kinds of elements in the initial fireball. These abundances, when compared with today's abundances and with the decay rates of different radioactive elements, allow us to estimate the time since element formation—the **age of elements.** The results can be adjusted for formation of additional heavy elements inside stars between the big bang and the present.

Models of uranium production show that if all uranium formed in the big bang, the universe should be about 7 billion years old. But if all uranium had formed slowly by heavy-element formation inside stars, the universe should be about 18 billion years old (Peebles, 1971). Since both processes probably occurred to some degree, the true answer should be between these two values.

From the rough agreement among these three lines of evidence, most astronomers draw the conclusion that the

Time since the big bang =
16 ± 4 billion years.

Some unique explosive event must have occurred around 16 billion years

ago, apparently creating matter and sending it flying out on its expanding journey. Many astronomers refer to this time as the **age of the universe,** marking the creation of the universe as we know it.

The term "explosive" may be misleading, because the big bang was not an ordinary explosion. According to many models, no one could have floated in nearby space and watched the fireball expand, because, paradoxically, the fireball always filled all space. Space itself is viewed as *contracted* at the beginning. An analog to these models would be the expanding balloon, whose surface represents space. An ant living on a spot on its surface could detect the expansion, and could detect other spots moving away from it, but could not get off the surface to look at the whole balloon. At the beginning, the surface of the balloon would have been "small," but it still would have constituted the whole universe, so there would have been no exterior point from which to observe the initial explosion.

The First Hour of the Universe

One of the astonishing things about the big-bang theory is that it lets us describe **initial conditions** during the first moments in the history of the universe. The assumptions are simple, because if we say that all the observable mass was once as concentrated as possible, then at the initial moment the density (g/cm³) should have been nearly infinite; it has been declining ever since. Packing all mass at high density would lead to extraordinary temperatures. A simple theoretical model developed around 1950 by George Gamow assumed infinite temperature and density at the zero

instant! After one second, the temperature was 15 billion degrees and the density about that of the air we are breathing. One success of this model is that it predicts a present temperature and density averaging about 20 K and 10^{-3} g/cm³, close to the observed values.

Predicting Abundances of Elements in Population II

Another success of the big-bang theory is that it describes how the primordial particles joined together to form atoms of the various elements, and it also predicts the abundances of those elements. In the high-temperature gas of the first seconds, matter was broken down into its simplest components—not grains or molecules or even atoms, but subatomic particles such as neutrons, protons, and electrons. Calculations based on known atomic physics show how these particles would interact under conditions after one second, two seconds, and so on.

The first step in making atoms was to make their nuclei, or inner cores. The outer electrons could be added only later, under cooler conditions. The nucleus of a hydrogen atom is simply a proton, so it could be said that the universe was already full of hydrogen nuclei, since protons were an abundant basic particle. The next-heaviest atomic nucleus would be that of deuterium, or heavy hydrogen, which consists of a joined proton and neutron. Gamow and his associates found that protons and neutrons could collide and form deuterium:

proton + neutron = deuterium + gamma radiation

The higher the temperature, the faster particles collide. During the first three minutes or so, collisions would have been so energetic that deuterium nuclei would have broken apart faster than they would have formed. After about three minutes, according to this model, deuterium began to accumulate and the stage was set for collisions that formed still heavier nuclei. Small amounts of helium, lithium, and heavier elements thus accumulated. Gamow concluded that most of the element-forming reactions had already occurred by the end of the first hour.

It may sound presumptuous to talk of events during "the first hour" 16 billion years ago, but the results can be checked. For example, Gamow

Figure 24·1 **Evidence for the big-bang theory. Observed abundances of elements (points) approximately match a theoretical prediction (solid curve) based on element-formation solely by nuclear reactions in the primeval fireball of the big bang. Modern results attribute most heavy elements to additional formation processes inside massive stars.** (After Gamow, 1952)

▼

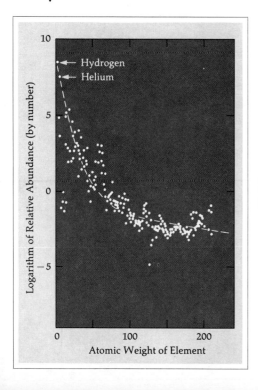

showed that relative abundances of elements predicted from this model are strikingly similar to actual observed abundances, as indicated in Figure 24·1. More detailed analysis reveals that the big-bang model alone probably does not account for enough heavy elements to match the universe that we now see. A recent calculation (Reeves et al., 1972) has given these abundances of nuclei at the end of the universe's first hour:

Hydrogen	^1H	75% by mass
Helium-4	^4He	25
Deuterium	^2H	0.1 ?
Helium-3	^3He	0.001 ?
Lithium 6 and 7	^6Li, ^7Li	0.000001 ?
Heavier nuclei		Negligible

Early theorists thought this lack of heavy nuclei was an argument against the big-bang model, but recent results reveal striking agreement with the real universe, because Population II stars show this pattern of abundances! These, remember, are the most ancient stars and must have been among the first formed. The heavier elements must have been formed later, as we will soon describe.

Radiation in the Early Universe

According to the big-bang model, radiation was more important than matter in the first few thousand years. According to Einstein's famous equation $E = mc^2$, energy, E, is equivalent to mass, m (c is the velocity of light). This means that each photon of radiation, carrying a certain energy, E, will have an equivalent mass, $m = E/c^2$. Thus, any given amount of radiation corresponds to an equivalent mass of material. At the billion-degree temperatures in the **primeval fireball,**

there must have been intense radiation, hence many photons per unit volume, and hence a large equivalent density. Under these conditions, the equivalent mass density of the radiation was greater than the density of the mass itself!

Today, on the other hand, radiation density is negligible. As Gamow pointed out, the mass of radiation in a cubic kilometer of air today is only about 10^{-14} g; and in the center of the sun or an exploding atomic bomb it is only about 10^{-3} g/cm³. But in the first tenth of a second it may have exceeded 1 g/cm³!

The importance of this result is twofold. First, the intense radiation agitated the particles of matter so violently that neither galaxies nor other clumps of matter could form. As time went on and everything rushed outward, the radiation density dropped rapidly. After some thousands of years, radiation density dropped to below the density of matter. Within the first billion years, recognizable masses of material could begin to form.

The second consequence of the high radiation density is that if we could look far enough into the past (by looking toward distant, high-red-shifted regions) we might see a different kind of radiation from that found near our own galaxy. It would be a trace of the dense radiation in the primeval fireball. The early universe was permeated by the kind of radiation that would come from material at a billion degrees or so. As matter flew outward, the radiation was diluted and red-shifted. The red-shifting would lengthen the apparent wavelength of the radiation, and, according to Wien's law, the peak energy would thus come at a wavelength corresponding to quite a low temperature. Starting as early as 1948, this temperature has been variously predicted to be about 1 K to 5 K. In other words, if the big-bang theory is right, the sky should be uniformly filled with faint radio radiation resembling radiation from an object at about 1 to 5 K.

This line of thought was foreseen by the father of the big-bang theory, Georges Lemaître, who wrote in 1931:

> The evolution of the universe can be compared to a display of fireworks that has just ended: some few red wisps, ashes, and smoke. Standing on a cooled cinder, we see the slow fading of the suns, and we try to recall the vanished brilliance of the origin of the worlds.

But why just try to recall it? Why not point detectors at the sky to see if there is long-wavelength radiation left over from the primeval fireball?

Discovering the Primeval Radiation

Indeed, there is direct observational evidence that this radiation exists. In the 1960s a research team at Princeton built a sensitive radio detector to search for it. The predicted radiation would be microwave radio noise. Even

Figure 24·2 **Observational detection of radiation attributed to the primeval fireball of the big bang. The dots and the dotted line are observed amounts of radiation at different wavelengths; the solid curve is the amount predicted for radiation emitted from material at 3 K.**

before the Princeton device could be applied, researchers using radio telescopes in experiments on the first Telstar communications satellite found puzzling microwave noise. First thought to be a problem in the instruments, it was identified in 1965 by Bell Laboratory and Princeton physicists as the predicted remnant radiation permeating the universe from the primeval fireball. The radiation corresponds to 3-K black-body emission and has been confirmed at many wavelengths, as shown in Table 24·1 and Figure 24·2.

This discovery is the strongest confirmation of the big-bang theory, especially because the **3-K radiation** seems to be uniform in all directions. Observations at many radio and X-ray wavelengths indicate no variations (within an accuracy of a few percent), as shown in Table 24·1. This is just

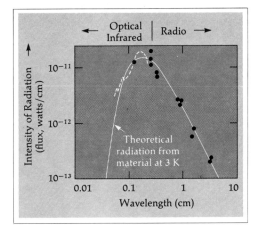

what the big-bang theory predicts; the primeval radiation should be seen equally at great red shifts in all directions.

Formation of Galaxies

According to the big-bang theory, the densities of both radiation and matter decreased as material rushed apart after the big bang. Perhaps within a few thousand years the density of matter became greater than that of radiation, allowing stronger gravitational forces between particles of matter. After a few million years, temperature and density were probably such that gravitational contraction could begin in the denser clouds of gas, as indicated in Figure 24·3. This could start **formation of galaxies.** Certain clouds or turbulent eddies had enough density and self-gravity to contract—perhaps to masses as large as entire superclusters of galaxies. Galaxies in large clusters show alignment of their rotation axes— they may be simply fragments of single, rotating, primitive clouds (Ozernoy, 1974). Thus, the galaxies themselves may be just one more case of the fragmentation of contracting clouds (as discussed in Chapter 15). Clouds in a hot gas can begin to contract as separate units if the density is high enough, and especially if tur-

Table 24·1 Some Observations of the 3-K Radiation[a]

Wavelength (cm)	Measured Thermodynamic Temperature (K)	Upper Limit on Variation in Different Directions (percent)
Radio 50–75	3.7 ± 1.2	
20	2.7 ± 0.6	
7	3.1 ± 1	
3.2	2.7 ± 0.1	0.5
0.86	2.6 ± 0.1	0.6
0.34	2.6 ± 0.2	3
0.26	2.8 ± 0.2	
0.05	3.0 ± 0.07	
X-ray 0.3–1.6 A		3

[a]See Webster, 1974.

bulence creates local clumping of matter. As suggested by Figure 24·4, cluster-size clouds may have appeared and eventually broken into individual galaxies. Inside the galaxies, globular clusters appeared, as illustrated by Figure 24·5. Some galaxies flattened to disks in which open clusters of individual stars formed.

Formation of Heavy Elements

The first galaxies and stars presumably formed from the gas produced in the big bang, about 75 percent hydrogen and 25 percent helium. This statement agrees with direct observations of the ancient Population II stars. A review of the theory of stellar evolution reveals the **origin of the heavy elements.** They were created inside stars by nuclear reactions such as the s-process and the r-process, described in Chapter 14. It is the massive stars, especially, that synthesize heavy elements in their high-pressure interiors and then blast them into space when they explode as supernovae. When the next generation of stars forms from the resulting interstellar clouds, these stars incorporate some of these ready-made heavy elements. Then, as they evolve, they will cook up still more heavy elements in their "stellar pressure cookers." As shown in Table 24·2, most heavy elements have probably been created in this way. The great stability of certain elements, such as iron, explains why they are especially abundant, since they rarely broke apart as other atoms joined and refragmented during atomic collisions inside stars.

Several other processes probably contributed to the present relative abundance of elements, as indicated in Table 24·2. For example, during the formation of galaxies, supermassive

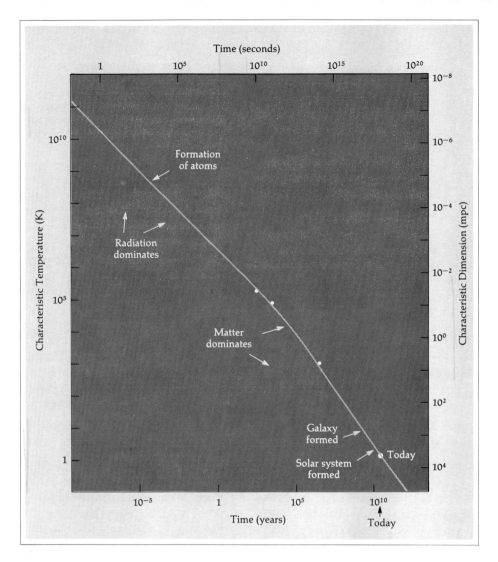

Figure 24·3 **A brief history of the universe, based on current research. Logarithmic time scale, bottom, expands the early history of the universe to fill most of the graph, allowing discrimination of early events. The curve shows declining temperature (left scale) and increasing distances to which internal observers could see (right scale).**

Radiation dominated during the early stages. Atoms formed by combinations of subatomic particles. Gravitational contraction of gas clouds led to the formation of galaxies. Characteristic temperatures corresponding to radiation have declined to 3 K at present.

objects of thousands of solar masses may have formed and soon exploded. Heavy elements from their interiors would be mixed into the gas that contributes to the galaxies. Few extremely massive stars would have lasted long enough to be observed today, but such stars might have contributed to heavy-element abundances in the past. Also, high-energy atomic particles, or cosmic rays, shoot through the galaxy and collide with atoms in space, splitting them into lighter nuclei such as lithium, beryllium, and boron. All these processes are still being studied.

◄ *Figure 24•4* **A remote cluster of galaxies, Abell 18. Almost every image is a complete galaxy. The arrangement resembles a loose globular cluster, but has a much larger scale. This may confirm that the process of galaxy-formation in clusters is similar to (but at a larger scale than) star-formation in clusters.** (Kitt Peak National Observatory)

Figure 24•5 **The giant elliptical galaxy M 87, 13 kpc across and 13,000 kpc away, suggests the hierarchical stages of formative processes in the universe. Faint starlike images surrounding the galaxy include a halo of more than 500 globular clusters, which in turn are made up of individual stars. The cloud that contracted to form the galaxy probably broke up first into globular clusters, then subfragmented into stars.** (Courtesy Malcom Smith and William E. Harris, Cerro Tololo Inter-American Observatory and McMaster University)
▼

Thus, the big-bang theory, combined with our knowledge of stellar processes, finds another major success in explaining the observed abundances of elements. As Figure 24•6 shows, the heavier elements have gradually increased in relative abundance from virtually none before about 12 billion years ago to a few percent today.

Before the Big Bang

What existed before the big bang? No one is sure that this question has meaning, or that any observations could reveal the answer.

According to classic big-bang theory, the mysterious explosive event at the beginning is *assumed* to have mixed matter and radiation in a primordial soup of near-infinite density, erasing any possible evidence of earlier environments. If so, science would have no way of answering the question.

Some astronomers suggest that the big bang was not unique, but only the

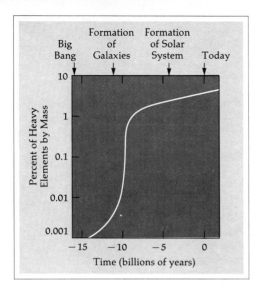

Figure 24·6 **The abundance of heavy elements as a function of time. Shortly after the big bang, heavy elements constituted virtually none of the total mass. They constituted about 1 percent of total mass soon after the formation of galaxies, because of production inside individual stars and dispersal by supernova explosions. They have since risen to several percent due to repeated generations of star-formation and disruption.** (After Reeves et al., 1972)

latest in a series of big bangs, as in curve D of Figure 23·5. This variant of big-bang cosmogony is called the hypothesis of the **oscillating universe.** In this view, after the primeval fireball explodes, gravity eventually pulls back the now-distant material and the universe collapses again, just as a rocket launched too slowly eventually slows and falls back to earth. Another high-density period would follow, then another expansion, and so on.

In a variant of the oscillating universe hypothesis, the high-density era was not the infinite-density epoch of the classical big bang, but only a period of maximum, finite density, during which some structures from a preceding era might have survived. At least one rotating neutron star has been interpreted as slowing its spin at a rate that may make it older than 16 billion years, and some astronomers have speculated that it might be a survivor from an earlier era.

In classic big-bang theory, the big bang is the true beginning of the universe as we know it, a "creation" like those in the creation myths of many cultures. The oscillating-universe model—that the big bang was not a true beginning, but only the beginning of the current cycle—resembles some Eastern mythologies.

Will the Universe Expand Forever?

Whatever its previous history, current observations suggest a **low probability that the universe will recollapse in the future.** If there is enough mass in the observable part of the universe, its total gravitational attraction ought to be able to slow down and stop the receding galaxies, and the universe would once again collapse. If not, the expansion should go on forever. Thus, the mean density of the universe is a critical factor in estimating whether the expansion will ever stop.

The density of material needed to make the galaxies fall back together is estimated at about 5×10^{-30} g/cm^3. However, many tests, such as measurement of the slowdown rate of galactic recession, indicate a density of material in the universe below that. For example, the observable, luminous material in the universe is calculated at a density of 5×10^{-32} g/cm^3—only about a percent of the required amount (Thorstensen and Partridge, 1975). Therefore, only large amounts of unseen mass would be enough to stop the expansion. Such mass does not exist within galaxies, and would thus have to occupy intergalactic space. Searches for such material—gas, dust, black holes, and so on—have been unsuccessful.

Thus, it now seems that the universe

Table 24·2 Probable Origins of Selected Elements

Element	Symbol	Big Bang	Supermassive Exploding "Pregalactic" Objects	Massive Unstable Stars	Ordinary Stars	Cosmic Ray Interactions
Hydrogen	^{1}H	Yes[a]				
Deuterium	^{2}H	Yes	?	?		
Helium-3	^{3}He	Yes		?	?	
Helium-4	^{4}He	Yes	?	?	?	
Lithium-6	^{6}Li					Yes
Lithium-7	^{7}Li	?	?	?	?	Some
Beryllium-9	^{9}Be		?			Yes
Boron-10	^{10}B					Yes
Boron-11	^{11}B	?	?			Yes
Heavier elements			?	Yes	Yes	

[a]"Yes" entries indicate strong mathematical evidence that the observed atoms formed in these environments. The entry "?" indicates the possibility or probability that some observed atoms formed. It is unlikely that significant numbers of atoms formed in the unmarked environments.

Source: Data adapted from Reeves et al. (1972)

will expand forever. This result suggests (although it does not *prove*) that the universe has not been oscillating in the past, because it will not oscillate in the future. Thus, many astronomers suspect that the big bang was indeed a unique event, and not one of a series of cycles.

Are There "Other Places"?

We have defined the universe as all matter and energy that exist anywhere. Until recent decades, most scientists had assumed that all parts of the universe had to be continuously connected, in the sense that light or a spaceship could go from one region to any other region and back again (even if the trip took a long time). However, if there are black holes there *might* be some regions permanently detached or unobservable from anywhere else. Some scientists speculate that there might be "other regions," such as regions inside black holes, in which events can occur but can never be seen by us.

A still more intriguing speculation is that we might be in a black hole in "someone else's" universe. Recent work suggests that—contrary to earlier theories that black holes could never emit anything—a black hole might in fact be able to explode in a burst of gamma ray radiation and subatomic particles. Thus, the big bang itself might have been some kind of black-hole explosion within some sort of larger universe.

Since black holes are thought to be permanently out of touch with each other and (so far) subject only to theoretical investigation, speculations on these problems are at the borders of what the scientific method can now deal with.

Why?

Why did the universe come into being? If the preceding problems were at the borders of the scientific method, this question is clearly beyond them. The scientific method is merely a procedure for analyzing observations and predicting phenomena slightly beyond those we know. Thus, people practicing the trade of science can only ask questions that can be answered by specific observations.

To questions of "why," astronomers' answers may be little better than anyone else's. One can always make up answers, but it is more interesting to admit that there are many things we do not understand—and things we do not even understand *how* to understand.

24

Summary

Three lines of astronomical evidence—ages of globular clusters, the expansion age calculated from Hubble's constant, and estimated ages of elements—indicate that the time elapsed since formative conditions existed in the universe has been roughly 16 ± 4 billion years. Most astronomers believe that an explosion-like event called the big bang occurred at that time. It takes some audacity to claim that we humans know something about this mysterious event 16 billion years ago, but there are several compelling observations:

1. Because galaxies are receding from each other, they must have once been concentrated in a much smaller volume, consistent with the big-bang model of initial high density.

2. The big-bang model explains the abundances of elements that we see both in Population II stars and, with the addition of later element-forming processes inside stars, in Population I stars as well.

3. The big-bang model predicted a pervasive weak radio radiation, known as the 3-K radiation, *later* found by radio astronomers.

4. Theories of galaxy formation, counts of distant galaxies, and observations of the most ancient galaxies and quasars, while not uniquely predicted by the big-bang theory, are consistent with it.

The big-bang theory takes us close to the limits of the ability of science to explain the properties of matter and energy in the universe. Questions such as why the universe came to exist are beyond the scope of science.

Concepts

cosmogony

ages of globular clusters

Hubble's constant

expansion age

age of elements

age of the universe

initial conditions

primeval fireball

3-K radiation

formation of galaxies

origin of heavy elements

oscillating universe

low probability that the
universe will recollapse
in the future

Problems

1. Which phrase describes the big-bang theory best?

a. *An assumption with logical consequences which turn out to be verified by observation.*
b. *A fact proven by repeated observation.*
c. *A revelation.*

2. Why was discovery of 3-K radiation from all over the sky heralded as strong evidence in favor of the big-bang theory? Does comparing the radiation from different parts of the sky give any evidence of asymmetry or inhomogeneity in the universe?

3. Why would planets such as the earth be unlikely to exist in globular clusters? (Hint: Consider the earth's composition.)

4. How would spectroscopic observations help reveal that a group of distant galaxies, such as Abell 18 (illustrated in Figure 24·4), is not a group of stars inside our own galaxy?

5. Why does the radiation from the primeval fireball as observed today have such long wavelength (radio waves) instead of the very short wavelengths (ultraviolet) that might be expected from the high temperatures theorized for the fireball?

6. Do you find any fundamental disagreement between the description of the universe's origin and structure, as described in this and the last chapter, and any philosophical or religious beliefs you may hold? If so, do you believe these might be clarified by further observations, or do you believe further observations are superfluous?

Advanced Problems

7. Hubble's constant, H, can be thought of as specifying the speed at which galaxies have traveled to reach any specified distance from the (arbitrarily chosen) central site of the big bang.

a. *Shown by logical deduction that $1/H$ ought to equal the age of the universe (the time since the big-bang).*
b. *Confirm that $1/H$ has the dimensions of time.*
c. *Confirm numerically that $1/H$ equals about 16 billions years. (Hint: One year approximately equals $\pi \times 10^7$ sec.)*

Alien Life in the Universe

One of the most intriguing questions in astronomy is whether planets elsewhere in the universe harbor what we are pleased to call "intelligent life." Extraterrestrial life must either exist or not exist. Either case has striking consequences. The nineteenth-century English writer Thomas Carlyle sardonically said that other worlds offer:

A sad spectacle. If they be inhabited, what a scope for misery and folly. If they be not inhabited, what a waste of space.

If *intelligent* aliens exist, our society (if it survives) is someday likely to be influenced by them, for better or worse. Anthropologist D.K. Stern notes that discovering such creatures "would irreversibly destroy man's self-image as the pinnacle of creation."

If alien life does not exist, we of earth would be the only living creatures in the universe—a remarkable alternative that would probably evoke dismay and likewise provoke a drastic revision of our myths.

What has astronomy to say about these two possibilities? Some scientists would argue that astronomy, lacking definitive data, has nothing to say. However, meteorites have supplied some evidence of organic (carbon-based) chemistry in space; the **organic molecules** (complex carbon-based molecules) recently found in interstellar gas are provocative; and Mars offers intriguing clues about biochemical possibilities on other planets. So a growing number of astronomers, allied with biologists, chemists, anthropologists, and theologians, are taking an interest in these questions.

In this chapter, we consider the issue in several steps. First, we discuss what we mean by "life," and what conditions life would require to exist on planets. Second, we review the long process that led to evolution of life on the earth, as best we understand it. Third, we ask whether there are planets that could support a similar evolution of life elsewhere in the universe. Finally, we try to estimate the probability of intelligent life actually existing elsewhere, and whether we might contact it, or it, us.

The Nature of Life

What do we mean by "life"? **Life** is not a status but a *process*—a series of chemical reactions using carbon-based molecules, by which matter is taken into a system and used to assist the system's growth and reproduction, with waste products being expelled. The system in which the processes occur is the cell. All known living things are composed of one or more cells.[1] A **cell** is in essence a container filled with an intricate array of organic and inorganic molecules (protoplasm). Codes for cellular processes are contained in very complex molecules (such as the famous DNA) located in a central body called the nucleus. The elements most prominent in the organic molecules are carbon, hydrogen, oxygen, and nitrogen—all very common in regions populated by evolved (Population I) stars. Phosphorus, also important to life (in small amounts), is widely available. Carbon is especially critical because it can combine to make long chains of atoms—large molecules that encourage the complicated chemistry of genetics, reproduction, and so on. That is why the term **organic chemistry** (as well as *organic molecules*) refers not specifically to life-forms but more generally to all chemistry (and molecules) involving carbon.

We often make the mistake of thinking of ourselves as static beings instead of dynamic systems. We casually assume that we are constant entities, as if our identity were solely dependent on the form of our bodies. But our bodies today are not the same ones we had seven years ago. Hardly a cell is still alive that was part of that body. This dynamic conception is a far cry from the conception of life only a few generations back, when bodies were thought of as semipermanent machines whose parts gradually wore out. Even our seemingly inert skeletons are living and changing, always replacing their cells. We *must* keep changing—the cells must keep processing new materials to stay alive.

---

[1] Viruses might be an exception. They are simpler than cells, yet can reproduce themselves using materials from host cells. Biologists disagree on whether or not viruses should be considered a form of life.

When the processing stops, we call it death.

The nature of living beings is illustrated in an analogy from the Russian biochemist A. I. Oparin (1962). Consider a bucket that has water pouring in at the top from a tap and flowing out at the same rate through a tap in the bottom. The water level in the bucket stays constant, and a casual observer would call it a "bucket of water." But it is not like an ordinary bucket standing full of water. The water at any instant is not the same water as at any other instant, yet the outward appearance is constant. We are like buckets with water and nutrients and air flowing through us, but with other, much more complex attributes, such as the ability to reproduce and to be affected by genetic changes that let us evolve from generation to generation.

The *dynamic nature of life* gives us some clue to the kinds of processes involved in the origin of life. We are not looking for a readymade machine. Instead, we are looking for a process in which complex carbon-based molecules can enter cell systems that draw from the incoming material to create new molecules, incorporate them into new structures, eject unused material, and reproduce themselves.

Scientists therefore usually choose to define "life" by these specific carbon-based processes. Often at this point people ask, "What about some unknown form of consciousness?[2] Or what about some unknown chemistry based on other elements, such as silicon, that can form big molecules?" The answer is that we have never

---

[2]For example, in *The Black Cloud* astronomer Fred Hoyle imagined an interstellar nebular cloud with matter and electromagnetic fields organized in such a way that it developed a consciousness or will of its own.

observed or experimented with such life-forms, so we can say nothing substantive about them. If we admit they are plausible, we simply increase the probability of "life" or "consciousness" in the universe. But researchers usually restrict their discussion to forms on which chemical and behavioral data are available.

Whatever other conceptions we invent—civilization, religion, technology, art, war, love, communication— it is the chemical processes of "life" that define us, just as they define the spiders, sea urchins, elephants, moths, amoebas, redwoods, and all the other incredibly varied living creatures around us. To judge whether life may exist on other planets—whether other planets are already "taken"— we must find out how those processes got started on the earth.

The Origin of Life on Earth

In addition to the life-forming elements—especially carbon (C), hydrogen (H), oxygen (O), and nitrogen (N)—water was crucial to the development of life on earth. Several facts indicate this:

1. Most of our body weight is made up of water. The percentage is even higher for plants.

2. Body fluids, like tears, have a saltiness similar to that of the oceans. The oceans are our heritage. As embryos, we are at first immersed in fluid and develop bodies more like fish than like mammals—for example, we have gills.

3. Organisms deprived of liquid water quickly die. And dead organisms are shriveled and dried.

4. Taking a more theoretical view, we realize that water provides a *medium* in which organic molecules can be suspended and interact—a likely place for life to begin.

So a good starting question is: What was the early history of C, H, O, N, and water on the earth? The nebula from which the planets formed, and the earth's atmosphere, must have been rich in compounds that would form from cosmic gases. As is clear from theory (Chapter 11) and from observation of gases preserved on Jupiter, these compounds would include hydrogen, ammonia, methane, water, and perhaps some nitrogen (respectively H_2, NH_3, CH_4, H_2O, and N_2). Note which atoms were involved: C, H, O, and N—the elements of life!

The earth formed about 4.6 billion years ago, as outlined in Chapters 3 and 5. Probably within a few hundred million years, heat from internal radioactivity and impacts melted parts of the earth's interior, unleashing volcanic activity. Volcanic activity released gases, especially steam (H_2O) and carbon dioxide (CO_2). Thus, even if there were no oceans at the beginning, by about 4.5 to 3.5 billion years ago there were probably bodies of surface water under an atmosphere that was rich in hydrogen, ammonia, methane, water, and nitrogen, and was becoming richer in oxygen.

Chemical reactions occurring naturally in such an environment *could* produce building blocks of life, as has been proved by laboratory experiments. In the 1950s, Chemist S. L. Miller put a gaseous mixture of hydrogen, ammonia, methane, and water vapor (to represent the primitive atmosphere) over a pool of liquid water (primitive seas) and passed electric sparks through it (simulating energy

Figure 25•1 **Sun and sea illustrate the earth's unusual environment favoring the origin of life. Our planet is far enough from its central star, the star is stable enough, and the orbit is circular enough so that the** temperature **remains fairly uniform and some water remains liquid. Energy, water, and the right chemical environment are believed necessary to initiate life.**

▲
Figure 25•2 **Fragments of one of several carbonaceous chondrite meteorites in which chemists have discovered extraterrestrial amino acids. Similar carbon-based compounds are important in living cells on earth. Although no fossils or life-forms have been found in the meteorite, its dark color is due to the abundance of carbon and carbon compounds, showing that building blocks of life can form outside the earth. This meteorite fell in France in 1964. (NASA)**

Later experiments have shown that many kinds of energy sources—including ultraviolet light from the sun, volcanic activity, and even meteorite impacts—would produce the same result. Thus we can conclude that building blocks of life *would* have formed on the early earth and in similar environments elsewhere. The conclusion has been confirmed by the discovery of extraterrestrial amino acids (but not fossils or life-forms) in several carbonaceous chondrite meteorites such as the one shown in Figure 25•2. These amino acids were proven extraterrestrial not only because they were in the meteorite, but because their molecular structures and symmetry differed from any observed on the earth.

From such results, researchers have concluded that molecular organic materials existed in the earth's primitive oceans, probably less than half an eon after the earth formed. Such material may have accumulated in

sources such as lightning). After several days, the pool began to darken. The water now contained a solution of **amino acids,** the class of molecules that join to form **proteins,** the huge molecules in cells. This so-called **Miller experiment** proved that the complex organic chemicals necessary for life could be built in a natural environment.

isolated tide-water ponds (Figure 25•3) and concentrated as water evaporated, leaving heavy molecules behind to interact and form complex substances.

Related discoveries have been made by Viking spacecraft on Mars. The Martian soil, unlike earth's soil, is exposed to solar ultraviolet rays, which produce unfamiliar reactive chemical states in minerals. This soil contains materials that can synthesize organic molecules from atmospheric carbon dioxide, and release gas when nutrient fluids are added. Because the Martian soil in its natural state contains virtually no organic molecules, most scientists believe these processes are not caused by Martian life-forms. In fact, laboratory experiments in 1977 duplicated most Viking results using simulated Martian soil (iron oxide

Figure 25•3 **Broths rich in amino acids and complex organic molecules formed in long-lived tidewater pools on the ancient** ▼ **earth. Evaporation of water could have concentrated the remaining organic materials, allowing complex reactions. The** **resulting products could "fertilize" oceans with living organisms or protoliving materials.**

minerals exposed to ultraviolet light in CO_2) without any organisms. Nonetheless, the processes may indicate how complicated chemical reactions create material from which life could form on other planets.

The next step is less certain. Florida biologist Sidney W. Fox has shown that simple heating of dry amino acids (as might happen on a dry planet such as present-day Mars) can create protein molecules. When water is added, these proteins assume the shape of round, cell-like objects called **proteinoids,** which take in material from the surrounding liquid, grow by attaching to each other, and divide. Though they are not considered "living," they resemble bacteria so much that experts have trouble telling the difference in microscope views.

Possibly related to proteinoids are objects discovered in the 1930s by Dutch chemist H.G. Bungenberg de Jong. When proteins are mixed in solution with other complex molecules, both sets of substances spontaneously accumulate into cell-sized clusters, called **coacervates.** The remaining fluid is almost entirely free of complex organic molecules.

The next step toward recognizable life-forms is still more uncertain, but many biologists around the world have theorized that cell-like structures in primeval pools of "organic broth" began reacting with fluids in the pools and with each other, accumulating more molecules and growing more complex, as suggested by Figure 25·4. Eventually these could have evolved into biochemical systems capable of reproducing.

Whatever the processes, microscopic cellular life must have arisen between 4.6 and 3.1 eons ago, because remains have been discovered in rocks after that period. In a sequence of rocks from Greenland, 3.7-eon-old rocks show no traces of ancient life, but rocks dating about 3 eons and younger do show traces. A number of other rocks younger than 3 eons, found in the 1960s and the 1970s, show microscopic structures believed to be fossil organisms. Most scientists believe these organisms gradually evolved through processes similar to those shown at the primitive level in the Miller and coacervate experiments.

The earth was not a passive background for these processes, but evolved as indicated in Chapter 3, with the atmosphere changing from hydrogen-rich to hydrogen-poor. Oxygen increased and formed the high-altitude **ozone** (O_3) **layer,** screening the surface from potent solar ultraviolet light, which could have broken up organic molecules on the surface.

Living things could hardly be unaffected by these changes. Whereas the earliest life-forms developed and existed without oxygen, life now had to adjust to oxygen. Such environ-

Figure 25·4 **The importance of fluid medium for primitive evolution is suggested by these photos of single-celled organisms engulfing**
▼
A

food from surrounding fluid. **A. An amoeba flows to surround a nearby food particle.** (Optical microscope photo; S.L. Wolfe)

B. The protozoan *Woodruffia* **ingests a** *Paramecium.* (Electron microscope photo; T.K. Golder)
B

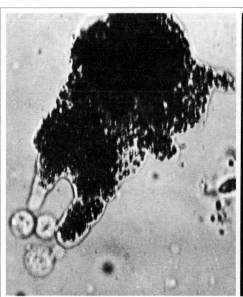

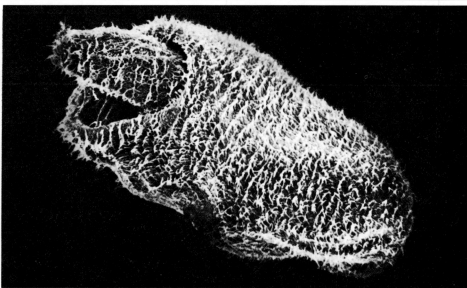

mental changes may have favored increases in the "rate" of evolution— when a mutation appeared that happened to be well-adapted to the new environment, it lived longer and had more offspring, promoting retention of the new trait. This promoted evolution of new species.

Yet by 2 eons ago, we still would scarcely have recognized our earth. The land was still barren. Some areas must have looked like today's deserts, or like Mars. Some areas were moist and washed by rains, but instead of luxurious forests there was only bare dirt, eroded gullies, and grand canyons. Brown vistas stretched to the sea coasts. Most life was still in the oceans, soft-bodied and rarely producing fossils. Such fossils as we have from that period suggest that algae, bacteria, and other simple forms were the dominant life. For a billion years, the life on land may have been no more complex than the lichens in Figure 25·5.

One evidence of the adaptability of life-forms, once they evolved, is the rapid proliferation of advanced species, as indicated in the geologic time scale shown in Table 3·1. While it took about half the available time to go from complex molecules to algae and bacteria, it took only the last 12 percent of earth history to go from the first hard-bodied sea creatures (such as trilobites, Figure 25·6) to humans. Biologists attribute evolution in general and this rapid proliferation in particular to **natural selection,** the greater production of offspring by those individuals best adapted to the ever-changing environment. Details of the natural selection process are uncertain. Prototypes of new species may have evolved in obscure ecological niches and then emerged rapidly when sudden environmental changes (such

▲
Figure 25·5 **Lichens, symbiotic combinations of algae and fungi, were probably some of the first organisms on land. They grow on rock surfaces, weathering the rock to** produce soil. **They are widely adaptable, existing both in arctic tundra and in hot deserts. Penny shows scale.**

Figure 25·6 **Fossil trilobite, a hard-bodied sea animal that appeared about 600 million years ago, providing diagnostic fossils for recognizing and dividing Cambrian time.** (Gayle Hartmann and Joyce Rehm)

as Ice Ages) caused extinction of earlier species better suited for earlier conditions.

Whatever the specific mechanism, we can say from the fossil record that the earth experienced a few-billion-year evolution from nonliving organic chemicals to small organisms, and that these evolved in less than a billion years to species with self-conscious intelligence.

▲
Figure 25·7 **Reptiles reached ascendency in the last 5 percent of geologic time. Many species of dinosaurs evolved rapidly in warm climates, but may have died out due to the onset of colder climates, to which they could not adjust. Warm-blooded mammals, which could survive larger temperature extremes, replaced them in the last 2 percent of geologic time.** (Smithsonian Institution)

Planets outside the Solar System?

From all we have just said, we conclude that if planetary surfaces with the necessary conditions—liquid water and C—H—O—N chemicals—exist anywhere, life is likely to evolve on them. And advanced species will probably eventually appear. But are there such planets?

In view of the hundred billion stars in our galaxy, not to mention the innumerable other galaxies, we should not limit our search for life to the solar system alone. There are four kinds of **evidence for planets near other stars.**

First, many or most newly formed stars are probably immersed in cocoon nebulae in which mineral grains have condensed. These grains may form planetary bodies. Production of such materials seems to be a natural by-product of star-formation.

Second, statistics of masses among

binary and multiple stars suggest the likelihood of many stars having low-mass companions that are planets.

Third, unseen low-mass companions have actually been detected around a few nearby stars. Barnard's star, for example, may have a companion about the size of the planet Jupiter. If so, then two of the four nearest stars have planets, as shown in Table 25·1. But chances for detection drop rapidly at greater distances. Of the 10 nearest systems, 2 may contain detectable planet-size objects, and a third has an object about 15 times more massive than Jupiter. A few stars still more distant may have companions around 10 times more massive than Jupiter.

Fourth, massive stars spin faster than smaller stars. Stars of about 1.5 solar masses (and the sun as well) may have lost angular momentum when some was transferred by magnetic forces to a surrounding nebula in which planets could have formed (Chapter 11). This finding has prompted speculation that most stars of less than 1.5 solar masses may have planets. (Huang, 1965).

To summarize, a contemporary astronomer's informed guess as to the percentage of stars accompanied by planets would be between one and thirty percent. Even if only one percent of the galaxy's stars had planets, that would give a billion planetary systems within the Milky Way!

Habitable Planets?

Even if planets exist near some other stars, there is no guarantee that they are habitable. Astronomers have proposed several conditions needed to make a planet habitable:

1. The central star should not be more than about 1.5 solar masses, so that it will last long enough for life to evolve (at least 2 eons), and so that it will not kill evolving life with too much ultraviolet radiation, which breaks down organic molecules.

2. The central star should have at least 0.3 solar masses, to be warm enough to create a reasonably large orbital zone in which a planet could retain liquid water.

3. The planet must orbit at the right distance from the star, so that liquid water will neither evaporate nor permanently freeze (Figure 25·8).

4. The planet's orbit must be near enough to circular to keep it that proper distance and prevent too drastic seasonal changes.

5. The planet's gravity must be strong enough to hold a substantial atmosphere.

The American physicist Stephen Dole (1964) reviewed these criteria and concluded that roughly 6 percent of the stars, mostly from 0.9 to 1.0 solar masses, have habitable planets. However, the figure remains highly uncertain.

Has Life Evolved Elsewhere?

If habitable planets exist, and if life evolves readily under habitable conditions, shall we immediately conclude that life must be abundant throughout the universe? There are several additional factors to consider. For one thing, planetary environments change with time, so that today's habitable planet may not be habitable tomorrow. This raises the question of the adaptability of life in response to change.

Table 25·1 Components of the Nearest Systems and Other Selected Systems

Name of Largest Star in System	Distance (psc)	(light-years)	10^{-4} (Jupiter)	10^{-3}	10^{-2}	10^{-1}	1 (Sun)	10	Tested for Radio Signals[a]
Nearest Stars									
Sun	0	0	x	x			x		
Alpha Centauri	1.3	4.3				x	x x		
Barnard's star	1.8	5.9	?	?		x			✔
Wolf 359	2.3	7.6				x			✔
+36°2147	2.5	8.1			x	x			✔
Sirius	2.6	8.6					x	x	
Luyten 726-8	2.7	8.9			x x				✔
Ross 154	2.9	9.4				x			✔
Ross 248	3.2	10.3				x			
Other Selected Systems									
61 Cygni	3.5	11.2			x		x x		
Krüger 60	3.9	12.9				x x			
Ross 614	4.0	13.0				x x			
B.D. +68°946	4.8	15.6			x		x		
B.D. + 20°2465	4.9	16.0			?		x		
Eta Cassiopeiae	5.9	19.2			x		x x		

[a]Check mark means search has been made for radio signals from near these stars that could have been made by intelligent beings. Results have so far been negative.
[b]A close star, recently identified. Its properties are uncertain.

Source: Data from Allen (1973); Berendzen (1973); and other sources.

Effects of Astronomical Processes on Biological Evolution

Basic planetary or stellar processes may be involved in encouraging or hindering biological evolution. For example, earth's fossil records indicate an episode called the "great dying," which witnessed simultaneous extinction of a significant fraction of land and sea species about 230 million years ago; many paleontologists believe widespread climate changes were involved. A similar episode saw the extinction of dinosaurs and other land and sea species about 70 million years ago, and there is evidence in this case for climatic cooling. Astronomical processes that have been suggested as responsible for these changes include:

1. Planetary convection, causing plate-tectonic crustal splitting and changing sea levels, ocean currents, wind patterns, and seasonal extremes. This process has been responsible for isolating land masses such as Australia and allowing evolution of different species there.

2. Volcanic eruptions may have spewed enough dust into the high atmosphere to reduce the sunlight reaching the surface. For example, widespread sunlight decreases of up to 25 percent occurred after the 1883 Krakatoa (Sumatra) and 1912 Katmai (Alaska) eruptions. Larger, rarer eruptions might have cut sunlight for some years, causing the decline of some species and ascendancy of new ones.

3. Slight changes in the tip of planetary axes to the plane of the ecliptic, caused by gravitational forces, are believed to have caused major climate changes such as the Ice Ages.

4. Slight changes in the sun's radia-

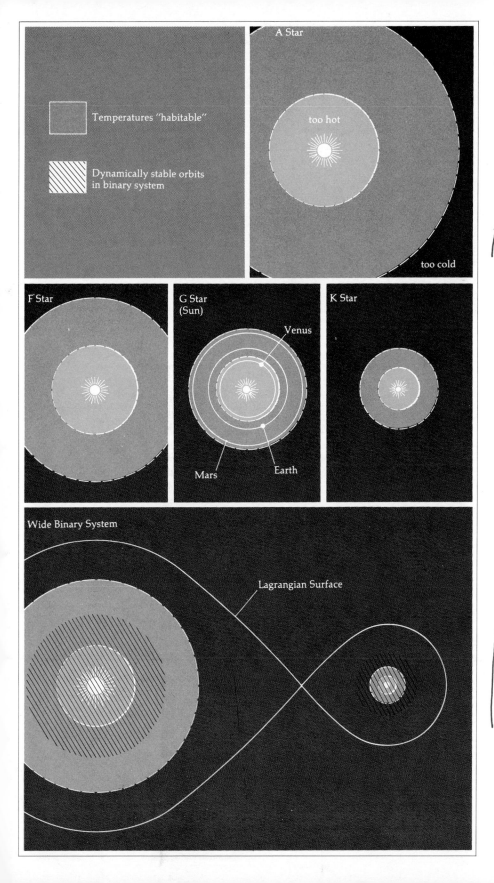

Figure 25·8 **Stippled zones show the relative sizes of temperate zones where water would remain liquid in planetary systems around stars of different spectral types. In a binary system (bottom), the situation is complicated because orbits outside hatched regions suffer large perturbations, causing evolution away from circular orbits, in turn causing temperature extremes.** (After Huang, 1960)

tion over various time scales may have changed climates.

Would these kinds of changes help or hinder biological evolution? One might suppose that any instabilities would hinder the chances for evolution of advanced life. But this may be true only for changes that are too extreme (for instance, boiling away of all water) or too rapid (an Ice Age developing within a year or two). Many biologists are coming to believe that changes of the magnitude that have occurred on earth, while causing the decline of some species, have promoted the emergence of advanced species. For example, if some aspects of intelligence had first appeared in a benign, constant environment where food was plentiful, these traits would have had little value. But if Ice Ages destroyed the mild environment, then the cleverer groups might have emerged from their previous obscurity.

Perhaps it is no coincidence, then, that the time scale of geologic change is comparable to the time scale of biological change. Life's response to changing environments is to change itself, on time scales as short as a million years.

Adaptability and Diversity of Life

The great variety of ancient and modern species on earth, and the variety of environments in which they have thrived, suggest that, given time, life could also have evolved to fit a wide range of conditions on other planets. Even humans have a remarkable adaptability. We thrive from equatorial wet jungles to dry deserts to arctic plains to Andean summits where air pressure is barely half that at sea level. In the past, during Ice Ages that glaciated New York, humans survived by migrating. We can survive a 3 percent variation in body temperature, from about 303 K to 313 K.

But humans are fine-tuned animals. The environmental range of simpler life-forms on the earth is more impressive. Microflora are known in supercooled Antarctic ponds that remain liquid at 228 K ($-49°F$) because of dissolved calcium salts. And bacteria are known in Yellowstone Hot Springs at temperatures of 363 K ($194°F$)—a 46 percent variation in temperature. In laboratory experiments, common bacteria have survived in liquid cultures at least 24 hours in CO_2 atmospheres at 433 K ($320°F$), extending the range to a 62 percent variation in temperature.

Habitable pressure regimes show an even greater range. Bacteria exist at altitudes where the atmospheric pressure is only about 0.2 atm, and more advanced organisms live in ocean depths with pressures of hundreds of atmospheres. These extremes range over a factor of a thousand in pressure. In general, the simpler the organism the wider the environmental range it can withstand.

Could any terrestrial organisms withstand the environments of other planets? For simple organisms, probably yes. Grass seeds can germinate in a variety of atmospheres quite different from the earth's, as long as they are composed of simple com-

Figure 25·9 **Tests with organisms in space included the flight of the spider Arabella, which flew with three astronauts during the 59-day Skylab 3 mission in 1973. The spider showed some disorientation because of weightlessness, but completed its web and survived the flight. The test was one of 25 experiments selected for Skylab from designs submitted by high school students.** (NASA)

pounds of the common elements C, H, O, and N. Eight species of insects (relatively complicated creatures) studied at different pressures behaved normally all the way down to 10 to 16 percent of normal atmospheric pressure on earth. Pressure could be dropped as low as 1 to 3 percent and permit survival if restored in a few minutes to 3 to 10 percent of normal. For comparison, the pressure at some places on Mars is somewhat over 1 percent our normal value in most regions (but the air is CO_2; Mars probably had more pressure, more oxygen, and higher temperature in the past).

Thus, bacteria or other simple organisms from the earth might survive in certain conditions found on other

Figure 25·10 **Electron microscope photo of single-celled bacteria discovered in 1973 in a California spring ten times more alkaline than the previous limit believed to permit living organisms. Some scientists have suggested this environment has some similarities to conditions that could exist on Jupiter, thus strengthening the chances for possible microscopic life-forms in the outer solar system. One bacterium is dividing into two separate organisms.** (NASA)

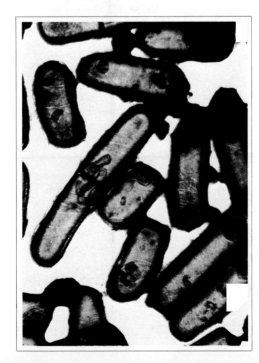

planets. For this reason, spacecraft have been at least partially sterilized to avoid contaminating the planets. Such contamination might not only alter planetary environments, it might ruin any chance for us to discover whether simple organisms or complex organic molecules ever formed independently on those planets.

It works the other way, too. Simple organisms from other planets might have devastating effects on the earth. There are historical examples of similar events. The plague caused by bacteria introduced into Europe in the 1300s killed about a quarter of all Europeans, and as much as three-quarters of the inhabitants in some areas. Diseases introduced into Hawaii after the first European contact in 1778 killed about half of all Hawaiians within 50 years. Some 95 percent of the natives of Guam were wiped out by disease within a century of continued European contact. For these reasons, early Apollo astronauts were quarantined until it was clear they carried no lunar organisms. And samples from Mars may be analyzed only in space labs, well above the earth's atmosphere.

With these facts as well as cultural competition in mind, anthropologist D.K. Stern (1975) has remarked that "It is likely that the meeting of two alien civilizations will lead to the subordination of one by the other." Thus, change and evolution in life populations are likely not only from life's adaptability to new environments, but also by the invasion and destruction of some populations by others.

In summary, natural selection seems to produce species capable of occupying any habitable environment. Thus, we should not be surprised if life has evolved on another planet. This life may look very strange to us. After all, if mushrooms and corals

and wooly mammoths and Venus fly-traps all evolved on one planet, how much greater may be the differences between life-forms on two different planets? Feathers and fur and sex and seeds and herd instinct and intelligence of the form we know may be products of earth only.

Effects of Technological Evolution on Biological Evolution

If we are right so far, that life will evolve when the right conditions exist, and that the conditions probably do exist elsewhere, then what can we predict about it? Should we predict intelligence and civilizations? What do these terms mean? Should we assume that other civilizations will achieve space flight, or might visit us some day? This raises the question of technology and its role in the evolution of life. Just as cosmologists have only one universe as an example, **exobiologists** (biologists concerned with possible life on other worlds) have only one inhabited world as an example, so the answers to these questions are uncertain. Many exobiologists have assumed that intelligence involves, among other things, use of tools to modify the environment; hence, technology.

But as we have seen, while limited environmental change can be helpful, environmental change that is too much or too fast can be fatal! As Pulitzer Prize-winning naturalist René Dubos points out, we are umbilical to earth, and if we alter our planet too much before acquiring an ability to leave it, we are finished. For this reason, consideration of our own case leads to the conclusion that development of technology may actually end civilization on some worlds. This is hardly wild speculation, since we see a few nomi-

nally moral, intelligent technologists on our own world spending entire careers devising weapons solely to deal death to our own species.

In the past, war did not threaten our whole species because conflicts involved only a small percent of the world. But today, nuclear, biological, and other types of weapons could involve the whole world. For example, the radioactive strontium-90 produced by nuclear explosions has a half-life of 28 years. It was blasted freely into the atmosphere before the Nuclear Test Ban Treaty of 1963. A year after the first H-bomb tests by the United States in the Pacific, strontium-90 deposits in American soil increased soil radioactivity by one-half percent. A few weeks after a 1976 Chinese nuclear test, airborne radioactive debris fell onto the United States, increasing radiation levels. Obviously, a sufficiently massive nuclear exchange could devastate not only civilization but also future forms of life, whose genetic pool would be exposed to high radiation levels for decades. Humanity has thus proved that a planetary culture could wipe itself out by conscious design of weapons, as irrational as that may seem.

We have also proved that this disaster *could* happen by mistake. As our technology assumes a planetary scale, so do our accidents. Problems as diverse as nuclear power and aerosol spray cans illustrate the issue. Many currently planned nuclear power plants will produce radioactive plutonium wastes, among the most toxic of known materials. Although safe when sealed, some kilograms of plutonium dust accidentally spilled into the air could devastate whole states. Yet many pounds are already being processed, and the policy of several governments, including the United States, promotes expanded dependence on nuclear power plants until other energy forms become available. Accidents are not the only danger: We have seen our society spawn terrorists and madmen. With some grams of stolen plutonium-238, such characters could threaten whole cities. Radioactivity presents a known technological danger that our society has allowed to develop in spite of knowing about it ahead of time.

The seemingly harmless aerosol can is a different story. In 1974, several scientists realized from theoretical calculations that when freon (the propellant gas used in such cans) is released into the air, it reaches the earth's protective ozone layer high in the stratosphere. Here its breakdown products act to destroy the ozone layer. Depletion of ozone would let more solar ultraviolet radiation reach the surface, increasing hazards of skin cancer. Even as more freon was being sold, planetary astronomers and chemists published refined calculations showing that to deplete the freon already in the atmosphere could take as long as a century. Thus, our generation's consumer gimmicks could be creating future health hazards. After a 1975 report by the National Academy of Sciences, confirming this danger, moves were made in 1976 and 1977 by American regulatory agencies to phase out freon aerosol propellants by 1979—an example of a technological danger that almost slipped up on us without our knowing it.[3]

If humanity is any example, long-term survival of a planetary culture is not assured. Although we have been around less than a hundredth of a percent of the age of the universe, we

are already having close brushes with global disaster. Thus, we can speculate that if evolution produces intelligent societies that remain tied to one planet, many of them may last only a fraction of a percent of the age of the universe—in which case there is little chance that a given culture will be around at the same time we are.

More optimistically, we have succeeded in identifying the **cultural hurdle** that we (and perhaps intelligent species on any planet) must surmount: the transition from scattered competing nation-states with the capability to damage the planet, to a stable global technology of intelligence and imagination. Perhaps we and some other cultures will cross this hurdle. After all, we have identified some perils in time. Perhaps some cultures have completed the transition from planetary cultures to interplanetary cultures spread across many planets, thus insuring their survival against ecological disaster on any one planet. Such cultures might last, and be detectable, for billions of years instead of thousands.

Our conclusion so far is that a fraction of the stars have planets; a fraction of the planets ought to be inhabited; a fraction of the inhabited planets ought to have either civilizations or relics of destroyed civilizations. The next question is whether any actual evidence for extraterrestrial carbon-based life as we know it exists today.

*Alien Life
in the Solar System?*

The ground rules of the search are that we must find a "habitable planet" that has (or has had) liquid water and an atmosphere in which complex carbon-based organic chemistry can

proceed—and where these conditions have lasted long enough for a specified degree of complexity to have evolved. If the specified degree is to be intelligence, or meter-scale animals made mostly out of liquid water (like ourselves), then, as we saw in the early chapters, we can almost certainly rule out every body in the solar system but the earth, since conditions elsewhere do not permit long-term bodies of liquid water.

But if we ask about primitive life-forms in the solar system, rather than intelligence, the situation is a little more promising. As noted, meteorites indicate that amino acids have been synthesized outside the earth. On Mars, ultraviolet light may have destroyed organic material on the surface, but native Martian soil has the ability to synthesize organic molecules from nutrients. Some Viking scientists leave open the possibility that Martian microbes exist below the surface.

In summary, there is almost certainly no extraterrestrial intelligent **life in the solar system** today, and probably no other life-forms big enough to be visible to the naked eye. There remains a low chance of microscopic, primitive life-forms elsewhere in the solar system, perhaps on Mars. It is certain that complex but nonliving organic materials have been created in other parts of the solar system. Evidence from our solar system favors the argument that life might have evolved on planets near other stars if those planets had enduring, temperate, moist environments.

Alien Life among the Stars?

The next step is to consider alien life on planets near other stars. There is no direct evidence, but American, Soviet, and other scientists in inter-

---

[3]But we should be haunted by the words of Harvard planetary physicist Michael McElroy, who performed some of the earliest freon calculations: "What the hell else has slipped by?"

national meetings have put together a method for considering the possibilities (Sagan, 1973). The logic is to try to estimate the various fractions—of stars having planets; of those planets that are habitable; of those where conditions remain favorable long enough for life to evolve; of those where life does evolve; of those where intelligence evolves; and of the planet's life during which intelligence lasts.

Table 25·2 shows "optimistic" and "pessimistic" estimates of the various fractions, and consequent estimates of the upper and lower limits on the fraction of stars that might have civilizations today. In the "optimistic" case, the answer comes out a few percent; in the "pessimistic" case, only one star in 10^{14} (one in 100 million million).

In these cases, how far away would the civilizations be? Figure 25·11 shows the answer, by plotting the radial distance required to include a given number of stars.

In the first case, the nearest civilization might be only 15 light years away—amazingly close. At the speed of light it might take only one generation to reach it. In the "pessimistic" case, the closest civilization would probably not be in our own galaxy, but roughly 10 million light years away in a distant galaxy. Nonetheless, in view of the innumerable galaxies, it is hard to avoid the conclusion—even with the most pessimistic view—that the likelihood of **life outside the solar system** amounts to millions or billions of technological civilizations. If our reasoning has been right,

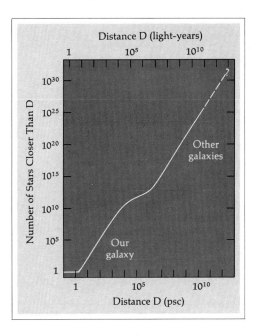

▲

Figure 25·11 **Distance in parsecs (bottom) and light-years (top) required to encounter the number of stars indicated at left. If one star in 10^9 has life-bearing planets, the nearest one might be within a few thousand parsecs.**

then at this moment, intelligent creatures may be pursuing their own ends in unknown places under unknown suns.

Where Are They?

One remaining observational fact bears on this question. American and Russian radio astronomers have listened for radio messages from alien civilizations and have found none (see Table 25·1). Nor do our skies seem to be overrun with alien visitors trying to contact us. Yet, if the "optimistic" estimate in Table 25·2 is anywhere near correct, aliens would be nearby. Why haven't we heard from them?

One answer might be that we *are* being visited—witness the "flying saucer" reports. But although some UFO reports are intriguing, verifiable

Table 25·2 Estimated Fraction of Stars with Planets That Have Intelligent Life

	Plausible Lower Limit	Plausible Upper Limit
Stars having planets	10^{-2}	0.3
Those stars ever having habitable conditions on at least one planet	10^{-1}	0.7
Those planets on which such conditions last long enough for life to evolve	10^{-1}	1
Those planets on which life does evolve	10^{-1}	1
Those planets on which habitable conditions last long enough for intelligence to evolve	10^{-1}	0.9
Those planets on which intelligence does evolve	10^{-1}	1
Those planets on which intelligent life endures	10^{-7}	10^{-1}
Product: Fraction of all stars with planets that bear intelligent life	10^{-14}	2×10^{-2}
Implication: Distance to nearest civilization	10^7 light years	15 light years

▲
Figure 25•12 **Astronomical data suggest that many planets may exist among the galaxies of the universe, but the question remains of how many are both habitable and inhabited. This imaginary earthlike world with a moonlike satellite lies in a small galaxy** orbiting a large spiral galaxy which dominates the evening sky. The planet's **temperature is too low to allow liquid water, so the abundant water remains frozen in snow fields and no life has evolved.** (Painting by Ron Miller)

evidence for actual alien spaceships is abysmally poor (for further discussion of UFOs see Enrichment Essay B). If any alien visits have really occurred they are so rare that we have no proof of them.

Another possible answer is that the "pessimistic" figures are right, and that the nearest civilizations are in distant galaxies. Even their radio messages, if any, could be 10 million years old by the time we receive them, and their spaceships would be unlikely to reach the earth if limited to speeds less than that of light, as current physics requires. But most investigators of the problem place the probability well above our "pessimistic" limit. At a 1971 Soviet–American conference on this problem, the favored estimate came out to be one civiliza-

tion per 10^5 stars. This would put a million civilizations in our galaxy and the nearest civilizations only a few hundred light years away. Why, then, no frequent visits to earth?

A third possible answer is that biological evolution need not produce creatures who have a desire to build "civilizations" or travel through space. Not even all human societies necessarily evolve toward technology. Are humans fated to be explorers, bridge-builders, and businessmen rather than artists, athletes, or daydreamers? Is the stereotyped aggressive Westerner more representative of the essence of humanity than the stereotyped contemplative Easterner? May not our aggressive technocracy be just one type of *cultural* activity rather than a universally achieved stage of *biological*

evolution? Historically, patterns we once assumed to be biologically imposed have turned out to be merely culturally imposed.

If humanity is not predestined to develop a technological civilization, how much less certain is the course of social development on other worlds. It is absurdly anthropocentric of us to suppose that beings on other planets would resemble us physically, psychologically, or socially. Consider again the variety of highly evolved life forms on our planet alone. Ants live in ordered societies that do not appear to regard individual survival as important. Dolphins communicate and have brains that seem almost comparable to ours, but they have no manipulative organs and hence no technology. Perhaps we cannot expect aliens to be motivated by emotions that mean much to us. We certainly cannot expect, as always happens in grade-C science fiction movies, that humanlike aliens will walk out of saucers and invite us to join their democratically constituted United Planets, a galactic organization structured by documents that are curiously reminiscent of the United States Constitution. So why assume that other civilizations might even care to try to visit or contact us?

Evolutionary Clocks
and the Explorative Interval

This brings us to a fourth and perhaps most significant answer: We may be farther from aliens in evolutionary time than in physical space. Biological evolution is so persistently experimental that even if another planet started evolving at exactly the same time as ours, and even if its biochemistry produced creatures like us, those creatures are not likely to be in a phase of evolution similar to ours.

If the evolutionary "clocks" on the two planets got only 0.02 percent out of synchronization, they would be one million years ahead of us or behind us. Thus, even in the unlikely event that other planets produce civilizations recognizable to us, we would have to contact one of those civilizations in a very narrow time interval in order to see any recognizable common interests.

Evolution may pass through only a brief **explorative interval** in which societies on one planet would care to reach other planets; beyond that stage communication or space exploration might be no more attractive than a national program on our part to communicate with chimpanzees, ants, or dolphins. To be sure, a few of our scholars try this, but they "contact" an infinitesimal fraction of these lower creatures. What fraction of the ant hills or dolphin schools have we humans tried to contact? By the same token our solar system might be ignored by advanced aliens. Aliens a million years ahead of us might be no more interested in us than we are in ants.

How long might an explorative interval last? We have used tools for about two million years, and it appears safe to assume that we will have progressed far beyond current technology in another million years, if we survive. Our explorative interval might be a few million years, then, or less than 0.1 percent of the history of the planet. Thus, if the last factor in Table 25·2 were interpreted as "time during which intelligence is recognizable to us and interested in communicating," this would revise the "optimistic" upper limit downward to three stars in 100,000, about the figure cited in the Soviet-American conference of 1971. If civilizations are at least 300 light years apart, then interstellar voyages and messages would take around half

a millennium. There might be little incentive for the effort. Any spaceships that did arrive on the earth would probably arrive many thousands or millions of years apart.

This brings us to a fifth possible answer to the question of alien visits: They may have happened in the remote past. This is the "ancient astronaut" hypothesis, popularized in several pseudo-science books (see Essay B). There is no good evidence for this possibility. The Soviet and American collaborators I. S. Shklovskii and Carl Sagan (1966) surveyed archeological and mythological literature even before the hypothesis was a popular fad, and found no compelling evidence for ancient astronauts. Neither earth, the moon, Mars, nor Venus

is littered with ancient alien artifacts, and not a single mysterious artifact has been advanced as physical evidence of ancient astronauts (though popular books have misrepresented some ancient artifacts belonging to known terrestrial cultures).

Radio Communication

Possibly our period of isolation may be nearing an end. For half a century we have been broadcasting radio communications among ourselves. Already our unintentional, but weak, alert is more than 50 light-years out from earth. Aliens may one day pick up our signals and send radio messages (or an expedition?) in return. Radio astron-

Figure 25·13 **A proposed facility for interstellar communication, "Project Cyclops," envisions an array of radio telescope antennae, each larger in diameter than a football field. Scale is indicated by** ▾

multi-story control building, right. Facility would listen for signals from alien civilizations. American and Soviet scientists have proposed similar facilities.

omers in both the United States and the Soviet Union are therefore still conducting modest programs to listen for such messages with large radio telescopes. Larger listening instruments have been proposed in both countries, as shown in Figure 25·13.

A message from 1,000 light years away would come from a civilization 1,000 years in the past, and no answers to our questions could come back for 2,000 years. Such communication would be unlike dialogue, but as physicist Philip Morrison has pointed out, more like our receipt of "messages" (books, letters, plays, art works) from ancient civilizations such as Greece.

Meanwhile, we have sent a few messages of our own. The Pioneer 10 spacecraft, which flew by Jupiter and left the solar system in 1973, carried a plaque designed to convey our appearance and location to possible alien discoverers of the derelict spacecraft sometime in the future, as shown in Figure 25·14. The first radio message was a test message sent from the large radio telescope at Arecibo, Puerto Rico, in 1974, beamed toward globular star cluster M 13, 27,000 light-years away.[4]

A

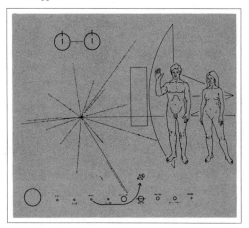

◄ Figure 25·14 Plaque carried out of the solar system on Pioneer 10. A. Plaque design showing scale of human figures in front of stylized spacecraft (right); path of spacecraft from earth past Jupiter (bottom); hydrogen atom as a unit of scale (top left); and map showing directions and frequencies of pulsars as measured from solar system (left center). Plaque could serve as indicator of place and time of origin of the spacecraft, and appearance of its builders, if found by alien creatures. B. Plaque installed on antenna support struts of Pioneer spacecraft. (NASA)

B

25

Summary

Questions of exobiology, and especially of intelligent life on other worlds, leave us with a mystery. Experimental evidence indicates rather strongly that life should start on other planets if liquid water, energy, and the right chemicals are present. Astronomical evidence suggests, but does not prove, that habitable planets ought to exist elsewhere in the universe. Biological evidence shows that life is adaptable and species can evolve to fit different environments, from ocean depths to low-pressure atmospheres of different composition.

While the limited evidence indicates that other life-forms should exist, there is no evidence that they do, or that they have tried to communicate with us. We can only speculate about the reasons. Perhaps they are too far

[4]Astronomers in these projects have been deluged with letters ranging from support to complaints about the nudity of the human figures on the Pioneer 10 plaque. The radio astronomers promptly received a telegram: "Message received. Help is on the way—M 13." Its authenticity might be questioned, because the round trip time to M 13 is 54,000 years.

away. Perhaps most civilizations destroy themselves before successfully exploring the universe. Perhaps evolution carries them beyond a stage where they would care to communicate with us. Perhaps they are unrecognizable.

Arthur C. Clarke has remarked that any technology much advanced beyond your own looks like magic to you. Perhaps we are too limited by our own concept of civilization. After all, one creature's civilization may be another's chaos, as shown by Mohandas Gandhi's remark when asked what he thought of Western civilization; he said it wouldn't be such a bad idea. It seems likely that our first contact with aliens (if they exist) might be as incomprehensible as the dramatized contact in the closing segment of Clarke's novel and the Kubrick-Clarke film, *2001.*

Clearly we have been reduced to speculation by lack of facts. Indeed, the whole field of exobiology has been criticized as a science without any subject matter. Exobiology recalls Mark Twain's comment, "There is something fascinating about science. One gets such wholesale returns of conjecture from such a trifling investment of fact." The only way to reduce the conjecture and increase the proportion of fact is to pursue research in many related fields—physics, chemistry, geology, meteorology, biology—and listen to the skies with radio telescopes. There may be surprises waiting out there.

Concepts

organic molecules

life

cell

organic chemistry

amino acids

proteins

Miller experiment

proteinoids

coacervates

ozone layer

natural selection

**evidence for planets
near other stars**

exobiologists

cultural hurdle

life in the solar system

life outside the solar system

explorative interval

Problems

1. Compare Alpha Centauri, Barnard's star, and Sirius in terms of the probability of detecting radio broadcasts from intelligent creatures.

2. Describe several ways in which earth's internal evolution has affected the evolution of life on earth.

3. Describe several ways in which extraterrestrial events, such as solar or stellar evolution, might have affected evolution of life on earth or other planets.

4. Describe ways technology could affect the survival of intelligent life on earth or on other planets. Construct scenarios of:
   a. *possible destruction of life;*
   b. *guaranteeing the survival of life.*
   (Hint: In case (b) consider the impact of space travel.)

5. In your own opinion, what would be the long-range consequences of:
   a. *Arrival of an alien spacecraft and*
visitors in a prominent place, such as the U.N. building.
   b. *Discovery of radio signals arriving from a planet about ten light-years distant and asking for two-way communication.*
   c. *Proof (by some unspecified means) that life existed* nowhere *else in the observable universe.*

6. Given the assumption that our technology has the potential for creating planet-wide changes in environment, defend the proposition that *if* life on other worlds produces technologies like ours, then that life is likely either to have become extinct or to be widely dispersed among many planets by means of space travel.

7. In view of the devastation wrought on many terrestrial cultures by contact with more technologically advanced cultures, which would you say is the safest course: (1) aggressive broadcasting of radio signals to show where we are in hopes of attracting friendly contacts; (2) careful listening with large radio receivers to see if there are any signs of intelligent life in space; (3) neither broadcasting nor listening, but just waiting to see what happens?
   a. *How would it affect the results of our listening program if other intelligent species had reached the first, second, or third conclusion?*
   b. *If we listen at many frequencies and pick up no artificial signals, does that prove that life has not evolved elsewhere in the universe?*

8. Why would native life be unlikely on a planet associated with:
   a. *an O or B star?*
   b. *a red giant star?*
   c. *a white dwarf?*
   d. *a pulsar?*

9. For each known extraterrestrial planet in the solar system, list several environmental factors believed to be adverse to the existence of advanced native life-forms.

10. According to present astronomical theory, the sun should eventually use all its hydrogen and turn into a red giant.

a. *When will this happen (see Chapter 12)?*

b. *How does this compare with the time scale of biological evolution?*

c. *Would you expect the human species to be recognizable by the time this happens?*

d. *In what way might life that originated on earth conceivably escape such events?*

Advanced Problems

---

11. If an earth-size planet (diameter 12,000 km) were circling a star one parsec away (3×10^{18} cm),

a. *What would be its angular diameter when seen from earth?*

b. *Could this be resolved by existing telescopes?*

c. *If the planet orbited one AU (1.5×10^{13} cm) from its star, what would be its maximum angular separation from the star as seen from earth?*

d. *Could this angle be resolved?*

12. Aside from its distance, why is the globular cluster M 13 a poor choice of target if we are really trying to send a message to inform carbon-based intelligent organisms, like ourselves, of our existence?

a. *Using the distance listed for M 13 in Table 19·1, confirm the round-trip travel time for radio waves given in the footnote on page 460.*

13. If an alien spacecraft happened to pass through the solar system at 99 percent of the speed of light, how long would it take to traverse the system (assumed to have the diameter of Pluto's orbit)?

a. *If it were broadcasting on the 21-cm radio wavelength of hydrogen, describe verbally how its radio transmissions would be received by us, taking into account Doppler shifts.*

b. *Why would this make detection difficult?*

The Cosmic Perspective

Just as a vacation in an interesting place can help us view workaday relationships and cares in a clearer perspective, our astronomical explorations to the ends of the known universe help us view earthly relationships and cares in a clearer, cosmic perspective. The cosmic perspective has already had deep impact. Astronomical experience has outmoded once-popular ideas, such as:

1. The idea that there are physical heavens and hells directly above or below the earth.

2. The idea that the earth is the universe's center around which all other bodies revolve.

3. The idea that humans are the inevitable lords of creation.

4. The idea that the earth is the sole and inexhaustible reservoir of raw materials.

What has the cosmic perspective given us in exchange for these cherished ideas? Some critics have said that science has only demeaned humans by putting us on a satellite of an average star on one side of an average galaxy. Certainly these ideas shocked people during the Renaissance. But as astronomer Harlow Shapley asked:

Are we debased by the greater speed of the sparrow, the larger size of the hippopotamus, the keener hearing of the dog, the finer detectors of odor possessed by insects? We can easily get adjusted to all of these. . . . We should also take the stars in stride. We should adjust ourselves to the cosmic facts.

Science has also been criticized for replacing outmoded ideas only with machines, which have brought us twin evils of pollution and a dehumanized existence. This growing criticism seems to me to confuse science (knowing) with technology (using knowledge and material). But there is a difference between knowing and using. To me there is something appealing about knowing as much as possible and using as little as necessary. The popular criticism of "science and technology" (lumped together as one) is really a criticism of how we have *used* knowledge and materials.

And it is justified. There is great irony in the fact that in spite of 10,000 years of effort to produce labor-saving devices, many people have trouble getting by on 40-hour-per-week jobs where they labor to make products which do not even interest them and which some people would label as "consumer gimmicks" or "wasteful." This is a failure of culture, rather than a failure of science or technology. The cosmic perspective is making this clearer, as more people begin to think in terms of our total planetary budget of materials, energy, and creative talent. This problem shows that the difficulties of maintaining a stable, workable civilization are at least as challenging as the problems of scientific exploration.

Besides giving us a new view of our place in the universe and our planetary society, the cosmic perspective gives us facts on which to base our actions. "Facts," of course, are merely conclusions drawn from observations, usually verified by generations of observers. They provide a foundation for practical life in the real world. They grade indistinguishably into hypotheses about how the universe acts and how it is put together.

What good are these? They allow us to act. They allow us to build bridges, insulate our homes, get to the moon, measure properties of remote galaxies. If a critic argues that we can never really know facts, a scientist can agree and argue that acting on hypotheses is like betting. If you are forced to put your money on one thing or another, you try to put it on what is most likely to be successful. The world forces us to act and we want to act on the basis of ideas most likely to be true. The scientific method of learning about the universe is simply a scheme for accumulating hypotheses on which to act until better information comes along. Hypotheses are useful because they suggest possibilities of new discoveries, new principles, new places to go. They suggest goals, tests, and experiments.

Some loosely connected ideas from the preceding chapters seem to me to be involved in the cosmic perspective:

1. For at least 5,000 years, astronomy has been one of the most successful and practical sciences, clarifying matters of navigation, agriculture, and

time-keeping, and revealing new worlds that we are just starting to explore.

2. A classic science-fiction plot deals with a vast interstellar spaceship, miles long, in which generations of men and women live and die, maintaining a stable environment and culture while en route from one part of the galaxy to another. This plot turns out to be the truth. We have only recently realized that we are on board just such a vehicle—spaceship earth—with finite area and finite resources.

3. Places in the universe must be very rare where we can walk naked away from our machines, breathe in the atmosphere, let the light of the nearest star fall on us, and find water to drink on the surface. While we can take technology with us to adapt to new places (as we have adapted to extremes of the earth), it would be a good idea to take care of the one known place in the universe where we can do these things.

4. Twentieth-century global society is not living within its planetary means. Certain resources such as metals, coal, and natural gas are being consumed at rates hundreds of thousands of times faster than their replacement rates. In the past, frontiers have swept west from China, east and west from India, out from the Mediterranean, and around the world from Europe. Humanity has lived by eating its frontier. But the earth is spherical and the frontier has closed on itself. We are now groping our way through the required period of adjustment, which may or may not be pleasant, depending on our ability to sense the proper direction and make the necessary changes. Global social changes are in the wind.

5. It seems within our technological competence to engineer a stable so-ciety, but it also seems within our competence to destroy civilization and all life on the earth, either by design or by accident.

6. We can describe with some accuracy the arrangement and history of the planetary and galactic systems in which we find ourselves, but we do not really understand the arrangement or history of the largest-scale clusters of galaxies. The most fundamental structure of the universe is still a mystery.

7. We may be prohibited, by the Hubble recession and the finite speed of light, from ever obtaining information about places beyond about 10,000 megaparsecs.

8. To some thinkers, the only reality is that directly witnessed by human observers. Yet three centuries of repeated, reliable observations give us the right to talk about real events that happened over a billion years ago, and real places where no person has ever walked. Time and distance can be objectively probed, giving results that remain consistent (within our limits of measurement) regardless of who does the measuring.

Certain places on Mars have looked the same whether photographed in 1971 by an American spacecraft or in 1973 by a Russian spacecraft. Thus, it is reasonable to suppose that there are real places where things happen, even though there are no humans to see them. In a way, it is astonishing to think that rocks roll down hills on the far side of the moon when no one is there; that winds churn dust across empty Martian landscapes just as black winds of water churn the bottoms of our seas; that gas heaves upward in stars; that starquakes crack crusts of unseen neutron stars; that unseen cataclysms play themselves out silently in space in front of no audiences; and that waves crashed on earth's beaches for billions of years before there were land animals to see them.

9. No place in the universe is permanent. The seemingly changeless panorama of stars in the sky will alter in a few thousand years as the stars move past the sun. The patterns of nebulae will change as the gas clouds are torn and stirred during their 250 million-year circuits around the galactic center, just as the mountains and continents of earth are torn and stirred by plate-tectonic processes in similar periods of time. Probably the sun will run out of hydrogen within the next 6 to 10 billion years, expanding to the red giant state, engulfing the earth, and finally collapsing to a dwarf. New stars will form near other stars, and perhaps other earthlike planets will form around some of these stars.

10. As the physical universe changes, the biological universe also changes. In relatively brief periods of millions of years, new creatures may evolve, just as humanity itself appeared only in the last few million years. A philosophy that sees mankind as a permanent fixture in an unchanging environment—while it may be practical for short periods such as human life-times or centuries—cannot be defended on a cosmic scale.

11. In spite of our seeming sophistication and ability to modify our local surroundings, the universe remains implacable. We humans live our lives on a small planet in the midst of forces and processes, some far greater than any we can create or modify. Known life and intelligence are adaptable and ingenious, but there are limits to their present ability to affect current events in the universe. Beyond our mortal processes there are greater

processes of death and birth which
we have to accept as part of nature.
Whole worlds are demolished and
whole worlds are created. Who are
we to say what sense there is behind
it?

The question of creators is a mystery—exciting, puzzling, beyond our
abilities to explain. Here astronomy
establishes a point of contact with
both old and new religious perceptions. Beyond this point, astronomers
are no better equipped than anyone
else to make definite statements.

12. Many mysteries remain, and
much space is left still to explore.
Such explorations would seem to be
an endeavor worthy of human energy—
worthier, for example, than warfare.
It is good to admit that there is still
mystery in the universe. There is more
to life than what we see on this particular planet at this particular time.

The Universe in Color

Our picture of the universe has changed, like our television picture, from black and white to color. Recent advances in color photography and television now bring us the universe's amazing variety of locales and landscapes in a color spectacular. At the same time, astronomers have learned how to use color as a clue to what is happening in space.

On the surfaces of the planets, different minerals in the rocks reflect different colors of sunlight, creating the reddish landscapes of Mars, the greys and browns of the moon, and the varied colors of earth. Gases, dust, and condensed particles in the planets' atmospheres interact with sunlight in various ways, creating the cloudy skies of Venus, the blue skies of earth, the red skies of Mars, and the yellow, brown, and red cloud-formations of Jupiter and Saturn. Color photographs are thus a key to surface and atmospheric conditions on the planets.

Similarly, images of the sun and stars photographed in light of different colors (wavelengths) reveal the different properties of their surfaces. From these properties astronomers get clues to internal structure and composition.

Gases in space can show an incredible variety of colors because of the many ways that gases interact with light. No one had seen these colors until recent decades because nebular gases were too dim to record on color films. Advances in color photography intensify both images and colors, showing emissions in the red light of hydrogen, other colored emissions from other gases, and the blue light scattered by dust particles. The colors are not only beautiful; they also tell us about composition, temperature, and other conditions in the gases.

Colors can also tell us about whole star-systems. If most of a system's stars are blue, stars are probably still forming there. Massive, hot, blue stars are the brightest stars of a forming system. Many red stars in a system often mean an older system where star-formation has ceased, the short-lived blue stars have burnt themselves out, and the brightest survivors are red giants near the end of their evolution. Thus color can help us to understand the evolution of systems as well as their present conditions.

Not the least of color's benefits are simple pleasure and wonder. The addition of color to the astronomer's view of the universe has increased not only our knowledge but our sense of awe of our cosmic environment. The photographs in the following section reflect both of these values. The photographs are grouped roughly in the order of topics in the text. The captions that follow briefly describe the photographs and refer the reader to the appropriate discussion in the text as well.

Color Photo 1 **This aerial view of Stonehenge, viewed from the north, shows earthworks and standing stones built around 2600–1800 B.C. The white pavement is a modern footpath. The Avenue leading toward summer solstice sunrise lies at lower left. See p. 16 ff.** (Courtesy of Georg Gerster)

Color Photo 2 **Summer solstice sunrise presents a spectacular view from the center of Stonehenge. Beyond the massive stone lintels, the distant heel stone marks the sun's rising point on this special date. See p. 16 ff.** (Courtesy of Georg Gerster)

Color Photo 3 **The Jantar Mantar Observatory at New Delhi, India, was built in 1710 and contained masonry structures designed to measure positions of the sun and other celestial bodies. This observatory was patterned after a still earlier, pre-telescopic observatory built in the 1400s at Samarkand, now in the Soviet Union. See p. 45.** (Photo by D. J. Forbert, courtesy Shostal Associates)

Color Photo 4 **Summer solstice sunset is framed in a cylindrical window in the western adobe wall of Casa Grande National Monument, Arizona. The four-story structure was built by Indians around A.D. 1350. Some evidence indicates that the windows were carefully designed to frame astronomically significant horizon points; the building was probably used as an astronomical observatory. See p. 20.** (Photo by author)

Color Photo 5 **The Apollo 11 lunar landing module returns to lunar orbit after the first manned landing on the moon in July, 1969. In the distance, the earth rises above the lunar horizon. See p. 92.** (NASA)

Color Photo 6 **Geologist-astronaut Harrison Schmitt examines a massive boulder dominating a lunar landscape in the hills at the edge of the Sea of Serenity. Lunar colors are muted by the mantle of lunar regolith, or soil layer of pulverized rock. See p. 92.**

Inset shows a portion of the lunar disk as photographed from earth. (NASA)

Color Photo 7 **An astronaut stands on the sloping rim of a modest-size lunar crater (background, left). This scene is backlighted by the sun, out of the frame beyond the astronaut. See Chapter 5.** (NASA)

Color Photo 8 **Weather on Mars.** In the floors of fault-produced canyons, frozen crystals of water create white clouds. Elsewhere in this view from the Viking orbiter, the cratered desert has the characteristic Martian red color. See p. 146 ff. (NASA)

Color Photo 9 **A large rock lies near the landing site of Viking 1 on the Chryse plain of Mars. It is about 3 m long and 8 m from the spacecraft. The rock's texture and grey color suggest it may be a lava boulder. The mantle of reddish dust on its top probably filtered out of the Martian atmosphere after the last local dust storm. Dust dunes surround the rock. Inset shows part of the disk of Mars. See p. 145.** (NASA)

Color Photo 10 **The Martian landscape at the Viking 2 landing site is a relatively uniform field of strewn boulders, colored red by iron oxide minerals. In the foreground are portions of the Viking spacecraft, including colored emblems which helped calibrate the true colors of the photographs sent back to earth. See Chapter 8.** (NASA)

Color Photo 11 **In the late afternoon landscape at the Viking 2 landing site in the Utopia plain of Mars, a low sun emphasizes scattered boulders and shallow depressions. Local time is about 15 minutes before sunset. See p. 146.** (NASA)

Color Photo 12 **This photograph in X-ray radiation by astronauts in the Skylab space station shows regions of intense activity on the sun. The brightest areas are flares emitting X-rays and other forms of energetic radiation. Since X-rays are not visible to the eye, the picture has been processed to appear in familiar yellowish-orange tones usually associated with sunlight. See p. 224 ff.** (NASA)

Color Photo 13 **When observed visually, Venus is a nearly featureless, cloud-covered disk. But when photographed at ultra-**

violet wavelengths, Venus shows cloud belt patterns, because some cloud regions reflect ultraviolet light whereas other regions absorb it. The blue color has been artificially added to represent the ultraviolet color of the photo, as the eye is insensitive to ultraviolet. See p. 128 ff. (NASA)

Color Photo 14 **Africa and Antarctica are prominent in this photograph of earth taken by Apollo 17 astronauts as they coasted out toward the moon. Coriolis forces from earth's rotation cause more pinwheel-shape cloud masses than on Venus (Color Photo 13). See Chapter 3. On this page, earth is shown in true proportion relative to Venus, moon, and Mars.** (NASA)

Color Photo 15 **The moon, shown at true scale relative to the earth (Color Photo 14), appears small and relatively colorless by comparison. See Chapter 5.** (NASA)

Color Photo 16 **Mars, shown at true scale relative to the earth (Color Photo 14), reveals a red color caused by oxidized iron minerals. The dark markings, which vary with seasons, were once thought to be vegetation, but are now believed to be windblown deposits of dust. The white polar cap can also be seen at the south pole (bottom), with light clouds at the north pole. See Chapter 8.** (Lunar and Planetary Laboratory, University of Arizona)

Color Photo 17 **Jupiter's Red Spot is a prominent feature in this photograph of the planet, made by the Pioneer 11 spacecraft as it passed by at a distance of 1.2 million km. The smallest features visible are cloud formations about 1,000 km across. To be in true proportion to the accompanying images of earth and the terrestrial planets, this image of Jupiter would have to be enlarged to a diameter of about 40 cm. See p. 163 ff.** (NASA Ames Research Center and University of Arizona)

Color Photo 18 **Saturn, photographed from earth with a 1.55 m telescope, shows the rings and yellowish cloud formations covering the disk. In this view the rings are inclined 27° to the line of sight.** (Lunar and Planetary Laboratory, University of Arizona)

Color Photo 19 **The Pleiades is an open cluster of relatively young stars, roughly 50 million years old. The cluster is about 127 parsecs away. Its brightest stars are massive, hot, blue stars that rapidly consume nuclear fuel. Dust clouds around the stars absorb red light and scatter blue light, creating wispy blue reflection nebulae. See Chapters 15, 17, 19.** (Copyright by California Institute of Technology and Carnegie Institution of Washington. Reproduced by permission from the Hale Observatories.)

Color Photo 20 **The Orion nebula is a massive emission nebula in the heart of a star-forming region about 500 parsecs away. A dark lane of dust is silhouetted in front of the central nebula, while outer regions glow with different colors of light emitted from atoms of different elements in different states of excitation. See Chapters 15, 17.** (Copyright Association of Universities for Research in Astronomy, Inc., Kitt Peak National Observatory)

Color Photo 21 **The Trifid nebula is an emission nebula roughly 1,000 parsecs away in the constellation Sagittarius. Dark lanes of dust are silhouetted in front of one gas cloud, which glows with red light of the hydrogen alpha spectral emission line. A neighboring cloud is a reflection nebula of dust, in which blue starlight is scattered toward us while red light is absorbed. See Chapter 17.** (Kitt Peak National Observatory)

Color Photo 22 **A spectacular young nebula and associated star cluster, NGC 6611 is a site of star-formation about 2,000 parsecs away. The nebula, about 6 parsecs across, contains dark clouds of dust silhouetted in front of red-glowing hydrogen gas. The complex is rich in hot, bluish-white, type O stars and is about 3 million years old. See page 285.** (Copyright Association of Universities for Research in Astronomy, Inc., Kitt Peak National Observatory)

Color Photo 23 **The Ring nebula, 0.2 parsecs in diameter, is a shell of gas blasted off the unstable old-age central star. It lies about 700 parsecs away in Lyra. The shell is expanding and has left a relatively hollow section in the center. Different colors are emitted by different kinds of gas, with hydrogen alpha red glow prominent on the outer rim. See Chapters 16, 17.** (Copyright Association of Universities for Research in Astronomy, Inc., Kitt Peak National Observatory)

Color Photo 24 **The Dumbbell nebula is a gas cloud 0.3 parsecs in diameter blown off an unstable star. It is about 220 parsecs away in the constellation Vulpecula. Gases at different distances from the central star are excited to different degrees and radiate in different colors. See Chapters 16, 17.** (Copyright by California Institute of Technology and Carnegie Institution of Washington. Reproduced by permission from the Hale Observatories.)

Color Photo 25 **The Crab nebula is the remnant of a supernova explosion recorded on earth in A.D. 1054. The central glow comes from synchrotron radiation caused by fast-moving electrons. Red-colored outer filaments are splatters of ionized hydrogen, glowing with the red light of hydrogen alpha emission. See Chapters 16, 17.** (Copyright by California Institute of Technology and Carnegie Institution of Washington. Reproduced by permission from the Hale Observatories.)

Color Photo 26 **This color image of spiral galaxy NGC 4535 was constructed from television and computer processing of separate images made in red, green, and blue light. The result reveals a reddish color for the central regions and a bluish color for the spiral arms. In most spiral galaxies, star-formation has ceased in the center, so that the brightest central stars are old-age red giants. In the arms, star-formation continues, and the brightest stars are young, bluish-colored, massive stars with high temperatures. See Chapter 21.** (Copyright Association of Universities for Research in Astronomy, Inc., Kitt Peak National Observatory)

Color Photo 27 **Spiral galaxy NGC 4603, like NGC 4535, shows reddish central regions and bluish spiral arms, as is believed to be the case for our own Milky Way galaxy. Scattered across the foreground are stars of different colors located in our galaxy. See Chapter 21.** (Copyright Association of Universities for Research in Astronomy, Inc., Kitt Peak National Observatory)

Color Photo 28 **The radio galaxy NGC 5128, also known as Centaurus A, is located about 4,400 kiloparsecs away. It strongly emits radio waves and has a peculiar form. Other data (Figure 22.5) indicate a jet of luminous gas extending from the center. The galaxy is believed to be the site of explosive activity of uncertain cause. See Chapter 22.** (Copyright Association of Universities for Research in Astronomy, Inc., Kitt Peak National Observatory)

Color Photo 29 **The H-R diagram is shown here in color, using the same format as found elsewhere in this book (for example, Figure 14.2). Colors of different star-types are illustrated. Comparative sizes are indicated, but scales are not exact, as this would necessitate much larger giant stars. See Chapter 14 for further discussion.**

Color Photo 30 **Schematic diagrams of spectra show different features studied by astronomers. Top diagram shows the array of colors unbroken by any absorption or emission lines; this is called a continuous spectrum. Wavelengths in Angstrom units (A) are identified. Center diagram shows spectrum with superimposed absorption lines. Bottom diagram shows spectrum composed only of emission lines, as produced by hot, glowing gas seen against a dark (or cooler) background. See Chapter 13.**

▲ 1 / Stonehenge   ▼ 2 / Stonehenge, summer solstice sunrise   ▲ 3 / Jantar Mantar Observatory, New Delhi   ▲ 4 / Casa Grande National Monument, Arizona

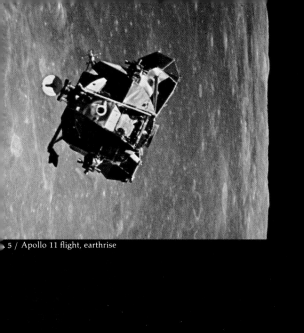

5 / Apollo 11 flight, earthrise

▼ 6 / The surface of the moon   ▲ 7 / Astronaut on the moon

8 / Aerial view of Mars ▼ 9 / The surface of Mars ▲ 10 / The Viking 2 lander on Mars ▲ 11 / A landscape on Mars

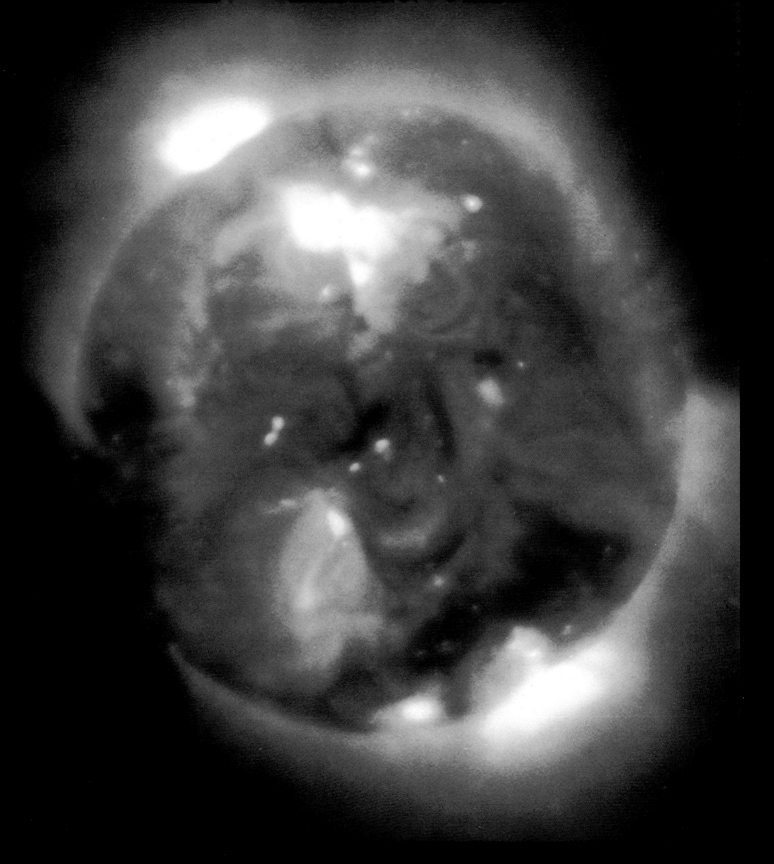

▲ 13 / Venus

▲ 14 / Earth from Apollo 17

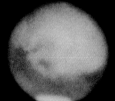

▲ 15 / The moon

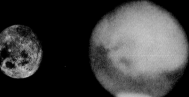

▲ 16 / Mars

▲ 17 / Jupiter

▼ 18 / Saturn

◄ 12 / X-ray photograph of the sun

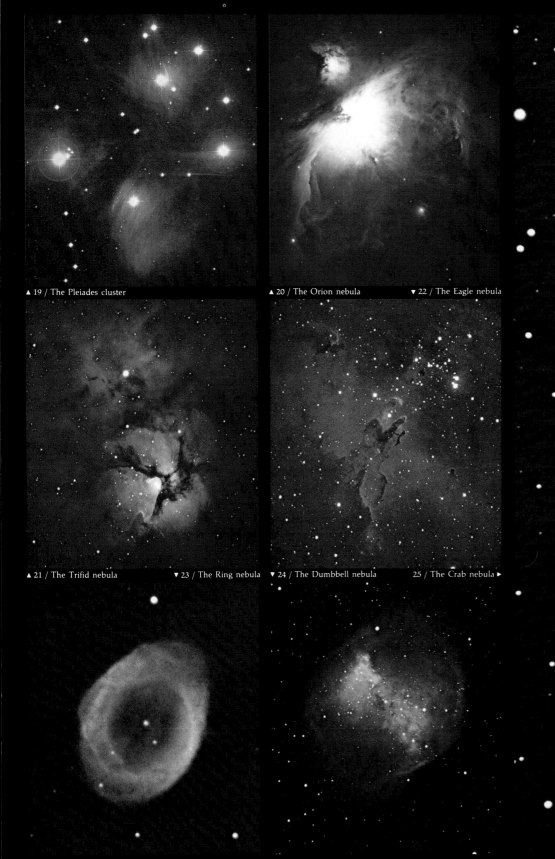

▲ 19 / The Pleiades cluster

▲ 20 / The Orion nebula   ▼ 22 / The Eagle nebula

▲ 21 / The Trifid nebula   ▼ 23 / The Ring nebula   ▼ 24 / The Dumbbell nebula   25 / The Crab nebula ▶

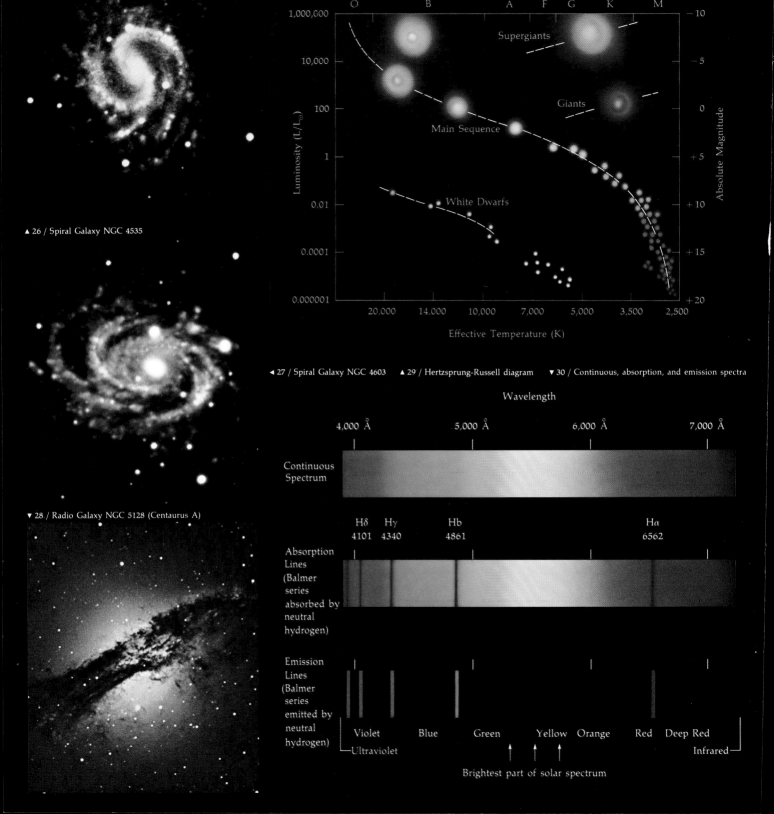

▲ 26 / Spiral Galaxy NGC 4535

◀ 27 / Spiral Galaxy NGC 4603   ▲ 29 / Hertzsprung-Russell diagram   ▼ 30 / Continuous, absorption, and emission spectra

▼ 28 / Radio Galaxy NGC 5128 (Centaurus A)

O   B   A   F   G   K   M

1,000,000   –10

10,000   –5
                        Supergiants

100   0
                                        Giants
           Main Sequence

1   +5

0.01   White Dwarfs   +10

0.0001   +15

0.000001   +20

20,000   14,000   10,000   7,000   5,000   3,500   2,500

Luminosity ($L/L_\odot$)   Absolute Magnitude

Effective Temperature (K)

Wavelength

4,000 Å   5,000 Å   6,000 Å   7,000 Å

Continuous
Spectrum

Hδ   Hγ   Hb   Hα
4101 4340   4861   6562

Absorption
Lines
(Balmer
series
absorbed by
neutral
hydrogen)

Emission
Lines
(Balmer
series
emitted by
neutral
hydrogen)

Violet   Blue   Green   Yellow   Orange   Red   Deep Red

Ultraviolet   Infrared

Brightest part of solar spectrum

h

Enrichment Essays

We have loved the stars too
fondly to be fearful of the night.

Inscription on the
crypt of John and
Phoebe Brashear,
Allegheny Observatory

Essay A

Telescopes and Observing

The purpose of an optical telescope is to focus light into as big an image as possible, and to collect as much light as possible. The more light, the brighter the image and the more it can be magnified.

Telescope Design

Two basic designs have been used. The first to be built, the *refractor*, uses a lens to bend, or refract, light rays to a focus, as in Figure A·1. Galileo first used this type astronomically in 1609. The second type, the *reflector*, uses a curved mirror to reflect light rays to a focus, as in Figure A·2. Isaac Newton built the first reflector in 1668. Several alternate reflector designs have since been constructed, but the simplest is Newton's,

sometimes called the *Newtonian reflector*. Recent years have seen new designs combining lenses and mirrors. Often called *compound* telescopes, these designs are often very compact and portable.

In any telescope, the main lens or mirror is called the *objective*. The distance from the objective to the place where the image is focused is called the *focal length* of the objective (marked F in the figures). The larger the focal length, the bigger the image. This principle is also used in cameras, which are merely small telescope systems with film at the focus.

Radio Telescopes

Radio telescopes serve the same function as optical telescopes. But because a 50-cm radio wave is a million times bigger than a visible light wave and carries less energy per photon, radio telescopes need larger surfaces to collect and focus enough energy to give strong signals. Radio telescopes are reflectors, with a large curved surface (the "dish") of metal or wire mesh, which reflects the radio waves to a focus where a smaller radio detector picks up the signal.

Functions of Visual Telescopes

A telescope designed to be looked through, as opposed to a radio telescope or a camera, may be called a *visual telescope*. The optical systems of Figures A·1 and A·2 can be converted into visual telescopes by adding an eyepiece (and usually tubing to keep out stray light), as shown in Figure A·3. The eyepiece is simply a lens or system of lenses designed to magnify the image still more and

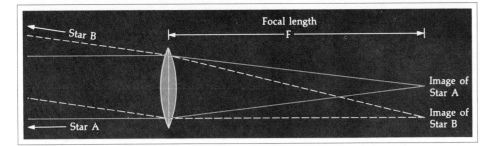

Figure A·1 **Cross section through a lens, showing principle of image formation in a refracting telescope (and most cam-** eras). **Light rays from two stars are focused into two images.**

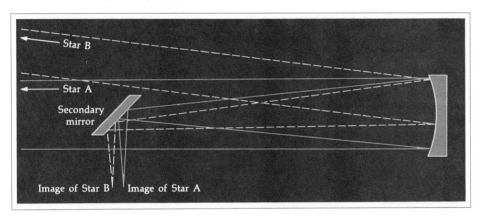

Figure A·2 **Cross section through a mirror system, showing principle of image formation in a reflecting telescope. Light rays strike curved mirror (right) and are reflected back toward focus. Focus would normally lie in an inconvenient position in front of mirror, but a secondary mirror is used to beam the light rays to one side for easier access to image.**

Figure A·3 **Cross sections showing how the systems of Figures A·1 and A·2 are converted to visual telescopes by the addition of an eyepiece. The eyepiece is a small magnifying lens (or several lenses mounted together) used to examine the image. Tubing from the objective to the eyepiece helps cut out stray light.**

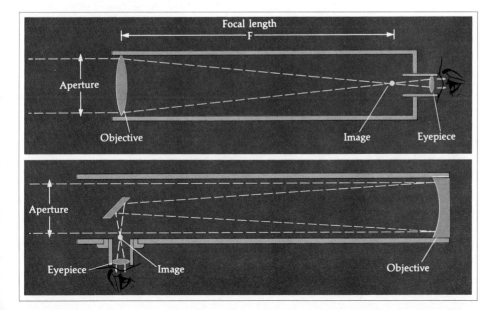

allow the eye to examine the image.

One of the first questions asked of backyard telescope enthusiasts is, "How far can you see with that thing?" This is not the right question to ask, because no telescope is limited by distance. Any optical system can see as far as there is an object large enough or bright enough to detect. The naked eye as well as the 200-inch Palomar telescope can see the Andromeda galaxy 19 billion billion kilometers away—but the telescope shows more detail and fainter regions. The two main functions of the telescope, then, are to enlarge the image in order to show small angular detail, and to gather enough light to reveal objects too faint to be otherwise seen. These two functions are called *magnifying power* and *light-gathering power*.

Magnifying Power

In any given telescope, the magnifying power, or *power,* as it is usually called, is controlled by the eyepiece. Suppose a distant object subtends an angle of one minute of arc (about the angular size of Jupiter when it is prominent). An eyepiece that makes the object appear 100 minutes across when seen through the telescope is said to give 100 power. The magnifying power is given by the formula,

$$\text{power} = \frac{\text{focal length of objective}}{\text{focal length of eyepiece}}$$

For instance, if a 1-cm eyepiece is inserted in a telescope of objective focal length 100 cm, we get 100 power, often written 100x.

In theory, any telescope can be made to give any magnifying power simply by insertion of an eyepiece of short enough focal length. In practice, however, three effects limit the magnifying power that can be used. First, because we look into space through turbulent air, higher powers magnify air turbulence, and too high a power causes the image to shimmer hopelessly. (The air quality, called *seeing*, varies from night to night, with occasional nights of good seeing occurring when the air is still.)

Second, as the magnification is increased, the image gets fainter and fainter because the light is spread out over a larger and larger area. Too high a power gives a hopelessly faint image.

Third, due to fundamental properties of light waves, a telescope of A cm aperture cannot resolve details smaller than $12/A$ seconds of arc. Thus, a 12 cm (5-inch) telescope can resolve about 1″, but nothing smaller. Therefore, too high a power gives a hopelessly fuzzy image.

The last two problems can be overcome by increasing the telescope aperture, giving more light and resolution of smaller angular detail. Nonetheless, the maximum useful magnifying power, for most telescopes on most nights, is about 20 power per cm of aperture. For example, a 5-cm (2-inch) telescope might be used at 100x; a 15-cm (6-inch) at 300x; a 30-cm (12-inch) at 600x.

Light-Gathering Power

The other important function of a telescope, gathering light, is controlled solely by the aperture size. Many astronomical objects, such as nebulae and galaxies, are very faint, and to see their details one needs light more than magnifying power. In fact, the best visual impression of faint nebulae comes from a large telescope (lots of light) used at *low* power (to concentrate light). For this purpose, one might

use only 4 power per cm of aperture; for instance, a 15-cm (6-inch) telescope at 60 power.

Generally, the amount of detail that can be seen on astronomical

objects depends ultimately on the stillness and clarity of the air and on the telescope aperture, since these control both the useful magnifying power and the light-gathering power.

Figure A·4 **A. The 2.2-m (88-inch) reflecting telescope at Mauna Kea Observatory, Hawaii. In this telescope design, light is focused through a hole in the center of the main mirror, where the astronomer adjusts**
▼

his electronic instruments to record data. Data are fed through cables to computer equipment, which performs initial analysis and stores data for further analysis. Dome

A

Photography with Telescopes

Instead of an eyepiece, the image could be formed on a piece of photo-graphic film. Development of properly exposed film then provides a perma-nent record of the observation. Many telescopes come equipped with photo-graphic attachments for this purpose. In the case of a bright object like the moon or a planet, the eye usually sees more than a photograph with the same telescope, because the eye can take advantage of moments of perfect seeing, whereas the photograph aver-ages moments of poor seeing that blur the image. On the other hand, faint objects like nebulae and galaxies are usually better shown in photos, be-cause light can be accumulated in long exposures, whereas the eye cannot "store" light.

Many photos of star fields and nebulae in this book were taken with small telescopes or ordinary cameras. The reader should consult picture captions, which in many cases describe the equipment and exposure used. In many cases, readers may be able to duplicate or improve the results with their own equipment.

In large, modern observatories, the recording instruments are often elec-tronic, instead of either photographic or visual, as shown in Figure A·4. Furthermore, computer-directed ma-chines can point the telescope almost precisely at any known celestial object. Thus, an astronomer may spend nearly the whole night observing without looking through the telescope, though he may look through a small telescope called a *finder* mounted parallel on the side to help find his target object, or he may view television screens that monitor the image.

Figure A·4 (cont.) **is open to night air in order to give clearest observing conditions; temperature inside dome is often below freezing during observing sessions. B. Schematic diagram of A.**

▼

B

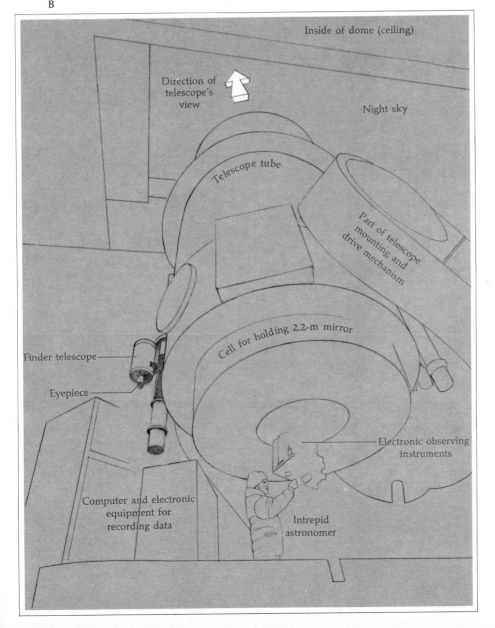

Solar Observing

A NORMAL TELESCOPE SHOULD NEVER, UNDER ANY CIRCUM-STANCES, BE POINTED AT THE SUN. An eye could be immediately burned or blinded, since the objective acts like a giant magnifying glass, concentrating solar light, heat, and ultraviolet radiation at a point near the eyepiece. Furthermore, the tele-scope is likely to be damaged, since the solar heat can crack the glass in eyepieces or secondary mirrors. Some telescopes can be modified for safe solar observing. One usually starts by putting an opaque screen over the objective with a new, small aperture as small as one or two centimeters, then adding additional filters at the eyepiece end. Instructors or telescope owners should be consulted before attempting this procedure.

The Pseudo-Science of Astrology, UFOs, Ancient Astronauts, and Astro-Catastrophes

One of the occupational hazards of being an astronomer is to be mistaken for an astrologer. The misnomer is annoying because astrology is not part of modern astronomy, but a pseudo-science associated with astronomy as it was practiced 3,000 years ago.

What Is Pseudo-Science?

When confronted with a hypothesis that sounds far-fetched, scientists may immediately label it pseudo-science, or even nonsense. This is unfair. Any hypothesis has as much right to vie for verification as any other. The true scientific method is to test hypotheses. To be worth anything, a hypothesis must predict some things about nature, and if the predictions are wrong, the hypothesis must be rejected or modified. Though a hypothesis can never be proven ultimately true, if experiments keep turning up consistent results it is considered more and more reliable, and comes to be an accepted theory, or law.

A pseudo-science is a body of hypotheses treated as if true (usually with commercial intent) but without any consistent body of supporting experimental or observational evidence. Pseudo-science may appear to be backed by the trappings of real science, such as quotations of evidence, but the evidence is often hearsay, the references are often to other poorly researched commercial books, and the work is rarely reviewed by other professional researchers before publication.

The dangers of pseudo-science are twofold. First, practitioners often bilk consumers of their money by falsely promising new discoveries or mystic knowledge. Second, it misrepresents real scientific discovery and often contributes to anti-intellectual attitudes that would exchange mysticism and magic for exploration and discovery. Furthermore, the exchange is a poor one because, as we hope we have shown in this book, real discoveries about the universe are as exciting as the erroneous claims of pseudo-science.

Astrology

Astrology is a pseudo-science which claims that events on earth are influenced by, and can be forecast from, positions of the sun, moon, and planets relative to the stars, or relative to the imaginary patterns of the constellations.

Origins of Astrology: Magic and Misunderstanding

Astrology can be traced back at least 3,000 years, when it flourished with other ancient magical beliefs. A common form of ancient magic was to associate patterns in nature with patterns of human events. Most people today would scoff at having their futures read from flight patterns of migrating birds or patterns in the bloody intestines of freshly sacrificed animals. Yet these were popular forms of divination in Babylon and Rome when astrology was flourishing, and they have the same basic logic. In astrology, the pattern is that of the planets and stars.

Archeological research suggests that early people used patterns in the sky in a practical way to foretell natural events, such as the change in seasons. Egyptian astronomer-priests, for example, designed their calendar to begin with the heliacal rising of Sirius, which was also the date of the Nile's flooding. Observation of the heliacal rising would help predict the time of the flood.

It was an easy mistake to confuse useful signs in the stars with the idea that the stars actually control events. Human minds are quick to assume that if one event follows another, the second is caused by the first. Sometimes this is true and sometimes not. Centuries ago, logicians named this error with

the Latin phrase, *post hoc, ergo propter hoc* (following this, therefore *because* of this). Historical evidence suggests that astrology grew out of this error. For example, if a king died a few days after an eclipse, some early observers reasoned that an eclipse causes or foretells the death of a king.

There is some evidence to suggest that early astronomical knowledge later degenerated into myth and pseudo-science. For example, the best astronomical alignments at Stonehenge (Chapter 1) were probably those constructed first, around 2600 B.C. Later construction, around 1800 B.C., muddled the design (Hoyle, 1971). In the nineteenth and twentieth centuries we see the monument used for mystical costumed rites by self-styled "Druids" and "witches." Here is an example of the loss of knowledge of original astronomical purpose.

In the same way, records of astronomical events, originally gathered by accurate observation, may have degenerated as later interpreters added mystical interpretations of astronomical events interspersed with historical incidents, describing them as if one controlled the other.

Historical Development

A Babylonian text of about 1600 B.C. describes motions of Venus interspersed with historical events, assuming that the same events will occur again in the future, whenever Venus presents the same patterns (paraphrased from Pannekoek, 1961):

If in the month of Abu, on the sixth day, Venus appears in the east, rains will be in the heavens and there will be devastation. Venus remains in the east until the tenth day of Nisannu and disappears on the eleventh. Three months she is not seen. On the eleventh of Duzu, Venus flares up in the west. Hostility will be in the land; the crops will prosper.

Similarly, a letter of an Assyrian astrologer to his king in 668 B.C. shows the same misconception still current a thousand years later (Pannekoek, 1961):

. . . when Jupiter appears in the third month the land will be devastated and corn will be expensive . . . when Jupiter enters Orion the gods will devour the land.

Mesopotamian astrology became a popular fad in the Greek world. Around 350 B.C. a Greek philosophical school, called *catasterism,* identified individual stars with certain of the human-like Greek gods. Of course, implicit in the concept of a god is the idea that divine will prevails over human will. So it was a short step to the idea that the stars arrange the fate of humans. The practice of forecasting personal fortunes spread from heads of state to sophisticated Greek households seeking diversion, and finally to commoners.

Astrology also spread rapidly eastward. In a Chinese example as recent as 1882, an unsuccessful astrological attempt to stave off the dire consequence of a comet is ascribed to official mistakes (Stephenson and Clark, 1977):

When a (comet) was seen last year, an imperial decree was written to the palace and court officers, ordering them to perform their respective duties conscientiously. In . . . this month, the comet was seen again in the southeast. This must be due to the frequent mistakes committed by those employed in the administration. . . .

Toward Modern Astrology

In medieval times, astrology was banned by the Church as magic, yet it persisted, deeply ingrained in people's thinking. For example, historian Owen Gingerich (1967) documents how astrology was apparently used to pick dates for laying cornerstones of palaces and churches, and how a palace built in 1563 was oriented toward sunset on the date of the battle it commemorated. Similarly, there was a tradition of orienting medieval cathedrals toward sunrise on their patron saints' days. These traditions suggest a holdover from the ancient patterns of astronomical temple-orientation at Stonehenge and in Egypt, but with the original purpose forgotten and replaced by ritual.

Astrology has continued to flourish alongside the development of scientific thought since the Renaissance. Today the ancient magic of 3,000 years ago arrives on our doorsteps in the astrology columns of our daily papers!

Problems with Astrology

A test of astrology is whether it is consistent with observations. The basic practice is to forecast the influences for a particular day by casting a *horoscope.* The horoscope is usually presented as a circular chart, showing the positions of the planets, sun, and moon with respect to the constellation signs at the moment being studied. For example, since Venus was the goddess of love, prominence of Venus might be used to predict a loving influence. This technique follows the rules laid down by astrologers centuries ago.

But application of the ancient rules creates an embarrassing problem for astrologers today. The original astrologers designated twelve signs, or portions of the zodiac corresponding to the twelve zodiacal constellations. Like the planetary gods and goddesses, each constellation and its sign were said to have a certain personality trait. In 1867 B.C., the point where the sun crossed the celestial equator on March 21 (spring equinox) and entered the constellation Aries was called the "first point of Aries." The twelve signs were measured from this point. Thus a person born between March 21 and April 19 was said to be "an Aries," since the sun lay in this constellation and in this sign. The original astrologers thought this pattern was fixed. But around 130 B.C., Hipparchus discovered precession, which shifts the first point of Aries relative to the constellations. Thus the signs no longer correspond to their appropriate constellations.

Astrologers during Ptolemy's time tried to patch up the scheme by claiming that the signs were more important than the actual constellations, and by defining signs as 30° intervals measured from the March equinox point. Astrologers still claim that a person born between March 21 and April 19 is "an Aries," but the sun during most of this period is really in Pisces! Thus astrology has become removed from the realities of the sky.

Another embarrassment is that we now know the stars are independently moving; thousands of years from now the constellation patterns themselves will have changed. Although one could patch up astrology further by inventing new mythological characteristics to cover this problem, the discrepancy shows that astrology as it is now practiced is not consistent with observed reality.

Does Astrology Work?

Another test of astrology is to examine the success of its predictions. It has failed in many spectacular cases. For example, a grouping of all known planets in Libra in 1186 led astrologers to predict a disastrous hurricane, because Libra was associated astrologically with the wind. People dug storm cellars in Germany; the Archbishop of Canterbury ordered fasting; the palace in Byzantium was walled up; people fled to caves in the Near East. But the conjunction of planets passed without incident. Similarly, in 1524, all known planets clustered in Aquarius, the water-carrier. This time astrologers predicted a second Deluge, but the month of the conjunction passed without disaster, and was reportedly drier than usual (Ashbrook, 1973).

Some statistical studies of astrological predictions have been reported, to see if they are more successful than random predictions. Jerome (1975) reviewed several of these studies and concluded:

> . . . legitimate statistical studies of astrology have found absolutely no correlation between the positions and motions of the celestial bodies and the lives of men.

Still another test would be to search for forces exerted by planets or stars by which astrological influences could be manifested. Some modern apologists for astrology argue that these forces might be like gravity. So far, no such mysterious forces have been detected. We can calculate the relative strengths of gravity on a newborn infant. The variations of force due to the local topography of earth (mountains vs. sea level) are greater than any variations due to sun, moon, or planet positions. Even the gravitational forces due to the position of the mother surpass the typical forces exerted by any extraterrestrial planet.

Thus, because its logic is identical to that of ancient magic, because of its inconsistency with the sky as observed by twentieth-century people, because of its predictive failures, and because no theoretical rationale for it has been advanced, astrology hardly seems to warrant our belief.

Pseudo-Science Based on Extraordinary Reports

Another form of pseudo-science has grown up around alleged eyewitness reports of strange events. Once again, it is perfectly permissible to hypothesize that strange events, such as an alien monster's visit to earth, might sometimes occur. We cannot dismiss the hypothesis, but we can ask if there is verifiable evidence, or if an intensive research program might turn up verifiable evidence. To answer this question, we need to understand the processes that lead to eyewitness reports.

Perception, Conception, and Reporting

One-time events in the sky are hard to confirm. It is easier to learn from repeatable experiments in the laboratory than from sporadic views of celestial phenomena, which is one reason why less occult lore accumulates around a science like hydraulics than around astronomy.

The first step in generating an eyewitness observation is *perception*—the observer's intake of external sensory stimuli. The problem is that this perception must be converted in the observer's brain into a *conception*, a step which involves subjective factors, such as the association the person may make between the object and concepts current in the culture. For example, witnesses almost invariably conceive meteors in terms of the distances of aircraft. They say "it landed just behind the barn," when the meteor may in fact be hundreds of miles away.

Reporting, the third step, transmits conceptions to other people. Throughout most of history, this process has been by word of mouth, with second-hand reports and hearsay blending into long-lived oral tradition, which may incorporate extraneous incidents or myth. Even today, it is hard to transmit conceptions accurately by words. There is a classic UFO anecdote about a man seeing a large orange object on the ground, with flashing red lights, rows of windows, and small people inside. The object was a school bus. In the context of UFO reports, simple, accurate words can lead many people to the wrong conception.

The quality of reporting cannot be overemphasized because it affects the beliefs of millions of people. We must retain a healthy skepticism, because many news or entertainment media thrive on sensational stories with minimal documentation. Astronomer Carl Sagan (Sagan and Agel, 1975) recounts an example of public over-acceptance. He saw a woman reading a pseudo-astronomy bestseller that contained many errors and flagrantly misquoted Sagan himself. Sagan asked the woman if she knew the book contained inaccuracies. "It couldn't," she said, "because they wouldn't let him publish it if it weren't true." Not so. Many books see print simply because they *will* sell, and many bookstands carry as much fantasy as fact in their so-called non-fiction selections.

In summary, the processes of perception, conception, and reporting of rare events could in principle be a source of scientific information, but in practice these processes often produce misleading data. Reports of weird, rare events are likely to circulate and be published whether these events really occurred or not.

UFOs

UFOs are any form of Unidentified Flying Object reported in the sky. Undeniably, there have been thousands of reports of different kinds of UFOs, ranging from alleged metal disks to amorphous glows to unusual objects detected by radar. The previous discussion shows why they are hard to study scientifically.

Most of the literature on UFOs is pseudo-science, not because of the subject, but because of the way it is treated. Most popular UFO books abound in exaggerated claims and distortions. For example, a 1968 University of Colorado study (in which I personally participated) established that many of the classic UFO photos which have been used on covers of UFO publications (and are still used) are fakes or photos of known natural phenomena.

A sociological, fad-like element in the UFO problem is shown by waves of UFO reports following space events such as the first Sputniks and the first photos from Mars (see Figure B·1), and by UFO hoaxes that followed within weeks of the first "flying saucer" report in 1947. Waves of UFO reports have even occurred in earlier eras, when the UFOs were reported to look, not like the saucers popularized in our movies, but like images popular in those times. A UFO wave in the 1890s reported "airships" looking like Victorian dirigibles, with fan-like propellers. One reportedly crashed into a windmill in Texas in 1897, killing its Martian occupants. A pro-saucer research organization in 1973 labeled this case a hoax.

When a satellite reentered the atmosphere and broke into a group of burning pieces in the night sky over the eastern U.S. in 1968, many people witnessed the phenomenon. Some correctly identified it, but many others misconceived it. Of a group of 30 extensive reports collected by the Air Force, 57 percent said that the objects were flying in formation, implying intelligent control. About 17 percent conceived that the glowing objects must be attached to an unseen black object, and they reported a "cigar-shaped" or "rocket-shaped" object that did not exist. Others conceived the glowing objects as windows, and reported a dark object with glowing windows. One totally erroneous report said, "It was shaped like a fat cigar. . . . It appeared to have rather square shaped windows along the side . . . the fuselage was constructed of many pieced or flat sheets . . . with

Figure B·1 **Examples of sudden increases in the number of UFO reports correlated with social factors such as the first satellite launches (top) and the first close-up photos** ▼ **of Mars (bottom). This correlation suggests that "waves" of sightings, discussed in many popular books, have social rather than physical causes.**

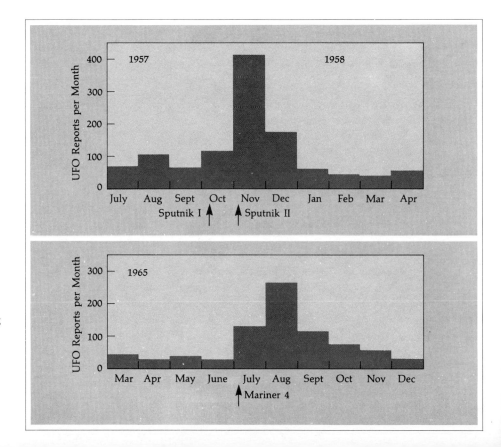

a 'riveted together look' . . ." (Condon et al., 1969).

The abundance of such misconceptions and misreporting does not, of course, prove that "flying saucers" or alien spaceships do not exist. But it does show that it is nearly impossible to prove that they do exist from the kinds of reports available. While a few unsolved UFO cases may suggest unusual atmospheric phenomena, there are none that give good evidence for spaceships. If there really are any spaceships around, one might predict, in view of increasing numbers of surveillance devices and tourist cameras, that good evidence should appear soon. So far, none has.

Ancient Astronauts

A similar area of pseudo-science is the literature on possible ancient astronauts. The basic technique in many of these books is to misrepresent astronomical or archeological evidence. One best-seller, for example, claims that studies of the 1908 Siberian meteorite impact (Chapter 10) "confirmed a nuclear explosion," which is not true. The same book presents the ancient lines in Peru (Figure 1·19) as a mystery requiring supertechnology to construct. I have visited them and they were made by simply pushing aside an inch-thick cover of dark stones. Several scientists examined the evidence for ancient alien visitors years before these books became popular, and found no convincing evidence (Shklovskii and Sagan, 1966).

Astro-Catastrophes

A more interesting case was a study of ancient myths by naturalist and psychiatrist Immanuel Velikovsky. He used a pseudo-scientific method of interpreting all ancient documents literally. For example, where the Book of Joshua says the sun stood still (10:13), Velikovsky assumes that the earth indeed suddenly stopped turning and then started again, ignoring modern evidence that this is unlikely. Compiling his results, Velikovsky (1950) concluded that two astronomical super-catastrophes occurred around 1500 B.C. and 750 B.C., during which Venus first appeared, passed near Mars, and then passed near earth, reaching its present orbit only after 750 B.C.

Velikovsky's work was interesting in its compilation of strange ancient myths, but when astronomers criticized his work, he was celebrated as an anti-establishment underdog and his books became best-sellers. However, his conclusions have been solidly refuted by scientists. For example, records of Venus' motions usually dated around 1600 B.C. (quoted earlier) and records of old eclipses show no signs of these disturbances; computer calculations can trace planetary motions back millions of years and show no indication of the hypothesized disturbances; the lunar lava flows that Velikovsky attributed to the catastrophes date from 3 billion years ago, not 3 thousand years ago.

Although Velikovsky was far too literal in interpreting old records, astronomical events such as meteorite falls may indeed have influenced early cultures and myths. Scientific methods can be applied to this problem. For example, Florida geologist Cesare Emiliani and his co-workers (1975) describe physical evidence of world-wide coastal flooding around 9600 B.C., which they believe may be the basis of the world-wide flood myth emphasized by Velikovsky.

What Should Be Done?

One might smile at pseudo-science as just another human foible, if it weren't for nagging fears that it threatens healthy civilization. For example, a 1975 Gallup poll indicated that about 12 percent of all Americans take astrology quite seriously. In an age of exploration, do we want to rely on ancient magic? Millions of low-income people seeking advice spend money to support astrologers, and, as astronomer Bart Bok (1975) points out, "The astrologer can refine his interpretations to any desired extent—the end product becoming increasingly more expensive as further items are added." Computers can be used to calculate the planetary positions, giving the operation a scientific look. It is thus easy for naive people to believe astrology has a scientific basis.

Newspapers have rejected suggestions that they drop astrology and other pseudo-science, because they are allegedly harmless entertainment that promotes sales. Of course, most advice in astrology columns is innocuous. It might be said of astrology, like fairy stories, that "if you believe it hard enough, it will work." For example, if you read morning advice to avoid frivolous expenses today, you may indeed be more prudent than usual. But one would hope we could find better sources of inspiration than ancient magic that is pseudo-science at best and fraudulent waste at worst.

Essay C

Astronomical Coordinates and Time-Keeping Systems

Two kinds of astronomical systems are related to the rotation of the earth. The first is the system of astronomical coordinates, whereby astronomers map positions of objects in the sky. The second is the time-keeping system employed around the world—an outgrowth of the early time-keeping and calendar systems described in Chapter 1.

Right Ascension and Declination Coordinates

We use the earth's spin to define coordinate systems for locating objects both on the ground and in the sky.

The ground system, the system of latitude and longitude, is based on the earth's spin, because latitude lines are perpendicular to the axis of rotation, whereas longitude lines intersect at the poles of rotation. (See Figure 3·5A, p. 54.)

A nearly identical system describes locations of stars and other celestial objects. It involves coordinates called *right ascension* and *declination* (often abbreviated R.A. and Dec.). *Right ascension and declination are merely longitude and latitude, respectively, projected from the center of the earth through the earth's surface and onto the sky, and fixed with respect to the stars.* The celestial equator (defined in Chapter 1) is the line of 0° declination, and the north celestial pole is at declination 90°N, usually written +90°. The south celestial pole is at −90°. (See Figure 3·5B.)

To visualize this system, imagine the earth as a giant nonrotating glass sphere with only the longitude and latitude lines engraved on it, and imagine yourself at the center of the sphere. You look out and see the longitude-latitude lines, by which you could locate features on the surface of the earth. But these lines are also lines of right ascension and declination, and you see them projected among the stars. The earth's equator overlaps the celestial equator. The earth's north pole overlaps the north celestial pole, and almost obscures the star Polaris.

As zero right ascension, astronomers have chosen the right ascension line running through the vernal equinox: the point where the sun appears to cross the celestial equator going north, once each year. Instead of measuring R.A. in degrees, they measure it in hours, because they are interested in the time required for apparent motions of stars across the sky. For example, if a star's R.A.

places it 1^h below the eastern horizon, astronomers know that it will rise in an hour. (The earth turns 15° in an hour, so 1 hr = 15°.)

Stars are cataloged by their right ascension and declination. Sirius, for example, lies at R.A. 6^h 41^m, Dec. −17°; Polaris at 1^h 23^m, +89°.

The Effect of Precession on Right Ascension and Declination

Now we can more clearly describe the effects of the precession Hipparchus discovered around 130 B.C. (Chapter 2). Because the R.A.-Dec. system is anchored on earth, and because the earth wobbles with respect to the stars every 26,000 years, the R.A.-Dec. system shifts among the stars. The coordinates of stars shift slightly every year, and astronomers must continually update the R.A. and Dec. positions listed in catalogs. Catalogs generally specify "epoch 1950" or "epoch 2000" as the year for which the positions are listed. The same shift of the celestial equator and vernal equinox invalidated the original system of astrology by shifting the R.A.-anchored astrological signs away from their original starry constellations, as described in the preceding essay. For an example of the changing position of the north celestial pole among the stars, see Figure 2·10 on pp. 40–41.

Another Celestial Coordinate System

A system useful for describing objects in the sky during both day or night is the *altitude-azimuth system*. *Altitude* is the number of degrees above the horizon. *Azimuth* is the number of degrees around the horizon, starting from the north point (0°)

and moving through east (90°). Thus, an object on the horizon due east would have altitude 0°, azimuth 90°. An object halfway up the sky in the west would have altitude 45°, azimuth 270°.

An object at the *zenith* has altitude 90°. The most important azimuth line is the *meridian* (see Chapter 1, p. 9).

Systems of Time-Keeping

Before the invention of clocks, time was kept by watching the apparent motion of the sun around the earth, caused by the earth's rotation. This was called *apparent solar time.* Noon was the moment when the sun crossed the meridian. AM. (ante-meridian) is the period before the sun crosses the meridian; P.M. (post-meridian), the period after. Once clocks were invented, it became clear that the sun's apparent motions were not uniform, because the combination of earth's rotation and orbital revolution produces days of somewhat different length. Later, more careful measurements revealed slight sudden speedups and slowdowns in earth's rotation due to earthquakes, tides, and winds.

For these reasons today's clocks run on *mean solar time*, a time system based on the uniform average rate of the sun's motion rather than its actual position in the sky. The most accurate calibration of clocks for scientific work is now based on an international definition of the second in terms of constant atomic processes: one second is 9,192,631,770 cycles in certain radiation from the cesium-133 atom. This standard is called *atomic time.*

For long-term prediction of planetary positions or eclipses, astronomers have defined an additional time system based on the motions of the planets, called *ephemeris time.*

Astronomers, nightly sighting their telescopes on the stars, need a star-based time system rather than the sun-based system of mean solar time by which most clocks run. This system is called *sidereal time* and is simply a clock-reading equal to the right ascension crossing the local meridian at any moment.

Toward the Modern Calendar

Just as time-keeping based on the earth's rotation has required improvement, so the calendar system based on the earth's orbital revolution has required improvement. Chapter 1 described how early people developed calendar systems and estimated the length of the year. The year was often officially started on an astronomically determined date, such as summer solstice. One problem with early calendars was accurately determining the number of days in the year. We now know that the sidereal year (earth's motion around the sun with respect to stars) is 365.256 days, and the tropical year (cycle of seasons) is 365.242 days.

If an early community or nation adopted an official calendar with, say, 365 days in a year, they would discover that the astronomically determined New Year's Day was one day in error every four years. This problem was dealt with by a process called *intercalation:* astronomers, government officials, or priests would announce extra days, often celebrated as holidays, every few years to keep the calendar in step with astronomical events.

As long as society was stable, this process could be kept up for many years. The Romans, for example, developed the ancestor of our calendar, beginning the year at spring equinox. The Latin names *September* through *December* represent month positions (7 through 10) in that calendar. The extra days were intercalated at the end of the last month, February, where we still insert our extra day every leap year.

However, by 46 B.C., local officials in different parts of the empire were inserting extra days as they pleased, and the calendar had become confused. Julius Caesar had the calendar revised in 44 B.C. under the Alexandrian astronomer Sosigenes, and the modern leap-year scheme was formalized. The fifth-month of the new *Julian* calendar (Quintilis) was renamed July in honor of Julius Caesar. August was named after Augustus Caesar, who made some additional improvements in 7 B.C.

The Julian calendar was so good that it worked well for more than a thousand years. But since the year is not exactly $365\frac{1}{4}$ days long, new errors crept in. In the 1200s, European scientists noted that there was a full week discrepancy in the date of the vernal equinox. After several earlier attempts, Pope Gregory XIII commissioned a new revision of the calendar in 1576–1603.

This Gregorian calendar was adopted quickly in Roman Catholic countries, more slowly in Protestant countries, and still more slowly in Orthodox and other countries. England and the colonies converted in 1752, dropping 11 days. (Riots and legislation against rent abuse followed because of the resulting 19-day month.) Russia converted during the 1917 Revolution and had to drop 13 days. Because of the different dates of conversion to the modern Gregorian calendar in different countries, scholars must beware of historical records of dates, and not compare different dates in differ-

ent countries without considering calendar evolution.

To clear up historical confusion about dates, astronomers have devised a system called the *Julian day-count,* which designates any date in history, anywhere in the world, by the number of days since January 1, 4713 B.C. On July 4, 1976, to pick a nonrandom date, the Julian day-count was 2,443,964. Computer-calculated dates of ancient eclipses and other astronomical events are expressed in this system.

Appendix 1

Powers of Ten

It is no accident that the word "astronomical" has become a synonym for "enormous, almost beyond conception." Astronomy is full of extraordinary numbers designating great ages and distances. In astronomy (and other sciences), therefore, a convenient shorthand system of writing numbers is favored. In this system, an exponent, or superscript, designates the number of factors of 10 that have to be multiplied together to give the desired quantity—that is, the number of zeros in the quantity. For example:

$$1 = 10^0$$
$$10 = 10^1$$
$$100 = 10^2$$
$$1,000 = 10^3$$
$$10,000 = 10^4, \text{ and so on.}$$

In this book the most often-used of these large numbers are:

$$\text{one thousand} = 1,000$$
$$= 10^3$$
$$\text{one million} = 1,000,000$$
$$= 10^6$$
$$\text{one billion}^1 = 1,000,000,000$$
$$= 10^9$$
$$\text{one trillion} = 1,000,000,000,000$$
$$= 10^{12}$$

The usefulness of this system is illustrated by the age of the earth:

$$4,600,000,000 \text{ years} = 4.6 \times 10^9 \text{ yr}$$

or the distance that light travels in a year:

$$6,000,000,000,000 \text{ miles} = 6 \times 10^{12} \text{ mi}$$

A similar system is used for very small numbers. Here the exponent is negative, and refers to the number of decimal places. Thus:

$$0.0001 = 10^{-4}$$
$$0.001 = 10^{-3}$$
$$0.01 = 10^{-2}$$
$$0.1 = 10^{-1}$$
$$1.0 = 10^{0}$$

The density of gas in interstellar space, which is:

$$0.000,000,000,000,000,000,000,002 \text{ g/cm}^3$$

can more conveniently be written:

$$2 \times 10^{-24} \text{ g/cm}^3$$

---

[1] Beware: The British system uses the term "billion" to refer *not* to 10^9 but to 10^{12}. This book uses "billion" to mean 10^9.

Appendix 2

Units of Measurement

Most scientists express distances and other measurements in metric units, and the world is in the process of converting to this system. This conversion will save time and money, since publications, quantities of goods, and machine parts will be in one uniform system. The metric system will also be easier to learn and use since it expresses units in multiples of ten (like our money system), instead of in irregular multiples such as 12 inches per foot, 5,280 feet per mile, and 16 ounces per pound.

Table A2·1 gives some common units in both systems. Since some scientists commonly use centimeters and grams from the metric system, while others use meters and kilograms, conversions are given to both types of units (labeled cgs and mks, respectively). In the text, we have stressed cgs measurements.

Since astronomers commonly must deal with measurements much greater than the ordinary, they use some larger units, which are listed at the right.

	Table A2·1 Units of Measurement		
	Metric System		
English System	cgs system (cm-g-sec)	mks system (m-kg-sec)	Astronomical Measurements
1 inch	2.54 cm	.025 m	
1 foot	30.5 cm	.305 m	
1 yard	91.4 cm	.914 m	
1 mile	1.609×10^5 cm	1,609 m	
1 pound	454 g	.454 kg	
	1.50×10^{13} cm	1.50×10^{11} m	1 Astronomical Unit
	9.46×10^{17} cm	9.46×10^{15} m	1 light-year
	3.08×10^{18} cm	3.08×10^{16} m	1 parsec
	3.08×10^{21} cm	3.08×10^{19} m	1 kiloparsec
	3.08×10^{24} cm	3.08×10^{22} m	1 megaparsec
	1.99×10^{33} g		1 solar mass ($1 \, M_\odot$)

Supplemental Aids in Studying Astronomy

Astronomical Calendar Published annually by Guy Ottewell; sponsored by Department of Physics, Furman University, Greenville, S.C. 29613, (803) 294-2207. $4.95; 50 or more copies for $2.97 each (1976 prices including tax and postage). Star maps for each month; lists of astronomical events for each month; explanation of phenomena and various terms; astronomical glossary.

Astronomy Magazine Published monthly; 411 East Mason St., 6th Floor, Milwaukee, Wis. 53202. $15/year (12 issues; 1976 price). Popular articles on astronomical phenomena and observing techniques; astronomical paintings and photographs; monthly star charts.

Griffith Observer Published monthly by the Griffith Planetarium; 2800 E. Observatory Road, Los Angeles, Calif. 90027. $5/year (12 issues; 1977 price). Popular articles about astronomy.

Journal of the Association of Lunar and Planetary Observers Published bimonthly; Box 3 AZ, University Park, N.M. 88003. $8.00/year membership (6 issues; 1976 price). Journal of an association of amateur astronomers devoted to reports of amateur research and observing projects on the moon, planets, and comets. Publishes drawings and reports by members and coordinates research projects.

Mercury Magazine (Journal of the Astronomical Society of the Pacific) Published bimonthly; Astronomical Society of the Pacific, 1244 Noriega St., San Francisco, Calif. 94122. $12.50/year membership (6 issues; 1976 price). Popular articles about astronomy and astronomical scene; news of the Society, which encourages astronomy and holds occasional public functions.

Sky and Telescope Magazine Published monthly; 49-50-51 Bay State Road, Cambridge, Mass. 02138. $12/year (12 issues; 1976 price). Popular and semi-technical articles on astronomical phenomena, history, and observing techniques; photographs; monthly star charts.

494

Glossary

A (see *Angstrom*)

AU (see *Astronomical Unit*)

absolute intrinsic; not dependent on the position or distance of the observer.

absolute magnitude the absolute brightness (luminosity) of a star expressed in the magnitude system. The sun's absolute magnitude is +5.

absolute brightness any measure of the intrinsic brightness or luminosity of a celestial object.

absorption line in a spectrum, a reduction in intensity in a narrow interval of wavelength, caused by absorption of light by atoms between the source and the observer.

achondrite a type of stony meteorite in which chondrules have been destroyed, probably by heating or melting.

active nuclei in galaxies bright central regions emitting with greater intensity than normal nuclei, and often variable.

age the time since origin. The age of earth, sun, and other planets is 4.6 billion years (4.6 × 10⁹ yr). The age of our galaxy is roughly 12 billion years. The age of the universe (since the postulated big-bang) is estimated to be 16 billion years.

airglow visible and infrared glow from the atmosphere produced when air molecules are excited by solar radiation.

Airy disk in the telescopic image of a star, a small disk caused by optical effects.

amino acid a complex organic molecule important in composing protein and called a "building block of life."

Alexandrian library The research institution and collection of ancient works preserved after the fall of Rome at Alexandria, Egypt. Alexandrian knowledge passed into Arab hands with the Arab conquest of Alexandria, and eventually back into Europe around A.D. 100–1500.

Andromeda galaxy the nearest spiral galaxy comparable to our own, about 660 kpc away.

Angstrom (A) 10⁻⁸ cm; a unit of length used in measuring spectral wavelengths. A hydrogen atom is about 1 A across; a human hair is about 500,000 A in diameter.

angular measure any measure of the size or separation of two objects, expressed in angular units (degrees, minutes of arc, or seconds of arc), but not linear units (such as kilometers, miles, or parsecs).

angular size the angle subtended by an object at a given distance.

annular solar eclipse an eclipse in which the light source is almost, but not quite, covered, leaving a thin ring of light at mid-eclipse.

anorthosite a type of igneous rock, lighter colored and more silica-rich than basalt, common in the lunar uplands and probably composing much of the lunar crust.

apogee the point in a circum-terrestrial orbit farthest from the earth.

Apollo asteroids asteroids that cross earth's orbit.

Apollo program the U.S. program to land humans on the moon, 1961–1972; first landing July 20, 1969.

apparent not intrinsic, but dependent on the position or distance of an observer.

apparent brightness the brightness of an object as perceived by an observer at a specified location (but not measuring the object's intrinsic, or absolute, brightness).

apparent magnitude apparent brightness of one star relative to another star as expressed in the magnitude system (see examples in Table 14·2, pp. 260–261).

apparent solar time time of day determined by the sun's actual position in the sky. Apparent solar noon occurs as the sun crosses the meridian. Apparent solar time is different at each different longitude.

apparition the period of a few weeks during which a planet is most prominent or best placed for observation from earth.

ashen light a reported airglow sometimes seen on the dark side of Venus.

association a loosely connected grouping of young stars.

asteroid a rocky or metallic interplanetary body (usually larger than 100 meters in diameter).

asteroid belt the grouping of asteroids orbiting between Mars and Jupiter.

astrometric binary a binary star system detectable from the orbital motion of a single visible component.

astrometry the study of positions and motions of stars.

Astronomical Unit (AU) the mean distance from earth to sun, about 150 million km.

astronomy the study of all matter and energy in the universe.

atom a particle of matter composed of a nucleus surrounded by orbiting electrons.

atomic time a time-keeping system that defines the second by atomic vibrations; it is the current standard system for scientific time measurement.

aurora glowing, often moving, colored light forms seen near the N and S magnetic poles of earth; caused by radiation from high-altitude air molecules excited by particles from the sun and Van Allen belts.

B-type shell star a star of spectral type B that occasionally blows off a cloud of gas, forming a gaseous shell around the star.

Balmer alpha line a brilliant red spectral line at wavelength 6563 A, caused by transition of the electron in the hydrogen atom from the third-level orbit to the second-level orbit.

barred spiral galaxy a spiral galaxy whose spiral arms attach to a barlike feature containing the nucleus.

basalt a type of igneous rock, often formed in lava flows, common on the moon and terrestrial planets.

basaltic rock igneous rocks (including basalt) with a composition resembling basalt and a relatively low content of silica (SiO_2).

basins large impact craters on the planets, usually several hundred km across, flooded with basaltic lava, and surrounded by concentric rings of faulted cliffs.

belts dark cloud bands on giant planets.

big-bang theory the hypothetical explosionlike event that initiated the universe as we know it, probably between 12 and 20 billion years ago.

binary star-system a pair of co-orbiting stars.

black dwarf a starlike object too small to achieve nuclear reactions in its center; any stellar object smaller than about 0.08 $M_\odot$.

blue globular clusters the hot, blue stars (which are massive and have short lifetimes) contained in the Magellanic clouds. These globulars must have formed less than a billion years ago, and others may be still forming.

blue shift a Doppler shift of spectral features toward shorter wavelengths, indicating approach of the source.

Bode's rule a convenient memory-aid for listing the planets' distances from the sun.

body tide a tidal bulge raised in the solid body of a planet.

Bok globule a relatively small, dense, dark cloud of interstellar gas and dust, usually silhouetted against bright clouds.

bolometric referring to the total of radiation at all wavelengths.

bolometric luminosity the total energy radiated by an object at all wavelengths, usually given in ergs/sec.

bulk density total mass divided by total volume, usually given in grams/cubic centimeter.

cgs system metric system of measurement using centimeters, grams, and seconds as the fundamental units.

calendar stick a primitive pole-shaped device carved with symbols to record time intervals or historic events.

Cambrian period a period from 570 to 500 million years ago which saw the first rapid proliferation of fossil-producing species of plants and animals.

canals alleged straight-line markings on Mars found not to exist after spacecraft visits to Mars.

capture theory a theory of origin of a planet-satellite or binary star-system in which one body captures another body by gravity.

carbon cycle a series of nuclear reactions in which hydrogen is converted to helium, releasing energy in stars more massive than about 1.5 $M_\odot$. Carbon is used as a catalyst.

carbonaceous chondrite a type of carbon-rich and volatile-rich meteorite believed to be nearly unaltered examples of some of the earliest-formed matter in the solar system.

Cassini's division the most prominent gap in Saturn's rings.

catastrophic theories theories invoking sudden or very short (cosmically or geologically speaking) energetic events to explain observed phenomena.

catastrophism an early scientific school which held that most features of nature formed in sudden events, or catastrophes, instead of by slow processes.

celestial equator the projection of earth's equator onto the sky.

celestial poles the projection of the two poles of earth's rotation onto the sky.

cell the unit of structure in living matter.

center of mass the imaginary point of any system or body at which all mass could be concentrated without affecting the motion of the system as a whole; the balance point.

Cepheid variable any of a group of luminous variable stars with periods of 5 to 30 days (depending on their population). The periods are correlated with luminosity, allowing distance estimates out to about 3 mpc.

Chandresekhar limit a mass of about 1.4 $M_\odot$, the maximum for white dwarfs; stars of greater mass have too great a central pressure, causing formation of a star type denser than a white dwarf.

channels riverbed-like valleys on Mars, possible sites of ancient Martian rivers.

chemical reactions reactions between elements or compounds, in which electron structures are altered; atoms may be moved from one molecule to another, but nuclei are not changed and thus no element is changed to another.

chondrites stony meteorites containing chondrules; they are believed to be little altered since their formation 4.6 billion years ago.

chondrules bb-size spherules in certain stony meteorites, believed among the earliest-formed solid materials in the solar system.

chromosphere a reddish-colored layer in the solar atmosphere, just above the photosphere.

circular velocity velocity of an object in circular orbit, measured with respect to the center of mass of the orbiting pair:

$$V = \sqrt{\frac{GM}{R}}$$

G = Newton's gravitational constant, 6.67×10^{-8}

M = Mass of central body

R = Distance of orbiter from center of central body

circumpolar zone the zone of stars centered on the celestial pole; they never set, as seen from a given latitude.

circumstellar nebula gas and dust surrounding a star.

closed universe a theoretical model of the universe with finite volume and curved space.

cluster of galaxies a relatively close grouping of galaxies, often with some members co-orbiting or interacting with each other.

coacervates cell-size, nonliving globules of proteins and complex organic molecules formed spontaneously in water solutions.

Coalsack nebula a prominent dark nebula about 170 kpc away silhouetted against the Milky Way.

cocoon nebula a dust-rich nebula enclosing and obscuring a star during its formation, but later shed.

coma (1) the diffuse part of the head of a comet, surrounding the nucleus; (2) a type of distortion in some telescopes and optical systems.

comet an ice-rich interplanetary body that releases gases when heated by the sun in the inner solar system, forming a bright head and diffuse tail. (see also *coma*)

comet nucleus the brightest starlike object near the center of a comet's head; the physical body (believed to be icy and a few km across) within a comet.

common condensation theory a theory of origin of a planet-satellite or binary star system by simultaneous growth of two co-orbiting bodies.

communication satellites artificial satellites used for communication on earth or in space.

comparative planetology an interdisciplinary field of astronomy and geology attempting to discover and explain differences from one planet to another in properties such as climate and interior structure.

condensation sequence the sequence in which chemical compounds condense to form solid grains in a cooling, dense nebula.

conduction one of three processes that transfer heat from hotter regions to colder regions; conduction occurs as fast-moving molecules in the hot region agitate adjacent molecules.

conjunction the period when a planet lies at zero or minimum angular distance from the sun, as seen from earth.

conservation of angular momentum a useful physical rule which states that the total angular momentum in an isolated system remains constant.

constellations imaginary patterns found among the stars, resembling animals, mythical heroes, etc.; different cultures map different constellations.

contact binary a co-orbiting pair of stars whose inner atmospheres or surfaces touch.

continental drift the motion of continents due to (convective?) motion of underlying material in the earth's mantle.

continental shield stable, ancient regions, usually flat and oval shaped, in continents.

continuous creation theory the hypothesis that matter is being created in interstellar or intergalactic space during the current era (and always).

continuous spectrum a spectrum consisting of radiation changing uniformly (continuously) with wavelength, with no absorption or emission lines.

continuum in a spectrum with absorption or emission lines, the background continuous spectrum.

convection one of three modes of transmission of heat (energy) from hotter regions to colder regions. Convection is the mode involving motions of masses of material.

core the densest inner region of earth, probably of nickel-iron composition; in other planets, similar high density central regions; in the sun or stars, a dense central region where nuclear reactions occur; in galaxies, the densest, brightest central regions.

coriolis force a force perceived by an observer in a rotating system caused by the rotation; coriolis effects in clouds were early evidence of the earth's rotation.

corona the outermost atmosphere of

the sun, with temperature about 1 to 2 million K.

coronograph an instrument permitting direct observation of the solar corona in the absence of eclipses.

cosmic fuels nonfossil energy sources provided by cosmic processes; for example, solar and geothermal energy.

cosmic rays high energy atomic particles (85 percent protons) which enter the earth's atmosphere from space. Many may originate in supernovae and pulsars.

cosmogony any theory of the origin of the universe or one of its component systems, such as galaxies or the solar system.

cosmological red shift any redward Doppler shift attributed to the mutual recession of galaxies or the expanding universe.

cosmology the study of the structure of the universe.

Crab nebula (M 1, NGC 1952) the expanding cloud of gas from the supernova in Taurus, about 2 kpc away; first witnessed by terrestrial observers in A.D. 1054.

crater (see *impact crater; volcanic crater*)

crust the outermost solid layer of a planet, with composition distinct from the mantle, and defined by a seismic discontinuity.

cultural hurdle the hypothetical survival requirement for a planetary culture between the time it achieves technology capable of quickly altering its planetary environment and the time it can establish viable bases off its planet; the uncertainty of the probability of survival affects our estimates of the probability of intelligent life elsewhere in space.

curved space space in which Euclidean solid geometry is not valid.

cusps the pointed horns of a crescent.

daughter isotope an isotope resulting from radioactive disintegration of a parent atom.

declination angular distance north or south of the celestial equator.

degenerate matter matter in a very high density state in which electrons are freed from atoms, and pressure is a function of density but not temperature.

density-wave theory the leading theory of formation of spiral arms in galaxies, by periodicities in star, dust, and gas motions.

deposition the accumulation of eroded materials in one place from another place.

differential rotation the differences in speed for stars at different distances from the center of the galaxy. Orbital velocities are actually slower at 5,000 parsecs from the center than at the sun's distance, which is 8,000 to 10,000 parsecs from the center.

differentiation any process that tends to separate different chemicals from their original mixed state and concentrate them in different regions.

diffraction the slight bending of light rays as they pass edges, producing spurious rays and rings in telescopic images of stars.

dimensions of the galaxy as early as 1935 astronomers agreed that the sun is about 8,000 to 10,000 parsecs from the center, and the overall diameter of our galaxy is about 30,000 parsecs.

dirty-iceberg model a theoretical description of a comet nucleus as a large icy body with bits of silicate "dirt" embedded in it.

Doppler shift the shift in wavelength of light or sound as perceived by the observer of an approaching or receding body. For speeds well below that of light the shift is given by:

$$\text{original wavelength} \times \frac{\text{radial velocity}}{\text{velocity of light}}.$$

dwarf stars (1) an unusually small or faint star, such as a white dwarf; (2) professional astronomers sometimes use this term to indicate any main-sequence star, but it is not so used in this book.

early intense bombardment a period of intense cratering as planetesimals were swept up in the last stages of planet formation.

earthquake vibration or rolling motion of the earth's surface accompanying the fracture of underground rock.

eclipse an event in which the shadow of one body falls on another body.

eclipsing binary a binary star system seen virtually edge-on so that the stars eclipse each other during each revolution.

ecliptic (1) the plane of the earth's orbit, and its projection in the sky as seen from earth; (2) approximately, the plane of the solar system.

effective temperature the temperature of an object as calculated from the properties of the radiation it emits.

ejecta blanket a layer of debris thrown out of a crater onto a planet's surface.

electron one of the negatively charged particles orbiting around the atomic nucleus, with mass $= 9.1 \times 10^{-28}$g.

element a chemical material with a specified number of protons in the nucleus of each atom. Atoms with one proton are hydrogen; with two protons, helium; and so on.

ellipse a closed oval-shaped curve (generated by passing a plane through a cone) describing the shape of the orbit of one body around another.

elliptical galaxies a galaxy of approximately elliptical cross section with no spiral arms. They generally have many red giant stars, no young stars, and little dust.

emission bands narrow wavelength intervals in which molecules emit light.

emission lines very narrow wavelength intervals in which atoms emit light.

emission nebula a nebula emitting energy, especially in the form of spectral emission lines.

energy in physics, a specific quality equal to work or the ability to do work. Energy may appear in many forms, including electromagnetic radiation, heat, motion, and even mass (according to the theory of relativity).

energy level the orbit of an electron around the nucleus of an atom.

English system a nondecimal system of units using pounds, inches, and seconds, now being replaced by the more convenient metric system. (see also *cgs* and *mks systems*)

ephemeris a table of predicted positions of a planet, asteroid, or other celestial body.

ephemeris time a time-keeping system based on the motions of planets, more regular than conventional systems based on earth's rotation.

epicycle a small circular motion superimposed on a larger circular motion.

epicycle theory an early theory (especially by Ptolemy) that the planets moved around the earth in epicycles.

equatorial zone on Jupiter, Saturn and possibly other giant planets, a bright cloud zone near the equator.

equinoxes the dates (twice a year) at which the sun passes through the earth's equatorial plane.

erg the unit of energy in the cgs metric system.

erosion removal of rock and soil by any natural process.

escape velocity the minimum speed needed to allow a projectile to move away from a planet and never return to its point of launch.

Eta Carinae nebula a nebula around a peculiar nova-like variable star about 2 kpc away.

event horizon the theoretical surface around a black hole, from which matter and energy do not escape.

evolutionary track the sequence of points on the H-R diagram occupied by a star as it evolves.

excitation the process of raising electrons in an atom to higher energy states.

exobiology study of life outside the earth.

expanding universe a term popularized by Eddington to describe the mutual recession of galaxies.

expansion age the age of a system (particularly the universe as a whole) calculated from present dimension and rate of expansion.

explorative interval the hypothetical interval of time during which a species actively engages in exploration of other planets.

extra-galactic standard of rest an assumed stationary frame of reference defined by using the nearby galaxies as reference objects.

fault a fracture along which displacement has occurred on the solid surface of a planet or other celestial body.

field star the general designation of random stars near any star of particular interest.

fission theory a theory of origin of a planet-satellite or binary star-system by breakup of a single original body.

flare (1) on the sun, a sudden, short-lived, localized outburst of energy, often ejecting gas at speeds exceeding 1,000 km/sec; (2) outbursts from certain types of variable stars, sometimes called *flare stars.*

focus (foci) one of the two interior points around which planets or stars move in an elliptical orbit.

forbidden lines spectral lines arising from metastable states in atoms.

force in physics, a specific phenomenon producing acceleration of mass. Forces can be generated in many ways, such as by gravity, pressure, and radiation.

Foucault pendulum a carefully constructed pendulum with a universal pivot, long suspension, and heavy weight. Swinging for several hours, it reveals the earth's rotation under it.

free fall motion under the influence of gravity only, without any other force or acceleration such as rocket firing.

free-fall contraction contraction of a cloud or system of particles by gravity only, unresisted by any other force.

frequency number of electromagnetic oscillations per second corresponding to electromagnetic radiation of any given wavelength.

funnel effect concentration of all evolutionary tracks into the giant region of the H-R diagram.

galactic equator the plane of the Milky Way galaxy projected on the sky.

galactic halo a spherical swarm of globular clusters "above" and "below" the galactic disk, centered on a point in the direction of the constellation Sagittarius.

galactic longitude and latitude angular distance from the galactic equator, and around the galactic equator from the galaxy's center, respectively.

galactic nucleus the center of the galaxy.

galaxy any of the largest groupings of stars usually of mass 10^8 to 10^{13} $M_\odot$.

Galilean satellites the four large satellites of Jupiter, discovered by Galileo.

geologic time scale the sequence of events, and their dates, in the history of the earth.

giant elliptical galaxy any of the largest elliptical galaxies comparable to or more massive than the Milky Way.

giant planets (1) Jupiter, Saturn, Uranus, and Neptune; (2) any planet much more massive than earth.

giant stars highly luminous stars larger than the sun. O and B main-sequence stars are sometimes called blue giants; evolved stars of extremely large radius are called red giants.

globular star cluster a dense spheroidal cluster of stars, usually old, with mass of 10^4 to 10^6 $M_\odot$.

grains small (usually microscopic) solid particles in space.

granitic rock a silica-rich, light-colored rock type common in earth's continents. Granites, being low in density, tend to float to the surfaces of planets that have had extensive melting in the outer layers.

granules convection cells 1,000 to 2,000 km across, rising from the subphotospheric layers of the sun. Each granule rises at a speed of 2 to 3 km/sec and lasts for a few minutes.

gravitational contraction slow contraction of a cloud, star, or planet due to gravity, causing heat and radiation.

gravitational red shifts red shifts caused by light emitted from regions of very high gravity, such as neutron stars or the environs of a black hole.

gravity the force by which all masses attract all other masses. (see also *Newton's law of gravitation*)

great circle any circle on the surface of a sphere (especially the earth or sky) generated by a plane passing through the center of the sphere; the shortest distance between two points on a sphere.

greenhouse effect heating of an atmosphere by absorption of outgoing infrared radiation.

Gregorian calendar essentially the modern calendar system, introduced around A.D. 1600 under Pope Gregory XIII, and containing the modern system of reckoning leap years.

ground state the lowest energy state of an atom, in which all electrons are in the lowest possible energy levels.

HI and HII regions interstellar regions in which hydrogen is predominantly neutral (HI) or ionized (HII).

H-R diagram a technique for representing stellar data by plotting spectral type (or color or temperature) against luminosity (or absolute magnitude), named after its early proponents, Hertzsprung and Russell.

half-life in any phenomenon, the time during which the main variable changes by half its original value; often used loosely to indicate the characteristic time scale of a phenomenon. In radioactive decay, the time for half the atoms in any system to disintegrate.

Hayashi track a sharply descending evolutionary track in the H-R diagram covering the early period of stellar evolution from the high luminosity phase to the main sequence.

heliacal risings and settings celestial events during morning or evening twilight; heliacal risings and settings of a star are its first rising and last setting visible during the yearly cycle.

helium flash runaway helium "burning" inside a star as it evolves off the main sequence and into the giant phase of evolution. It occurs when degenerate gas at the star's center reaches a temperature of about 10^8 K.

Helmholtz contraction slow contraction of a cloud or system of particles by force of gravity, retarded by outward gas pressure and the limited rate at which radiation can escape.

Herbig-Haro object compact or star-like nebula of variable brightness believed to contain newly forming stars.

Hertzsprung gap a relatively unpopulated region in the upper part of the H-R diagram, from spectral classes A0 to about F5, crossed quickly by variable stars during post-giant evolution.

high-luminosity phase a star's short-lived stage of maximum brightness during pre-main-sequence evolution.

high velocity stars stars with high velocity, generally associated with the galactic halo.

homogeneous uniform in composition throughout the volume considered.

hour angle the number of hours since a star (or other body) last crossed the local meridian.

Hubble's constant the ratio between a galaxy's recession speed and its distance, measured to be about 57 km/sec per mpc.

Hubble's relation expresses how the recession velocity increases with distance.

hydrogen alpha line the designation of hydrogen's red spectral line at 6563 A, more properly called hydrogen Balmer alpha.

hydrogen Balmer series the series of all hydrogen spectral lines from 3646 A to 6563 A caused by electron transitions between the second and higher energy levels.

hyperbola the orbital curve followed by any free-falling body moving faster than escape velocity.

hypothesis a proposed explanation of an observed phenomenon, or a proposal that a certain observable phenomenon occurs.

IC number the catalog number of a cluster, nebula, or galaxy in the *Index Catalog*.

igneous rocks rocks crystallized from molten material.

impact crater a roughly circular depression of any size (known examples range from microscopic to diameters greater than 1,000 km) caused by impact of a meteorite.

inferior planets Mercury and Venus.

infrared light radiation of wavelength too long to see, usually about 0.0001 to 0.01 cm.

infrared star a star detected primarily by infrared light.

instability strip the Hertzsprung gap, or a portion of it, in which certain types of variable stars exist.

interacting galaxies pairs of galaxies in contact or connected by gas filaments.

interstellar atoms atoms of gas in interstellar space.

interstellar grains microscopic solid grains in interstellar space; interstellar dust.

interstellar molecules molecules of gas in interstellar space.

interstellar obscuration absorption of starlight by interstellar dust, causing distant objects to appear fainter.

interstellar reddening loss of blue starlight due to interstellar dust, causing distant objects to appear redder and fainter.

interstellar snowballs hypothetical interstellar particles larger than interstellar grains (perhaps icy).

inverse square law the relation describing any entity, such as radiation or gravity, that varies as $1/r^2$, where r is the distance from the source.

ion a charged atom or molecule.

ionization the process of removing at least one electron from an atom or molecule.

iron meteorites meteorites composed of a nearly pure alloy of nickel and iron.

irregular galaxy a galaxy of amorphous shape. Most have relatively low mass (10^8–10^{10} $M_\odot$).

irregular variable a variable star that fluctuates in brightness irregularly.

isotope a form of an element with a specified number of neutrons in the nucleus. Each element may have many possible isotopic forms, but only a few are stable.

isotropic appearing uniform no matter what the direction of view.

Julian date the date based on a running tabulation of days, starting January 1, 4713 B.C.

Kepler's laws the three laws of planetary motion that describe how the planets move, show that the sun is the central body, and allow accurate prediction of planetary positions.

kiloparsec 1,000 parsecs, or 3×10^{21} cm.

Kirchhoff's laws laws describing conditions that produce emission, absorption, and continuum spectra.

$L_\odot$ the luminosity of the sun, 4×10^{33} ergs/sec.

LSR (see *local standard of rest*)

Lagrangian points in an orbiting system with one large and one small body, an array of five points where a still smaller body would retain a fixed position with respect to the other two.

Lagrangian surface an imaginary surface with a figure-8 cross section, surrounding two co-orbiting bodies in circular orbits and constraining motions of particles within the system.

lava molten rock on the surface of a planet.

librations apparent wobbles in the moon's (or other satellite's) rotation, caused partly by gravitational forces and partly by geometric effects.

life a process in which complex carbon-based materials organized in cells take in additional material from their environment, replicate molecules, reproduce, and do other weird things like writing books.

light-year the distance light travels in a year (9.5×10^{17} cm).

limb the apparent edge of a celestial object.

line of nodes a line formed by the intersection of an orbit and some other reference plane, such as the plane of the solar system.

lithosphere the solid rocky layer in a partially molten planet.

local group the cluster of galaxies to which the Milky Way and about 20 other nearby galaxies belong.

local standard of rest (LSR) a frame of reference moving with the average velocity of the nearby stars (out to about 50 psc from the sun).

luminosity The total energy radiated by a source per second. The luminosity of the sun ($L_\odot$) $= 4 \times 10^{33}$ ergs/sec.

$M_\odot$ the mass of the sun, 2×10^{33}g.

M giants giant stars of spectral class M.

Magellanic clouds the two galaxies nearest the Milky Way, irregular in form and visible to the naked eye in the southern hemisphere.

Magellanic stream gas filaments connecting the Magellanic clouds to the Milky Way.

magma underground molten rock.

magnetic braking the slowing of rotation of a star or planet by interaction of its magnetic field with surrounding ionized material.

main sequence the group of stars defined on the H-R diagram that have a relatively stable interior configuration and are consuming hydrogen in nuclear reactions.

mantle a region of intermediate density surrounding the core of planets.

mare (maria) a dark-colored region on a planet or satellite; a region of basaltic lava flow on the moon.

Mars-crossing asteroids asteroids whose orbits cross that of Mars.

maser (Microwave Amplification by Simulated Emission of Radiation) (1) a device which amplifies microwave radio waves through special electronic transitions in atoms; (2) an interstellar cloud that acts in this way.

mass (1) material; (2) the amount of material.

mass/luminosity ratio the number of grams per unit of light or total radiation emitted from an object such as a galaxy.

mass-luminosity relation the relation between the mass of a main-sequence star and its total radiation rate; the more massive, the greater the luminosity.

mass of the galaxy applying the circular velocity equation gives a result of 2×10^{44} grams, or 10^{11} solar masses.

Maunder minimum the interval from 1645 to 1715, when solar activity was minimal.

Mayan Sacred Round a cycle of 260 days in the Mayan calendar, probably used in eclipse prediction.

mean solar time time shown by conventional clocks, determined by the sun's mean rate averaged over the year.

megaparsec 1,000,000 parsecs, or 3×10^{24} cm.

megaregolith a layer of meteoritically pulverized rubble at least several km deep, hypothesized as the primeval surfaces of planets.

Mercury's orbital precession shift of Mercury's line of nodes due to relativistic effects.

meridian (1) a north-south line on a planet, moon, or star; (2) a great circle through celestial pole and zenith.

Messier number the catalog number of a nebula, star cluster, or galaxy in *Messier's Catalog*.

metallic hydrogen a high-pressure form of hydrogen with free electrons.

metastable state in an atom, a configuration of electrons which is relatively long-lived, but is rarely found on earth because it is disrupted by collisions with other atoms; it may be found in interstellar atoms, creating forbidden spectral lines.

meteor a rapidly moving luminous object visible for a few seconds in the night sky (a "shooting star").

meteor shower a concentrated group of meteors, seen when the earth's orbit intersects debris from a comet.

meteorite an interplanetary rock or metal object that strikes the ground.

meteoritic complex the total swarm of all interplanetary particles, from microscopic to 1,000 km in diameter.

meteoroid a particle in space, generally implied to be smaller than a few meters across.

meter 39.4 inches.

Milky Way galaxy the spiral galaxy in which we live.

Miller experiment an experiment in which amino acids are created in laboratory conditions simulating the early earth.

minerals chemical compounds, usually in the form of crystals, that constitute rocks.

mks system a metric system of units expressing length in meters, mass in kilograms, and time in seconds. (see also *cgs system*)

multiple scattering redirection of electromagnetic radiation (such as light waves) by repeated interaction with atoms, molecules, or dust grains in space or in an atmosphere.

multiple star-system a system of three or more stars orbiting around each other.

mutual recession of galaxies all distant galaxies are moving away from us, and the farther away they are, the faster they are receding.

natural selection the theory in which those individuals best adapted to the ever-changing environment produce a greater number of offspring.

NGC number the catalog number of a nebula, cluster, or galaxy in the *New General Catalog.*

neap tide tides that occur near the time of first-quarter or last-quarter lunar phases.

nebula a cloud of denser-than-average gas and/or dust in interstellar space or surrounding a star.

neutron one of the two major particles constituting the atomic nucleus; it has zero charge and mass about 1.6749×10^{-24} g.

neutron star a star with a core composed mostly of neutrons, with density 10^{13} to 10^{15} g/cm^3.

Newtonian–Euclidean static cosmology a hypothetical model of the universe with infinite volume, no expansion, and Euclidean geometry.

Newton's law of gravitation the inverse square law, giving the force of gravity at any distance from a given mass.

Newton's laws of motion three rules describing motion and forces. Briefly, (1) a body remains in its state of motion unless a force acts on it; (2) force equals mass times acceleration; (3) for every action there is an equal and opposite reaction.

node either of two points where an orbit crosses a reference plane.

noncosmological red shift a hypothetical red shift of distant galaxies *not* caused by the Doppler effect.

non-Euclidean geometries hypothetical geometries in which Euclid's relations are not true; geometries of curved space.

nonthermal radiation radiation *not* due to heat of the source; example: synchrotron radiation.

north star (1) Polaris; (2) any bright star that happens to be within a few degrees of the north celestial pole during a given era.

nova (1) a new star, visible for a few weeks; (2) an exploding star expending about 10^{44} ergs, believed to be a member of a binary system and exploding as mass is transferred onto it from the other member.

nuclear reactions reactions involving the nuclei of atoms, in which a nucleus changes mass.

nucleus (1) the matter at the center of an atom, composed of protons and neutrons; (2) the central bright core (or solid body) of a comet; (3) the bright central core of a galaxy.

O association an association of O-type stars.

oblate spheroid the shape assumed by a sphere deformed by rotation.

obliquity the angle by which a planet's rotation axis is tipped to its orbit.

Occam's razor the principle that the simplest hypothesis, with the fewest assumptions, is most likely to be correct. Named after its use to cut away false hypotheses.

Olbers' paradox the problem of why the sky is dark at night if the universe is filled with stars.

Oort cloud the swarm of comets surrounding the solar system.

opacity the extent to which gaseous (or other) material absorbs light.

open star cluster a grouping of relatively young Population I stars (usually young Population I stars (usually 10^2–10^3 M$_\odot$), sometimes called a galactic cluster.

open universe a universe with infinite volume and no boundaries.

opposition the period when a superior planet lies in an opposite direction from the sun as seen from earth appearing in the midnight sky, well placed for observation.

optical double star a pair of stars that have small angular separation but are not co-orbiting.

organic molecules molecules based on the carbon atom, usually large and complex.

Orion a major constellation which lies in the direction of a star-forming region about 500 psc away, rich in bright young stars.

oscillating universe a hypothetical model of the universe with cycles of contraction and expansion.

outgassing emission of gas from the interior of a planet during volcanic activity.

ozone layer an atmospheric layer rich in ozone (O_3), created by interaction between oxygen molecules (O_2) and solar radiation. On earth, its altitude is about 20–60 km.

parabola (1) the curved trajectory followed by a particle moving at escape velocity; (2) the curve of a Newtonian telescope's primary mirror.

parallax angular shift in apparent position due to observer's motion. More specifically, small angular shift in a star's apparent position due to earth's motion around the sun. Stellar parallax, used to measure stellar distance, is defined as the angle subtended by the radius of earth's orbit as seen from the star.

parent body a body in which a meteorite formed and later broke off as a fragment.

parent isotope a radioactive isotope that disintegrates and forms a daughter isotope.

parsec a distance of 206,265 AU, 3.26 light-years, or 3.09×10^{13} km; defined as the distance corresponding to a parallax of one second of arc.

partial solar eclipse an eclipse in which the light source, as seen by a specified observer, is not totally obscured.

Pauli exclusion principle a principle of subatomic physics, specifying that no two electrons in a very small volume have exactly the same properties of energy, motion, or other properties.

peculiar velocity a star's velocity with respect to the local standard of rest.

penumbra (1) the outer, brighter part of a shadow, from which the light source is not totally obscured; (2) the outer, lighter part of a sunspot.

perfect cosmological principle an assumption made by some researchers that the universe—on the large scale—looks the same not only in each region of space, but also in different eras of time.

perigee the point in an orbit around the earth at which the orbit is closest to earth.

permafrost semi-permanent underground ice.

photon the quantum unit of light, having some properties of a wave. For each wavelength of radiation the photon has a different energy.

photosphere the light-emitting surface layer of the sun.

physical binary system two stars

orbiting around their common center of mass.

Planck's law a formula that describes the amount of energy/sec radiated at each wavelength by a body of specified temperature.

planet a solid (or partially liquid) body orbiting around a star, but too small to generate energy by nuclear reactions.

Planet X hypothetical tenth planet beyond Pluto, for which no current evidence exists.

planetary nebula any of a group of nebulae thrown off exploding stars and having roughly dish-like shape, resembling faint planets in small telescopes; they have no physical relation to planets.

planetesimals small bodies from which planets formed, usually ranging from micro-meters to kilometers in diameter.

planetology the study of planets' origins, evolution, and conditions.

plasma a high-temperature gas consisting entirely of ions, instead of neutral atoms or molecules. Because of the high temperature, the atoms strike each other hard enough to keep at least the outer electrons knocked off.

plate tectonics motions of the earth's crust in large plates the size of continents, creating major ocean basins and mountain ranges.

Polaris the north star.

Population I stars with a few percent heavy elements (heavier than He) found in the disks of spiral galaxies and in irregular galaxies.

Population II stars composed of nearly pure hydrogen and helium, found in the halo and center of spiral galaxies, in elliptical galaxies, and to a limited extent in irregulars.

powers of ten the number of times tens must be multiplied together to give a specific number; the exponent of ten. (Example: $10^2 = 100$; the power, or exponent, is 2; see also Appendix 1.)

precession the wobble in the position of a planet's rotation axis, caused by

external forces. Also, the change in a coordinate system (tied to any planet) caused by such a wobble.

pressure line broadening broadening of spectral lines due to high pressure in the gas emitting or absorbing the line.

primeval fireball the hypothetical expanding initial cloud of high-temperature plasma during the big bang.

primitive atmosphere the atmosphere of a planet (if any) just after the planet formed. (see also *secondary atmosphere*)

principle of relativity the principle that observers can measure only relative motions, since there is no absolute frame of reference in the universe by which to specify absolute motions.

prograde motion revolution or rotation from west to east, the most common type of motion in the solar system.

prominence a radiating gas cloud extending from the solar surface into the thinner corona.

proper motion the angular rate of motion of a star or other object across the sky (most stars have proper motions less than a few seconds of arc per year).

proteins any of several types of complex organic molecules made from amino acids inside plants and animals, essential in living organisms.

proteinoids cell-like, nonliving spheroids of protein molecules created in the laboratory by heating amino acids and adding water; a possible step in the evolution of life.

proton one of the two basic particles in an atomic nucleus, with positive charge and mass of 1.6726×10^{-24}g.

proton-proton cycle a series of thermonuclear reactions that convert hydrogen nuclei to helium nuclei, converting a tiny amount of mass into energy.

protoplanet a planet shortly before its final formation. Sometimes hypothesized to have a massive atmosphere and greater mass than in its present state.

pseudo-science research which has the external trappings of science but does not follow the scientific method, usually lacking review and repetition of observations by independent researchers.

Ptolemaic model the ancient earth-centered model of the solar system, with the sun, moon, and other planets moving in epicycles.

pulsar a rapidly rotating neutron star with a strong magnetic field, observed to emit pulses of radiation.

QSOs quasi-stellar objects; faint bluish stars with spectra similar to those of quasars.

quantized orbit orbits of electrons inside atoms; unlike orbits in the solar system, they can exist only in certain patterns or energy levels.

quantum a small indivisible unit of some quantity such as energy or mass.

quasar any of a group of starlike, faint celestial objects with very large red shifts. Many astronomers believe they are extremely distant galaxies of unusually energetic form.

r-process reactions rapid reactions, probably occurring inside supernovae, in which heavy elements are formed as atomic nuclei capture neutrons. (see also *s-process reactions*)

radial velocity the component velocity along the line of sight, toward or away from an observer. Recession is positive; approach is negative.

radiant the direction in the sky from which meteor showers appear to come.

radiation (1) any electromagnetic waves or atomic particles that transmit energy across space; (2) one of three modes of transmission of heat (energy) through stars or planets from warm regions to cooler regions.

radiation pressure an outward pressure on small particles exerted by electro-magnetic radiation in a direction away from the light source.

radio galaxy a galaxy that emits unusually large amounts of energy in the form of radio radiation.

radioactive atom any atom whose nucleus spontaneously disintegrates.

radioisotopic dating dating of rock or other material by measuring amounts of parent and daughter isotopes.

ray a bright streak of material ejected from craters on the moon or other planets.

red shift of galaxies Doppler shift toward longer wavelengths.

Red Spot a large, reddish, oval, semi-permanent cloud formation on Jupiter.

refractory elements elements least likely to be driven out of a material by heat. They are usually concentrated in the last components to melt when a material such as rock is heated.

regolith a powdery soil layer on the moon and some other bodies, caused by meteorite bombardment.

relativistic moving at speeds near that of light.

relativity (see *principle of relativity*)

retrograde revolution or rotation from east to west contrary to the usual motion in solar system.

retrograde motion of Mars an apparent reversal of Mars' motion among the stars, due to the faster terrestrial motion past Mars at the time of opposition.

right ascension longitude lines projected onto the celestial sphere.

rille a type of lunar valley.

Roche's limit the distance within which tidal forces would disrupt a satellite.

rotation curve orbital velocity as a function of distance from the center of a galaxy.

rotational line broadening broadening of spectral lines due to rotation of the source.

RR Lyrae star a type of variable star similar to the Cepheids that has been found associated with Population II and not Population I.

runaway star a star rapidly moving away from a region of recent star-formation.

Russell-Vogt theorem the equilibrium structure of a star is determined by its mass and chemical composition.

s-process reactions slow reactions in massive giant stars, in which heavy elements are built up as atomic nuclei capture neutrons. (see also *r-process reactions*)

Sagittarius A the radio source marking the center of our galaxy.

saros cycle an interval of 6,585 days (about 18 years) separating cycles of similar eclipses, used by ancient people to predict eclipses.

satellite any small body orbiting around a larger body.

scientific method the method of learning about nature, based on making observations, formulating hypotheses, and constructing observational or experimental tests to see if the hypotheses are accurate.

secondary atmosphere a planet's atmosphere after modification by outgassing and other processes. (see also *primitive atmosphere*)

sedimentary rocks formed from sediments.

seismology study of vibrational waves passing through planets, revealing internal structure.

selection effect any effect that systematically biases observations or statistics away from a correct result.

Seyfert galaxy a type of galaxy with a bright, bluish nucleus, possibly marking a transition between ordinary galaxies and quasars.

sidereal referring to stars.

sidereal time time measured by the apparent motion of the stars (instead of the sun), used by astronomers to point telescopes toward celestial targets; it is the right ascension that is on the meridian at any given location.

significant figures the number of digits known for certain in a quantity.

small angle equation the equation giving the relation between the distance D of an object, its diameter d, and its angular size α:

$$\frac{\alpha''}{206265} = \frac{d}{D}$$

solar apex the direction toward which the sun is moving relative to nearby stars.

solar constant the amount of energy reaching a sun-facing square centimeter at the top of the earth's atmosphere per unit time; 1.39×10^6 ergs/cm²/sec.

solar cycle 22-year cycle of solar activity.

solar nebula the cloud of gas around the sun during the formation of the solar system.

solar system the sun and all bodies orbiting around it.

solar wind an outrush of gas past the earth and beyond the outer planets. Near the earth, the solar wind travels at velocities near 600 km/sec, and sometimes reaches 1,000 km/sec.

solstices the two dates when the sun reaches maximum distance from the celestial equator.

solstitial orientation orientation of a building or structure so as to provide sight-lines to measure the date of solstice.

space velocity a star's velocity with respect to the sun.

spectral class a class to which a star belongs because of its spectrum, which in turn is determined by its temperature. The spectral classes are O, B, A, F, G, K, and M, from hottest to coolest.

spectral line strength measure of the total energy absorbed or emitted in a spectral line.

spectrograph an instrument for recording a photographic image of a spectrum.

spectroheliograph an instrument for observing the sun in certain specified wavelengths.

spectrometer an instrument for tracing the intensity of a spectrum at different wavelengths; the result is a graph.

spectroscope an instrument for visually studying spectra.

spectroscopic binary a binary star revealed by varying Doppler shifts in spectral lines.

spectroscopy study of spectra, especially as revealing the properties of the light source.

spectrum (1) the array of all electromagnetic radiation in order of wavelength, (2) an array of the visible colors of radiation from blue to red, in order of wavelength. (see Color Photo 30)

spectrum binary a binary revealed by mixture of two spectral classes in the spectrum.

spectrum-luminosity diagram the H-R diagram.

speed of light often designated as c, the speed of light is about 300,000 km/sec and is constant as perceived by all observers.

spicules narrow jets of gas extending out of the solar chromosphere, with lifetimes of about 5 minutes.

spiral arms in spiral galaxies, the arms lying at an angle to the sun-center line. The arms contain open clusters, O and B stars, and nebulae.

spiral galaxy a disk-shaped galaxy with a spiral pattern, typically containing 10^{10}–10^{12} $M_\odot$ of stars, dust, and gas.

spring tide a tide at new or full moon.

standard time solar time appropriate to the given local time zone.

star a mass of material, usually wholly gaseous, massive enough to initiate (or to have once initiated) nuclear reactions in its central regions.

steady-state cosmology a hypothetical cosmology in which the large-scale properties of the universe remain the same indefinitely.

Stefan-Boltzmann law a law giving the total energy E radiated from a surface of area A and temperature T per second: $E = \sigma T^4 A$. Sigma (σ), the Stefan-Boltzmann constant, $= 5.7 \times 10^{-5}$.

stellar evolution evolution of every star from one form to another forced by changes in composition as nuclear reactions proceed.

Stonehenge a prehistoric English ruin with built-in astronomical alignments.

stony-iron meteorites stony meteorites that probably come from deep within the parent body, where melted stony and metallic material coexisted.

stony meteorites meteorites of primarily rock composition.

Stromgren sphere an idealized conception of the region of ionized interstellar gas around a hot star.

subfragmentation breakup of a contracting cloud into smaller condensations.

subtend to have an angular size equal to a specified angle

sunspot a magnetic disturbance on the sun's surface, cooler than the surrounding area.

supercluster a hypothetical cluster of clusters of galaxies.

supergiant star an extremely luminous star in the uppermost part of the H-R diagram.

supergranulation large-scale (15,000–30,000 km diameter) convective cell patterns in the solar photosphere.

superior planets all planets with orbits outside earth's orbit.

supernova a very energetic stellar explosion expending about 10^{49} to 10^{51} ergs and blowing off most of the star's mass, leaving a dense core.

symbiotic stars pairs of stars whose evolutions are affecting each other, especially by mass transfer.

synchronous rotation any rotation such that a body keeps the same face toward a co-orbiting body.

synchrotron radiation radiation emitted when electrons move at nearly the speed of light in a magnetic field.

synodical month one complete cycle of lunar phases, 29.53 days.

T association an association of T Tauri stars.

T Tauri star a type of variable star, often shedding mass, believed to be still forming and contracting onto the main sequence.

tangential velocity the velocity component perpendicular to the line of sight.

Tarantula nebula (30 Doradus) a huge HII emission nebula in the Large Magellanic cloud.

tectonics disruption of planetary or satellite surfaces by large-scale mass movements, such as faulting.

temperature a measure of the average energy of a molecule of a material.

terminator the dawn or dusk line separating night from day on a planet or satellite.

terra (terrae) light-colored hilly uplands on the moon and Mercury.

terrestrial planets (1) Mercury, Venus, Earth, and Mars; (2) planets primarily composed of rocky material.

theory a body of hypotheses, often with mathematical backing and having passed some observational tests; often implying more validity than the term "hypothesis."

thermal escape escape of the fastest-moving gas atoms or molecules from the top of a planet's atmosphere by means of their thermal motion.

3-K radiation radio radiation coming uniformly from all over the sky, believed to be a red-shifted remnant of the big-bang radiation.

3-kiloparsec arm an inner arm of our galaxy expanding from the center at about 53 km/sec.

thrust the force generated by a high-speed discharge, as from a rocket or airplane.

tidal recession recession of the moon (or other satellite) from the earth (or other planet) caused by tidal forces.

tide a bulge raised in a body by the gravitational force of a nearby body.

total solar eclipse (1) an eclipse in which the light source is totally obscured from a specified observer; (2) an eclipse in which a body is entirely immersed in another's shadow. (see also *eclipse*)

transit (1) passage of a planet across the sun's disk; (2) any passage of a body with small angular size across the face of a body with large angular size.

Triangulum galaxy (M 33, NGC 598) the second nearest spiral galaxy.

triple-alpha process a nuclear reaction in which helium is transformed into carbon in red giant stars.

Trojan asteroids asteroids caught near the Lagrangian points in Jupiter's orbit, 60° ahead of and 60° behind the planet.

tsunami a large ocean wave generated by earthquake or volcanic activity (the correct name for a tidal wave).

"tuning fork" diagram a classification scheme for galaxies.

turbulent line broadening broadening of spectral lines by turbulence in the source gas.

21-cm emission line the important radio radiation at 21 cm wavelength from interstellar neutral atomic hydrogen.

21-cm radio waves produced by neutral hydrogen, these waves are especially useful for galactic mapping, because they allow us to detect HI clouds, which are concentrated in the spiral arms.

ultrabasic rocks describing rocks of high density, low silica content, and high iron content, in many cases derived from the upper mantle of a planet or satellite.

ultraviolet light radiation of wavelength too short to see, but longer than that of X-rays.

umbra (1) the dark inner part of a shadow, in which the light source is totally obscured; (2) the dark inner part of a sunspot.

universe everything that exists.

Urey reaction reaction by which earth's carbon dioxide was concentrated in carbonate rocks after dissolving in sea water.

Van Allen belts doughnut-shape belts around the earth (and other planets with strong magnetic fields) in which energetic ions from the sun are trapped.

variable star a star that varies in brightness.

Viking missions the first automated spacecraft missions to land successfully on Mars, in 1976.

visual binary star a binary in which both components can be seen.

volatile elements elements easily driven out of a material by heating.

volcanic crater a circular depression caused by volcanic processes such as explosion or collapse.

volcanism eruption of molten materials at the surface of a planet or satellite.

W Ursa Majoris star a contact binary star.

wavelength (1) the length of the wave-like characteristic of electromagnetic radiation; (2) in any wave, the distance from one maximum to the next.

white dwarf stars a planet-size star of roughly solar mass and very high density (10^5–10^8 g/cm^3) produced as a terminal state after nuclear fuels have been consumed.

Wien's law a formula giving the wavelength W at which the maximum amount of radiation comes from a body of temperature T. The formula is $W = 0.290T$ (cgs units).

Wolf-Rayet star a type of very hot star ejecting mass.

X-ray electromagnetic radiation of wavelength about 0.1 to 100 A.

X-ray source celestial objects emitting X-rays; many are probably binary systems where mass is transferred.

zenith the point directly overhead.

zodiac a band around the sky about 18° wide, centered on the ecliptic, in which the planets move.

zodiacal light a glow barely visible to the eye, caused by dust particles spread along the ecliptic plane.

zone of avoidance a band around the sky, centered on the Milky Way, in which galaxies are obscured by the Milky Way's dust.

zones light cloud bands on giant planets.

References

Asterisked references are the less technical references, more readily available and recommended for general reading.

Chapter 1

Aveni, A.F. "Possible Astronomical Orientations in Ancient Mesoamerica," in *Archaeoastronomy in Pre-Columbian America,* ed. A.F. Aveni. Austin: University of Texas Press, 1975.

Aveni, A.F., S.L. Gibbs, and H. Hartung. "The Astronomical Significance of the Caracol of Chichén Itzá." *Science, 188* (1975), 977.

Baity, E.C. "Archaeoastronomy and Ethnoastronomy So Far." *Current Anthropology, 14* (1973), 389.

*****Carlson,** J.B. "Lodestone Compass: Chinese or Olmec Primacy?" *Science, 189* (1975), 753.

Castetter, E.F., and W.H. Bell. *Pima and Papago Indian Agriculture.* Albuquerque: University of New Mexico Press, 1942.

Crommelin, A. "The Ancient Constellation Figures," in *Splendour of the Heavens,* ed. T. Phillips and W. Steavenson. New York: McBride, 1925.

*****Doig,** P. *A Concise History of Astronomy.* London: Chapman and Hall, 1950.

Eddy, J.A. "Astronomical Alignment of the Big Horn Medicine Wheel." *Science, 184* (1974), 1035.

*****Harber,** H.E. "Five Mayan Eclipses in Thirteen Years." *Sky and Telescope, 37* (1969), 72.

Hartner, W. "The Earliest History of the Constellations in the Near East, and the Motif of the Lion-Bull Contest." *Journal of Near Eastern Studies, 24* (1965), 1.

————. "Eclipse Periods and Thales' Prediction of a Solar Eclipse: Historic Truth and Modern Myth." *Centaurus, 14* (1969), 60.

*****Hawkins,** G., and J.B. White. *Stonehenge Decoded.* New York: Doubleday, 1965.

*****Hoyle,** F. *From Stonehenge to Modern Cosmology.* San Francisco: W.H. Freeman, 1972.

Jacobsen, T. "Mesopotamia," in *Before Philosophy,* ed. H. Frankfort et al. Baltimore: Penguin Books, 1946.

Lockyer, J.N. *The Dawn of Astronomy.* Cambridge, Mass.: MIT Press, 1964.

Lowell, P. *Mars and Its Canals,* 2nd ed. New York: Macmillan, 1906.

Luce, G.G. "Trust Your Body Rhythms." *Psychology Today* (Apr. 1975), 52.

*****Marshack,** A. *The Roots of Civilization.* New York: McGraw-Hill, 1972.

Neugebauer, O. *The Exact Sciences in Antiquity.* Providence, R.I.: Brown University Press, 1957.

Ovenden, M. "The Origin of Constellations." *Philosophical Journal, 3* (1966), 1.

Owen, N.K. "The Use of Eclipse Data to Determine the Maya Correlation Number," in *Archaeoastronomy in Pre-Columbian America,* ed. A.F. Aveni. Austin: University of Texas Press, 1975.

*****Pannekoek,** A. *A History of Astronomy.* London: George Allen and Unwin, 1961.

Smiley, C.H. "The Solar Eclipse Warning Table in the Dresden Codex," in *Archaeoastronomy in Pre-Columbian America,* ed. A.F. Aveni. Austin: University of Texas Press, 1975.

...nom. *Megalithic*
... Oxford: Oxford
...ss, 1971.

...tonehenge." *Journal of the*
...y of Astronomy, 5 (1974), 71.

Wilson, J.A. *The Culture of Ancient Egypt.* Chicago: University of Chicago Press, 1951.

*——. "Egypt," in *Before Philosophy,* ed. H. Frankfort et al. Baltimore: Penguin Books, 1961.

Chapter 2

***Durant,** W. *The Story of Philosophy.* New York: Simon and Schuster, 1926.

Einstein, A. *Relativity.* New York: Crown, 1961.

Gingerich, O. "Musings on Antique Astronomy." *American Scientist, 55* (1967).

Gribbon, J. "Did Chinese Cosmology Anticipate Relativity?" *Nature, 256* (1975), 619.

***Lewis,** D. *We, the Navigators.* Honolulu: University of Hawaii Press, 1973.

***Mozans,** H.J. *Woman in Science.* New York: D. Appleton, 1913.

Needham, J. *Science and Civilization in China,* Vol. 3. London: Cambridge University Press, 1959.

Neugebauer, O. *The Exact Sciences in Antiquity.* Providence, R.I.: Brown University Press, 1957.

***North,** J.D. "The Astrolabe." *Scientific American, 230* (Jan. 1974), 96.

Pannekoek, A. *A History of Astronomy.* London: George Allen and Unwin, 1961.

Pritchard, J.B., ed. *Ancient Eastern Texts Relating to the Old Testament,* 2nd ed. Princeton: Princeton University Press, 1955.

***Wilson,** J.A. *The Culture of Ancient Egypt.* Chicago: University of Chicago Press, 1951.

Chapter 3

***Davies,** G.L. *The Earth in Decay.* New

York: American Elsevier, 1969.

***Dewey,** J.F. "Plate Tectonics." *Scientific American, 226* (May 1972), 56.

Dunbar, C.O. *Historical Geology.* New York: John Wiley, 1963.

Einstein, A. *Relativity.* New York: Crown, 1961.

Engel, A. et al. "Crustal Evolution and Global Tectonics: A Petrogenic View." *Geological Society of America. Bulletin, 85* (1974), 843.

***Hallam,** A. "Alfred Wegener and the Hypothesis of Continental Drift." *Scientific American, 232* (Feb. 1975), 88.

Hargraves, R.B. "Precambrian Geologic History." *Science, 193* (1976), 363.

***Hurley,** P.M. "The Confirmation of Continental Drift." *Scientific American, 218* (Apr. 1968), 52.

Jeffreys, H. *The Earth,* 5th ed. New York: Cambridge University Press, 1970.

Jordan, T.H. "The Continental Tectosphere." *Reviews of Geophysics and Space Physics, 13* (1975), 1.

***Nelkin,** D. "The Science-Textbook Controversies." *Scientific American, 234* (Apr. 1976), 33.

Pepin, R.O. "The Formation Interval of the Earth." *Abstracts Seventh Lunar Science Conference,* Houston, 1976.

***Renfrew,** C. "Carbon-14 and the Prehistory of Europe." *Scientific American, 225* (Oct. 1971), 63.

Schneider, S.H., and R. Dickinson. "Climate Modeling." *Reviews of Geophysics and Space Physics, 12* (1974), 447.

Schopf, J.W. "The Age of Microscopic Life." *Endeavour, 34* (1975), 51.

***Tazieff,** H. *When the Earth Trembles.* London: Hart-Davis, 1964.

Wyllie, P.J. *The Dynamic Earth.* New York: John Wiley, 1971.

*——. "The Earth's Mantle." *Scientific American, 232* (Mar. 1975), 50.

Chapter 4

Ball, W.W.R. *An Essay on Newton's 'Principia.'* New York: Johnson Reprint Collection, 1972. (First published 1893.)

***Clarke,** A.C. *The Exploration of Space.*

New York: Harper & Bros., 1951.

Dyson, F. "Human Consequences of the Exploration of Space." *Bulletin of Atomic Scientists, 25* (Sept. 1969), 8.

Heinlein, R. *The Man Who Sold the Moon.* New York: New American Library, 1950.

***Lewis,** R.S. *Appointment on the Moon.* New York: Viking, 1969.

***Logsdon,** J. *The Decision to Go to the Moon.* Cambridge, Mass.: MIT Press, 1970.

Newton, I. *Principia.* A. Motte, trans. Berkeley: University of California Press, 1962.

Nicholson, M. *Voyages to the Moon.* New York: Macmillan, 1949.

***Sagan,** C., and J. Agel. *The Cosmic Connection.* New York: Doubleday, 1973.

Swenson, L., J. Grimwood, and C. Alexander. *This New Ocean.* Washington, D.C.: NASA, Special Publication 4201, 1966.

Verne, J. *From the Earth to the Moon.* New York: Didear, 1949. (First published 1865.)

Wells, H.G. *The First Men in the Moon.* Bridgeport, Conn.: Airmont, N.D. (First published 1901.)

Chapter 5

Abetti, G. *The History of Astronomy.* New York: Henry Schuman, 1952.

Cortright, E.M., ed. *Exploring Space with a Camera.* Washington, D.C.: NASA, 1968. Pub. no. SP-168.

Darwin, G.H. *The Tides and Kindred Phenomena in the Solar System.* San Francisco: W.H. Freeman, 1962. (First published 1898.)

El-Baz, F. "The Moon after Apollo." *Icarus, 25* (1975), 495.

Fairbridge, R.W. *Encyclopedia of Geochemistry and Environmental Sciences.* New York: Van Nostrand, 1972.

***Goldreich,** P. "Tides and the Earth-Moon System." *Scientific American, 226* (Apr. 1972), 42.

Hartmann, W.K., and D. Davis. "Satellite-sized Planetesimals and Lunar Origin." *Icarus, 24* (1975), 504.

Kaula, W.K., and A. Harris. "Dynamics of Lunar Origin and Orbital Evolution." *Review of Geophysics and Space Physics, 13* (1975), 363.

Lammlein, D. et al. "Lunar Seismicity, Structure, and Tectonics." *Review of Geophysics and Space Physics, 12* (1974), 1.

*****Mason,** B., and W.G. Melson. *The Lunar Rocks.* New York: Wiley-Interscience, 1970.

*****Mutch,** T. *Geology of the Moon,* 2nd ed. Princeton: Princeton University Press, 1973.

*****Page,** T. "Notes on the Fourth Lunar Science Conference." *Sky and Telescope, 45* and *46* (June and July/August 1973).

Ringwood, A.E. "Some Comparative Aspects of Lunar Origin." *Physics of Earth and Planetary Interiors, 6* (1972), 366.

Shoemaker, E. et al. "Lunar Regolith at Tranquillity Base." *Science, 167* (1970), 452.

Taylor, S.R. "Geochemistry of the Lunar Highlands." *Moon, 7* (1973), 181.

———. *Lunar Science: A Post-Apollo View.* New York: Pergamon Press, 1975.

Toksoz, M.N. et al. "Structure of the Moon." *Review of Geophysics and Space Physics, 12* (1974), 539.

Wood, J. "Thermal History and Early Magmatism in the Moon." *Icarus, 16* (1972), 229.

Chapter 6

Allen, C.W. *Astrophysical Quantities.* Cambridge, Mass.: Athlone Press, 1973.

*****Chapman,** Clark R. *The Inner Planets.* New York: Charles Scribner's Sons, 1977.

*****de Santillana,** Georgio. *The Crime of Galileo.* New York: Time, 1962.

Gingerich, O. *"Crisis" versus Aesthetic in the Copernican Revolution.* Cambridge, Mass.: Smithsonian Astrophysical Observatory, 1973.

*———. "Copernicus and Tycho." *Scientific American, 229* (Dec. 1973), 86.

Lerner, L.S., and E.A. Gosselin. "Giordano Bruno." *Scientific American, 228* (Apr. 1973), 86.

Morrison, D., and D.P. Cruikshank. "Physical Properties of the Natural Satellites." *Space Science Review, 15* (1974), 641.

*****Pannekoek,** A. *A History of Astronomy.* New York: Interscience, 1961.

*****Rosen,** E. "Copernicus' Place in the History of Astronomy." *Sky and Telescope, 45* (1973), 72.

Williams, H.S. *The Great Astronomers.* New York: Simon and Schuster, 1930.

Chapter 7

Connes, P. et al. "Traces of HCl and HF in the Atmosphere of Venus." *Astrophysical Journal, 147* (1967), 1230.

*****Cruikshank,** D.P., and C.R. Chapman. "Mercury's Rotation and Visual Observations." *Sky and Telescope, 34* (1967), 24.

Duncombe, R. "Report on Numerical Experiment on the Possible Existence of an 'Anti-Earth'," in *Scientific Study of Unidentified Flying Objects,* ed. E.U. Condon. New York: Bantam, 1969.

Goldreich, P., and S. Peale. "Spin-orbit Coupling in the Solar System, II. The Resonant Rotation of Venus." *Astronomical Journal, 72* (1967), 662.

Goldstein, R. "Review of Surface and Atmosphere Studies of Venus and Mercury." *Icarus, 17* (1972), 571.

Kerzhanovich, V., M.Y. Marov, and M. Rozhdestvensky. "Data on Dynamics of the Subcloud Venus Atmosphere from Venera Spaceprobe Measurements." *Icarus, 17* (1972), 659.

Lewis, J. "An Estimate of the Surface Conditions of Venus." *Icarus, 8* (1968), 434.

Marov, M.Y. "Venus: A Perspective at the Beginning of Planetary Exploration." *Icarus, 16* (1972), 415.

*****Moore,** P. *A Guide to the Planets.* New York: Norton, 1954.

Morrison, D. "Thermophysics of the Planet Mercury." *Space Science Review, 11* (1970), 271.

*****Murray,** Bruce C. "Mercury." *Scientific American, 233* (Sept. 19.

Sagan, C. "Structure of the Lower Atmosphere of Venus." *Icarus, 1* (1962), 151.

Sill, G.T. "Sulfuric Acid in the Venus Clouds." *Communications of the Lunar and Planetary Laboratory, University of Arizona, 9* (1973), 191.

Singer, S.F. "How Did Venus Lose Its Angular Momentum?" *Science, 170* (1970), 1196.

Young, A.T. "Are the Clouds of Venus Sulfuric Acid?" *Icarus, 18* (1973), 564.

*———, and Louise Young. "Venus." *Scientific American, 233* (Sept. 1975), 70.

Chapter 8

Bradbury, R. *The Martian Chronicles.* New York: Doubleday, 1950.

Burns, J., and M. Harwit. "Towards a More Habitable Mars—or—The Coming Martian Spring." *Icarus, 19* (1973), 126.

Burroughs, E.R. *The Chessmen of Mars.* New York: Doubleday, 1972. (First published 1922.)

*****Hartmann,** W.K., and O. Raper. *The New Mars.* Washington, D.C.: NASA, 1974.

Lowell, P. *Mars and Its Canals,* 2nd ed. New York: Macmillan, 1906.

*****Murray,** B.C. "Mars from Mariner 9." *Scientific American, 228* (Jan. 1973), 48.

Sagan, C., O. Toon, and P. Gierasch. "Climatic Change on Mars." *Science, 181* (1973), 1045.

Wells, H.G. *The War of the Worlds.* London: W. Heinemann, 1898.

*****Note:** Early results from Viking are reported in special issues of *Science,* 27 August 1976 and 1 October 1976.

Chapter 9

*****Alexander,** A.F.O.D. *The Planet Saturn.* London: Faber and Faber, 1962.

Bodenheimer, P. "Contraction Models for the Evolution of Jupiter." *Icarus, 29* (1976), 165.

*Chapman, C.R. "The Discovery of Jupiter's Red Spot." *Sky and Telescope,* 35 (1968), 276.

Cruikshank, D.P., and R.E. Murphy. "The Post-eclipse Brightening of Io." *Icarus,* 20 (1973), 7.

*Cruikshank, D.P., and D. Morrison. "The Galilean Satellites of Jupiter." *Scientific American,* 234 (May 1976), 108.

Cruikshank, D.P., C. Pilcher, and D. Morrison. "Pluto: Evidence for Methane Frost." *Science,* 194 (1976), 835.

Grosser, M. *The Discovery of Neptune.* Cambridge: Harvard University Press, 1962.

*Hartmann, W.K. *Moons and Planets.* Belmont, Calif.: Wadsworth, 1972.

Hubbard, W.B. "Structure of Jupiter: Chemical Composition, Contraction, and Rotation." *Astrophysical Journal,* 162 (1970), 687.

*Hunten, D.M. "The Outer Planets." *Scientific American,* 233 (Sept. 1975), 131.

Joyce, R., R. Knacke, and T. Owen. "An Upper Limit on the 4.9-Micron Flux for Titan." *Astrophysical Journal,* 183 (1973), L31.

Khare, B.N., and C. Sagan. "Red Clouds in Reducing Atmospheres." *Icarus,* 20 (1973), 311.

Kuiper, G.P. "Comments on the Galilean Satellites." *Communications of the Lunar and Planetary Laboratory, University of Arizona,* 10 (1973), 28.

Lebofsky, L., T. Johnson, and T. McCord. "Saturn's Rings: Spectral Reflectivity and Compositional Implications." *Icarus,* 13 (1970), 226.

Lewis, J., and R.G. Prin. "Jupiter's Clouds: Structure and Composition." *Science,* 169 (1970), 472.

*Lyttleton, R.A. *Mysteries of the Solar System.* New York: Oxford University Press, 1968.

Minton, R.B. "The Red Polar Caps of Io." *Communications of the Lunar and Planetary Laboratory, University of Arizona,* 10 (1973), 35.

Morrison, D. "Determination of the Radii of Satellites and Asteroids from Radiometry and Photometry." *Icarus,* 19 (1973), 1.

Owen, T.C. "The Atmosphere of Jupiter." *Science,* 167 (1970), 1675.

*Peek, B.M. *The Planet Jupiter.* New York: Macmillan, 1958.

Pollack, J. "Greenhouse Models of the Atmosphere of Titan." *Icarus,* 19 (1973), 43.

Sinton, W.M. "Does Io Have an Ammonia Atmosphere?" *Icarus,* 20 (1973), 284.

Soter, S. Untitled paper presented at Satellite Colloquium, Cornell University, Aug. 18–21, 1974.

Tombaugh, C.W. "The Trans-Neptunian Planet Search," in *Planets and Satellites,* eds. G. Kuiper and B. Middlehurst. Chicago: University of Chicago Press, 1961.

Trafton, L. "On the Possible Detection of H_2 in Titan's Atmosphere." *Astrophysical Journal,* 175 (1972), 285.

Veverka, J. "Titan: Polarimetric Evidence for an Optically Thick Atmosphere?" *Icarus,* 18 (1973), 657.

*Wolfe, J.H. "Jupiter." *Scientific American,* 233 (Sept. 1975), 118.

Zellner, B. "On the Nature of Iapetus." *Astrophysical Journal,* 174 (1972), L107.

Chapter 10

*Chapman, C.R. "The Nature of Asteroids." *Scientific American,* 232 (Jan. 1975), 24.

*Chapman, C.R., and D. Morrison. "The Minor Planets: Size and Mineralogy." *Sky and Telescope,* 47 (1974), 92.

Chapman, C.R., D. Morrison, and B. Zellner. "Surface Properties of Asteroids." *Icarus,* 25 (1975), 104.

Gehrels, T. "Physical Parameters of Asteroids and Interrelations with Comets," in *From Plasma to Planet,* ed. A. Elvius. New York: John Wiley, 1972.

*Hartmann, W.K. *Moons and Planets.* Belmont, Calif.: Wadsworth, 1972.

*———. "The Smaller Bodies of the Solar System." *Scientific American,* 233 (Sept. 1975), 143.

Krinov, E.L. *Giant Meteorites.* New York: Pergamon Press, 1966.

*Marsden, B. "The Recovery of Apollo." *Sky and Telescope,* 46 (1973), 155.

McCord, T. "Asteroid Surface Materials." Abstracts of Lunar Science Conference, Houston, 1977.

Millman, P.M. "Observational Evidence of the Meteorite Complex," in *The Zodiacal Light and the Interplanetary Medium,* ed. J. Weinberg. Washington, D.C.: NASA AP-150, 1967, 399.

Nininger, H.H. "Meteorite Collecting among Ancient Americans." *American Antiquity,* 4 (1938), 39.

Oort, H.H. "Empirical Data on the Origin of Comets," in *The Moon, Meteorites, and Comets,* eds. B. Middlehurst and G.P. Kuiper. Chicago: University of Chicago Press, 1963.

*Sagan, C. "Kalliope and the Kaa'ba: The Origin of Meteorites." *Natural History,* 84 (1975), 8.

Shapley, H., and H. Howarth, eds. *A Source Book in Astronomy.* New York: McGraw-Hill, 1929.

van Houten, C., I. van Houten-Groenveld, and T. Gehrels. "Minor Planets and Related Objects, V., The Density of Trojans Near the Preceding Lagrangian Point." *Astronomical Journal,* 75 (1970), 659.

Wasson, J.T. *Meteorites.* New York: Springer-Verlag, 1974.

Williams, J.G. "Proper Elements, Families, and Belt Boundaries," in *Physical Studies of Minor Planets,* ed. T. Gehrels. Washington, D.C.: NASA SP-267, 1971.

*Wood, J.A. *Meteorites and the Origin of Planets.* New York: McGraw-Hill, 1968.

Chapter 11

Cameron, A.G.W. "Accumulation Processes in the Primitive Solar Nebula." *Icarus,* 18 (1973), 407.

*———. "The Origin and Evolution of the Solar System." *Scientific American,* 233 (Sept. 1975), 32.

Clayton, R., L. Grossman, T. Mayeda. "A Component of Primitive Nuclear Composition in Carbonaceous Chondrites." *Science,* 182 (1973), 485.

Elvius, A., ed. *From Plasma to Planet.* New York: Wiley-Interscience, 1972.

Fairbridge, R.W. *The Encyclopedia of Geochemistry and Environmental Sciences.* New York: Van Nostrand Reinhold, 1972.

*Grossman, L. "The Most Primitive Objects in the Solar System: Carbonaceous Chondrites." *Scientific American,* 232 (Feb. 1975), 30.

*Lewis, J.S. "The Chemistry of the Solar System." *Scientific American,* 230 (Mar. 1974), 50.

*Mason, B., and W.G. Melson. *The Lunar Rocks.* New York: Wiley, 1970.

*Page, T.L. "Notes on the 4th Lunar Science Conference." *Sky and Telescope,* 46 (1973), 88.

Papanastassiou, D., and G. Wasserburg. "Initial Strontium Isotopic Abundance and the Resolution of Small Time Differences in the Formation of Planetary Objects." *Earth and Planetary Science Letters,* 5 (1968), 361.

Safranov, V.S. *Evolution of the Protoplanetary Cloud and Formation of the Earth and Planets.* Springfield, Va.: National Technical Information Service, 1972.

Schmidt, O. *A Theory of the Origin of the Earth.* Moscow: Foreign Languages Publishing House, 1958.

Shapley, H., and H. Howarth, eds. *A Source Book in Astronomy.* New York: McGraw-Hill, 1929.

Taylor, S.R. "Chemical Evidence for Lunar Melting and Differentiation." *Nature,* 245 (1973), 203.

Urey, H.C. *The Planets: Their Origin and Development.* New Haven, Conn.: Yale University Press, 1952.

*Wood, J.A. *Meteorites and the Origin of Planets.* New York: McGraw-Hill, 1968.

Chapter 12

*Abetti, G. *The Sun,* J. Sidgwick, trans. London: Faber and Faber, 1957.

Babcock, H.W. "Topology of the Sun's Magnetic Field and the 22-year Cycle." *Astrophysical Journal, 133* (1961), 572.

*Bahcall, J.N. "Neutrinos from the Sun." *Scientific American, 221* (July 1969), 28.

Cameron, A.G.W. "Major Variations in Solar Luminosity?" *Review of Geophysics and Space Physics, 11* (1973), 505.

*Clark, W. "How to Harness Sunpower and Avoid Pollution." *Smithsonian, 2* (Nov. 1971), 14.

Eddy, J.A. "The Maunder Minimum." *Science, 192* (1976), 1189.

Gibson, E.G. *The Quiet Sun.* Washington: NASA SP-303, 1973.

*Hubbert, M. "The Energy Resources of the Earth." *Scientific American, 224* (Sept. 1971), 61.

*Kummel, B. *History of the Earth.* San Francisco: W.H. Freeman, 1970.

Leighton, R.B. "A Magneto-Kinematic Model of the Solar Cycle." *Astrophysical Journal, 156* (1969), 1.

*Pasachoff, J.M. "The Solar Corona." *Scientific American, 229* (Oct. 1973), 68.

Schneider, S.H., and C. Mass. "Volcanic Dust, Sunspots, and Temperature Trends." *Science, 190* (1975), 741.

Shapley, H., and H. Howarth, eds. *A Source Book in Astronomy.* New York: McGraw-Hill, 1929.

*Snell, J.E., P. Achenbach, and S. Peterson. "Energy Conservation in New Housing Design." *Science, 192* (1976), 1305.

Ulrich, R.K. "Solar Neutrinos and Variations in the Solar Luminosity." *Science, 190* (1975), 619.

*Wilcox, J.M. "Solar Structure and Terrestrial Weather." *Science, 192* (1976), 745.

Williams, H.S. *The Great Astronomers.* New York: Simon and Schuster, 1930.

Wood, C.A., and R.R. Lovett. "Rainfall, Drought, and the Solar Cycle." *Nature, 251* (1974), 594.

Chapter 13

Allen, C.W. *Astrophysical Quantities,* 3rd ed. London: Athlone Press, 1973.

*Aller, L.H. *Atoms, Stars, and Nebulae.* Cambridge: Harvard University Press, 1971.

Shapley, H., ed. *Source Book in Astronomy, 1900–1950.* Cambridge: Harvard University Press, 1960.

Shapley, H., and H. Howarth, eds. *A Source Book in Astronomy.* New York: McGraw-Hill, 1929.

Young, A.T. "Television Photometry: The Mariner 9 Experience." *Icarus, 21* (1974), 262.

Chapter 14

Allen, C.W. *Astrophysical Quantities,* 3rd ed. London: Athlone Press, 1973.

*Aller, L.H. *Atoms, Stars, and Nebulae.* Cambridge: Harvard University Press, 1971.

Boesgaard, A., and W. Hagen. "The Age of Alpha Centauri." *Astrophysical Journal, 189* (1974), 85.

Chandrasekhar, S. *An Introduction to the Study of Stellar Structure.* New York: Dover, 1957.

Eddington, A.S. *The Internal Constitution of the Stars.* Cambridge: Harvard University Press, 1926.

Iben, I. "Stellar Evolution: Within and Off the Main Sequence." *Annual Review of Astronomy and Astrophysics, 4* (1967), 171.

Larson, R.B. "Numerical Calculations of the Dynamics of a Collapsing Proto-Star." *Monthly Notices of the Royal Astronomical Society, 145* (1969), 271.

Schwarzschild, M. *Structure and Evolution of the Stars.* Princeton: Princeton University Press, 1958.

Shapley, H., ed. *Source Book in Astronomy, 1900–1950.* Cambridge: Harvard University Press, 1960.

Swihart, T. *Astrophysics and Stellar Astronomy.* New York: John Wiley, 1968.

Westbrook, C., and C.B. Tarter. "On Protostellar Evolution." *Astrophysical Journal, 200* (1975), 48.

Chapter 15

Aannestad, P.A. "Absorptive Properties of Silicate Core-Mantle Grains." *Astrophysical Journal, 200* (1975), 30.

Arny, T., and W. Weissman. "Interaction of Proto-stars in a Collapsing Cluster." *Astronomical Journal, 78* (1973), 309.

Becklin, E., G. Neugebauer, and C. Wynn-Williams. "On the Nature of the Infrared Point Source in the Orion Nebula." *Astrophysical Journal, 182* (1973), L7.

Bodenheimer, P. "Contraction Models for the Evolution of Jupiter." *Icarus, 29* (1976), 165.

*Bok, B.J. "The Birth of Stars." *Scientific American, 227* (Aug. 1972), 49.

Bok, B.J., C. Cordwell, and R. Cromwell. "Globules," in *Dark Nebulae, Globules, and Protostars,* ed. B. Lynds. Tucson: University of Arizona Press, 1970.

Bok, B.J., and E. Reilly. "Small Dark Nebulae." *Astrophysical Journal, 105* (1947), 255.

Fowler, W., and F. Hoyle. "Star Formation." *Royal Observatory Bulletin, No. 67* (1963).

Hartmann, W.K. "On the Nature of the Infrared Nebula in Orion." *Astrophysical Journal, 149* (1967), L87.

Hayashi, C. "Stellar Evolution in the Early Phases of Gravitational Contraction." *Publications of the Astronomical Society of Japan, 13* (1961), 450.

Henyey, L., R. LeLevier, and R. Levée. "The Early Phases of Stellar Evolution." *Publications of the Astronomical Society of the Pacific, 67* (1955), 154.

*Herbig, G. "The Youngest Stars." *Scientific American, 223* (Aug. 1970), 30.

————. "The Structure and Spectrum of R Monocerotis." *Astrophysical Journal, 152* (1968), 439.

————. "Introductory Remarks," *Proceedings of the Liège Colloquium.* Mémoires de la Société Royale des Sciences de Liège, *19* (1970), 13.

————. "The Spectrum and Structure of 'Minkowski's Footprint:' M 1-92." *Astrophysical Journal, 200* (1975), 1.

Iben, I. "Stellar Evolution I. The Approach to the Main Sequence." *Astrophysical Journal, 141* (1965), 993.

Kuhi, L.V. "T Tauri Stars: A Short Review." *Journal of the Royal Astronomical Society of Canada, 60* (1966), 1.

Larson, R.B. "Numerical Calculations of the Dynamics of a Collapsing Protostar." *Monthly Notices of the Royal Astronomical Society, 145* (1968), 271.

————. "The Evolution of Spherical Protostars with Masses 0.25 to 10 $M_\odot$." *Monthly Notices of the Royal Astronomical Society, 157* (1972), 121.

Layzer, D. "On the Fragmentation of Self-gravitating Gas Clouds." *Astrophysical Journal, 137* (1963), 351.

Low, F.J. "Galactic Infrared Sources," in *Dark Nebulae, Globules, and Protostars,* ed. B. Lynds. Tucson: University of Arizona Press, 1970.

Low, F.J., and B. Smith. "Infrared Observations of a Preplanetary System." *Nature, 212* (1966), 675.

Lynds, B.T., ed. *Dark Nebulae, Globules, and Protostars.* Tucson: University of Arizona Press, 1970.

Mendoza, V.E. "Infrared Excesses in T Tauri Stars and Related Objects." *Astrophysical Journal, 143* (1968), 1010.

Mestel, L. "Problems of Star Formation I." *Royal Astronomical Society Quarterly Journal, 6* (1966), 161.

*Neugebauer, G., and E. Becklin. "The Brightest Infrared Sources." *Scientific American, 228* (Apr. 1973), 28.

Ney, E., and D. Allen. "The Infrared Sources in the Trapezium Region of M 42." *Astrophysical Journal, 155* (1969), L193.

Penston, M. "Photoelectric UBV Photometry of Stars in Ten Selected Areas." *Monthly Notices of the Royal Astronomical Society, 164* (1973), 121.

Poveda, A. "The H-R Diagram of Young Clusters and the Formation of Planetary Systems." *Boletin de los Observatorio Tonantzintla y Tacubaya* (Mexico City), *4* (1965), 15.

Roberts, M.S. "Upper Limit to the Neutral Hydrogen Density in the Halo Regions of Spiral Galaxies." *Physics Review, Letters, 17* (1966), 1203.

Sim, M. "A Search for Globules in OB Clusters and Associations." *Monthly Notices of the Royal Astronomical Society, 145* (1969), 375.

Spitzer, L., Jr. "The Dynamics of the Interstellar Medium, II. Radiation Pressure." *Astrophysical Journal, 94* (1941), 232.

Stockton, A., D. Chesley, and S. Chesley. "Spectroscopy of R Monocerotis and NGC 2261." *Astrophysical Journal, 199* (1975), 406.

Strom, S. et al. "The Nature of the Herbig Ae and Be-type Stars Associated with Nebulosity." *Astrophysical Journal, 173* (1972), 353.

*Strom, S., and K. Strom. "The Early Evolution of Stars." *Sky and Telescope, 45* (1973), 279, 359.

Walker, M. "Studies of Extremely Young Clusters, VI." *Astrophysical Journal, 175* (1972), 89.

Weaver, W. "The Coalsack III: A Search for T Tauri Stars." *Astrophysical Journal, 189* (1974), 263.

Wolstencroft, R., and T. Simon. "The Variable Circular Polarization of V1057 Cygni." *Astrophysical Journal, 199* (1975), L169.

Wright, A. "Results of a Computer Program for Gravito-Gas Dynamic Collapse." *Proceedings of the Liège Colloquium.* Mémoires de la Société Royale des Sciences de Liège, *19* (1970), 75.

Wynn-Williams, C., E. Becklin, and G. Neugebauer. "Infrared Sources in the HII Region W3." *Monthly Notices of the Royal Astronomical Society, 160* (1972), 1.

Chapter 16

Aller, L.H. *Atoms, Stars, and Nebulae.* Cambridge: Harvard University Press, 1971.

Baade, W., and F. Zwicky. "Cosmic Rays from Super-novae." *Proceedings of the National Academy of Science, 20* (1934), 259.

Bessell, M. "2U 1700–37: Another Black Hole?" *Astrophysical Journal, 187* (1974), 355.

Bregman, J. et al. "On the Distance to Cygnus X-1." *Astrophysical Journal*, *185* (1973), L117.

Bodenheimer, P., and J. Ostriker. "Do Pulsars Make Supernovae? II." *Astrophysical Journal*, *191* (1974), 465.

*****Bova**, B. "Obituary of Stars: A Tale of Red Giants, White Dwarfs, and Black Holes." *Smithsonian*, *4* (1973), 54.

Christy-Sackman, J., and K. Despan. "An Interpretation of the Puzzling Observations of FG Sagittae." *Astrophysical Journal*, *189* (1974), 523.

Cowley, A., and D. Macconnel. "Spatial Coincidences of X-ray Sources and Supernova Remnants." *Astrophysical Journal*, *11* (1972), 217.

Gamow, G. "Physical Possibilities of Stellar Evolution." *Physics Review*, *55* (1939), 718.

Glasby, J.S. *Variable Stars*. London: Constable, 1968.

Gold, T. "Rotating Neutron Stars as the Origin of Pulsating Radio Sources." *Nature*, *218* (1968), 731.

Hawking, S.W. "The Quantum Mechanics of Black Holes." *Scientific American*, *236* (Jan. 1977), 34.

Hewish, A. et al. "Observation of a Rapidly Pulsating Radio Source." *Nature*, *217* (1968), 709.

Hoxie, D. "The Structure and Evolution of Stars of Very Low Mass." Ph.D. Dissertation, University of Arizona, 1969.

Iben, I. "Stellar Evolution: Within and Off the Main Sequence." *Annual Reviews of Astronomy and Astrophysics, 5* (1967), 571.

Kraft, Robert. "Binary Stars among Cataclysmic Variables, III." *Astrophysical Journal*, *139* (1964), 457.

Kumar, S. "On the Nature of Planetary Companions of Stars." *Zeitschrift für Astrophysic*, *58* (1964), 248.

Langer, G., R. Kraft, and K. Anderson. "FG Sagittae: the s-Process Episode." *Astrophysical Journal*, *189* (1974), 509.

Leach, R., and R. Ruffini. "On the Masses of X-Ray Sources." *Astrophysical Journal*, *180* (1973), L15.

Morgan, W., R. White, and J. Tapscott. "A New Shell Phase in Pleione."

Astronomical Journal, 78 (1973), 302.

Motz, L., and A. Duveen. *Essentials of Astronomy*. Belmont, Calif.: Wadsworth, 1968.

Norman, C.A., and D. ter Haar. "On the Black Hole Model of Galactic Nuclei." *Astronomy and Astrophysics, 24* (1973), 121.

Novikov, I., and K. Thorne. "Astrophysics of Black Holes," in *Black Holes*, eds. C. DeWitt and B.S. DeWitt. Gordon and Breach, 1973.

Oppenheimer, J.R., and R. Serber. "On the Stability of Stellar Neutron Cores." *Physics Review, 54* (1938), 540.

Peery, B. "Spectroscopic Observations of VV Cephei." *Astrophysical Journal, 144* (1966), 672.

*****Penrose**, R. "Black Holes." *Scientific American, 226* (May 1972), 38.

*****Ruderman**, M. "Solid Stars." *Scientific American, 224* (Feb. 1971), 24.

————. "Pulsars: Structure and Dynamics." *Annual Review of Astronomy and Astrophysics, 10* (1972), 427.

Schwarzschild, M. *Structure and Evolution of the Stars*. Princeton, N.J.: Princeton University Press, 1958.

Shapley, H., ed. *Source Book in Astronomy, 1900–1950*. Cambridge: Harvard University Press, 1960.

Strohmeier, W. *Variable Stars*. New York: Pergamon Press, 1972.

Thorne, K.S. "The Death of a Star." Adapted from *Science Year: World Book Science Annual*. New York: Field Enterprises, 1968.

*****Wade**, N. "Discovery of Pulsars: A Graduate Student's Story." *Science, 189* (1975), 359.

Walker, M. "Nova D.Q. Herculis (1934): An Eclipsing Binary with Very Short Period." *Publications of the Astronomical Society of the Pacific, 66* (1954), 230.

Wickramasinghe, D. et al. "2U 0900-40: A Black Hole?" *Astrophysical Journal, 188* (1974), 167.

Witten, T.A. "Compounds in Neutron-Star Crusts." *Astrophysical Journal, 188* (1974), 615.

Zwicky, F. "On the Theory and Ob-

servation of Highly Collapsed Stars." *Physics Review, 55* (1939), 726.

Chapter 17

Allen, C.W. *Astrophysical Quantities*, 3rd ed. London: Athlone Press, 1973.

*****Aller**, L. *Atoms, Stars and Nebulae*. Cambridge: Harvard University Press, 1971.

Bomen, I.S. "The Origin of the Chief Nebular Lines." *Publications of the Astronomical Society of the Pacific, 39* (1927), 295.

Bok, B., and C. Cordwell. *A Study of Dark Nebulae*. Tucson, Ariz.: Steward Observatory, 1971.

Deharveng, L., and M. Maucherat. "Optical Study of the Carina Nebula." *Astronomy and Astrophysics, 41* (1975), 27.

Dombrovsky, W.A. "On the Nature of the Crab Nebula Radiation." *Doklady Academy of Sciences USSR, 94* (1954), 1021.

Erickson, E. et al. "Infrared Spectrum of the Orion Nebula between 55 and 300 Microns." *Astrophysical Journal, 183* (1973), 535

Gaustad, J.E. "The Composition of Interstellar Dust," in *Dark Nebulae, Globules, and Protostars*, ed. B.T. Lynds. Tucson: University of Arizona Press, 1970.

*****Gehrz**, R., and E. Ney. "The Core of Eta Carinae." *Sky and Telescope, 44* (1972), 4.

Greenberg, J.M. "Interstellar Grains," in *Nebulae and Interstellar Matter*, eds. B. Middlehurst and L. Aller. Chicago: University of Chicago Press, 1968.

————. "The Interstellar Depletion Mystery, or Where Have All These Atoms Gone?" *Astrophysical Journal, 189* (1974), L81.

Hartmann, J.F. "Investigations on the Spectrum and Orbit of δ Orionis." *Astrophysical Journal, 19* (1904), 268.

Hartmann, W.K. "Growth of Planetesimals in Nebulae Surrounding Young Stars," in *Evolution Stellaire avant la Séquence Principale*. Mémoires de la

Société Royale des Sciences de Liège. Collection in −8°, 5th sér., 19 (1970), 215.

Herbig, G.H. "Introductory Remarks," in *Evolution Stellaire avant la Séquence Principale.* Mémoires de la Société Royale des Sciences de Liège. Collection in −8°, 5th sér., 19 (1970), 13.

*———. "Interstellar Smog." *American Scientist*, 62 (1974), 200.

Johnson, H.L. "Interstellar Extinction," in *Nebulae and Interstellar Material*, eds. B. Middlehurst and L. Aller. Chicago: University of Chicago Press, 1968.

*Lovi, G. "Rambling through May Skies." *Sky and Telescope*, 43 (1972), 306.

Lynds, B.T., ed. *Dark Nebulae, Globules and Protostars.* Tucson: University of Arizona Press, 1971.

*Maran, S. "The Gum Nebula." *Scientific American*, 224 (Dec. 1971), 20.

Maran, S., J. Brandt, and T. Stecher, eds. *The Gum Nebula and Related Problems.* Washington, D.C.: NASA SP-332, 1973.

Matthews, W.G. "The Time Evolution of an HII Region." *Astrophysical Journal*, 142 (1965), 1120.

Menon, T.K. "Interstellar Structure of the Orion Region." *Astrophysical Journal*, 127 (1958), 28.

*Miller, J.S. "The Structure of Emission Nebulae." *Scientific American*, 231 (Oct. 1974), 34.

Neugebauer, G., and J. Westphal. "Infrared Observations of Eta Carinae." *Astrophysical Journal*, 152 (1968), L89.

Ney, E.P. "Infrared Excesses in Supergiant Stars: Evidence for Silicates." *Publications of the Astronomical Society of the Pacific*, 84 (1972), 613.

Oort, J., and H.C. van de Hulst. "Gas and Smoke in Interstellar Space." *Bulletin of the Astronomical Society of the Netherlands*, 10 (1946), 187.

Penston, M. "Multicolor Observations of Stars in the Vicinity of the Orion Nebula." *Astrophysical Journal*, 183 (1973), 505.

Poveda, A. "G.W. Orionis, a 20,000 Year Old T Tauri Star?" *Boletin de los Observatorio Tonantzintla y Tacubaya*, 4 (1965), 77.

Scoville, N.Z. "Molecular Clouds in the Galaxy." *Astrophysical Journal*, 199 (1975), L105.

Shapiro, P.R. "Interstellar Polarization: Magnetite Dust." *Astrophysical Journal*, 201 (1975), 151.

Shlovsky, I.S. "On the Nature of Radio Radiation of Nebular Masses." *Astronomical Journal of the USSR*, 31 (1954), 533.

Simon, T., and H.M. Dyck "Silicate Absorption at 18 μm in Two Peculiar Infrared Sources." *Nature*, 253 (1975), 101.

Steffey, P.C. "A Kinematical Study of Field Supergiants near the Sun." Ph.D. Dissertation, Tucson: University of Arizona, 1964.

Stromgren, B. "The Physical State of Interstellar Hydrogen." *Astrophysical Journal*, 89 (1939), 526.

Trumpler, R.J. "Preliminary Results on the Distances, Dimensions, and Space Distributions of Open Star Clusters." *Lick Observatory Bulletin*, 14 (1930), 154.

*Turner, B.E. "Interstellar Molecules." *Scientific American*, 228 (Mar. 1973), 50.

*Vandervoort, P.O. "The Age of the Orion Nebula." *Scientific American*, 213 (Feb. 1965), 90.

*Wickramasinghe, R.M. "Interstellar Molecules." *Scientific American*, 216 (Mar. 1967), 50.

Zuckerman, B. "A Model of the Orion Nebula." *Astrophysical Journal*, 183 (1973), 863.

Chapter 18

Abt, H., C. Bolton, and S. Levy. "1C 4665, A Cluster of Binaries." *Astrophysical Journal*, 171 (1972), 259.

Abt, H., and S. Levy. "Multiplicity among Solar-Type Stars." *Astrophysical Journal* Supplement, in press.

Arny, T., and P. Weissman. "Interaction of Proto-stars in a Collapsing Cluster." *Astronomical Journal*, 78 (1973), 310.

*Batten, A.H. *Binary and Multiple Systems of Stars.* Oxford: Pergamon Press, 1973.

Blaauw, A., and T.S. van Albada. "Exploration of Statistical Properties of Early Type Spectroscopic Binaries," in *The Determination of Radial Velocities and Their Applications*, eds. A. Batten and J. Heard. London: Academic Press, 1967.

*Eggen, O.J. "Stars in Contact." *Scientific American*, 218 (June 1968), 34.

Feibelman, W.A. "Stars near Beta Lyrae." *Astrophysical Journal*, 137 (1963), 701.

Gatewood, G., and H. Eichhorn. "An Unsuccessful Search for a Planetary Companion of Barnard's Star." *Astronomical Journal*, 78 (1973), 769.

Hershey, J.L. "Astrometric Analysis of the Triple Star BD +66°34." *Astronomical Journal*, 78 (1973), 935.

Jensen, O.G., and T. Ulrych. "An Analysis of the Perturbations on Barnard's Star." *Astronomical Journal*, 78 (1973), 1104.

Kopal, Z. "The Classification of Close Binary Systems." *Annals of Astrophysics*, 18 (1955), 379.

———. *Close Binary Systems.* London: Chapman and Hall, 1959.

———. "The Eclipsing System of Epsilon Aurigae and Its Possible Relevance to the Formation of a Planetary System." *Astrophysics and Space Science*, 10 (1971), 332.

Kraft, R.P. "Cataclysmic Variables as Binary Stars." *Advances in Astronomy and Astrophysics*, 2 (1964), 43.

———. "Binary Systems as X-ray Sources: A Review," in *Gamma-Ray Astronomy*, eds. H. Bradt and R. Giaconni. Dordrecht, Netherlands: Reidel, 1973.

Kuiper, G.P. "On the Interpretation of Beta Lyrae and Other Close Binaries." *Astrophysical Journal*, 93 (1941), 133.

Kumar, S. "On the Formation of Double Stars." *Astrophysics and Space Science*, 17 (1972), 453.

Lindenblad, E.W. "Multiplicity of the Sirius System." *Astronomical Journal*, 73 (1973), 205.

Radzievskii, V., and E. Radzievskaya. "Coplanar System of Binary Stars in Aquila." *Soviet Astronomy*, 17 (1973), 239.

Standish, E.M., Jr. "The Dynamical

Evolution of Triple Star Systems: A Numerical Study." *Astronomy and Astrophysics, 21* (1972), 185.

*Struve, O. "The Duplicity of Nova Herculis." *Sky and Telescope, 14* (1955), 275.

van de Kamp, P. "Astrometric Study of Barnard's Star with Plates Taken with the Sproul 24-inch Refractor." *Astronomical Journal, 68* (1963), 515.

———. "Alternate Dynamical Analysis of Barnard's Star." *Astronomical Journal, 74* (1969), 757.

———. "Parallax and Mass Ratio of the Visual Binary *61 Cygni*." *Astronomical Journal, 78* (1973), 1099.

———. "Astrometric Analysis of Barnard's Star from Plates Taken with the Sproul Observatory 61-cm. Refractor." *Astronomical Journal, 80* (1975), 658.

Wanner, J.F. "The Visual Binary Krüger 60." *Sky and Telescope, 33* (1967), 16.

*Warner, B. "Six Ultra-short Period Binary Stars." *Sky and Telescope, 44* (1972), 358.

Wilson, R.E. "The Secondary Component of Beta Lyrae." *Astrophysical Journal, 189* (1974), 319.

Chapter 19

Ambartsumian, V.A. *Stellar Evolution and Astrophysics.* U.S.S.R.: Erevan, 1947.

———. "Expanding Stellar Associations," in *Source Book in Astronomy, 1900–1950,* ed. H. Shapley. Cambridge: Harvard University Press, 1960.

*Iben, I. "Globular-Cluster Stars." *Scientific American, 223* (July 1970), 26.

Jones, K G. *Messier's Nebulae and Star Clusters.* New York: American Elsevier, 1969.

*Metz, W.D. "Astronomy from Space: New Class of X-ray Sources Found." *Science, 189* (1975), 1073.

Milhalas, D., and P. Routly. *Galactic Astronomy.* San Francisco: W.H. Freeman, 1968.

Ostriker, J., L. Spitzer, and R. Chevalier. "On the Evolution of Globular Clus-

ters." *Astrophysical Journal, 176* (1972), L51.

Peebles, P., and R. Dicke. "Origin of the Globular Star Clusters." *Astrophysical Journal, 154* (1968), 891.

Ruzmaikina, T. "On the Cosmological Origin of Globular Clusters." *Astronical' Zhurnal Nauk SSSR, 49* (1972), 1229.

Schlesinger, B.M. "The Main Sequences of Synthetic Clusters with Finite Formation Times." *Astronomical Journal, 77* (1972), 584.

Shapley, H. *Star Clusters.* Cambridge: Harvard University Press, 1930.

Trumpler, R. "Absorption of Light in the Galactic System." *Publications of the Astronomical Society of the Pacific, 42* (1930), 214.

Chapter 20

Allen, C.W. *Astrophysical Quantities.* London: Athlone Press, 1973.

Balick, B., and R. Sanders. "Radio Fine Structure in the Galactic Center." *Astrophysical Journal, 191* (1974), 325.

Becker, W., and G. Contopoulos, eds. "The Spiral Structure of Our Galaxy." *International Astronomical Union. Symposium 38.* Dordrecht, Netherlands: Reidel, 1970.

Blaauw, A., and M. Schmidt, eds. *Galactic Structure.* Chicago: University of Chicago Press, 1965.

Bok, B.J., and P. Bok. *The Milky Way,* 4th ed. Cambridge: Harvard University Press, 1974.

Dicke, R.H. "The Age of the Galaxy from the Decay of Uranium." *Astrophysical Journal, 155* (1969), 123.

Huang, S.-S. "Disk and Ring Structure in the Universe." *Sky and Telescope, 43* (1972), 225.

Iaki, S.L. *The Milky Way, an Elusive Road for Science.* New York: Science History Publications, 1972.

Kapitzky, J., and W. Dent. "A High-Resolution Map of the Galactic-Center Region," *Astrophysical Journal 188* (1974), 27.

Kapteyn, O.C. "On the Parallaxes and Motions of the Brighter Galactic

Helium Stars." *Astrophysical Journal, 47, 107, 146,* 255.

Kerr, F.J. "The Large-Scale Distribution of Hydrogen in the Galaxy." *Annual Review of Astronomy and Astrophysics, 7* (1969), 39.

Klare, G., and T. Neckel. "Polarization of Southern OB Stars," in *The Spiral Structure of Our Galaxy,* eds. W. Becker and G. Contopolos. Dordrecht, Netherlands: Reidel, 1970.

Kraft, R.P. "Pulsating Stars and Cosmic Distances." *Scientific American, 200* (July 1959), 48.

Lin, C.C. "Theory of Spiral Structure," in *Highlights in Astronomy.* Dordrecht, Netherlands: Reidel, 1971, 2, 88.

Lindblad, B. *Handbuch der Physik,* ed. A.A. Winkelmann. *53* (1959), 21.

Maraschi, L., A. Treves, and M. Tarenghi. "Accretion by Neutron Stars at the Galactic Center." *Astronomy and Astrophysics, 25* (1973), 153.

Moffat, F.J., and N. Vogt. "An Up-to-Date Picture of Galactic Spiral Features Based on Young Open Star Clusters." *Astronomy and Astrophysics, 23* (1973), 317.

O'Connell, D.J.K., ed. *Stellar Populations.* New York: Wiley-Interscience, 1958.

Oort, J., and G. Rougoor. "The Position of the Galactic Center, V." *Monthly Notices of the Royal Astronomical Society, 121* (1960), 171.

Rieke, G., and F.J. Low. "Infrared Maps of the Galactic Nucleus." *Astrophysical Journal, 184* (1973), 415.

Rougoor, G., and J. Oort. "Distribution and Motion of Interstellar Hydrogen in the Galactic System with Particular Reference to the Region within 3 kpc of the Center." *Proceedings of the National Academy of Science, 46* (1960), 1.

Sanders, R., and K. Prendergast. "The Possible Relation of the 3-kpc Arm to Explosions in the Galactic Nucleus." *Astrophysical Journal, 188* (1974), 489.

*Saunders, J. "The Globular Cluster Omega Centauri." *Sky and Telescope, 26* (1963), 133.

Shapley, H. *Star Clusters.* Cambridge: Harvard University Press, 1930.

Shapley, H., and H. Howarth, eds. *A Source Book in Astronomy.* New York: McGraw-Hill, 1929.

*Struve, O. "A Historic Debate about the Universe." *Sky and Telescope, 19* (1960), 398.

Verschuur, G.L. "Interstellar Neutral Hydrogen and Its Small-Scale Structure," in *Galactic and Extra-Galactic Radio Astronomy,* eds. G. Verschuur and K. Kellermann. New York: Springer-Verlag, 1974.

Walborn, N.R. "The Space Distribution of the O Stars in the Solar Neighborhood." *Astronomical Journal, 78* (1973) 1067.

Whitney, C.A. *The Discovery of Our Galaxy.* New York: Alfred A. Knopf, 1971.

Wielen, R. "The Density-Wave Theory of the Spiral Structure of Galaxies." *Publications of the Astronomical Society of the Pacific, 86* (1974), 341.

Chapter 21

Allen, C.W. *Astrophysical Quantities.* London: Athlone Press, 1973.

Bok, B.J. "Magellanic Clouds." *Annual Review of Astronomy and Astrophysics,* 4 (1966), 95.

de Vaucouleurs, G. "Structure, Dynamics, and Statistical Properties of Galaxies," in *The Formation and Dynamics of Galaxies,* ed. R. Shakeshaft. Boston: Reidel, 1974.

Gott, J.R. "Dynamics of Rotating Stellar Systems: Collapse and Violent Relaxation," in *The Formation and Dynamics of Galaxies,* ed. R. Shakeshaft. Boston: Reidel, 1974.

Harris, W.E., and J.E. Hesser. "NGC 5694: A Globular Cluster Escaping from the Galaxy?" *Publications of the Astronomical Society of the Pacific, 88* (1976), 377.

*Hodge, P.W. *The Physics and Astronomy of Galaxies and Cosmology.* New York: McGraw-Hill, 1966.

Hubble, E. "N.G.C. 6822, a Remote Stellar System." *Astrophysical Journal, 62* (1925), 409.

Light, E., R. Danielson, and M. Schwarzschild. "The Nucleus of M 31." *Astrophysical Journal, 194* (1974), 257.

Mathewson, D., M. Cleary, and J. Murray. "The Magellanic Stream." *Astrophysical Journal, 190* (1974), 291.

Oort, J.H. "Summary and Desiderata," in *The Magellanic Clouds,* ed. A. Muller. Boston: Reidel, 1971.

*Rubin, Vera C. "The Dynamics of the Andromeda Nebula." *Scientific American, 228* (June 1973), 30.

Sandage, A. *The Hubble Atlas of Galaxies.* Washington, D.C.: The Carnegie Institution, 1961.

Sandage, A., and G.A. Tammann. "Steps toward the Hubble Constant, III." *Astrophysical Journal, 194* (1974), 223.

Shapley, H. *The Inner Metagalaxy.* New Haven, Conn.: Yale University Press, 1957.

——. *Galaxies.* Cambridge: Harvard University Press, 1961.

Smith, E., and K. Jacobs. *Introductory Astronomy and Astrophysics.* Philadelphia: W.B. Saunders, 1974.

Thackeray, A.D. "Survey of Principal Characteristics of the Magellanic Clouds," in *The Magellanic Clouds,* ed. A. Muller. Boston: Reidel, 1971.

——. "Colour-Magnitude Arrays of Brightest Stars," in *The Magellanic Clouds,* ed. A. Muller. Boston: Reidel, 1971.

van den Bergh, S. "Search for Faint Companions to M31." *Astrophysical Journal, 171* (1972), L35.

*Whitney, C.A. *The Discovery of Our Galaxy.* New York: Alfred A. Knopf, 1971.

Chapter 22

Allen, C.W. *Astrophysical Quantities.* London: Athlone Press, 1973.

Arp, H. "Evidence for Non-velocity Redshifts—New Evidence and Review," in *The Formation and Dynamics of Galaxies,* ed. J. Shakeshaft. Dordrecht, Netherlands: Reidel, 1974.

Balkowski, C. et al. "Observational Evidence for Non-velocity Redshift in Stephan's Quintet," in *The Formation and Dynamics of Galaxies,* ed. J. Shakeshaft. Dordrecht, Netherlands: Reidel, 1974.

de Vaucouleurs, G. "Structure, Dynamics, and Statistical Properties of Galaxies," in *The Formation and Dynamics of Galaxies,* ed. J. Shakeshaft. Dordrecht, Netherlands: Reidel, 1974.

Eddington, A. *The Expanding Universe.* Cambridge: Cambridge University Press, 1933.

*Hodge, P.W. "Some Current Studies of Galaxies." *Sky and Telescope, 44* (1972), 23.

*Hoyle, F. *From Stonehenge to Modern Cosmology.* San Francisco: W.H. Freeman, 1972.

Hubble, E.P. "A Relation between Distance and Radial Velocity among Extra-galactic Nebulae," in *Source Book in Astronomy, 1900–1950,* ed. H. Shapley. Cambridge: Harvard University Press, 1960.

——. "The Velocity Distance Relation among Extra-galactic Nebulae." *Astrophysical Journal, 74* (1931), 43.

——. *The Realm of the Nebulae.* New York: Dover, 1958.

Khachikian, E.Y., and D. Weedman. "An Atlas of Seyfert Galaxies." *Astrophysical Journal, 192* (1974), 581.

Miley, G., H. van der Laan, and K. Wellington. "Recent Westerbork Observations of Head-Tail Galaxies," in *The Formation and Dynamics of Galaxies,* ed. J. Shakeshaft. Dordrecht, Netherlands: Reidel, 1974.

Oort, J.H. "Recent Radio Studies of Bright Galaxies," in *The Formation and Dynamics of Galaxies,* ed. J. Shakeshaft. Dordrecht, Netherlands: Reidel, 1974.

Petrosian, V. "The Hubble Relation for Non-standard Candles and the Origin of the Redshift of Galaxies." *Astrophysical Journal, 188* (1974), 443.

Sandage, A., and G. Tamman. "Steps Toward the Hubble Constant, IV." *Astrophysical Journal, 194* (1974), 559.

Saslaw, W.C. "Theory of Galactic Nuclei," in *The Formation and Dynamics of Galaxies,* ed. J. Shakeshaft. Dordrecht, Netherlands: Reidel, 1974.

Shapley, H., ed. *Source Book in Astronomy, 1900–1950.* Cambridge: Harvard University Press, 1960.

*Toomre, A., and J. Toomre. "Violent Tides Between Galaxies." *Scientific American,* 229 (Dec. 1973), 38.

*Weymann, R.J. "Seyfert Galaxies." *Scientific American,* 220 (Jan. 1969), 28.

*Whitney, C.A. *The Discovery of Our Galaxy.* New York: Alfred A. Knopf, 1971.

Wolfe, A.M. "On the Steady Flow of Gas from the Nuclei of Seyfert Galaxies." *Astrophysical Journal,* 188 (1974), 243.

Chapter 23

Bondi, H. *Cosmology.* Cambridge: Cambridge University Press, 1960.

Brown, G.S., and B. Tinsley. "Galaxy Counts as a Cosmological Test." *Astrophysical Journal,* 194 (1974), 555.

*de Vaucouleurs, G. "The Case for a Hierarchical Cosmology." *Science,* 167 (1970), 1203.

Einstein, A. "Cosmological Considerations on the General Theory of Relativity," in *The Principle of Relativity.* New York: Dover, 1952. (First published 1917.)

Gamow, G. *The Creation of the Universe.* New York: Viking, 1952.

*Gaster, T.H. *The Oldest Stories in the World.* Boston: Beacon Press, 1952.

Gott, J.R., J. Gunn, D. Schramm, and B. Tinsley. "An Unbound Universe?" *Astrophysical Journal,* 194 (1974), 543.

Harrison, E.R. "Why the Sky Is Dark at Night." *Physics Today,* 27 (1974), 5.

*Hodge, P.W. *Concepts of the Universe.* New York: McGraw-Hill, 1969.

*Hoyle, F. *From Stonehenge to Modern Cosmology.* San Francisco: W.H. Freeman, 1972.

Kant, I. *Kant's Cosmology,* W. Hastie, trans. New York: Greenwood Publishing, 1968.

North, J.D. *The Measure of the Universe.* Oxford: Oxford University Press, 1965.

Peebles, P.J.E. *Physical Cosmology.* Princeton, N.J.: Princeton University Press, 1971.

Peterson, B.A. "The Distribution of Galaxies in Relation to Their Formation and Evolution," in *The Formation and Dynamics of Galaxies,* ed. J. Shakeshaft. Dordrecht, Netherlands: Reidel, 1974.

Ryle, M. "The Counts of Radio Sources." *Annual Review of Astronomy and Astrophysics,* 6 (1968), 249.

*Schatzman, E.S. *The Origin and Evolution of the Universe.* New York: Basic Books, 1965.

*Wilson, J.A. "Egypt," in *Before Philosophy,* eds. H. Frankfort and H.A. Frankfort. Baltimore: Penguin Books, 1949.

Chapter 24

Alpher, R.A., and R.C. Herman. "Evolution of the Universe." *Nature,* 162 (1948), 774.

Dicke, R.H. et al. "Cosmic Black Body Radiation." *Astrophysical Journal,* 142 (1965), 414.

Gamow, G. *The Creation of the Universe.* New York: Viking, 1952.

*Hawking, S.W. "The Quantum Mechanics of Black Holes." *Scientific American,* 236 (Jan. 1977), 34.

*Morrison, D., and N. Morrison. "An Infinitely Expanding Universe." *Mercury,* 4 (Sept. 1975), 27.

Ozernoy, L.M. "Dynamics of Superclusters as the Most Powerful Test for Theories of Galaxy Formation," in *The Formation and Dynamics of Galaxies,* ed. J. Shakeshaft. Dordrecht, Netherlands: Reidel, 1974.

Peebles, P.J.E. *Physical Cosmology.* Princeton, N.J.: Princeton University Press, 1971.

Penzias, A.A., and R.W. Wilson. "A Measurement of Excess Antenna Temperature at 4080 Mc/s." *Astrophysical Journal,* 142 (1965), 419.

Reeves, H. et al. "On the Origin of Light Elements." Cal Tech Preprint OAP-296, 1972.

Sandage, A., and G. Tammann. "Steps toward the Hubble Constant, IV." *Astrophysical Journal,* 194 (1974), 559.

Smirnov, Y.N. "Hydrogen vs. He^4 Formation in the Prestellar Gamow Universe." *Soviet Astronomy,* 8 (1964), 864. (English trans.)

Thorstensen, J.R., and R.B. Partridge. "Can Collapsed Stars Close the Universe?" *Astrophysical Journal,* 200 (1975), 527.

*Webster, A. "The Cosmic Background Radiation." *Scientific American,* 231 (Aug. 1974), 26.

Chapter 25

Ambartsumian, V.A. "Introduction," in *Extraterrestrial Civilizations,* ed. S.A. Kaplan, trans. Israel Program for Scientific Translations. Springfield, Va.: U.S. Department of Commerce, National Technical Information Service, 1971.

Berendzen, R., ed. *Life Beyond Earth and the Mind of Man.* Washington, D.C.: NASA SP-328, 1973.

Berkner, L.V., and L.C. Marshall. "On the Origin and Rise of Oxygen Concentration in the Earth's Atmosphere." *Journal of Atmospheric Science,* 22 (1965), 225.

Brown, H. "Planetary Systems Associated with Main-Sequence Stars." *Science,* 145 (1964), 1177.

*Dole, S.H. *Habitable Planets for Man.* Waltham, Mass.: Ginn/Blaisdell, 1964.

*Dubos, R. *So Human an Animal.* New York: Scribners, 1968.

*Gould, S.J. "The Great Dying." *Natural History,* 83 (Oct. 1974), 22.

*———. "An Unsung Single-Celled Hero." *Natural History,* 83 (Nov. 1974), 33.

Hartmann, W.K. "Star Formation in Clusters and the Stellar Mass Function," in *Evolution Stellaire avant la Séquence Principale.* Mémoires de la Société Royale des Sciences de Liège. Collection in −8°, 5th ser., 19 (1970), 215.

Holland, H.D. "Model for Evolution of the Earth's Atmosphere," in *Petrologic Studies.* New York: Geological Society of America, 1962.

Horowitz, N.H., G. Hobby, and J. Hubbard. "The Viking Carbon Assimilation Experiments: Interim Report." *Science*, 194 (1976), 1321.

*Hoyle, F. *From Stonehenge to Modern Cosmology*. San Francisco: W.H. Freeman, 1972.

*Huang, S.-S. "Life Outside the Solar System." *Scientific American*, 202 (Apr. 1960), 55.

————. "Rotational Behavior of the Main-Sequence Stars and Its Plausible Consequences Concerning Formation of Planetary Systems." *Astrophysical Journal*, 141 (1965), 985.

Kuiper, G.P. "Discourse Following Award of Kepler Gold Medal." *Communications of the Lunar and Planetary Laboratory, University of Arizona*, 9 (1971), 403.

*Kurten, B. "Continental Drift and Evolution." *Scientific American*, 220 (Mar. 1969), 54.

Kvenvolden, K. et al. "Evidence for Extraterrestrial Amino Acids and Hydrocarbons in the Murchison Meteorite." *Nature*, 228 (1970), 923.

Levin, Gilbert V., and Patricia Ann Straat. "Viking Labeled Release Biology Experiment: Interim Results." *Science*, 194 (1976), 1322.

*Maruyama, M., and A. Harkins. *Cultures Beyond the Earth*. New York: Vintage Books, 1975.

Miller, S.L. "Production of Some Organic Compounds under Possible Primitive Earth Conditions." *Journal of the American Chemical Society*, 77 (1955), 2351.

*Oparin, A.I. *Life: Its Nature, Origin, and Development*. New York: Academic Press, 1962.

*Sagan, C. "Life," in *Encyclopaedia Britannica*. Chicago: Encyclopaedia Britannica, Inc., 1970.

————. ed. *Communication with Extraterrestrial Intelligence*. Cambridge: M.I.T. Press, 1973.

Shklovskii, I.S., and C. Sagan. *Intelligent Life in the Universe*. San Francisco: Holden-Day, 1966.

*Siegel, S.M. "Experimental Biology of Extreme Environments and Its Significance for Space Bioscience—2" *Spaceflight*, 12 (1972), 256.

Stern, D.K. "First Contact with Non-human Cultures." *Mercury*, 4 (Sept./Oct. 1975), 14.

*Underwood, J. *Biocultural Interactions and Human Variation*. Dubuque, Iowa: W.C. Brown, 1975.

Urey, H.C. *The Planets: Their Origin and Development*. New Haven, Conn.: Yale University Press, 1952.

————. "Biological Material in Meteorites: A Review." *Science*, 151 (1966), 157.

Van Valen, L., and R. Sloan. "Ecology and the Extinction of Dinosaurs." *Twenty-fourth International Geological Congress Abstracts*, 1972.

Enrichment Essay B

*Abetti, Giorgio. *The History of Astronomy*. New York: Schuman, 1951.

*Ashbrook, J. "Astronomical Scrapbook." *Sky and Telescope*, 46 (1973), 300.

Bjorkman, Judith K. "Meteors and Meteorites in the Ancient Near East." *Meteoritics*, 8 (1973), 91.

Bok, B., and M. Mayall. "Scientists Look at Astrology." *Scientific Monthly*, 52 (1941), 233.

*Bok, B. "A Critical Look at Astrology." *The Humanist*, 35 (Sept./Oct. 1975), 5.

*Condon, E.U. et al. *Scientific Study of Unidentified Flying Objects*. New York: Dutton, 1969.

Emiliani, C. et al. "Paleoclimatological Analysis of Late Quaternary Cores from the Northeastern Gulf of Mexico." *Science*, 189 (Sept. 1975), 1083.

Farrington, Benjamin. "Astrology," in *Encyclopaedia Britannica*, vol. 2. Chicago: Encyclopaedia Britannica, Inc., 1973, 640.

Gingerich, Owen. "Musings on Antique Astronomy." *American Scientist*, 55 (Jan./Feb. 1967), 88.

Gleadow, Rupert. *The Origin of the Zodiac*. New York: Atheneum, 1963.

*Gould, S.J. "Velikovsky in Collision." *Natural History*, 84 (Feb. 1975), 20.

Jerome, L.E. "Astrology: Magic or Science?" *The Humanist*, 35 (Sept./Oct. 1975), 10.

Mathur, K.D. "Indian Astronomy in the Era of Copernicus." *Nature*, 251 (1974), 283.

*Minnaert, M. *Light and Colour in the Open Air*. New York: Dover, 1954.

Neugebauer, O. "History of Ancient Astronomy." *Journal of Near Eastern Studies*, 4 (1945), 1.

————. *The Exact Sciences in Antiquity*. Providence, R.I.: Brown University Press, 1957.

*Sagan, C., and J. Agel. *Other Worlds*. New York: Bantam Books, 1975.

*Shklovskii, I.S., and C. Sagan. *Intelligent Life in the Universe*. San Francisco: Holden-Day, 1966.

*Stephenson, F.R., and D.H. Clark. "Ancient Astronomical Records from the Orient." *Sky and Telescope*, 53 (1977), 84.

*Story, R. *The Space Gods Revealed*. New York: Harper & Row, 1976.

Velikovsky, I. *Worlds in Collison*. New York: Doubleday, 1950.

von Däniken, E. *Chariots of the Gods?* New York: Bantam Books, 1971.

Index of Names

Index of Terms

Key *Concepts* and the page numbers on which they first appear are set in **boldface** type.

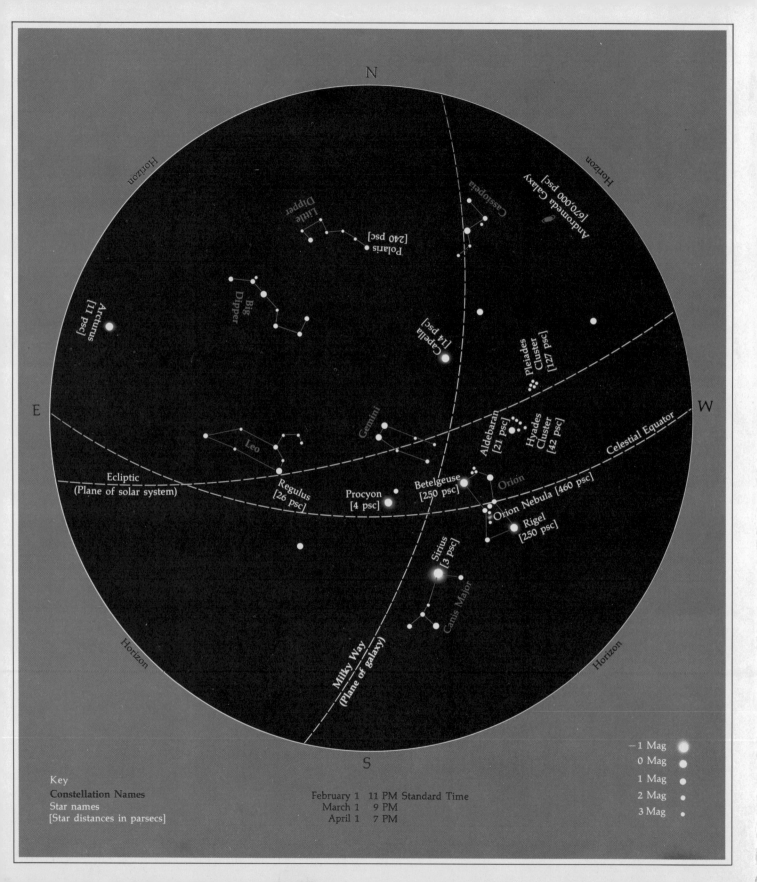